Genome Editing for Crop Improvement: Theory and Methodology

Genome Editing for Crop Improvement: Theory and Methodology

Edited by

Jameel M. Al-Khayri
Department of Agricultural Biotechnology, King Faisal University, Al-Ahsa, Saudi Arabia

Muhammad N. Sattar
Central Laboratories, King Faisal University, Al-Ahsa, Saudi Arabia

Sudhir K. Sopory
International Centre for Genetic Engineering and Biotechnology, New Delhi, India

Shri Mohan Jain
Department of Agricultural Sciences, University of Helsinki, Helsinki, Finland

CABI

CABI is a trading name of CAB International

CABI
Nosworthy Way
Wallingford
Oxfordshire OX10 8DE
UK

Tel: +44 (0)1491 832111
E-mail: info@cabi.org
Website: www.cabi.org

CABI
200 Portland Street
Boston
MA 02114
USA

T: +1 (617)682-9015
E-mail: cabi-nao@cabi.org

A catalogue record for this book is available from the British Library, London, UK.

ISBN-13: 9781800622494 (hardback)
9781800622500 (ePDF)
9781800622517 (ePub)

DOI: 10.1079/9781800622517.0000

Commissioning Editor: Rebecca Stubbs
Editorial Assistant: Emma McCann
Production Editor: Shankari Wilford

Typeset by Exeter Premedia Services Pvt Ltd, Chennai, India
Printed and bound by CPI Group (UK) Ltd, Croydon, CR0 4YY

Contents

Contributors

Khadija Aaliya, Centre of Excellence in Molecular Biology (CEMB), University of the Punjab, 87-West Canal Bank Road, Thokar Niaz Baig, Lahore, Pakistan. Email: aaliya.khadija20@gmail.com

Oumer Abdie, Department of Biotechnology, College of Natural and Computational Science, Wolkite University, Wolkite, Ethiopia. Email: oumer.abdie@aau.edu.et; oumar.abdie@gmail.com

Muhamed Adem, Department of Forestry, College of Agriculture and Natural Resources, Madda Walabu University, Bale Robe, Ethiopia. Email: muhamed.adem@aau.edu.et; muhamedadem@gmail.com

Farooq Ahmad, Department of Plant Production and Forest Resources, University of Valladolid, Avda. de Madrid, 34004, Palencia, Spain. Email: Farooq.Ahmad@uva.es

Sallah A. Al-Hashedi, Central Laboratories, King Faisal University, 31982 Al-Ahsa, Al-Hofuf, Saudi Arabia. Email: sahmad@kfu.edu.sa

Zulfiqar Ali, Islamic Organization for Food Security, Astana, Republic of Kazakhstan. Email: zulfiqar@iofs.org.kz

Alekya Allu, School of Bioengineering and Biosciences, Lovely Professional University, Jalandhar-144411, India. Email: alekyaallu333@gmail.com; jjaskani@uaf.edu.pk

Asadullah, Department of Plant Breeding and Genetics, Directorate of Agriculture Research Wayaro Farm, Lasbela, Pakistan. Email: asadkhanagri24@gmail.com

Habib-ur-Rehman Athar, Institute of Botany, Bahauddin Zakariya University, Multan 60800, Pakistan. Email: habibathar@yahoo.com

Farrukh Azeem, Department of Bioinformatics and Biotechnology, Government College University, Faisalabad 38000, Pakistan. Email: farrukh@gcuf.edu.pk

Anu Bansal, School of Bioengineering and Biosciences, Lovely Professional University, Jalandhar-144411, India. Email: anu.21971@lpu.co.in

Sherien Bukhat, Institute of Molecular Biology and Biotechnology, Bahauddin Zakariya University, Multan 60800, Pakistan. Email: sherien.bukhat@gmail.com

Mallesham Bulle, Institute of Genomics for Crop Abiotic Stress Tolerance (IGCAST), Department of Plant and Soil Science, Texas Tech University, Lubbock, Texas, USA. Email: mbulle@ttu.edu

Jahnavi Chivukula, Department of Life Sciences, School of Sciences, CHRIST (Deemed to be University), Hosur Road, Bangalore – 560 029, India. Email: ch.jahnavi@mbty.christuniversity.in

Rupesh Deshmukh, Department of Biotechnology, Central University of Haryana, Mahendragarh, India. Email: rupesh0deshmukh@gmail.com

Vinay Kumar Dhiman, Department of Basic Sciences, Dr. Yashwant Singh Parmar University of Horticulture and Forestry, Solan, Himachal Pradesh, India. Email: vinaykrdhiman4@gmail.com

Vivek Kumar Dhiman, Centre for Herbal Pharmacology and Environmental Sustainability, Chettinad Hospital and Research Institute, Chettinad Academy of Research and Education, Kelambakkam 603103, Tamil Nadu, India.

Kaushik Ghose, Institute of Genomics for Crop Abiotic Stress Tolerance (IGCAST), Department of Plant and Soil Science, Texas Tech University, Lubbock, Texas, USA. Email: kaushik.ghose@ttu.edu

Ambreen Gul, Centre of Excellence in Molecular Biology (CEMB), University of the Punjab, 87-West Canal Bank Road, Thokar Niaz Baig, Lahore, Pakistan. Email: ambreen.gul@cemb.edu.pk; ambreengulpu@yahoo.com

Sehrish Iftikhar, Microfluidics Innovation Centre, Elvesys, 75011, Paris, France. Email: sehrish.iftikhar05@gmail.com

Sumera Iqbal, Department of Botany, Lahore College for Women University, Jail Road, Lahore, Pakistan. Email: sumeraiqbal2@yahoo.com

Khajista Jabeen, Department of Botany, Lahore College for Women University, Jail Road, Lahore, Pakistan. Email: khajista_1@hotmail.com

Phanikanth Jogam, Department of Biotechnology, Kakatiya University, Warangal-506009, India. Email: phanikanth.jogam@gmail.com

Sana Khalid, Department of Botany, Lahore College for Women University, Jail Road, Lahore, Pakistan. Email: sanakhalidpu@gmail.com

Dipanshu Kumar, Department of Life Sciences, School of Sciences, CHRIST (Deemed to be University), Hosur Road, Bangalore – 560 029, India. Email: dipanshu.kumar@mbty.christuniversity.in

Hamid Manzoor, Institute of Molecular Biology and Biotechnology, Bahauddin Zakariya University, Multan, Pakistan. Email: dr.hamid@bzu.edu.pk

Altafhusain Nadaf, Department of Botany, Savitribai Phule Pune University, Pune, Maharashtra, India. Email: abnadaf@unipune.ac.in

Muddesra Naeem, Department of Botany, Lahore College for Women University, Jail Road, Lahore, Pakistan. Email: Muddesra508@gmail.com

Nimra Naeem, Department of Botany, Lahore College for Women University, Jail Road, Lahore, Pakistan. Email: nimranaeem245@gmail.com

Praveen Nagella, Department of Life Sciences, School of Sciences, CHRIST (Deemed to be University), Hosur Road, Bangalore – 560 029, India. Email: praveen.n@christuniversity.in

Kiran Nawaz, RNAissance Ag. TechAccel, 1005 North Warson Road, Ste 375, Saint Louis, MO 63132, USA. Email: Kiran@techaccel.net

Himanshu Pandey, Department of Agriculture, Khalsa College, Amritsar, India. Email: hpandey1990@gmail.com

Gunvant B. Patil, Institute of Genomics for Crop Abiotic Stress Tolerance (IGCAST), Department of Plant and Soil Science, Texas Tech University, Lubbock, Texas, USA. Email: gunvant.patil@ttu.edu

Blessy Ponnachan, Department of Life Sciences, School of Sciences, CHRIST (Deemed to be University), Hosur Road, Bangalore – 560 029, India. Email: blessy.ponnachan@mbty.christuniversity.in

Lubna Rasool, Department of Chemistry, University of Sahiwal, Sahiwal 57000, Pakistan. Email: lubnarasool@uosahiwal.edu.pk

Maria Rasul, Department of Environment and Energy, Sejong University, Seoul 05006, South Korea. Email: mariarasul@ymail.com

Sumaira Rasul, Institute of Molecular Biology and Biotechnology, Bahauddin Zakariya University, Multan 60800, Pakistan. Email: dr.sumaira@bzu.edu.pk

Abdur Rauf, Department of Chemistry, University of Sahiwal, Sahiwal 57000, Pakistan. Email: drarauf@uosahiwal.edu.pk

Fozia Saeed, Institute of Molecular Biology and Biotechnology, Bahauddin Zakariya University, Multan 60800, Pakistan. Email: fozia.saeed12@yahoo.com

Muhammad Naeem Sattar, Central Laboratories, King Faisal University, Box 31982 Al-Ahsa, Al-Hofuf, Saudi Arabia. Email: naeem.sattar1177@gmail.com; mnsattar@kfu.edu.sa

S.R.U. Shah, Department of Soil and Environment, Swedish University of Agricultural Sciences, Uppsala, Sweden. Email: syed.shah@slu.se

Shabnum Shaheen, Department of Botany, Lahore College for Women University, Jail Road, Lahore, Pakistan. Email: shabnum_shaheen78@hotmail.com

Muhammad Naveed Shahid, Department of Botany, Division of Science and Technology, University of Education, Lahore, Pakistan. Email: naveed.shahid@ue.edu.pk

Avinash Sharma, Faculty of Agricultural Sciences Arunachal University of Studies, Namsai 792103, India. Email: avinashcau@gmail.com

Hemendra Shinde, K.V.N. Naik Arts, Commerce and Science College, Nashik, Maharashtra, India. E-mail: shindehemant79@gmail.com

Baraka Siddique, Centre of Excellence in Molecular Biology (CEMB), Lahore, Pakistan. Email: barkah.siddique@gmail.com

Devendra Singh, Faculty of Biotechnology, Institute of Biosciences and Technology, Shri Ramswaroop Memorial University, Barabanki-225003, Uttar Pradesh, India. Email: devendrasingh.ibst@srmu.ac.in

Nidhi Singh, Medi-caps University, Indore, India. Email: nidhi.singh@medicaps.ac.in

Renuka Suravajhala, Amrita School of Biotechnology, Amrita Vishwa Vidyapeetham, Clappana PO 690525, Kerala, India. Email: rainprash@gmail.com

Muhammad Nouman Tahir, Department of Plant Protection, Ghazi University, Dera Ghazi Khan, Pakistan. Email: mntahir@gudgk.edu.pk

Tooba Wajid, Department of Botany, Lahore College for Women University, Jail Road, Lahore, Pakistan. Email: wajidtooba22@gmail.com

Sana Zahra, Department of Plant Protection, Ghazi University, Dera Ghazi Khan, Pakistan. Email: zahrasana492@gmail.com

Preface

The past 50 years have seen remarkable growth in food crop productivity. Notably, advancements in crop improvement techniques, like those introduced during the Green Revolution (1966–1985), significantly increased yields for staple crops like wheat and rice in developing countries. Later, recombinant DNA technology led to the development of highly efficient genetically modified (GM) crops, pioneered by researchers like Mary-Dell Chilton who codeveloped a key transformation method. However, GM crops face limitations in cost and public acceptance in some regions. Additionally, conventional breeding struggles to keep pace with the combined challenges of population growth and climate change. For instance, the current rate of yield increase for major crops like wheat, rice, maize, and soybean needs to double to meet projected demands by 2050. These challenges necessitate the exploration of novel breeding technologies that hold the potential to revolutionize agriculture.

One such transformative technology is genome editing. It allows precise modification of an organism's DNA sequence, often by creating targeted double-stranded breaks (DSBs) at specific locations. Sequence-specific nucleases (SSNs) act as molecular scissors, engineered to induce these DSBs. Different SSN types, including zinc-finger nuclease (ZFN), transcription activator-like effector nuclease (TALEN), and the CRISPR (clustered regularly interspaced short palindromic repeats)-Cas systems, have been successfully used for efficient genome editing in various plant species. Among these, CRISPR-Cas has emerged as the preferred method due to its ease of customization. CRISPR-Cas utilizes a guide RNA molecule designed to recognize and cleave the target DNA, significantly simplifying the engineering process.

Developed in 2012 and applied to eukaryotic cells in 2013, CRISPR-Cas has become a revolutionary tool in plant biology. It empowers researchers to delve deeper into the genetics of non-model plants and offers a powerful breeding technology for crop improvement. The use of CRISPR in plant research has seen a dramatic rise in recent years. However, choosing the most appropriate system for genome editing in plants and achieving the desired outcome can be challenging for newcomers to the field. This book aims to bridge this gap by providing comprehensive coverage of genome editing systems and their applications in plants.

The book offers a structured journey through genome editing in plants. Chapters 1 and 2 explore the history and core concepts, whereas chapters 3 through 9 dive into specific editing techniques and their essentials. Chapters 10 through 17 showcase practical applications for crop improvement, and chapter 18 discusses the regulatory, biosafety, and ethical considerations of genome editing. This resource is valuable for researchers, students, and professionals in crop science and biotechnology.

We extend our appreciation and gratitude to the contributing authors for their tremendous efforts and diligence and to the CABI staff for allowing us to publish this book.

Jameel M. Al-Khayri
Al-Ahsa, Saudi Arabia

Muhammad N. Sattar
Al-Ahsa, Saudi Arabia

Sudhir K. Sopory
New Delhi, India

Shri Mohan Jain
Helsinki, Finland

Acknowledgement

The editors extend their appreciation to the support of the Deanship of Scientific Research, Vice Presidency for Graduate Studies and Scientific Research, King Faisal University, Saudi Arabia (Project No. GrantA0000).

1 Historical Perspectives of Cutting-edge Genome Editing Techniques

Farooq Ahmad[1,2*]

[1]Department of Plant Production and Forest Resources, University of Valladolid, Palencia, Spain; [2]Sustainable Forest Management Research Institute, University of Valladolid & INIA, Palencia, Spain

Abstract

Humans have always been curious about the heredity of traits from one generation to another. For a large part of history, however, their curiosities only led to the development of mysteries and philosophies. Gregor Mendel developed testable hypotheses in genetics and tested these hypotheses using his famous pea plants in the 1860s. After his discoveries, it took another century to properly understand the heredity and functioning of genes. Once the molecular basis of gene functions was understood in greater details, the race was on to change the genome and produce organisms with the desired traits. Humans have been engineering genomes for centuries, unintentionally, through classical breeding. DNA discovery opened the path to genetic engineering, and several tools have become available over the years. Mutagenesis was one of the first tools available to edit plant genomes and prepare the plants of the future. For this purpose, chemical or physical mutagens were used to induce mutations in the plants and select mutants with desirable traits. Classical breeding and mutagenesis were successfully used to prepare crops with desirable characteristics such as high yield, disease resistance, and resistance to environmental stress. However, breeding techniques were limited, as it was only possible to breed plants of a single or related species. Mutagenesis was also a painstakingly slow process and looked like searching for a needle in a haystack. By the end of the 20th century, scientists discovered several other ways to edit the DNA sequence precisely and change the genomes. The more prominent methods included *Agrobacterium*-mediated gene transformation and zinc-finger nucleases (ZFNs). The 21st century brought more advanced and precise technologies, such as transcription activator-like effector nucleases (TALENs) and clustered regularly interspaced short palindromic repeats (CRISPR). The latter has become a revolutionary technology in genome editing and is widely used these days to edit plant genomes. The ethical concerns of genome editing persist, thus slowing the development of plants with new and desirable traits through this technology. Several countries have banned the use of genetically modified organisms because of ethical and safety concerns. Others have made strict laws, making it virtually impossible to commercialize genome-edited plants. However, modern CRISPR-based methods are becoming increasingly acceptable in some countries because of their effectiveness and precision. In the near future, we expect to see a few vegetables and fruits in the market that were modified using modern genome-editing tools.

Keywords: Plant genome editing, History, Modern technology, Zinc-finger nucleases (ZFNs), Transcription activator-like effector nucleases (TALENs), Clustered regularly interspaced short palindromic repeats (CRISPR)

*Corresponding author: Farooq.Ahmad@uva.es

DOI: 10.1079/9781800622517.0001

1.1 Introduction

Several millennia ago, humans became fascinated with observations of inheriting traits from parents to offspring. They were curious about family resemblance and similarities in their physical and behavioral traits. This sense of inheritance was the basis of the family system and royal dynasties in the old times. Around 6000 years ago, humans living in Babylonia stored the pedigree of horses and their traits (Donaghy, 2014). The selective breeding of animals started in early nomadic tribes when humans started using desirable characteristics and started the domestication of wild animals (Marshall *et al.*, 2014). Thanks to selective plant breeding, humans began growing crops, transforming our civilization from hunter-gatherers to growers. Some old carvings were found to represent the pollination process in date palm trees (DeMason and Sekhar, 1988). Although heredity was known to exist, no mechanism explained heredity and it remained a mystery.

Several philosophers came forward to explain heredity in different ways. Heredity philosophies were based on abstract ideas and logical thinking without scientific evidence. Nevertheless, these ideas were crucial to providing testable hypotheses to 19th- and 20th-century scientists. Hippocrates and Aristotle proposed the most influential early theories of heredity. The concept of Hippocrates was more similar to Darwins' theory than Aristotle and he suggested in his book 'On Generation' that heredity material comes from throughout the body (Zirkle, 1935). Aristotle, however, said that traits are transmitted through semen (which he thought to be a purified form of blood) and the mother's menstrual blood (Zirkle, 1935). According to him, semen and menstrual blood interact in the womb to direct the early development of an organism. Several other philosophers and scientists proposed theories about heredity, but no one could develop a testable hypothesis of heredity until 1856. Between 1856 and 1865, Mendel developed scientific hypotheses about genetics and tested them with his genius experiments. Because of his contributions to the field, he is known as the father of genetics (Schwarzbach *et al.*, 2014).

The term 'genetics' was first used in 1905 by an English scientist, William Bateson (Bateson, 2002). He referred to the study of heredity and the inheritance of traits from one generation to the next. The term was derived from the Greek word 'genetikos', which means 'originating' or 'causing to be born'. Even though genetics as a field was introduced by Bateson, the term to refer to hereditary properties already existed. For example, the word 'gene' was already used by a Danish botanist Wilhelm Johannsen in 1909 (Churchill, 1974).

Decades before Bateson, Gregor Mendel made a groundbreaking discovery in 1865 regarding the inheritance of traits from parents to offspring. However, his findings went unnoticed and unappreciated due to the scientific community's focus on Charles Darwin's influential work, 'The Origin of Species by Natural Selection'. It was not until 35 years later, in 1900, that Mendel's laws were independently rediscovered by three botanists: Hugo de Vries in the Netherlands, Carl Korens in Germany, and Erich von Chermak in Austria (Corcos and Monaghan, 1990). Through their experiments involving crossbreeding various plant types, these experiments independently confirmed and emphasized the profound significance of Mendel's conclusions. The rediscovery and validation of Mendel's laws in the 1900s was a critical point in genetics, establishing it as a fundamental and vital scientific discipline. This period from 1900 to 1912 was the triumph of Mendelism, as his laws were recognized as universally applicable to all organisms.

The second milestone in genetics was the construction of the chromosomal theory of inheritance. In 1900s, Boveri and Sutton proposed the hypothesis that chromosomes serve as carriers of hereditary factors (O'Connor and Miko, 2008). Thomas Hunt Morgan tested this theory on fruit flies, demonstrating that genes are, in fact, located on chromosomes and play a crucial role in heredity (O'Connor and Miko, 2008). Morgan collaborated with his students (A. Sturtevant, K. Bridges, and H. Mueller) to establish the chromosomal theory of heredity. This theory provided an explanation of experiments of Hugo de Vries and others. It also enabled the construction of chromosome maps indicating gene locations. In 1913, Sturtevant created the first chromosome map for a

Drosophila (fruit fly) chromosome (Sturtevant, 1951). The chromosomal theory also facilitated the understanding and proof of the chromosomal mechanism of sex determination, as explored by Thomas H. Morgan, the German geneticist Richard B. Goldschmidt, and the American cytologist, Edmund B. Wilson. This period of genetic research witnessed significant advancements in agriculture-related fields, such as establishing inheritance patterns for quantitative traits and recognizing the importance of heterosis, as exemplified by G. Nilsson. The Swedish scientist Wilhelm Johannsen raised the question of the search for the biochemical nature of the gene (Roll-Hansen, 2010).

The period between 1925 and 1940 was characterized mainly by the discovery of several processes that cause mutations. Spontaneous mutations have long been known, but the causes were not well studied. Some researchers, such as Weissman and Hugo de Vries, believed that mutations occur in the body spontaneously under the influence of internal causes and do not depend on external forces (Endersby, 2013). Several experiments rejected this hypothesis by inducing mutations artificially. For the first time in 1925, G. Nadson and G. Filipov caused mutations in yeast by radiation and in 1927, H. Mueller caused mutational changes in Drosophila with X-rays (Inge-Vechtomov *et al.*, 2019). These experiments marked the beginning of radiation genetics and the emergence of mutation breeding.

The next stage in the advancement of genetics was from 1940 to 1955. This era saw an increased work on physiological and biochemical traits of living organisms and the inclusion of microorganisms in genetic research. The following findings were particularly important: genetic transformation of bacteria (Méthot, 2016), DNA as a carrier of genetic information (Avery *et al.*, 1944), and the discovery of the structure of the DNA molecule (Watson and Crick, 1953). These findings were of great importance for the development of molecular genetics and molecular biology. During this stage, several advancements were also achieved in genetics and cytological studies of genetic diseases in humans (Ford, 1960). After all these discoveries, it became clear that humans can use genetics not only to study but also to improve their health. Medical genetics emerged to investigate the cause of genetic diseases and identify humans at risk of developing diseases (Böök, 1955). Research in the field of population genetics continued and included human subjects. On the other hand, chemical mutagenesis also continued, especially in plants. Crops that were better suitable to climatic conditions and had higher yields were developed through these technologies (Monteith, 1977). Classical breeding techniques were slowly overtaken by radiation breeding and chemical mutagenesis (Van Harten, 1998).

The development of modern genetics dates from 1955 to the present day. It is characterized by the study of genetic phenomena at the molecular level. With the work of Watson and Crick, it was clear that genes are parts of the DNA molecule and that hereditary information is localized in the chemical structure of genes. In 1969, H.G. Khorana artificially synthesized a large number of copies of parts of a gene for the first time, which marked the beginning of a new branch in genetics: genetic engineering and its subdivision cell engineering (Khorana, 1978). The year 1991 proved to be the most significant in the history of DNA when the number of publications mentioning the word 'DNA' in their title, abstract, or keywords more than doubled compared with the previous year (Fig. 1.1). This was probably due to the collective impact of the human genome project (Watson and Cook-Deegan, 1991), new polymerase chain reaction (PCR) tool (Hayashi, 1991), and advancements in genetic engineering tools (Martuza *et al.*, 1991).

1.2 Genome Editing

With advancements in genetics, humans realized the possibility of changing genetic information and producing organisms with desired traits. Therefore, breeders started looking for efficient and safe technologies to precisely modify genomes. Breeders realized that humans have been editing genomes unknowingly for centuries through breeding practices. However, classical breeding methods for crop improvement have their limitations. First, organisms belonging to a single species (or closely related species) could be bred with each other (Poehlman, 2013). Second, the breeding process did not

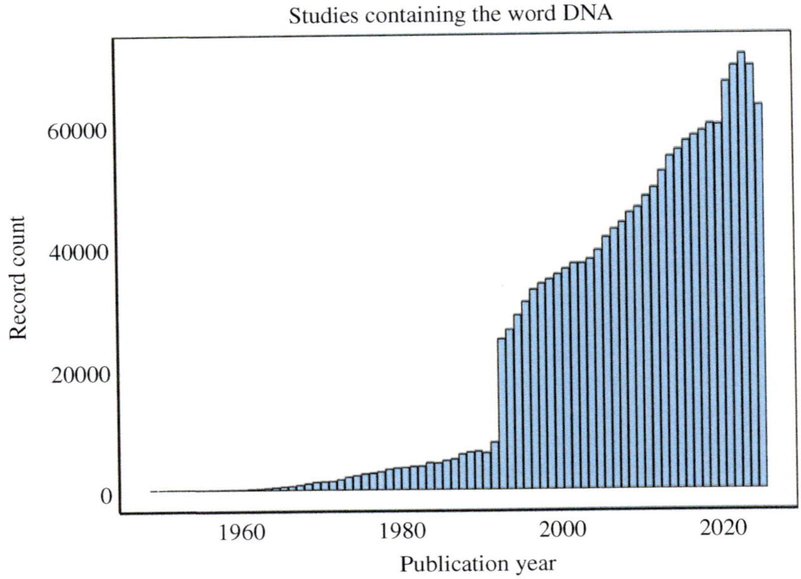

Fig. 1.1. The number of printed publications mentioning the word 'DNA' in their titles, abstracts, or keywords available in the database of Web of Science (accessed: 18 March 2024). Author's own figure.

allow precise genetic modifications (Poehlman, 2013). A better understanding of DNA helped overcome these challenges and precise editing of DNA revolutionized plant sciences (Chen *et al.*, 2019). These advancements allowed targeted modifications in plant genomes, improving crop traits and tackling global issues like food security and environmental sustainability. Genome-editing techniques will be vital for sustaining the ever-increasing population's food demands. Below, different genome-editing strategies are discussed in detail, starting from classical breeding to modern genetic engineering tools.

1.3 Classical Breeding

The history of plant genome editing can be traced back to traditional breeding practices, which involved mating plants with desirable traits to create new varieties. This approach, known as classical breeding, has been extremely useful in crop improvement for centuries. Early plant breeders selected plants with desired traits such as high yield, disease resistance, and nutritional content and used them as parents to produce the next generation of plants (Ulukan, 2011). Over time, this process led to the development of improved crop varieties.

Classical breeding techniques date back thousands of years. The earliest evidence of intentional plant breeding can be traced back to ancient civilizations, where early farmers selectively saved seeds from plants with desirable traits for the next planting season (Hawkes, 1983). This practice of selecting and propagating plants with beneficial characteristics laid the foundation for developing formal breeding programs. Classical breeding gained significant momentum during the 18th and 19th centuries (Davidson, 2008). Plant breeders began to adopt more systematic approaches to enhance crop traits and adapt plants to different environmental conditions. This era saw the emergence of notable plant breeders who made substantial contributions to the field. The work of Luther Burbank on plant hybridization and selection was particularly significant (Stansfield, 2006).

Selective breeding has been crucial in developing improved crop varieties and emerged as a core aspect of classical breeding techniques. Early plant breeders recognized the influence of selective breeding on plants' characteristics and actively sought to enhance desirable traits (Allard, 1999). The practice of selective breeding evolved gradually through the understanding of genetics. In the early 20th century, breeders began to employ scientific methods to improve crops systematically (Kingsbury, 2011). Shortly after, breeders could identify genetic traits and develop statistical tools to analyze breeding data that contributed to the refinement of selective breeding techniques. Breeders could also correlate phenotypic traits with specific genetic markers, further enhancing the efficiency of selective breeding programs (Holland, 2004). The techniques of crossbreeding and hybridization also started to improve with time (Kingsbury, 2011). These techniques involved crossing different plant varieties to combine their favorable traits.

A practical example of such breeding programs is the period of the Green Revolution, which revolutionized global agriculture through the development of high-yielding varieties of crops in the 1950s (Borlaug, 2002). One significant example is the development of semi-dwarf wheat varieties by a scientist called Norman Borlaug (Borlaug *et al.*, 1969). New varieties developed under this program played a crucial role in increasing wheat yields, particularly in Mexico and India. These varieties did not only give higher yields but were also resistant to lodging (falling over due to the weight of the grains). The Green Revolution's impact extended beyond wheat to crops like rice, maize, and soybeans, significantly boosting global food production and alleviating hunger in many parts of the world (Borlaug, 2002).

The knowledge about plant reproductive biology was also improving, contributing to even better breeding techniques. Pioneers like Joseph Goss, Thomas Fairchild, and Charles Naudin contributed significantly to understanding and applying hybridization techniques (Rieseberg, 1997). Their work demonstrated that crossing two genetically distinct parental lines could result in offspring with desirable traits, often exhibiting hybrid vigor. Scientists began to investigate the principles underlying hybrid vigor and found that the high yield is due to the beneficial effects of genetic diversity in hybrid offspring (Kingsbury, 2011). This knowledge generated by scientists enabled breeders to strategically utilize hybridization and develop superior crop varieties with higher yield, disease resistance, and other desirable traits (Kingsbury, 2011). An important example is the hybridization of maize pioneered by George H. Shull and Edward M. East (Shull, 1911). They produced hybrids that displayed remarkable vigor and uniformity by crossing inbred lines of different traits. Hybrid vigor has since been applied to a wide range of crops, including tomatoes, rice, and cotton, contributing significantly to global food production (Kingsbury, 2011).

1.4 Inbreeding

Inbreeding, in contrast to hybrid vigor, is a technique to mate individuals within the same line or variety that have a common ancestry (Blouin and Blouin, 1988). It is a type of classical breeding that has been practiced for centuries. Early farmers used inbreeding by saving seeds from the best-performing plants and using these seeds to prepare future plants. Over time, this led to the development of locally adapted landraces and heirloom varieties. In the 19th and 20th centuries, plant breeders actively employed inbreeding techniques to stabilize desirable traits and eliminate undesirable ones. This involved repeated self-pollination or crosses between closely related individuals. Notable historical figures, such as the Danish botanist Wilhelm Johannsen, helped in understanding inbreeding and coined the term 'pure line' to describe genetically uniform populations resulting from inbreeding (Bonneuil, 2016). Inbreeding was among the first ways humans could control plant traits. However, it is a slow and time-consuming process that could take years or even decades to complete. Therefore, scientists were looking for faster ways to achieve the same or better results through modern technologies.

Inbreeding has several advantages; it leads to exposure and elimination of harmful recessive genes (Jain, 1976). It also helps in the accumulation of better genes, especially from the inbred line (Jain, 1976). However, inbreeding also has some limitations; prolonged inbreeding often leads to a decrease in genetic diversity instead of increasing it. This is particularly the case in natural populations, for example Amos and

Harwood (1998). One historical example of the consequences of inbreeding was the case of the European royal families, where marriages among close relatives led to increased rates of genetic disorders. In agriculture, continuous inbreeding can result in decreased yields, susceptibility to diseases, and reduced adaptability to changing environmental conditions (Jain, 1976). Thus, while inbreeding can be a useful tool in certain breeding programs to establish uniformity or fix specific traits, it must be carefully managed to avoid the negative consequences associated with decreased genetic diversity.

1.5 Mutation Breeding and Mutagenesis

The systematic application of mutagenesis techniques created mutation breeding programs in the mid-20th century. Mutation breeding is to use radiation or chemical mutagens to induce genetic variations in crops and combine it with classical selection, that is, inbreeding and other breeding techniques (Das *et al.*, 2014). The rigorous screening and selection processes of classical breeding helped improve the development of crops with desirable traits. This involved evaluating the mutant population for traits like improved yield, disease resistance, abiotic stress tolerance, nutritional quality, and other agronomically important traits (Das *et al.*, 2014). Selected mutants with improved characteristics were further bred and developed into new crop varieties. A significant success story of mutation breeding was the development of new wheat varieties with improved resistance to diseases such as stem rust and leaf rust (Alemu, 2019). Breeders employed gamma radiation to induce mutations in wheat in the 1960s and 1970s, resulting in the development of high-yielding, disease-resistant cultivars that helped to increase the world food supply dramatically.

Mutagenesis involves using chemical- or radiation-induced mutations and introducing genetic variability (Datta, 2014). In this technique, plants are exposed to different mutagens (either physical or chemical) that induce random changes in their DNA and generate mutant phenotypes with altered traits. The mutants are screened for desirable characteristics and used as breeding parents to develop new crop varieties. Mutagenesis techniques played a significant role in plant sciences by contributing to the development of improved crop varieties. This laid the foundation for later advancements in genome editing by demonstrating the importance of genetic variation in crop improvement.

The understanding of mutagenesis began with the observation of spontaneous mutations in plants (Datta, 2014). Over the years, mutagenesis techniques have continued to evolve, with advancements in radiation and chemical mutagenesis. The discovery of new mutagens and improved protocols for mutagen treatment enhanced the efficiency and precision of mutagenesis. Therefore, mutation breeding became a more systematic and targeted approach in plant breeding programs. Fig. 1.2 shows new mutant varieties of different crops that were developed around the world using technologies like mutagenesis.

1.6 Radiation Mutagenesis Versus Chemical Mutagenesis

The discovery of X-rays by W. Roentgen in 1895 and radioactivity by H. Becquerel in 1896 opened new possibilities for mutagenesis research (Kardamakis *et al.*, 2023). Researchers found out that exposing plants to X-rays and radioactive substances could induce changes in plant DNA, leading to the development of new traits. The work of Hermann Muller in the 1920s and 1930s was significant in this regard (Pontecorvo, 1968). He demonstrated that X-rays could induce mutations in fruit flies and could be used to create heritable genetic variations. This discovery was done in an animal, but it led to the possibility of the deliberate use of radiation as a mutagenic tool in plant breeding.

Parallel to the developments in radiation mutagenesis, scientists were trying to use chemicals to induce plant mutations. The discovery of chemical mutagens in the 1940s, such as ethyl methanesulfonate and N-methyl-N-nitrosourea, was a significant step toward using chemicals to cause mutation in DNA (Mullins *et al.*, 2021). Researchers found out that these chemicals could cause changes in DNA structure, leading to genetic alterations. Chemical mutagens

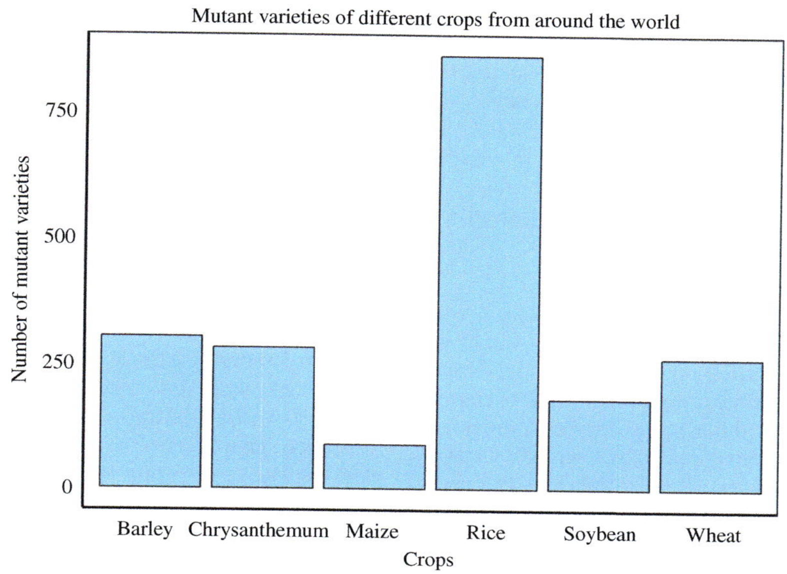

Fig. 1.2. Number of reported mutant crop varieties according to the International Atomic Energy Agency's (IAEA) online database (International Atomic Energy Agency, 2024).

offered several advantages over radiation, including ease of use and the ability to control dosage. Researchers can precisely administer the mutagens to plants, leading to a controlled induction of mutations.

1.7 Genetic Engineering

Before the discovery of DNA structure, it was a blackbox for the breeders who did not know how it worked. Breeders could modify it and cause mutations but were not accurate about what a mutation was at the structure level. The Watson and Crick model provided a deeper understanding of how genetic material is organized in a double-helix form (Watson and Crick, 1953). A code-like arrangement of nucleotides showed that a change in a nucleotide might also change the phenotype. The double-helix structure of DNA revealed the basis for heredity and the transfer of genetic information. This breakthrough sparked scientists' interest in manipulating DNA to change the traits of living organisms. Breeders no longer had to rely on using mutagens to cause random mutation if they could find a way to change the sequence of nucleotides. This looked impossible until the 1970s when some scientists successfully transferred genetic material between organisms through recombinant DNA technology (Sinsheimer, 1977).

Recombinant DNA technology involved cutting DNA fragments from one organism and inserting them into the genome of another. Genetically modified plants were produced using this technology by introducing foreign genes into plant genomes (Sinsheimer, 1977). A significant shift in plant breeding approaches was under way, as scientists could introduce specific genes into plants to induce desired traits, such as herbicide or insect resistance (Khan *et al.*, 2016). Recombinant DNA technology revolutionized plant sciences by making the modifications of plant genomes much more precise than classical breeding and mutation breeding. This technology has profoundly improved crops and agricultural productivity and contributed to solving global challenges like food security and environmental sustainability (Khan *et al.*, 2016). Key historical milestones include: Stanley Cohen and Herbert Boyer successfully fused DNA from different sources to create the first recombinant DNA molecule (Cohen *et al.*,

1973). This breakthrough laid the groundwork for genetic engineering. Another landmark achievement was the invention of PCR in 1983 by Kary Mullis, enabling the rapid amplification of specific DNA sequences (Mullis, 1990).

1.8 Agrobacterium-mediated Transformation

The discovery of *Agrobacterium tumefaciens* as a natural genetic engineer was a significant milestone to genetically engineer plants (Nester, 2015). Researchers found that the soil bacterium *Agrobacterium*, known for causing crown gall disease in plants naturally transfers its DNA portions to plants. These transferred DNA parts, known as the T-DNA, were exploited to develop *Agrobacterium*-mediated transformation for introducing foreign genes into plants. The T-DNA was modified to remove the disease-causing genes and replace them with desired genes of interest. This enabled the production of genetically modified plants with novel traits, such as herbicide resistance, insect resistance, or improved nutritional content (Khan *et al*., 2016). In the past decades, we have been seeing a continuous increase in genetic engineering research shown by the number of publications that mentioned the words 'genetic engineering' in either title, abstract, or keywords (Fig. 1.3). The figure shows a constant increase in the articles published, showing the scientific community's potential interest in the topic over the years.

The first transgenic plants were created in the early 1980s using *Agrobacterium*-mediated transformation (Gelvin, 2003). Plant biologists expressed genes from bacteria and viruses in plants to demonstrate the feasibility of introducing foreign genes into plants. Some tobacco plants expressed a bacterial gene for antibiotic resistance and marked a significant milestone in plant genetic engineering (Fraley *et al*., 1983). For the first time in the history of genetics, crop improvement broke through traditional breeding and mutagenesis methods. Researchers could now introduce genes of interest to induce desirable traits, such as enhanced pest resistance or tolerance to environmental stresses in plants (Fraley *et al*., 1983). Moreover, the expression of genes encoding enzymes for metabolic pathways enabled the production of plants with improved

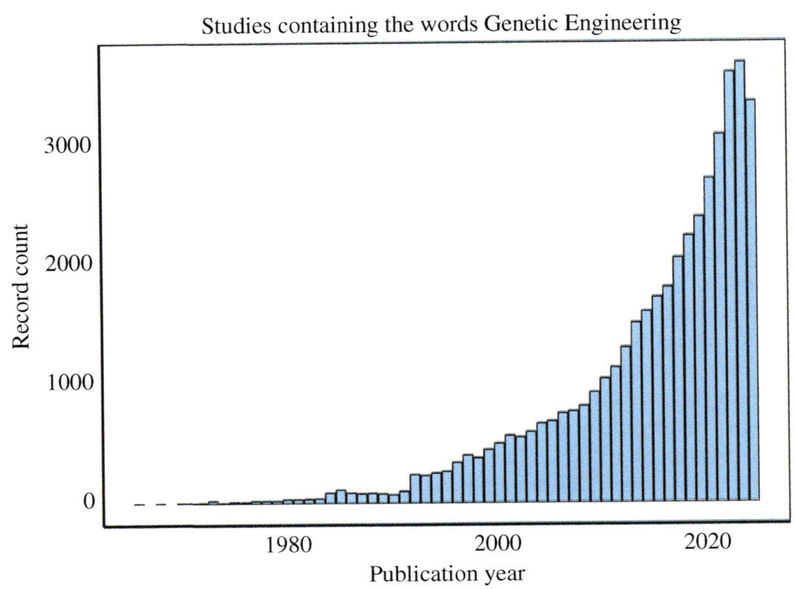

Fig. 1.3. The number of printed publications mentioning the words 'Genetic Engineering' in their titles, abstracts, or keywords available in the database of Web of Science (accessed: 18 March 2024). Author's own figure.

nutritional value or the synthesis of valuable compounds (Galili *et al.*, 2002).

1.9 Transformation Systems and Genome-editing Tools

Researchers developed alternative methods for delivering foreign DNA into plant cells and expanded the range of plant species amenable to genetic transformation. Biolistic or particle bombardment techniques were developed in the late 1980s to make a direct delivery of DNA-coated particles into plant cells (Taylor and Fauquet, 2002). This approach facilitated the alteration of crop species that were previously difficult to manipulate through *Agrobacterium*-mediated transformation.

1.10 Zinc-finger Nucleases (ZFNs)

Zinc-finger nucleases (ZFNs) are engineered proteins that recognize specific DNA sequences and induce double-stranded breaks at desired locations in the genome (Durai *et al.*, 2005). The induced double-stranded cuts stimulate the plant's natural DNA repair mechanisms and modify that part of the genome. The modification usually results in gene silencing, as the repair mechanisms add other nucleotides that will make the gene non-functional (Durai *et al.*, 2005). ZFNs are designed by fusing a zinc-finger DNA-binding domain with a nuclease domain to recognize and cut specific DNA sequences. ZFN is a promising genome-editing tool that allowed scientists to modify plant genomes precisely without introducing a foreign gene. ZFNs were successfully used to create targeted mutations or insertions in plant genes (Townsend *et al.*, 2009). However, ZFNs had limitations in terms of specificity and ease of design (Gaj *et al.*, 2013). The development of more versatile and efficient genome-editing techniques was needed.

The use of ZFNs as a genetic engineering tool can be traced back to the discovery of zinc-finger proteins (ZFPs). These proteins are DNA-binding proteins characterized by their ability to recognize specific DNA sequences. In the 1980s, scientists unraveled the structural and functional properties of ZFPs that lead to the development of engineered zinc-finger domains (ZFDs) with custom DNA-binding specificity. However, it wasn't until the early 2000s when researchers realized the potential of combining engineered ZFPs with the nuclease activity of a restriction enzyme called FokI (Gaj *et al.*, 2013). This fusion created the first ZFNs to introduce targeted DNA double-stranded breaks at specific loci (Townsend *et al.*, 2009).

ZFNs were a breakthrough in the field of targeted genome editing. Unlike earlier gene-editing techniques, which relied on random DNA integration, ZFNs could precisely modify specific DNA sequences in the genome without introducing a foreign gene (Chandrasegaran and Carroll, 2016). ZFNs function as dimers, with each ZFD recognizing a particular triplet of DNA. When the two ZFN monomers bind adjacent target sites, the FokI nuclease domains are brought into proximity, resulting in the cleavage of the DNA double helix. This targeted DNA cleavage triggered cellular repair mechanisms, such as non-homologous end joining (NHEJ) or homology-directed repair (HDR). These mechanisms could be exploited to introduce specific genetic modifications.

In the past years, researchers have utilized ZFNs to introduce targeted mutations, knock-outs, gene replacements, or plant insertions (Gaj *et al.*, 2013). These changes had several applications in agriculture. For example, they helped improve several crop traits, such as disease resistance and nutritional content (Barbadikar *et al.*, 2019). One significant example is the development of wheat that was resistant against powdery mildew, a fungal disease that poses a significant threat to wheat production worldwide. For this purpose, researchers precisely edited the susceptibility gene Mildew Locus O (MLO) of the wheat (Wang *et al.*, 2014). This achievement demonstrated the potential of ZFNs in conferring durable resistance against pathogens. ZFNs have also been utilized to enhance agronomically important characteristics, such as yield and drought tolerance (Davies *et al.*, 2017). ZFNs helped develop crops with improved nutritional value, including increased iron or zinc content, addressing nutrient deficiencies in vulnerable and malnourished populations (Barbadikar *et al.*, 2019).

Over the years, advancements in ZFNs technology have improved efficiency and specificity. Researchers refined the design of ZFDs to enhance DNA-binding specificity and minimize off-target effects (Pattanayak *et al.*, 2011). Moreover, the modular assembly methods and better computational tools facilitated the engineering of customized ZFNs to target desired genomic sites more precisely (Periwal, 2017). However, it is important to note that the design and optimization of ZFNs were labor-intensive and technically challenging so that limited their widespread adoption (Pattanayak *et al.*, 2011). The large size of ZFNs and the requirement for two monomers to function in a dimeric fashion have also posed challenges in their delivery into plant cells (Gaj *et al.*, 2013; Ain *et al.*, 2015). Therefore, scientists were looking to overcome these challenges through new tools of genomic editing.

1.11 Transcription Activator-like Effector Nucleases (TALENs)

Transcription Activator-like Effector Nucleases (TALENs) emerged as an alternative genome-editing tool to ZFNs. TALENs are designer enzymes that use a DNA-binding domain derived from TAL effectors to recognize specific DNA sequences (Carlson *et al.*, 2012). TALENs induce double-stranded breaks and trigger DNA repair processes, just like ZFNs do. However, they showed improved specificity compared to ZFNs that led to improved and precise genome-editing applications (Carlson *et al.*, 2012).

Transcription Activator-like Effectors (TALEs) led to the discovery of TALENs and their use in genetic engineering in the early 2000s. TALEs are naturally occurring DNA-binding proteins of some plant-pathogenic bacteria, such as *Xanthomonas* species. TALEs consist of a repetitive array of modular DNA-binding domains called repeat units (Morbitzer *et al.*, 2011). Each repeat unit is responsible for recognizing a specific DNA base pair. This modular nature of TALEs allowed for the customization and engineering of DNA-binding domains to target specific DNA sequences (Morbitzer *et al.*, 2011).

The fusion of TALE DNA-binding domains with nuclease domains was the next major milestone in TALEN development. For this purpose, the researchers created TALENs capable of introducing site-specific DNA-double-stranded breaks by fusing TALEs with the FokI nuclease domain. TALENs function as dimers, each monomer recognizing a specific DNA sequence (Aouida *et al.*, 2014). When the TALEN monomers bind adjacent target sites, the FokI nuclease domains are brought into proximity, leading to the cleavage of the DNA double helix (Aouida *et al.*, 2014). This targeted DNA cleavage triggers cellular repair mechanisms, such as NHEJ or HDR, facilitating precise genetic modifications (Shamshirgaran *et al.*, 2022).

The development of TALENs offered a versatile tool for targeted genome editing. They have been successfully employed in various organisms, including plants, animals, and human cells (Becker and Boch, 2021). In plant sciences, TALENs have been utilized in inducing precise genetic modifications in crops. Researchers have used TALENs to create knockout mutants, insert or replace specific genes, and cause targeted gene expression changes (Becker and Boch, 2021). This has led to advancements in crop improvement, such as enhanced disease resistance, improved nutritional content, and increased yield (Davies *et al.*, 2017; Gao, 2021).

Over the years, advancements in TALEN technology have improved efficiency and specificity. Scientists developed user-friendly methods for engineering TALENs that allowed for customizing DNA-binding domains to target specific genomic sequences. Introducing modular assembly methods and optimized design strategies has also facilitated the construction of TALENs with high specificity and reduced off-target effects (Gaj *et al.*, 2013). Additionally, new advancements in delivery methods (using viral vectors or nanoparticles) have greatly enhanced the efficiency of TALEN delivery into cells and tissues (Becker and Boch, 2021).

TALENs have been partially overshadowed by the more advanced CRISPR-Cas9 system, which revolutionized the field of genome editing (Becker and Boch, 2021). The CRISPR-Cas9 system offers several advantages over TALENs, including ease of design, multiplexing capabilities, and a wider range of targetable sequences (Doudna and Charpentier, 2014). However, TALENs continue to be utilized in specific applications where their unique features, such as longer target sites and potentially lower off-target effects, are advantageous (Shamshirgaran *et al.*, 2022).

1.12 Clustered Regularly Interspaced Short Palindromic Repeats (CRISPR-Cas)

The discovery of CRISPR-Cas (CRISPR-associated) proteins in the early 2010s revolutionized genome editing across various fields. In the beginning, it appeared just another naturally occurring immune system in bacteria and archaea (Rath *et al.*, 2015). However, its application as a powerful genome-editing tool changed it into an extraordinary discovery.

In the 1980s and 1990s, scientists observed some repetitive DNA sequences (CRISPR sequences) in bacteria and archaea of viral origin but did not fully understand their purpose (Ishino *et al.*, 2018). Almost two decades later in the year 2005, researchers found out that the CRISPR system is a unique and adaptive defense system in bacteria and archaea against viral infections (Rath *et al.*, 2015). CRISPR sequences in organisms enabled them to capture and retain snippets of viral DNA to have a memory of these viral infections (Ishino *et al.*, 2018). In the early 2010s, groundbreaking studies led by J. Doudna, E. Charpentier, and their research teams elucidated the mechanism of the CRISPR-Cas system (Doudna and Charpentier, 2014). Doudna and Charpentier identified the Cas9 protein as a key component in DNA cleavage and demonstrated its potential for targeted genome editing. The work of J. Doudna and E. Charpentier was such a breakthrough in genome editing that they were awarded a Nobel Prize in chemistry in the year 2020 (Gostimskaya, 2022).

CRISPR-Cas systems consist of two main elements: the guide RNA (gRNA) and the Cas nuclease (Doudna and Charpentier, 2014). The gRNA is a molecular guide and is complementary to the target DNA sequence. The gRNA directs the Cas nuclease to the precise genomic location, where it attaches and cleaves the DNA (Doudna and Charpentier, 2014). Cas9 is an RNA-guided endonuclease that can be programmed with a customized gRNA to target specific DNA sequences within the genome. The Cas9-gRNA complex identifies and binds to its target DNA sequence. After binding, it induces a double-stranded break in the DNA at the precise location. The cell's repair machinery then repairs the break by either NHEJ or HDR. These modifications can perform gene knockout, insertion, replacement, or precise nucleotide substitutions (Doudna and Charpentier, 2014).

CRISPR-Cas9 has transformed genome editing and provided a powerful tool for targeted genome modifications in various organisms (Gostimskaya, 2022). Its simplicity, versatility, and cost-effectiveness have made it popular in research laboratories and industrial applications (Ahmad *et al.*, 2024). CRISPR-Cas9 has been extensively used in multiple fields, including agriculture, medicine, and biotechnology. In plant sciences, CRISPR-Cas9 has been employed to develop crops with improved traits such as disease resistance, enhanced nutritional value, increased yield, and tolerance to environmental stresses (Gostimskaya, 2022). Since its discovery, the CRISPR-Cas9 system has undergone rapid advancements and refinements (Ahmad *et al.*, 2024). Scientists have significantly improved the specificity, efficiency, and versatility of CRISPR-Cas9 (Gostimskaya, 2022). They addressed the concerns of off-target effects and expanded their applications to several fields (Vakulskas and Behlke, 2019). New Cas proteins, such as Cas12a and Cas13, have expanded the CRISPR toolbox, allowing for more diverse functionalities, such as base editing, transcriptional regulation, and RNA editing (Pickar-Oliver and Gersbach, 2019). These advancements hold great potential for further progress in genetic engineering and therapeutic applications (Gostimskaya, 2022).

A significant example of CRISPR-Cas9 in plant genome editing is its role in engineering crops with improved resistance to bacterial blight (Oliva *et al.*, 2019). Bacterial blight is a devastating disease of rice (*Oryza sativa*) that is caused by the bacterium *Xanthomonas oryzae* pv. *oryzae* (Sharma *et al.*, 2017). Using CRISPR-Cas9, the susceptibility genes OsSWEET (Sugar Will Eventually be Exported Transporter) such as 11, 13, and 14 in rice were edited which led to the production of rice varieties with enhanced resistance to bacterial blight (Oliva *et al.*, 2019). The targeted gene editing through CRISPR-Cas9 significantly reduced the vulnerability of rice crops to this destructive disease. Even though many other plants were developed using CRISPR-Cas technologies, different ethical and safety issues persist (Ahmad *et al.*, 2021). Therefore, commercialization of these plants has not been possible in almost all countries (Ahmad *et al.*, 2024). Different countries are trying to tackle this issue by introducing new legislation. A good example is a proposal by the European Commission that was recently approved by the EU parliament (European Parliament, 2024). The aim of this proposal is to consider many new genomic techniques equivalent to conventional breeding methods and exempt them from the legislation on genetically modified organisms (European Parliament, 2024). Meanwhile, scientists are trying to develop and optimize technologies that can change phenotypes by altering genetic expressions without making any change in the genome. Technologies like virus-induced gene silencing and spray-induced gene silencing (VIGS and SIGS, respectively) are some examples (Robertson, 2004; Bocos-Asenjo *et al.*, 2022).

1.13 Conclusions

The history of genome editing started centuries ago when philosophers came up with different ways to explain heredity. Mendel's breakthrough work on pea plants changed that and converted philosophies of heredity to a new branch of biology called genetics. Testable hypotheses were developed later to further explain heredity, such as the chromosomal theory of inheritance developed by Thomas H. Morgan. Most of the 20th century was spent exploring genetics to better understand different mechanisms that govern heredity. However, the ultimate aim of the geneticists was to be able to change genetic information. Breeders had been doing classical breeding to improve plant traits without knowing the mechanistic understanding of genetics for centuries. Classical breeding was a slow and laborious process, like finding a needle in a haystack, and could take up to decades to develop a new variety. Another drawback of classical breeding was that it depended only on the natural variability of plants and combining traits from different varieties that already existed.

In the mid-20th century, scientists found that some chemical and physical materials cause mutations and change plant phenotypes. That was the first instance of genome editing, where scientists could artificially induce gene mutations using mutagens. The new approach was significantly better than the classical breeding practices. However, scientists still could not cause the mutations at the point of interest; they could only do so at random locations. This resulted in many screening programs that would last several years. Nevertheless, mutation breeding successfully created plants with desirable traits that went on toward the Green Revolution. Even though overall production increased during the Green Revolution, so did the global population. Intensive agriculture and high yields brought new challenges of pests and pathogens. Therefore, it became essential to use more effective tools to improve plant traits and ensure food security.

By the end of the 20th century, the Green Revolution turned into the gene revolution. The growing interest in genetics due to projects like the Human Genome Project also affected plant sciences. A breakthrough discovery in the tumor-inducing plant pathogen *Agrobacterium tumefaciens* enabled scientists to develop a precise genome-editing system. This further led to developing economically important plant varieties that boosted food production. One disadvantage of *Agrobacterium*-mediated gene editing is the potential for unwanted plasmid recombination (Sustiprijatno *et al.*, 2022). Another disadvantage is the activation of plant immune responses, such as pattern-triggered immunity, which can limit transformation efficiency (Aliu *et al.*, 2020). Due to these disadvantages, there was a search for more precise and efficient genome-editing techniques.

The 21st century became the century of genome editing, where several new and advanced tools were developed. It started with ZFNs and TALENs, eventually leading to the CRISPR-Cas system. These tools are currently used depending on the system's requirements, but the CRISPR-Cas system has become the most famous tool for editing plant genomes. ZFNs were reported to have a high probability of off-target editing. While ZFNs and TALENs have been effective in targeted genome editing, the CRISPR-Cas system offers simplicity, high efficiency, and multiplex loci editing advantages over these traditional tools. Overall, understanding and minimizing off-target effects are crucial for the safe and precise application of gene-editing technologies in plants. Several countries are struggling to legalize CRISPR-Cas gene editing, making it easier to use and develop new varieties. It is predicted that in the upcoming years, the new regulations will increase the acceptability of genome-edited plants and their use around the globe. Nevertheless, several ethical issues of genome editing exist, such as biodiversity loss by promoting monoculture farming and potentially reducing the number of available plant varieties. The potential long-term effects of editing plant genomes on soil quality or other environmental factors are also unknown.

References

Ahmad, A., Ghouri, M.Z., Munawar, N., Ismail, M., Ashraf, S. *et al.* (2021) Regulatory, ethical, and social aspects of CRISPR crops. In: Ahmad, A., Khan, S.H. and Khan, Z. (eds) *CRISPR Crops: The Future of Food Security*. Springer, Singapore, pp. 261–287.

Ahmad, A., Arif, A. and Munir, A. (2024) CRISPR technology commercialization and biosafety. In: *Global Regulatory Outlook for CRISPRized Plants*. Academic Press, Elsevier, pp. 461–514.

Ain, Q.U., Chung, J.Y. and Kim, Y.-H. (2015) Current and future delivery systems for engineered nucleases: ZFN, TALEN and RGEN. *Journal of Controlled Release* 205, 120–127.

Alemu, G. (2019) Wheat breeding for disease resistance. *Journal of Microbiology & Biotechnology* 4(2), 141.

Aliu, E., Azanu, M.K., Wang, K. and Lee, K. (2020) Generation of thymidine auxotrophic *Agrobacterium tumefaciens* strains for plant transformation. *BioRxiv*. DOI: 10.1101/2020.08.21.261941.

Allard, R.W. (1999) *Principles of Plant Breeding*. John Wiley & Sons, New York, USA.

Amos, W. and Harwood, J. (1998) Factors affecting levels of genetic diversity in natural populations. *Philosophical Transactions of the Royal Society of London. Series B: Biological Sciences* 353(1366), 177–186.

Aouida, M., Piatek, M.J., Bangarusamy, D.K. and Mahfouz, M.M. (2014) Activities and specificities of homodimeric TALENs in *Saccharomyces cerevisiae*. *Current Genetics* 60, 61–74.

Avery, O.T., Macleod, C.M. and McCarty, M. (1944) Studies on the chemical nature of the substance inducing transformation of pneumococcal types: Induction of transformation by a desoxyribonucleic acid fraction isolated from pneumococcus type III. *The Journal of Experimental Medicine* 79, 137–158.

Barbadikar, K.M., Aglawe, S.B., Mangrauthia, S.K., Madhav, M.S. and Kumar, S.P.J. (2019) Genome editing: New breeding technologies in plants. In: Banerjee, R., Kumar, G.V. and Kumar, J. (eds) *OMICS-Based Approaches in Plant Biotechnology*. John Wiley & Sons, New York, USA, pp. 245–285.

Bateson, P. (2002) William Bateson: A biologist ahead of his time. *Journal of Genetics* 81, 49–58.

Becker, S. and Boch, J. (2021) TALE and TALEN genome editing technologies. *Gene and Genome Editing* 2, 100007.

Blouin, S.F. and Blouin, M. (1988) Inbreeding avoidance behaviors. *Trends in Ecology & Evolution* 3(9), 230–233.

Bocos-Asenjo, I.T., Niño-Sánchez, J., Ginésy, M. and Diez, J.J. (2022) New insights on the integrated management of plant diseases by RNA strategies: Mycoviruses and RNA interference. *International Journal of Molecular Sciences* 23(16), 9236.

Bonneuil, C. (2016) Pure lines as industrial simulacra: A cultural history of genetics from Darwin to Johannsen. *Heredity Explored: Between Public Domain and Experimental Science, 1850-1930* 213–242.

Böök, J.A. (1955) Heredity counseling: Medical genetics and counseling practices. *Eugenics Quarterly* 2(3), 174–183.

Borlaug, N.E. (2002) The green revolution revisited and the road ahead. Nobelprize. org, Stockholm. Available at: https://static1.squarespace.com/static/5a372ca9f9a61ed6e86178a7/t/5d1656f71c9a720001483a68/1561745144191/The+Green+Revolution+Revisited+and+The+Road+Ahead.pdf (accessed 15 June 2024).

Borlaug, N.E., Narvaez, I., Aresvik, O. and Anderson, R.G. (1969) A green revolution yields a golden harvest. *Columbia J. World Business* 4, 9–19.

Carlson, D.F., Fahrenkrug, S.C. and Hackett, P.B. (2012) Targeting DNA with fingers and TALENs. *Molecular Therapy-Nucleic Acids* 1, e3.

Chandrasegaran, S. and Carroll, D. (2016) Origins of programmable nucleases for genome engineering. *Journal of Molecular Biology* 428(5), 963–989.

Chen, K., Wang, Y., Zhang, R., Zhang, H. and Gao, C. (2019) CRISPR/cas genome editing and precision plant breeding in agriculture. *Annual Review of Plant Biology* 70, 667–697.

Churchill, F.B. (1974) William Johannsen and the genotype concept. *Journal of the History of Biology* 5–30.

Cohen, S.N., Chang, A.C., Boyer, H.W. and Helling, R.B. (1973) Construction of biologically functional bacterial plasmids *in vitro*. *Proceedings of the National Academy of Sciences* 70(11), 3240–3244.

Corcos, A.F. and Monaghan, F.V. (1990) Mendel's work and its rediscovery: A new perspective. *Critical Reviews in Plant Sciences* 9(3), 197–212.

Das, P., Mishra, M., Lakra, N., Singla-Pareek, S.L. and Pareek, A. (2014) Mutation breeding: A powerful approach for obtaining abiotic stress tolerant crops and upgrading food security for human nutrition. In: Tomlekova, N.B., Kozgar, M.I. and Wani, M.R. (eds) *Mutagenesis: Exploring Novel Genes and Pathways*. Wageningen Academic Publishers, Wageningen, Netherlands, pp. 615–621.

Datta, S.K. (2014) Induced mutagenesis: Basic knowledge for technological success. In: Tomlekova, N.B., Kozgar, M.I. and Wani, M.R. (eds) *Mutagenesis: Exploring Genetic Diversity of Crops*. Wageningen Academic Publishers, Wageningen, Netherlands, pp. 19–39.

Davidson, J. (2008) *Breeding: A Partial History of the Eighteenth Century*. Columbia University Press, New York.

Davies, J.P., Kumar, S. and Sastry-Dent, L. (2017) Use of zinc-finger nucleases for crop improvement. *Progress in Molecular Biology and Translational Science* 149, 47–63.

DeMason, D.A. and Sekhar, K.N.C. (1988) The breeding system in the date palm (*Phoenix dactylifera* L.) and its recognition by early cultivators. *Advances in Economic Botany* 6, 20–35.

Donaghy, T. (2014) *Horse Breeds and Breeding in the Greco-Persian World: 1st and 2nd Millennium BC*. Cambridge Scholars Publishing, Newcastle upon Tyne, UK.

Doudna, J.A. and Charpentier, E. (2014) The new frontier of genome engineering with CRISPR-Cas9. *Science* 346(6213), 1258096.

Durai, S., Mani, M., Kandavelou, K., Wu, J., Porteus, M.H. *et al.* (2005) Zinc finger nucleases: Custom-designed molecular scissors for genome engineering of plant and mammalian cells. *Nucleic Acids Research* 33(18), 5978–5990.

EFSA Panel, on G. M. O. (GMO), Mullins, E., Bresson, J.-L., Dalmay, T. *et al.* (2021) *In vivo* and *in vitro* random mutagenesis techniques in plants. *EFSA Journal* 19(11), e06611.

Endersby, J. (2013) Mutant utopias: Evening primroses and imagined futures in early twentieth-century America. *Isis* 104(3), 471–503.

European Parliament (2024) Amendments adopted by the European Parliament on 7 February 2024. Available at: https://www.europarl.europa.eu/doceo/document/TA-9-2024-0067_EN.html (accessed 19 March 2024).

Ford, C.E. (1960) Human cytogenetics: Its present place and future possibilities. *American Journal of Human Genetics* 12(1), 104.

Fraley, R.T., Rogers, S.G., Horsch, R.B., Sanders, P.R., Flick, J.S. *et al.* (1983) Expression of bacterial genes in plant cells. *Proceedings of the National Academy of Sciences* 80(15), 4803–4807.

Gaj, T., Gersbach, C.A. and Barbas, C.F. (2013) ZFN, TALEN, and CRISPR/Cas-based methods for genome engineering. *Trends in Biotechnology* 31(7), 397–405.

Galili, G., Galili, S., Lewinsohn, E. and Tadmor, Y. (2002) Genetic, molecular, and genomic approaches to improve the value of plant foods and feeds. *Critical Reviews in Plant Sciences* 21(3), 167–204.

Gao, C. (2021) Genome engineering for crop improvement and future agriculture. *Cell* 184(6), 1621–1635.

Gelvin, S.B. (2003) Agrobacterium-mediated plant transformation: The biology behind the "gene-jockeying" tool. *Microbiology and Molecular Biology Reviews* 67(1), 16–37.

Gostimskaya, I. (2022) CRISPR–Cas9: A history of its discovery and ethical considerations of its use in genome editing. *Biochemistry (Moscow)* 87(8), 777–788.

Hawkes, J.G. (1983) *The Diversity of Crop Plants*. Harvard University Press, Cambridge, MA, USA. DOI: 10.4159/harvard.9780674183551.

Hayashi, K. (1991) PCR-SSCP: A simple and sensitive method for detection of mutations in the genomic DNA. In: Hawkes, J.G. (ed.) *The Diversity of Crop Plants*. Harvard University Press, Cambridge, Massachusetts.

Holland, J.B. (2004) Implementation of molecular markers for quantitative traits in breeding programs-challenges and opportunities. In: *4th International Crop Science Congress in conjunction with the 12th Australian Agronomy Conference and the 5th Asian Crop Science Conference, Brisbane, Australia, 26 September-1 October 2004.*

Inge-Vechtomov, S., Zhouravleva, G. and Golubkova, E. (2019) The 100th anniversary of the Department of Genetics and Biotechnology, St. Petersburg State University. *Biological Communications* 64(2), 77–88.

International Atomic Energy Agency (2024) Mutant variety database. Available at: https://nucleus.iaea.org/sites/mvd/SitePages/Home.aspx (accessed 19 March 2024).

Ishino, Y., Krupovic, M. and Forterre, P. (2018) History of CRISPR-Cas from encounter with a mysterious repeated sequence to genome editing technology. *Journal of Bacteriology* 200(7), 10–1128.

Jain, S.K. (1976) The evolution of inbreeding in plants. *Annual Review of Ecology and Systematics* 7(1), 469–495.

Kardamakis, D., Baatout, S., Bourguignon, M., Foray, N. and Socol, Y. (2023) History of radiation biology. In: Baatout, S. (ed.) *Radiobiology Textbook*. Springer, Cham, Switzerland, pp. 1–24.

Khan, S., Ullah, M.W., Siddique, R., Nabi, G., Manan, S. *et al*. (2016) Role of recombinant DNA technology to improve life. *International Journal of Genomics* 2016(2), 1–14.

Khorana, H.G. (ed.) (1978) Studies on nucleic acids: Total synthesis of a biologically functional gene. *Bioorganic Chemistry* 7(3), 351–393.

Kingsbury, N. (2011) *Hybrid: The History and Science of Plant Breeding*. University of Chicago Press, Chicago.

Marshall, F.B., Dobney, K., Denham, T. and Capriles, J.M. (2014) Evaluating the roles of directed breeding and gene flow in animal domestication. *Proceedings of the National Academy of Sciences* 111(17), 6153–6158.

Martuza, R.L., Malick, A., Markert, J.M., Ruffner, K.L. and Coen, D.M. (1991) Experimental therapy of human glioma by means of a genetically engineered virus mutant. *Science* 252(5007), 854–856.

Méthot, P.O. (2016) Bacterial transformation and the origins of epidemics in the interwar period: The epidemiological significance of fred griffith's "Transforming experiment.". *Journal of the History of Biology* 49(2), 311–358.

Monteith, J.L. (1977) Climate and the efficiency of crop production in Britain. *Philosophical Transactions of the Royal Society of London. B, Biological Sciences* 281(980), 277–294.

Morbitzer, R., Elsaesser, J., Hausner, J. and Lahaye, T. (2011) Assembly of custom TALE-type DNA binding domains by modular cloning. *Nucleic Acids Research* 39(13), 5790–5799.

Mullis, K.B. (1990) The unusual origin of the polymerase chain reaction. *Scientific American* 262(4), 56–65.

Nester, E.W. (2015) Agrobacterium: Nature's genetic engineer. *Frontiers in Plant Science* 5, 730.

O'Connor, C. and Miko, I. (2008) Developing the chromosome theory. *Nature Education* 1(1), 44.

Oliva, R., Ji, C., Atienza-Grande, G., Huguet-Tapia, J.C., Perez-Quintero, A. *et al*. (2019) Broad-spectrum resistance to bacterial blight in rice using genome editing. *Nature Biotechnology* 37(11), 1344–1350.

Pattanayak, V., Ramirez, C.L., Joung, J.K. and Liu, D.R. (2011) Revealing off-target cleavage specificities of zinc-finger nucleases by *in vitro* selection. *Nature Methods* 8(9), 765–770.

Periwal, V. (2017) A comprehensive overview of computational resources to aid in precision genome editing with engineered nucleases. *Briefings in Bioinformatics* 18(4), 698–711.

Pickar-Oliver, A. and Gersbach, C.A. (2019) The next generation of CRISPR–Cas technologies and applications. *Nature Reviews Molecular Cell Biology* 20(8), 490–507.

Poehlman, J.M. (2013) *Breeding Field Crops*. Springer Science & Business Media, New York.

Pontecorvo, G. (1968) Hermann Joseph Muller, 1890-1967. *Biographical Memoirs of Fellows of the Royal Society* 14, 348–389.

Rath, D., Amlinger, L., Rath, A. and Lundgren, M. (2015) The CRISPR-Cas immune system: Biology, mechanisms and applications. *Biochimie* 117, 119–128.

Rieseberg, L.H. (1997) Hybrid origins of plant species. *Annual Review of Ecology and Systematics* 28(1), 359–389.

Robertson, D. (2004) VIGS vectors for gene silencing: Many targets, many tools. *Annual Review of Plant Biology* 55, 495–519.

Roll-Hansen, N. (2010) Eugenics and the science of genetics. In: Bashford, A. and Levine, P. (eds) *Oxford Handbook of the History of Eugenics*. Oxford University Press, New York, pp. 80–97.

Schwarzbach, E., Smýkal, P., Dostál, O., Jarkovská, M. and Valová, S. (2014) Gregor J. Mendel - genetics founding father. *Czech Journal of Genetics and Plant Breeding* 50(2), 43–51.

Shamshirgaran, Y., Liu, J., Sumer, H., Verma, P.J. and Taheri-Ghahfarokhi, A. (2022) Tools for efficient genome editing; ZFN, TALEN, and CRISPR. *Applications of Genome Modulation and Editing* 2495, 29–46.

Sharma, P., Bora, L.C., Puzari, K.C., Baruah, A.M., Baruah, R. *et al.* (2017) Review on bacterial blight of rice caused by *Xanthomonas oryzae* pv. *oryzae*: Different management approaches and role of *Pseudomonas fluorescens* as a potential biocontrol agent. *International Journal of Current Microbiology and Applied Sciences* 6(3), 982–1005.

Shull, G.H. (1911) The genotypes of maize. *The American Naturalist* 45(532), 234–252.

Sinsheimer, R.L. (1977) Recombinant DNA. *Annual Review of Biochemistry* 46(1), 415–438.

Stansfield, W.D. (2006) Luther Burbank: Honorary member of the American breeders' association. *Journal of Heredity* 97(2), 95–99.

Sturtevant, A.H. (1951) A map of the fourth chromosome of *Drosophila melanogaster*, based on crossing over in triploid females. *Proceedings of the National Academy of Sciences* 37(7), 405–407.

Sustiprijatno, S., Waluyo, S. and Suharsono, S. (2022) Transformation of csp gene into tobacco plant mediated by *Agrobacterium tumefaciens*. In: *AIP Conference Proceedings* (Vol. 2462, No. 1), AIP Publishing, Bogor, Indonesia, p. 040007.

Taylor, N.J. and Fauquet, C.M. (2002) Microparticle bombardment as a tool in plant science and agricultural biotechnology. *DNA and Cell Biology* 21(12), 963–977.

Townsend, J.A., Wright, D.A., Winfrey, R.J., Fu, F., Maeder, M.L. *et al.* (2009) High-frequency modification of plant genes using engineered zinc-finger nucleases. *Nature* 459(7245), 442–445. DOI: 10.1038/nature07845.

Ulukan, H. (2011) The use of plant genetic resources and biodiversity in classical plant breeding. *Acta Agriculturae Scandinavica Section B - Soil and Plant Soil* 61(2), 97–104.

Vakulskas, C.A. and Behlke, M.A. (2019) Evaluation and reduction of CRISPR off-target cleavage events. *Nucleic Acid Therapeutics* 29(4), 167–174.

Van Harten, A.M. (1998) *Mutation Breeding: Theory and Practical Applications*, Vol. 1. Cambridge University Press, New York.

Wang, Y., Cheng, X., Shan, Q., Zhang, Y., Liu, J. *et al.* (2014) Simultaneous editing of three homoeoalleles in hexaploid bread wheat confers heritable resistance to powdery mildew. *Nature Biotechnology* 32(9), 947–951.

Watson, J.D. and Crick, F.H.C. (1953) The structure of DNA. *Cold Spring Harbor Symposia on Quantitative Biology* 18, 123–131.

Watson, J.D. and Cook-Deegan, R.M. (1991) Origins of the human genome project. *The FASEB Journal* 5(1), 8–11.

Zirkle, C. (1935) The inheritance of acquired characters and the provisional hypothesis of pangenesis. *The American Naturalist* 69(724), 417–445.

2 Classification and Nomenclature of Different Site-specific Nucleases for Genome Editing

Sallah A. Al-Hashedi[1], Muhammad Naeem Sattar[1], Baraka Siddique[2] and Zulfiqar Ali[3]*

[1]Central Laboratories, King Faisal University, Al-Hofuf, Saudi Arabia; [2]Centre of Excellence in Molecular Biology (CEMB), Lahore, Pakistan; [3]Islamic Organization for Food Security, Astana, Republic of Kazakhstan

Abstract

Clustered regularly interspaced short palindromic repeats (CRISPR) and CRISPR-associated systems (Cas) in bacteria and archaea provide adaptive immunity with diverse effector mechanisms. Their inherent adaptability through RNA guides has fueled their application in therapeutics and diagnostics. These RNA-guided systems rely on specific targeting and interference by effectors. Class 1 CRISPR-Cas systems depend on effector complexes composed of multiple subunits, whereas Class 2 systems employ single proteins with various functional domains for their effector function. The compact nature of Class 2 systems, particularly Cas9 nucleases, has revolutionized molecular biology and biotechnology due to their wide applicability in genome editing (GE). Beyond Cas9, computational analysis of genomes and metagenomes has revealed a vast diversity of Class 2 effector enzymes, including Cas12 and Cas13 variants. This expansion has led to the development of versatile molecular tools with distinct properties, including varied protospacer adjacent motifs (PAMs), improved editing specificity, RNA targeting capabilities, smaller crRNAs, and diverse cleaving mechanisms. Notably, the promiscuous RNase activity of Cas13 enables highly sensitive nucleic acid detection. Expressing and delivering multi-protein Class 1 effectors pose significant challenges. The diverse array of CRISPR enzymes has significantly expanded the GE toolbox, enabling multiple applications from gene knockout to transcriptional modulation, and even RNA editing. The plant molecular biology and biotechnology toolkit powered by GE is constantly expanding due to advancements in rational design and the inherent versatility of CRISPR and related bacterial RNA-guided systems. Artificial intelligence (AI) is being used to design new and improved gene editors from scratch, showing promise for increased precision and versatility. However, this technology is still under development in human genomics and needs further work before it can be applied in fields like agriculture.

Keywords: Genome editing, CRISPR-Cas, Classification, Class 1 effectors, Class 2 effectors

2.1 Introduction

Bacteria and archaea possess a remarkable adaptive immune system called Clustered Regularly Interspaced Short Palindromic Repeats (CRISPR) and CRISPR-associated (CRISPR-Cas) protein, which employs a one-of-a-kind defense strategy. It combines an array of repeated DNA sequences

*Corresponding author: zulfiqar@iofs.org.kz

 Genome Editing for Crop Improvement: Theory and Methodology (eds J.M. Al-Khayri *et al.*)
DOI: 10.1079/9781800622517.0002

(CRISPR repeats) alongside Cas genes to target and neutralize invading viruses and plasmids (Hille *et al.*, 2018; Wang *et al.*, 2022). This system integrates snippets of foreign DNA (CRISPR spacers) into the CRISPR array. These spacer sequences, along with the flanking repeats, are transcribed into CRISPR RNAs (crRNAs). These crRNAs then guide Cas enzymes in targeting specific locations. These enzymes, armed with crRNAs, precisely target and cleave corresponding sequences in the invader's DNA or RNA, effectively disarming it. CRISPR-Cas engages in an ongoing evolutionary arms race with parasites, driving a remarkable diversification within the system itself. While the core mechanism doesn't require strict synchronization with parasite genes, a significant and expanding diversity exists in CRISPR composition, organization, and sequence variation (Koonin *et al.*, 2023). This highlights the system's remarkable adaptability in the face of ever-evolving threats. CRISPR's exceptional ability to manipulate genomes has fueled extensive research across various biotechnological applications (Medina-de la Cruz *et al.*, 2024). Initially, Cas9, the signature enzyme of Type II CRISPR systems (particularly from *Streptococcus pyogenes* – SpCas9), dominated genome editing (GE) (Jinek *et al.*, 2012). However, despite its utility, SpCas9 presents significant limitations. First, while CRISPR boasts remarkable programmability, it faces a hurdle in the form of a protospacer adjacent motif (PAM) requirement (Sternberg *et al.*, 2014). This specific sequence, needed for Cas9-guide RNA recognition adjacent to the target site, restricts targeting to a limited portion of the genome where the PAM sequence is present. Second, the specificity of Cas9 can be compromised, leading to unintended binding and cleavage at sites other than the intended target (Hsu *et al.*, 2013). This hinders precise editing, especially when targeting unique locations in complex genomes. Finally, using Cas9 for editing necessitates the creation of a unique single guide RNA (sgRNA) for each target (Jinek *et al.*, 2012). This often involves assembling a customized crRNA with a generic trans-acting CRISPR RNA (tracrRNA), a cumbersome process.

It is worth noting that the Cas9 proteins initially described, such as SpCas9, are notably large, typically comprising 1250–1400 amino acids (Chylinski *et al.*, 2013). This size presents significant challenges, hindering the packaging of Cas9 and Cas9-based technologies, complicating mRNA synthesis, and making the delivery of Cas9 into plant cells substantially more difficult (Medina-de la Cruz *et al.*, 2024). Given the immense potential of CRISPR tools and the limitations imposed by Cas9, efforts to diversify and improve CRISPR tools began shortly after Cas9 was first adapted for GE. Scientists have achieved significant progress through two main approaches: utilizing the inherent variation within CRISPR systems and rationally designing and engineering CRISPR tools (Koonin *et al.*, 2023). This perspective primarily focuses on the former approach, sometimes in conjunction with the latter. Initially, various CRISPR variants were discovered serendipitously during genome and metagenome sequence analyses (Haft *et al.*, 2005). Recent advancements in search algorithms have enabled researchers to efficiently identify new CRISPR system variants within vast genomic and metagenomic databases. This has led to a substantial increase in our understanding of the diversity within CRISPR systems (Koonin *et al.*, 2023; Han *et al.*, 2024). In the succeeding sections, we will examine the rich variation within CRISPR systems and how scientists are harnessing the potential of existing CRISPR variants to develop powerful molecular tools. We will also explore the possibilities offered by less-studied CRISPR systems.

2.2 Basic Components of the CRISPR-Cas System

The development of CRISPR-Cas9 has opened new avenues for engineering plants with desirable traits, offering a precise and efficient tool for GE. This technology allows for precise and efficient modification of targeted genes within a plant's genome. Initially discovered in archaea and bacteria as part of their adaptive immune system (Jinek *et al.*, 2012), CRISPR-Cas9 has been successfully adapted for plant research (Medina-de la Cruz *et al.*, 2024). CRISPR-Cas9, a revolutionary gene-editing technique, relies on two key components: Cas9 protein and a sgRNA molecule. These elements combine to form a ribonucleoprotein (RNP) complex. The sgRNA, typically 18–21 nucleotides long, guides Cas9 to specific target locations within the genome called protospacers. These targeted DNA segments (protospacers) have a specific sequence rich in guanine (G) located right before them. This sequence is called the PAM (Bhatia and Yadav, 2023). Scientists have engineered sgRNA by combining crRNA and tracrRNA functionalities. The crRNA portion contains a sequence complementary to the target DNA, while the tracrRNA forms a stem-loop structure that interacts with the Cas9 protein. This combined sgRNA molecule directs the Cas9-mediated cleavage of both DNA strands at the targeted site (Jinek *et al.*, 2012). Following target recognition by sgRNA through base pairing, the Cas9-sgRNA complex induces a double-stranded break (DSB) in the DNA (Makarova *et al.*, 2015; Shamshirgaran *et al.*, 2022). The cell repairs this break via the error-prone non-homologous end-joining (NHEJ) mechanism, often resulting in insertions or deletions (InDels) or substitutions near the PAM site. These mutations can disrupt the gene's coding sequence, leading to either loss or gain of function in the resulting organism. This innovative approach has empowered researchers to gain a deeper understanding of plant biological processes (Medina-de la Cruz *et al.*, 2024).

2.3 Structural Comparison of CRISPR-Cas9 Tools with ZFNs and TALENs

Genome-editing techniques are revolutionizing our understanding of gene function and hold immense potential for advancements in biotechnology. Some of the unique features of different nucleases used for GE are listed in Table 2.1 for comparison. Zinc-finger nucleases (ZFNs) are engineered proteins that act like molecular scissors, targeting specific locations in DNA for precise modification. These remarkable tools combine two key functionalities: recognition and cleavage (Liu *et al.*, 2024). A series of 3–6 zinc-finger (ZF) motifs form the recognition domain of ZFNs. Each ZF motif, roughly 30 amino acids long, folds into a unique structure stabilized by a bound zinc ion (Li *et al.*, 2020). The remarkable ability of these motifs assists them in recognizing and binding the specific short DNA sequences (roughly three base pairs) in the major groove of DNA. This allows scientists to design ZFNs for new targets by identifying ZFs with affinity for the desired sequence. Efforts are underway to create modular libraries encompassing all possible DNA triplets, although achieving high specificity remains challenging (Katayama *et al.*, 2024). The second domain of

Table 2.1. Comparative analysis of CRISPR-Cas9 with ZFNs and TALENs. Author's own table.

Technique	DNA binding	Endonuclease	gRNA length	Specificity	Positive potential	Challenges
CRISPR-Cas9	crRNA/sgRNA	Cas9	18–21	Variable	High accuracy, high efficiency, and multi-site targeting	PAM sequences constraint for target selection; high off-target activity
ZFNs	Zinc-finger protein, FokI1018	FokI	18–36	Low	Targeted editing with minimal off-targeting	Low efficacy; edits one genomic site; low precision
TALENs	Transcription activator-like effector	FokI	30–40	High	High on-target specificity, low off-target editing events	Low efficiency; targets only one site; hindered by DNA methylation

ZFNs is derived from the FokI enzyme, a naturally occurring DNA cutter from *Flavobacterium okeanokoites*. This domain lacks specific DNA recognition but possesses powerful cleavage activity. However, for FokI to function, two ZFN molecules need to bind to opposite strands of DNA, forming a dimer. This ensures precise targeting and prevents unintended cleavage. By combining specific DNA recognition and targeted cleavage, ZFNs induce DSBs at defined locations (Guha and Edgell, 2017). These DSBs possess unique 3-nucleotide overhangs, distinguishing them from breaks created by other GE tools like Cas9 (blunt ends) or Cas12a (5-nucleotide overhangs). This specificity allows for precise control over DNA repair processes within the cell. Despite their immense potential, designing highly specific ZFNs remains a complex task. Optimizing ZFN activity for efficient and accurate targeting is a time-consuming process. However, ongoing research continues to refine ZFN design and improve its efficiency and ease of use.

Unlike other tools, Transcription Activator-like Effector Nucleases (TALENs) feature a highly predictable one-to-one relationship between their repeat units and the targeted nucleotides on the DNA strand, reducing the need for extensive screening (Liu *et al.*, 2024). Each TALEN molecule is assembled from several essential elements. One key component is a DNA-binding domain, which contains a variable number of unique repeats, typically ranging from 12 to 28. Additionally, a nuclear localization signal (NLS) is crucial for directing the molecule to the appropriate location within the cell, an acidic activation domain, and the Fok1 nuclease enzyme (Bhardwaj and Nain, 2021). The DNA-binding region features a consistent 33–35 amino acid repeat sequence with variable di-residues (RVDs) at specific positions within each repeat (Boch *et al.*, 2009). These RVDs determine the target nucleotide binding specificity. Research indicates a direct interaction between the specific amino acid positioned at 13 and a nucleotide situated within the major groove of the target DNA, while the amino acid at position 12 stabilizes this interaction (Mak *et al.*, 2012). Notably, a short, duplicated sequence of just 20 amino acids is present at the 3′ terminus of the DNA-binding site. In a typical TALEN system, two DNA-binding domains (DBDs) originating from TALE proteins flank a Fok1 nuclease catalytic domain on either side. The Fok1 nuclease demands dimerization to become active. When both TALENs bind their specific DNA targets on opposite strands, the Fok1 domains dimerize, inducing a DSB in the DNA. This DSB activates the cell's DNA repair machinery, which can be harnessed for GE applications (Bhardwaj and Nain, 2021).

CRISPR-Cas9 technology has revolutionized GE due to its simplicity, efficacy, and precise target cleavage compared to ZFNs and TALENs (Liu *et al.*, 2024). Unlike the others, CRISPR-Cas9 leverages programmable guide RNAs (sgRNAs) for targeting. This allows multiple sgRNAs to work with a single Cas9 protein, enabling simultaneous editing of diverse targets. The CRISPR-Cas9 system utilizes a Cas9 nuclease enzyme guided by an RNA complex. This approach demonstrates significantly higher efficacy in gene editing compared with ZFNs and TALENs. Moreover, CRISPR-Cas9's reliance on sgRNAs for target identification, rather than direct DNA binding, enhances its reliability. While further research is needed to fully understand CRISPR-Cas9's specificity, it already exhibits a lower frequency of off-target mutations compared to physical and chemical mutagenesis methods.

2.4 Unveiling the Diversity and Classification of CRISPR Systems

CRISPR classification presents a unique challenge due to the system's inherent diversity and rapid gene evolution. Unlike traditional organism classification, which relies heavily on universally conserved genes and phylogenetic approaches, CRISPR classification employs a multifaceted strategy (Koonin *et al.*, 2023). This strategy considers several key factors: the presence of distinctive genes defining CRISPR types, subtypes, and variants; the arrangement of genes within CRISPR-Cas loci; sequence-based clustering; and, importantly, the functional characteristics of the system. As of 2020, the latest classification scheme categorizes CRISPR systems into two classes, encompassing six types and 33 subtypes (Table 2.2) (Makarova *et al.*, 2020). Notably, researchers have recently discovered potential new subtypes of CRISPR-Cas systems (Figs 2.1 and 2.2). While both classes share core modules

Table 2.2. CRISPR/Cas toolkit for application in GE. Author's own table.

Class	Type	Subtypes	Effector unit	tracrRNA	Essential protein	Substrate
1	I	A, B, C, D, E, F, G	Heterogeneous	Not present	Cas3	DNA
	III	A, B, C, D, E, F	Heterogeneous	Present	Cas10	DNA or RNA
	IV	A, B, C	Heterogeneous	Present	Undefined	Undefined
2	II	A, B, C	Unimodular	Present	Cas9	DNA
	V	A, B, C, D, E, F, G, H, I, K, U	Unimodular	Present	Cas12	DNA or RNA
	VI	A, B, C, D	Unimodular	Not present	Cas13	RNA

for adaptation, their effector modules – responsible for targeting and manipulating foreign nucleic acids – exhibit distinct architectures. Class 1 systems feature complex effector machinery composed of multiple Cas genes. These genes encode protein subunits or standalone enzymes that work together to achieve their function. In contrast, Class 2 systems streamline the process by condensing the effector module into a single, multidomain protein.

Focusing on Class 1, it encompasses types I, III, and IV, further divided into 16 subtypes, with four discovered since 2015 (Makarova *et al.*, 2015) (Fig. 2.1). These systems likely share a common ancestor based on the general organization of their effector complexes. These complexes typically involve proteins from the Cas5 and Cas7 families, which often contain highly specialized RNA recognition motifs (RRMs). Additionally, the Cas6 family plays a crucial role in most Class 1 systems, with its standalone ribonucleases responsible for processing pre-crRNA molecules. Finally, Class 1 CRISPR systems utilize effector complexes formed by two similar proteins, typically designated as a large subunit and a small subunit, alongside the RRM proteins. While the specific composition and structure of these complexes can vary across different types, their conserved architecture across Class 1 highlights their shared ancestry.

2.4.1 Class 1 CRISPR systems: Diverse functions and evolution

Comprising roughly 90% of all identified CRISPR loci in bacteria and archaea, Class 1 systems exhibit a surprising degree of functional diversity beyond their primary role in defense. While complete CRISPR systems are most common, researchers have uncovered various 'derived forms' lacking the ability for interference. These derivatives serve intriguing non-defense functions. Notably, subtypes I-F, I-B, and I-E harbor interference-defective CRISPR derivatives found within numerous bacterial Tn7-like transposons (Klompe *et al.*, 2022). These derivatives enable a remarkable process known as RNA-guided transposition. The functions of many other Class 1 CRISPR derivatives with substantial sequence variation are still unknown, presenting intriguing opportunities for future exploration.

One particularly interesting feature of Class 1 systems is subtype III-E. Unlike its Class 1 counterparts, subtype III-E employs a single effector protein, resembling Class 2 systems (Özcan *et al.*, 2021). The evolution of Class 1 CRISPR-Cas systems from subtype III-D to III-E reveals a distinct shift. This change involved the fusion of four Cas7 proteins with a Cas11-like protein, resulting in the formation of a single, multifunctional protein called Cas7-11. Additionally, III-E systems lack other Cas genes, such as the Cas10-encoded DNA endonuclease, specializing them for RNA targeting only (Özcan *et al.*, 2021). Type III-E systems evolved alongside several proteins: Csx29, a protease with Csx30 as its substrate, the accessory protein Csx31, and the sigma factor RpoE (Makarova *et al.*, 2020; Özcan *et al.*, 2021). These acquisitions formed the basis for a remarkable RNA-triggered protease system (Hu *et al.*, 2022).

2.4.1.1 Type I CRISPR-Cas system

The hallmark of the type I CRISPR-Cas system is the Cas3 protein. This multifunctional protein exhibits both helicase and DNase activities. In some subtypes, the Cas3 helicase domain

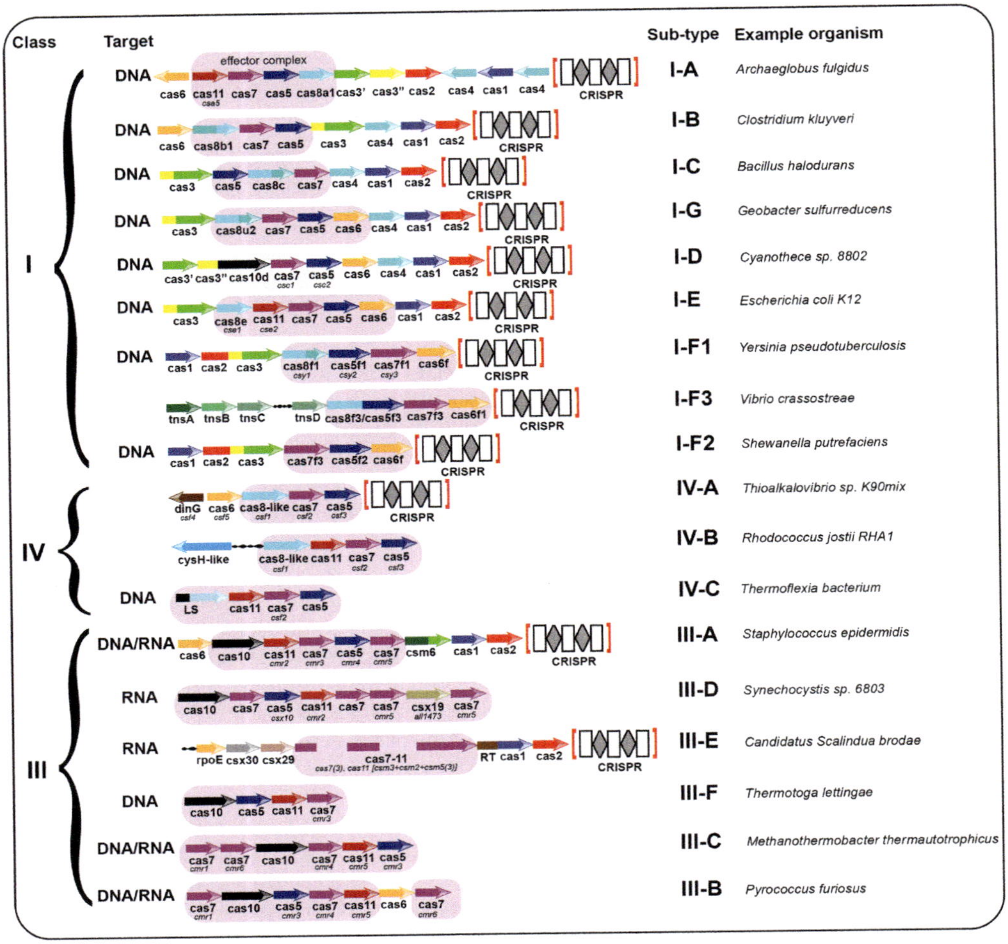

Fig. 2.1. This figure showcases the diversity of Class 1 CRISPR systems. It presents the CRISPR-Cas loci for each subtype, highlighting their unique gene arrangements. Labels on the right list examples of organisms for their origin. Homologous genes across the subtypes are color coded. The nomenclature of CRISPR system components reflects historical classifications. The small subunit comprises genes encoding Csm2, Cmr5, Cse2, Csa5, and a collection of related genes designated as Cas11. Notably, in subtypes III-A and III-E, Cas1 and Cas2 genes appear to be non-essential. CRISPR arrays are indicated at the end of each subtype, wherever present. The currently adopted gene names are written in bold text, with their previous names displayed separately for clarity. The figure was created in Adobe Illustrator 2020. Abbreviations: RT: reverse transcriptase; TPR: tetratricopeptide repeat domain.

combines with the HD domain, responsible for endonuclease activity, within the HD family. This HD domain resides at the beginning (N-terminus) of Cas3. It is encoded by a distinct gene within the same Cas3 locus. In such cases, the helicase domain is referred to as Cas3′ and the HD domain as Cas3″ (Bhatia and Yadav, 2023). The type I system typically encompasses Cas1 and Cas2 genes that encode proteins crucial for the effector complex, alongside larger and smaller subunits. Additionally, Cas5, Cas6, and Cas7 contribute to pre-crRNA transcript processing within this system. Each subtype displays a unique operon organization pattern with subtype-specific genes (Jackson *et al.*, 2014). Type IA is distinguished by the presence

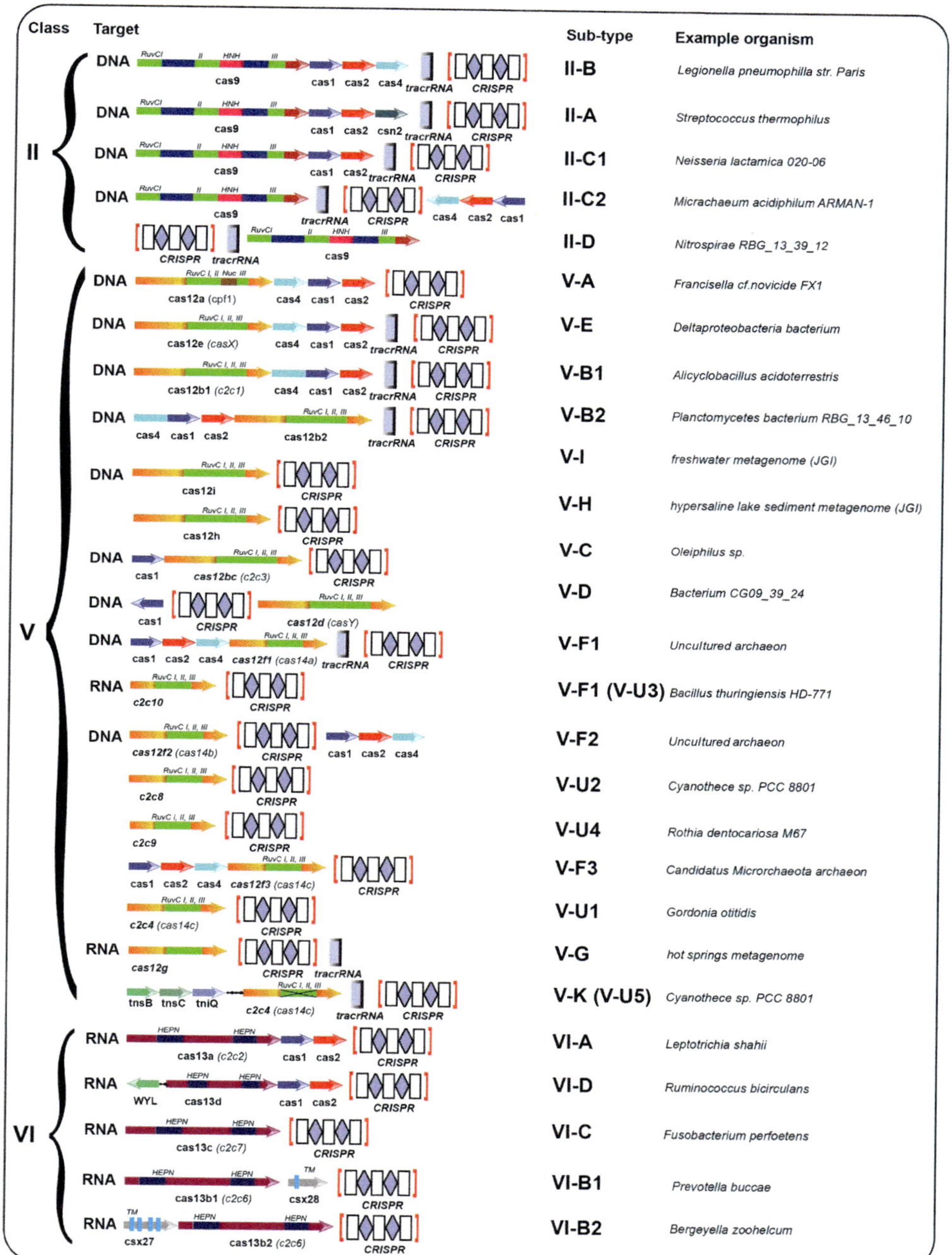

Fig. 2.2. Diversity of Class 2 CRISPR systems. The representative CRISPR-Cas loci for each subtype are depicted, highlighting the existence of distinct variants within each subtype. On the right, labels showcase exemplary host names. Genes with similar functions across subtypes are color coded for easy comparison. These gene names reflect previous classifications of CRISPR systems, with modern names displayed prominently (in bold) alongside their older counterparts for clarity. The figure was created in Adobe Illustrator 2020.

of Cas8a2 and Cas5 genes. Notably, Cas3 in this subtype consists of two separate subunits: Cas3′ (helicase) and Cas3″ (HD domain). Additionally, the csa5 gene encodes the smaller Cas3 subunit. Type IB and type IC are differentiated by their signature genes Cas8b and Cas8c, respectively. Type ID features Cas10d, as its signature element. Conversely, type IE and IF lack the Cas4 gene. However, they contain unique signature genes, such as cse1, cse2, cys1, cys2, cys3, and Cas6f, respectively (Fig. 2.1). Notably, in type I-F, Cas2 fuses with Cas3, eliminating the need for a separate gene encoding the smaller Cas3 subunit. This subunit may be entirely absent or potentially fused to the larger subunit (Makarova *et al.*, 2015). Among subtypes IA, IB, and IC, Cas8 is the only universally shared protein. Type IA and IC likely diverged from type IB due to rearrangements in their gene order. Notably, type IB systems exhibit a distinct gene order, with Cas genes positioned in a specific sequence: cas1-cas2-cas3-cas4-cas5-cas6-cas7-cas8 (Makarova *et al.*, 2015).

2.4.1.2 *Type III CRISPR-Cas system*

The hallmark of the type III CRISPR-Cas system is its Cas10 protein, a multidomain structure with intriguing functionalities. Recent structural studies revealed several key domains within Cas10 (Kolesnik *et al.*, 2021). The N-terminal domain exhibits a resemblance to cyclases or polymerases, although lacking enzymatic activity in this context. The palm domain harbors the protein's catalytic activity, while the helical domain contains a zinc-binding motif (Li *et al.*, 2024). Finally, the C-terminal domain shares similarities with the thumb domain found in DNA polymerases and the Cmr5 protein specific to certain type III systems (Zhu and Ye, 2012). Cas10 forms the larger subunit of the type III effector complex, collaborating with smaller Cas5 RAMP proteins and multiple Cas7 RAMP proteins. Interestingly, type III systems often differ from other CRISPR types by lacking Cas1 and Cas2 genes. Instead, they may rely on crRNA produced from type I or II CRISPR arrays for their immune response (Makarova *et al.*, 2015). Cas10 displays significant sequence variation across different type III systems. The complete type III locus also includes genes encoding a smaller Cas5 subunit and multiple Cas7 proteins. The first classification of type III systems divided them into subtypes III-A and III-B. Subtype III-A, alternatively referred to as the Mtube or Csm family, is characterized by the presence of the csm2 gene, which encodes a smaller subunit protein. This subtype retains Cas1, Cas2, and Cas6 genes, enabling DNA cleavage activity. Conversely, type III-B also referred to as the Cmr subtype family or RAMP module type, utilizes Cmr5 as its signature gene and lacks Cas1, Cas2, and Cas6 genes. Consequently, this subtype targets RNA instead of DNA (Koonin *et al.*, 2023; Schwartz *et al.*, 2024). The immense diversity within type III systems has led to the identification of additional subtypes, including III-C, III-D, III-E, and III-F. Type III-C is unique due to an inactive Cas10 protein. Type III-D possesses an active Cas10 gene but lacks Cas1 and Cas2 genes (Koonin *et al.*, 2023). Type III-E systems share characteristics with type III-D, including the presence of multiple Cas7 proteins and proteins similar to the Cas11 (Csm-2) subunit, hinting at a close evolutionary link. In contrast, type III-F features a single Cas7 protein with a unique structure. This protein combines an HD nuclease domain with a subunit resembling Cas10, enabling DNA cleavage (Makarova *et al.*, 2020).

2.4.1.3 *Type IV CRISPR-Cas system*

Most CRISPR-Cas loci fall within the three previously discussed classes based on their evolutionary relationships. However, the CRISPR-Cas system found in *Acidithiobacillus ferrooxidans* ATCC 23270 exhibits unique features that prevent its classification within any of these established categories. Consequently, it has been designated as a distinct type IV class (Makarova *et al.*, 2015; Pinilla-Redondo *et al.*, 2020).

This type IV system notably lacks genes for Cas1 and Cas2, which are commonly found in other classes. Some organisms exclusively rely on the type IV system for CRISPR-Cas functionality, lacking any other detectable CRISPR cassettes. The type IV effector complex is characterized by a streamlined architecture. This subtype typically features a smaller and less intricate large subunit. It also includes two genes encoding RAMP proteins – one belonging to the Cas5 family and another to the Cas7 family. Additionally, a gene encoding the smaller subunit is present. The type IV class itself encompasses further diversity. Type IV-A systems are distinguished by the presence of a DinG or csm4 helicase gene, acting as a signature element. Unlike type IV-A systems, type IV-B lacks the DinG helicase gene. However, they retain a gene encoding a smaller subunit, typically a small alpha-helical protein. Similar to type III systems, type IV systems can utilize crRNAs derived from other CRISPR arrays present within the genome, if available (Makarova *et al.*, 2020; Pinilla-Redondo *et al.*, 2020). Unlike type IV-A and type IV-B systems, type IV-C possesses an HD nuclease domain in addition to the shared Cas5 and Cas7 proteins. This distinct feature suggests that type IV-C might have the ability to cleave DNA, potentially setting it apart functionally from its closest relatives (Pinilla-Redondo *et al.*, 2020; Bhatia and Yadav, 2023).

2.4.2 Class 2 CRISPR systems: Powerhouses of genome editing

Class 2 systems, encompassing CRISPR types II, V, and VI, are defined by the presence of a single, large, multidomain effector protein (Fig. 2.2, Table 2.3). This characteristic makes them ideal tools for various applications, particularly GE. Class 2 systems constitute a smaller portion (around 10%) of bacterial CRISPR systems and are even rarer in archaea. Previously, researchers recognized only three subtypes within type II systems. These subtypes all rely on Cas9 as their primary effector protein. However, a significant breakthrough occurred in 2015 with the introduction of type V, featuring the distinct Ca12 effector protein (Makarova *et al.*, 2015). This discovery, coupled with the realization of the potential of Class 2 systems in GE, spurred a systematic effort to identify novel variants in genomic and metagenomic sequences. This effort involved pinpointing CRISPR-Cas loci and searching for large, uncharacterized proteins encoded within the remaining loci, considered potential candidates for new Class 2 systems. As a result, type V was significantly expanded, and a completely new type – VI – was discovered (Abudayyeh *et al.*, 2016). The use of similar computational approaches leads to the continuous identification of new subtypes within both types V and VI. Given the immense potential of diverse Class 2 CRISPR variants as molecular tools, a detailed exploration of these systems follows.

2.4.2.1 *Type II CRISPR systems*

Their hallmark lies in the unique domain organization of their effector protein, Cas9. Cas9 possesses two nuclease domains: HNH and RuvC-like (Jinek *et al.*, 2012). These domains cleave opposite strands of the target DNA, creating a DSB with staggered ends (Anders *et al.*, 2014). Additionally, Cas9 possesses a large recognition (REC) lobe crucial for target binding. The presence of a small tracrRNA is a unique feature of the type II system. Within type II CRISPR loci, tracrRNA is essential for processing precursor CRISPR RNA (pre-crRNA) through its interaction with RNase III. Notably, a protein called IscB, encoded by a specific subfamily of IS200/IS605 transposons, shares a similar domain structure, lacking only the REC lobe. This suggests IscB as a possible evolutionary precursor to Cas9, before Cas9 gained additional domains for its specific function (Kapitonov *et al.*, 2016). The association of IscB with CRISPR across various branches suggests independent evolutionary acquisitions. Functionally, IscB likely involves RNA-guided endonuclease activity. This suggests a potential role in RNA-directed transposition, similar to mechanisms observed in other IS200/IS605 transposons (Altae-Tran *et al.*, 2021).

Type II systems exhibit a high degree of uniformity in gene composition and locus organization. Three subtypes (II-A, II-B, and II-C) are differentiated based on the presence of specific signature genes. Subtype II-A includes an additional gene, Csn2, encoding a non-functional P-loop ATPase potentially involved in adaptation (Chylinski *et al.*, 2014). Subtype II-B possesses the Cas4 nuclease, also associated with adaptation. Finally, subtype II-C comprises the core Cas1, Cas2, and Cas9 genes. A potential new subtype, II-D, has been proposed, consisting of a minimalist version with only a small Cas9 protein and a CRISPR array (Altae-Tran *et al.*, 2021).

2.4.2.2 *Type V CRISPR systems*

Type V CRISPR-Cas systems share some organizational similarities with type II systems. Both typically have CRISPR-Cas loci encoding a

Table 2.3. List of single-protein Class 2 CRISPR effectors for GE. Author's own table.

Subtype/effector	Approximate protein size	TracrRNA	Target	Cut structure	Collateral activity
II-A/Cas9a	1150–1350	Yes	dsDNA	Blunt DSBs	No
II-B/Cas9b	1300–1600	Yes	dsDNA	Blunt DSBs	No
II-C/Cas9c	1050–1400	Yes	dsDNA	Blunt DSBs	No
II-D/Cas9d	700	Yes	dsDNA	Blunt DSBs	No
V-A/Cas12a (Cpf1)	1200–1500	No	dsDNA	Asymmetric DSBs	ssDNA
V-B/Cas12b (C2c1)	1300	Yes	dsDNA	Asymmetric DSBs	ssDNA
V-C/Cas12c (C2c3)	1209–1330	Yes	dsDNA	Asymmetric DSBs	No
V-D/Cas12d (CasY)	1200	Yes	dsDNA	Asymmetric DSBs	ssDNA
V-E/Cas12e (CasX)	986	Yes	dsDNA	Asymmetric DSBs	ssDNA
V-F/Cas12f	400–700	Yes	dsDNA	Asymmetric DSBs	ssDNA
V-G/Cas12g	720–830	Yes	ssRNA	N/A	ssRNA/ssDNA
V-H/Cas12h	870–924	No	dsDNA	dsDNA and ssDNA	ssDNA
V-I/Cas12i	1033–1093	No	dsDNA	Asymmetric DSBs or nick	ssDNA
V-J/Cas12j	700–800	No	dsDNA	Asymmetric DSBs	Yes
V-K/Cas12k	640–643	Yes	dsDNA	60 bp downstream of PAM	No
VI-A/Cas13a (C2c2)	1228	No	ssRNA	Multiple cleavage event	ssRNA
VI-B/Cas13b	1128	No	ssRNA	Multiple cleavage event	ssRNA
VI-C/Cas13c	1118	No	ssRNA	Multiple cleavage event	ssRNA
VI-D/Cas13d	928	No	ssRNA	Multiple cleavage event	ssRNA
III-E/Cas7-11	1367–1602	No	ssRNA	Duplex cleavage events with six nucleotides distance	No

large effector protein. However, type V systems often include an additional adaptation module and sometimes even more accessory proteins. While both Cas9 (type II) and Cas12 (type V) proteins possess similar RuvC-like nuclease domains, their overall architectures differ significantly. Cas9 utilizes two nuclease domains, one for each DNA strand, whereas Cas12 relies on a single RuvC-like domain for consecutive cleavage of both target DNA strands (Swarts *et al.*, 2017). Unlike the uniformity observed in type II CRISPR systems, type V systems boast a remarkable degree of diversity. Despite their relative scarcity in bacteria, they exhibit a wide range in size, architecture, and even target specificity of their effector proteins. This diversity is reflected in the current classification scheme (Makarova *et al.*, 2020), which recognizes 12 type V subtypes with more likely to be discovered in the future. Many newly identified variants belong to the provisional subtype V-U. These systems encode RuvC-like proteins that are smaller than those found in typical Cas12 enzymes. The encoded proteins also exhibit sequence similarity to TnpB proteins associated with IS605-like transposons (Altae-Tran *et al.*, 2021). Notably, like the IscB protein associated with type I-E CRISPR systems, these TnpB proteins function as RNA-guided nucleases, potentially mediating RNA-guided transposon homing.

The TnpB with CRISPR association suggests multiple independent acquisitions of TnpB by CRISPR systems, leading to the emergence of diverse type V subtypes. This is evident in phylogenetic analyses of both TnpB and Cas1 proteins (Altae-Tran *et al.*, 2021). Similar to the fusion event that created Cas9, the evolution of CRISPR effector proteins within each subtype likely involved the gradual acquisition of new protein domains alongside the diversification of the RuvC-like nuclease domain. The observed variations in size and sequence similarity of TnpB proteins within type V CRISPR systems likely reflect distinct stages in the evolutionary history of CRISPR. This diversity is likely due to numerous independent interactions between TnpB and CRISPR, facilitated by the prevalence of TnpB in bacterial genomes and its association with mobile transposons. These interactions make the classification of type V CRISPR systems challenging. These new variants showcase remarkable differences in interference specificity. For instance, Cas12f (originally named Cas14) demonstrates single-stranded DNA cleavage, although some studies report double-stranded activity as well (Harrington *et al.*, 2018). Conversely, Cas12g acts as an RNA-guided RNase with additional collateral RNase and single-stranded DNase activities (Yan *et al.*, 2019). The pre-crRNA processing mechanisms also vary significantly across type V subtypes. Subtype V-B relies on tracrRNA and the enzyme RNase III for processing (Kato *et al.*, 2022), while subtype V-A utilizes Cas12 itself through a separate catalytic site for pre-crRNA maturation. Subtype V-K, located in cyanobacterial Tn7-like transposons, possesses an inactive Cas12k protein homologous to the RuvC nuclease. This protein lacks the essential residues required for catalytic activity (Faure *et al.*, 2019). Experimental evidence suggests that Cas12k interacts with subunits of Tn7-like transposases, enabling RNA-guided transposition. This interaction is similar to those observed in interference-deficient variants of other CRISPR subtypes (Strecker *et al.*, 2019).

2.4.2.3 *Type VI CRISPR systems*

Unlike most CRISPR systems, type VI systems specialize in targeting RNA. These systems, exemplified by Cas13 proteins, belong to the Class 2 effector category. This class is distinguished by large, single-protein effectors compared to the multi-protein complexes found in Class 2 and 5 systems (Abudayyeh *et al.*, 2016). Cas13 proteins harbor two distinct HEPN ribonuclease (RNase) domains. These domains primarily target transcripts derived from invading DNA genomes, offering defense against foreign genetic material. Additionally, laboratory studies suggest their potential role in protecting bacteria from RNA phages (Meeske *et al.*, 2019). Notably, Cas13 proteins possess collateral RNase activity, triggered upon target recognition. This activity can induce dormancy in virus-infected bacteria, hindering viral replication. Similar to Cas12 proteins in certain type V subtypes, Cas13 possesses a unique active site for processing pre-crRNA molecules (Fig. 2.2). The type VI category encompasses four subtypes, with some variants incorporating accessory proteins to modulate their activity. For instance, subtype VI-B may include membrane proteins, while subtype VI-D utilizes WYL domain proteins (Yan *et al.*, 2018). Interestingly, while some VI-A and VI-D variants possess an adaptation module for acquiring new spacer sequences, most type VI systems lack this module. This suggests that they rely on coexisting CRISPR-Cas systems for spacer acquisition. Notably, one VI-B system has been observed to utilize a subtype II-C adaptation module for this purpose (Hoikkala *et al.*, 2021). Finally, the two HEPN nuclease domains of Cas13 share significant sequence similarity with those found in abortive infection (ABI) systems. This suggests that type VI systems likely evolved from ABI systems through duplication or fusion events, followed by the acquisition of additional functional domains (Koonin *et al.*, 2023).

2.5 Expanding the Diverse CRISPR Systems and Future Perspectives

Traditionally, researchers have identified diverse CRISPR systems by mining genomic and metagenomic data for elements commonly associated with these systems. This approach is then followed by a detailed biochemical analysis of the identified candidates and the development of new enzymes for plant GE. However, this strategy can be broadened by expanding the search beyond typical CRISPR markers like Cas1 or CRISPR arrays (Shmakov *et al.*, 2019). By analyzing the vast and ever-growing databases of microbial genomic and metagenomic sequences for any distinctive markers, researchers can uncover a wider range of potential CRISPR systems. This approach has already yielded significant results, leading to the discovery of numerous novel CRISPR subtypes and variants. It may even reveal entirely new classes of these systems. As we continue to explore these expanding databases, we can expect to uncover even more diverse CRISPR variants. These may include new single-protein or multi-subunit effector molecules, alongside novel enzymatic activities beyond nucleases, potentially encompassing proteases or ATPases associated with CRISPR effectors. The discovery of OMEGA systems, which share characteristics with both type II and type V CRISPR systems, suggests the existence of further RNA-guided mechanisms in prokaryotes (Altae-Tran *et al.*, 2021; Karvelis *et al.*, 2021). These mechanisms likely await discovery through further co-evolutionary exploration.

This book chapter explores the expanding CRISPR-Cas toolkit for plant GE. Discoveries like CasU, a recently identified Cas protein significantly smaller than Cas9 (Pausch *et al.*, 2020), showcase the remarkable diversity within CRISPR-Cas systems. These novel systems offer unique properties compared to established tools like Cas9 and Cas12a (Table 2.4). However, their suitability for plant applications remains to be fully explored. For instance, the transposon-associated CRISPR-Cas system, which can insert DNA fragments, has so far only been shown to function in bacteria (Strecker *et al.*, 2019; Klompe *et al.*, 2022). Compared to Cas9 and Cas12a, these newer systems haven't undergone extensive optimization for plant use, suggesting significant potential for improvement.

Efficient plant GE requires more than just new tools. Plants present unique challenges for CRISPR-Cas9 gene editing. These include the requirement for tissue culture, the inefficiency of homology-directed repair (HDR), and the extended timeframe needed to create homozygous lines, particularly in non-model or polyploid plant species. Codon and vector optimization are often necessary depending on the plant species. In polyploids, mutating multiple gene copies is crucial for knockouts. Furthermore, generating plants with the desired edits without introducing foreign genes (null segregants) is often favored. This approach minimizes potential off-target effects and regulatory hurdles (Kocsisova and Coneva, 2023). Recent advancements like *de novo* meristem induction for plant editing, improved targeting in *Arabidopsis*, and GE to induce haploid production, offer promising solutions to these challenges (Kelliher *et al.*, 2019). By combining these technological leaps, plant GE is on track to become faster, more efficient, and more precise.

2.6 Conclusions

The discovery of Cas9 ignited a revolution in genome engineering. However, the true potential lies within the vast natural repertoire of CRISPR effector enzymes. Class 2 CRISPR systems, along with their ancestral forms and engineered derivatives, offer a wealth of versatile and orthogonal (independent) molecular tools. Remarkably, in just 7 years since the identification of the first non-type II systems, scientists have explored all 17 recognized subtypes within types II, V, and VI. This exploration has yielded a treasure trove of potential applications, with most subtypes having ongoing or developed functionalities. Notably, researchers have tested multiple orthologs (genetically similar counterparts) within each subtype, but only a select few have proven effective for GE and other purposes. These diverse CRISPR variants are highly sought after due to several advantages. Their compact size facilitates efficient delivery into cells. Additionally, they often exhibit high editing efficiency and specificity.

Table 2.4. Summary of recent CRISPR-Cas technologies for plant GE. Author's own table.

Type	Primary origin	Crop	PAM or PFS	Results	Reference
I-E CRISPR-Cas3	*S. thermophilus*	Maize	AA	Gene activation was achieved with comparable efficacy to dCas9-CBF1 in *Zea mays*.	Young *et al*., 2019
I-D CRISPR-Cas10 (TiD)	*M. aeruginosa*	Tomato	GTT, GTC	TiD resulted in small insertions and deletions (up to 7.5 kb) at targeted locations in the tomato genome.	Osakabe *et al*., 2020
V-B CRISPR-Cas12b	*A. acidoterrestris, A. acidiphilus, B. thermoamylovorans and B. hisashii*	Rice	VTTV (V:A,C,G)	The study employed AaCas12b for targeted mutagenesis in rice, a monocot plant. This approach resulted in highly specific edits with larger deletion sizes (4–14 bp) compared to those achieved using Cas9.	Ming *et al*., 2020
V-B CRISPR-Cas12b	*A. acidoterrestris*	Upland cotton	TTN	Cotton plants were edited using the AacCas12b enzyme, resulting in increased resistance to high temperatures.	Wang *et al*., 2020
V-B CRISPR-Cas12b	*Bacillus* sp. V3-13 and *B. hisashii*	*Arabidopsis*	ATTA, ATTG	*Arabidopsis* genome was successfully edited using the BvCas12b and BhCas12b v4 enzymes.	Wu *et al*., 2020
V CRISPR-CasU	Bacteriophage genomes	*Arabidopsis*	TBN (B:G,T, and C)	CasU treatment resulted in mutations within the target genes of *Arabidopsis* protoplasts.	Pausch *et al*., 2020
VI CRISPR-Cas13a	*L. wadei*	Rice	No	LwaCas13 effectively reduced the target gene in rice.	Abudayyeh *et al*., 2017

Continued

Table 2.4. Continued

Type	Primary origin	Crop	PAM or PFS	Results	Reference
VI CRISPR-Cas13a	*L. shahii*	Tobacco; *Arabidopsis*	A, U, and C	Tobacco and *Arabidopsis* plants expressing LshCas13a exhibited interference against TuMV-GFP.	Aman *et al.*, 2018
VI CRISPR-Cas13a	*L. shahii*	Rice; Tobacco	Not defined	LshCas13a, a CRISPR-Cas effector, effectively inhibited viral infection in both monocot and dicot plant species.	Zhang *et al.*, 2018
VI CRISPR-Cas13d	*R. flavefaciens*	Tobacco	Not defined	CasRx demonstrated highly specific interference in tobacco.	Mahas and Mahfouz, 2018

Notably, some variants possess relaxed or complementary PAM requirements, allowing them to target a broader range of DNA sequences. Furthermore, unique properties like the collateral RNase or DNase activity of Cas13 and Cas12 enable novel applications such as single-molecule nucleic acid detection. While Class 2 systems have dominated research so far, efforts are underway to explore the potential of Class 1 CRISPR systems. Although challenges exist in expressing and delivering these multi-protein complexes, they offer the possibility of surpassing Class 2 effectors in terms of activity and specificity. By combining the power of rational design for both effector proteins and RNA guides, alongside creative strategies like fusing inactive effectors with other DNA or RNA modifying enzymes, researchers are harnessing the natural diversity of CRISPR systems. This paves the way for a significantly expanded toolkit in molecular biology and biotechnology.

CRISPR-dCas9 (dead Cas9) has emerged as a game-changer in plant science. Beyond its established role in editing genomes, it now offers exciting possibilities for enhancing plant resilience to both environmental stresses (abiotic) and disease/pest threats (biotic). Additionally, it allows for live-cell monitoring of chromatin dynamics, providing valuable insights into gene regulation. By fine-tuning dCas9 proteins and their associated effectors, researchers are expanding the capabilities of CRISPR-dCas9. The use of multiple guide RNAs (sgRNAs) and transcriptional regulatory tools further broaden its applications. One promising area of exploration is the precise control of plant secondary metabolite production, which has implications for medicine, industry, and agriculture. Despite its potential, CRISPR-dCas9 faces challenges. Public concerns surrounding the safety of consuming CRISPR-edited crops necessitate rigorous research and transparent communication. Regulatory frameworks for these crops might also need adjustments to ensure both safety and innovation. While further research is needed, CRISPR-dCas9 undoubtedly holds immense promise for plant GE. As the technology matures, it is poised to become a cornerstone for developing crops that meet the growing demands of a changing world.

CRISPR-based gene editors offer powerful tools for molecular biology. However, they often face limitations when adapted to non-native environments, such as human cells. Artificial intelligence (AI) offers a solution by designing gene editors that bypass these evolutionary constraints, achieving optimal performance. These powerful AI models use massive biological data to design entirely new gene editors from scratch. However, the application of these large language models (LLMs) has been limited to the human genome until now. Some of these AI-designed editors have shown equal or better performance than the standard Cas proteins, despite being significantly different in sequence. These AI-designed editors outperform natural ones and have the potential to be highly specific and versatile. However, translating this technology to practical applications in agriculture requires further development.

References

Abudayyeh, O.O., Gootenberg, J.S., Konermann, S., Joung, J., Slaymaker, I.M. *et al.* (2016) C2c2 is a single-component programmable RNA-guided RNA-targeting CRISPR effector. *Science* 353(6299), aaf5573.

Abudayyeh, O.O., Gootenberg, J.S., Essletzbichler, P., Han, S., Joung, J. *et al.* (2017) RNA targeting with CRISPR-Cas13. *Nature* 550(7675), 280–284. DOI: 10.1038/nature24049.

Altae-Tran, H., Kannan, S., Demircioglu, F.E., Oshiro, R., Nety, S.P. *et al.* (2021) The widespread IS200/IS605 transposon family encodes diverse programmable RNA-guided endonucleases. *Science* 374(6563), 57–65.

Aman, R., Ali, Z., Butt, H., Mahas, A., Aljedaani, F. *et al.* (2018) RNA virus interference via CRISPR/Cas13a system in plants. *Genome Biology* 19, 1.

Anders, C., Niewoehner, O., Duerst, A. and Jinek, M. (2014) Structural basis of PAM-dependent target DNA recognition by the Cas9 endonuclease. *Nature* 513(7519), 569–573.

Bhardwaj, A. and Nain, V. (2021) TALENs – an indispensable tool in the era of CRISPR: A mini review. *Journal of Genetic Engineering and Biotechnology* 19(1), 125.

Bhatia, S. and Yadav, S.K. (2023) CRISPR-Cas for genome editing: Classification, mechanism, designing and applications. *International Journal of Biological Macromolecules* 238, 124054.

Boch, J., Scholze, H., Schornack, S., Landgraf, A., Hahn, S. *et al.* (2009) Breaking the code of DNA binding specificity of TAL-type III effectors. *Science* 326(5959), 1509–1512.

Chylinski, K., Le Rhun, A. and Charpentier, E. (2013) The tracrRNA and Cas9 families of type II CRISPR-Cas immunity systems. *RNA Biology* 10(5), 726–737.

Chylinski, K., Makarova, K.S., Charpentier, E. and Koonin, E.V. (2014) Classification and evolution of type II CRISPR-Cas systems. *Nucleic Acids Research* 42(10), 6091–6105.

Faure, G., Shmakov, S.A., Yan, W.X., Cheng, D.R., Scott, D.A. *et al.* (2019) CRISPR–Cas in mobile genetic elements: Counter-defence and beyond. *Nature Reviews Microbiology* 17(8), 513–525.

Guha, T.K. and Edgell, D.R. (2017) Applications of alternative nucleases in the age of CRISPR/Cas9. *International Journal of Molecular Sciences* 18(12), 2565.

Haft, D.H., Selengut, J., Mongodin, E.F. and Nelson, K.E. (2005) A guild of 45 CRISPR-associated (Cas) protein families and multiple CRISPR/Cas subtypes exist in prokaryotic genomes. *PLoS Computational Biology* 1(6), e60.

Han, Y., Liu, G., Wu, Y., Bao, Y., Zhang, Y. *et al.* (2024) Crisprstitch: Fast evaluation of the efficiency of CRISPR editing systems. *Plant Communications* 5(3), 100783.

Harrington, L.B., Burstein, D., Chen, J.S., Paez-Espino, D., Ma, E. *et al.* (2018) Programmed DNA destruction by miniature CRISPR-Cas14 enzymes. *Science* 362(6416), 839–842.

Hille, F., Richter, H., Wong, S.P., Bratovič, M., Ressel, S. *et al.* (2018) The biology of CRISPR-Cas: Backward and forward. *Cell* 172(6), 1239–1259.

Hoikkala, V., Ravantti, J., Díez-Villaseñor, C., Tiirola, M., Conrad, R.A. *et al.* (2021) Cooperation between different CRISPR-Cas types enables adaptation in an RNA-targeting system. *MBio* 12(2). DOI: 10.1128/mbio/03338-03320.

Hsu, P.D., Scott, D.A., Weinstein, J.A., Ran, F.A., Konermann, S. *et al.* (2013) DNA targeting specificity of RNA-guided Cas9 nucleases. *Nature Biotechnology* 31(9), 827–832.

Hu, C., van Beljouw, S.P., Nam, K.H., Schuler, G., Ding, F. *et al.* (2022) Craspase is a CRISPR RNA-guided, RNA-activated protease. *Science* 377(6612), 1278–1285.

Jackson, R.N., Lavin, M., Carter, J. and Wiedenheft, B. (2014) Fitting CRISPR-associated Cas3 into the helicase family tree. *Current Opinion in Structural Biology* 24, 106–114.

Jinek, M., Chylinski, K., Fonfara, I., Hauer, M., Doudna, J.A. *et al.* (2012) A programmable dual-RNA–guided DNA endonuclease in adaptive bacterial immunity. *Science* 337(6096), 816–821.

Kapitonov, V.V., Makarova, K.S. and Koonin, E.V. (2016) ISC, a novel group of bacterial and archaeal DNA transposons that encode Cas9 homologs. *Journal of Bacteriology* 198(5), 797–807.

Karvelis, T., Druteika, G., Bigelyte, G., Budre, K., Zedaveinyte, R. *et al.* (2021) Transposon-associated TnpB is a programmable RNA-guided DNA endonuclease. *Nature* 599(7886), 692–696.

Katayama, S., Watanabe, M., Kato, Y., Nomura, W. and Yamamoto, T. (2024) Engineering of zinc finger nucleases through structural modeling improves genome editing efficiency in cells. *Advanced Science* 11(23), 2310255.

Kato, K., Okazaki, S., Kannan, S., Altae-Tran, H., Esra Demircioglu, F. *et al.* (2022) Structure of the IscB–ωRNA ribonucleoprotein complex, the likely ancestor of CRISPR-Cas9. *Nature Communications* 13(1), 6719.

Kelliher, T., Starr, D., Su, X., Tang, G., Chen, Z. *et al.* (2019) One-step genome editing of elite crop germplasm during haploid induction. *Nature Biotechnology* 37(3), 287–292.

Klompe, S.E., Jaber, N., Beh, L.Y., Mohabir, J.T., Bernheim, A. *et al.* (2022) Evolutionary and mechanistic diversity of type IF CRISPR-associated transposons. *Molecular Cell* 82(3), 616–628, .

Kocsisova, Z. and Coneva, V. (2023) Strategies for delivery of CRISPR/Cas-mediated genome editing to obtain edited plants directly without transgene integration. *Frontiers in Genome Editing* 5.

Kolesnik, M.V., Fedorova, I., Karneyeva, K.A., Artamonova, D.N. and Severinov, K.V. (2021) Type III CRISPR-Cas systems: Deciphering the most complex prokaryotic immune system. *Biochemistry (Moscow)* 86, 1301–1314.

Koonin, E.V., Gootenberg, J.S. and Abudayyeh, O.O. (2023) Discovery of diverse CRISPR-Cas systems and expansion of the genome engineering toolbox. *Biochemistry* 62(24), 3465–3487.

Li, H., Yang, Y., Hong, W., Huang, M., Wu, M. *et al.* (2020) Applications of genome editing technology in the targeted therapy of human diseases: Mechanisms, advances and prospects. *Signal Transduction and Targeted Therapy* 5(1), 1.

Li, X., Han, J., Yang, J. and Zhang, H. (2024) The structural biology of type III CRISPR-Cas systems. *Journal of Structural Biology* 216(1), 108070.

Liu, Q., Sun, Q. and Yu, J. (2024) Gene editing's sharp edge: Understanding Zinc Finger Nucleases (ZFN), Transcription Activator-Like Effector Nucleases (TALEN) and Clustered Regularly Interspaced Short Palindromic Repeats (CRISPR). *Transactions on Materials, Biotechnology and Life Sciences* 3, 170–179.

Mahas, A. and Mahfouz, M. (2018) Engineering virus resistance via CRISPR-Cas systems. *Current Opinion in Virology* 32, 1–8. DOI: 10.1016/j.coviro.2018.06.002.

Mak, A.N.-S., Bradley, P., Cernadas, R.A., Bogdanove, A.J. and Stoddard, B.L. (2012) The crystal structure of TAL effector PthXo1 bound to its DNA target. *Science* 335(6069), 716–719.

Makarova, K.S., Wolf, Y.I., Alkhnbashi, O.S., Costa, F., Shah, S.A. *et al.* (2015) An updated evolutionary classification of CRISPR–Cas systems. *Nature Reviews Microbiology* 13(11), 722–736.

Makarova, K.S., Wolf, Y.I., Iranzo, J., Shmakov, S.A., Alkhnbashi, O.S. *et al.* (2020) Evolutionary classification of CRISPR–Cas systems: A burst of class 2 and derived variants. *Nature Reviews Microbiology* 18(2), 67–83.

Medina-de la Cruz, R.O., Gonzalez-Avila, L.U., Saldaña-Padilla, A., Mora-Piña, C.J., Hernández-Cortez, C. *et al.* (2024) CRISPR-based techniques and their application in plants. In: Abd-Elsalam, K.A., Ahmad, A. and Zhang, B. (eds) *CRISPRized Horticulture Crops*. Elsevier, Amsterdam, Netherlands, pp. 91–104.

Meeske, A.J., Nakandakari-Higa, S. and Marraffini, L.A. (2019) Cas13-induced cellular dormancy prevents the rise of CRISPR-resistant bacteriophage. *Nature* 570(7760), 241–245.

Ming, M., Ren, Q., Pan, C., He, Y., Zhang, Y. *et al.* (2020) CRISPR–Cas12b enables efficient plant genome engineering. *Nature Plants* 6(3), 202–208.

Osakabe, K., Wada, N., Miyaji, T., Murakami, E., Marui, K. *et al.* (2020) Genome editing in plants using CRISPR type ID nuclease. *Communications Biology* 3(1), 648.

Özcan, A., Krajeski, R., Ioannidi, E., Lee, B., Gardner, A. *et al.* (2021) Programmable RNA targeting with the single-protein CRISPR effector Cas7-11. *Nature* 597(7878), 720–725.

Pausch, P., Al-Shayeb, B., Bisom-Rapp, E., Tsuchida, C.A., Li, Z. *et al.* (2020) CRISPR-casφ from huge phages is a hypercompact genome editor. *Science* 369(6501), 333–337.

Pinilla-Redondo, R., Mayo-Muñoz, D., Russel, J., Garrett, R.A., Randau, L. *et al.* (2020) Type IV CRISPR–Cas systems are highly diverse and involved in competition between plasmids. *Nucleic Acids Research* 48(4), 2000–2012.

Schwartz, E.A., Bravo, J.P., Ahsan, M., Macias, L.A., McCafferty, C.L. *et al.* (2024) RNA targeting and cleavage by the type III-Dv CRISPR effector complex. *Nature Communications* 15(1), 3324.

Shamshirgaran, Y., Liu, J., Sumer, H., Verma, P.J. and Taheri-Ghahfarokhi, A. (2022) Tools for efficient genome editing; ZFN, TALEN, and CRISPR. In: Verma, P.J., Sumer, H. and Liu, J. (eds) *Applications of Genome Modulation and Editing*. Humana, New York, pp. 29–46.

Shmakov, S.A., Faure, G., Makarova, K.S., Wolf, Y.I., Severinov, K.V. *et al.* (2019) Systematic prediction of functionally linked genes in bacterial and archaeal genomes. *Nature Protocols* 14(10), 3013–3031.

Sternberg, S.H., Redding, S., Jinek, M., Greene, E.C. and Doudna, J.A. (2014) DNA interrogation by the CRISPR RNA-guided endonuclease Cas9. *Nature* 507(7490), 62–67. DOI: 10.1038/nature13011.

Strecker, J., Ladha, A., Gardner, Z., Schmid-Burgk, J.L., Makarova, K.S. *et al.* (2019) RNA-guided DNA insertion with CRISPR-associated transposases. *Science* 365(6448), 48–53.

Swarts, D.C., Oost, J. and Jinek, M. (2017) Structural basis for guide RNA processing and seed-dependent DNA targeting by CRISPR-Cas12a. *Molecular Cell* 66(2), 221–233. .

Wang, J.Y., Pausch, P. and Doudna, J.A. (2022) Structural biology of CRISPR–Cas immunity and genome editing enzymes. *Nature Reviews Microbiology* 20(11), 641–656.

Wang, Q., Alariqi, M., Wang, F., Li, B., Ding, X. *et al.* (2020) The application of a heat-inducible CRISPR/Cas12b (C2c1) genome editing system in tetraploid cotton (*G. hirsutum*) plants. *Plant Biotechnology Journal* 18(12), 2436–2443.

Wu, F., Qiao, X., Zhao, Y., Zhang, Z., Gao, Y. *et al.* (2020) Targeted mutagenesis in *Arabidopsis thaliana* using CRISPR-Cas12b/C2c1. *Journal of Integrative Plant Biology* 62(11), 1653–1658.

Yan, W.X., Chong, S., Zhang, H., Makarova, K.S., Koonin, E.V. *et al.* (2018) Cas13d is a compact RNA-targeting type VI CRISPR effector positively modulated by a WYL-domain-containing accessory protein. *Molecular Cell* 70(2), 327–339.

Yan, W.X., Hunnewell, P., Alfonse, L.E., Carte, J.M., Keston-Smith, E. *et al.* (2019) Functionally diverse type V CRISPR-CAS systems. *Science* 363(6422), 88–91.

Young, J.K., Gasior, S.L., Jones, S., Wang, L., Navarro, P. *et al.* (2019) The repurposing of type IE CRISPR-Cascade for gene activation in plants. *Communications Biology* 2(1), 383.

Zhang, T., Zheng, Q.F., Yi, X., An, H., Zhao, Y.L. *et al.* (2018) Establishing RNA virus resistance in plants by harnessing CRISPR immune system. *Plant Biotechnology Journal* 16(8), 1415–1423.

Zhu, X. and Ye, K. (2012) Crystal structure of Cmr2 suggests a nucleotide cyclase-related enzyme in type III CRISPR-Cas systems. *Febs Letters* 586(6), 939–945.

3 Structure and Mechanism of Zinc-finger Nucleases-mediated Genome Editing in Plants

Jahnavi Chivukula, Blessy Ponnachan, Dipanshu Kumar and Praveen Nagella*
Department of Life Sciences, School of Sciences, CHRIST (Deemed to be University), Bangalore, India

Abstract
The chapter provides a comprehensive exploration of zinc-finger nucleases (ZFNs), intricate molecular tools designed to integrate genes of interest into target sites, thereby instigating both phenotypic and genotypic transformations. The focal point of this genetic manipulation lies in the activity of the FokI endonuclease enzyme, which, through the formation of homodimers, orchestrates precise cleavage at integration sites. Leveraging the wealth of genomic information across diverse organisms, the chapter elucidates the mechanistic underpinnings of ZFNs. A particular emphasis is placed on the modular assembly of ZFNs, unravelling the formation of ββα motifs and explicating the nuanced mechanisms governing their actions. The application of this methodology in plant engineering holds paramount significance, particularly in the realm of augmenting stress tolerance and nutritional value. The chapter systematically examines the gene of interest for various plants, including tobacco, soybean, *Arabidopsis thaliana*, rice, and wheat, elucidating the corresponding ZFN mechanisms. The amalgamation of sophisticated genetic tools with a detailed understanding of plant genomes presents a promising avenue for tailoring crops to meet diverse agricultural challenges. This work not only contributes to the fundamental understanding of ZFNs but also underscores their practical implications in advancing crop improvement strategies for sustainable agriculture.

Keywords: Zinc-finger nucleases, Genome editing, Endonuclease enzyme, Crop plants

3.1 Introduction

Zinc-finger nucleases (ZFNs) are user friendly and depict gene-targeted functions. After the discovery of genomics, it became clear that there are two types of genes. The first is accountable for driving developmental and physiological changes within an organism, whereas the second serves as the regulatory element, determining the extent to which a specific gene's expression will be manifested (Durai *et al.*, 2005).

ZFNs, synthetically designed enzymes, are used to target genes and create favorable mutations. They are used in generating site-directed mutagenesis (SDM), where they are responsible for creating indels (insertions and deletions). These are coupled enzymes made up of zinc-finger domains (ZFDs), which have specificity to binding nucleotides and a restriction enzyme, FokI. The ZFDs are known for their DNA-binding

*Corresponding author: praveen.n@christuniversity.in

DOI: 10.1079/9781800622517.0003

feature and FokI for its activity. The key feature that enables the use of ZFNs is that the enzyme functions as a nuclease only upon dimerization (Bitinaite *et al.*, 1998; Xiao *et al.*, 2011).

Mutations are random changes, which occur in the genome. This mutated genome can result in damaged DNA, which can alter cellular functioning. Mutations occur due to several factors like exposure to mutagens, genetic inheritance, lifestyle, and occupation. Mutations can be broadly classified as DNA damage at nucleotide level, physical or chemical, and chromosomal aberration. The chemical DNA damage at nuclear level is of several forms like transitions and transversions. There is a major form of DNA damage usually occurring during cell cycle, which is physical damage of DNA-single-stranded breaks or double-strand breaks (Hershberg, 2015).

These mutations are generally repaired by the cell itself. The physical mutations, which generally occur during cell division, are addressed by homologous recombination (HR) or non-homologous end-joining (NHEJ) mechanisms (Bibikova *et al.*, 2001; Chang *et al.*, 2017). In NHEJ, once the cell recognizes DNA-double-stranded breaks (DSBs) during cell division, the ends of the DNA are broken resulting in blunt or mismatched ends, and this process is called end processing. The processed ends of DNA are then ligated together by DNA ligase, a process that may result in the creation of some indels. NHEJ is the go-to repair mechanism due to its energy-friendly nature. Active throughout the cell cycle, especially during the G1 phase (Chang *et al.*, 2017), NHEJ safeguards the integrity of the genome by reattaching the broken DNA strands. If NHEJ repair is not successful, it may lead to errors or incomplete repair. Unsuccessful repair can trigger apoptosis (Lieber, 2010).

Alternatively, HR involves a series of well-coordinated steps, beginning with the recognition of breaks and followed by end resection, which generates single-stranded DNA tails. Subsequently, the single-stranded DNA invades a homologous region in the sister chromatid or homologous chromosome, forming a D-loop during DNA strand invasion and synthesis. This intricate mechanism relies heavily on HR repair mechanisms after DNA replication in the S-phase, where accurate repair is paramount (Meselson and Radding, 1975). However, during the G1 phase, when the sister chromatid is yet to be duplicated, the feasibility of HR is reduced. The resolution and repair of the recombination intermediate ultimately ensure the fidelity of the genetic material (Holliday, 1964; Lieber, 2010; Poetsch, 2020).

Gene manipulation works effectively only when the genome of the organism is entirely known along with the phenotypic effect of each gene. Model organisms such as mice, Drosophila, and yeast have a well-understood genome; hence, addition and deletion of a gene can be easily studied. All organisms resort to repair mechanisms in case of breaks that are caused in the DNA, which are referred to as NHEJ repair mechanism and HR (Smith *et al.*, 2000). Drawing from their understanding, scientists applied this knowledge to initiate the inaugural genome editing in *Drosophila melanogaster* (Urnov *et al.*, 2010). In case of introduction of an exogenous sequence by the donor in ZFNs with the help of homology-directed repair, new traits can be introduced. Plants, insects, roundworms, and human cells are among the subjects that have been subjected to gene modification with the help of ZFNs that have reported 18% higher endogenous gene correction rates (Miller *et al.*, 2007).

3.2 ZFN Structure Elucidation History

ZFNs were identified within the DNA-binding domain of a widely distributed protein class in eukaryotes known as transcription factors, a discovery credited to Klug and colleagues in 1986. These DNA-binding domains, numbering approximately 4000 and found in over 700 proteins, make up roughly 2% of the human genome (Porteus and Carroll, 2005). Notably, the native particle nomenclature refers to the 7S particle present in *Xenopus laevis*, characterized by the presence of 7 to 11 zinc atoms. Miller *et al.* (1985) revealed that zinc-binding domains are recurring features in these proteins, shedding light on the prevalence and significance of zinc-mediated structural motifs (Miller *et al.*, 1985).

The structure of the zinc finger (ZF) and the transcription factor IIA (TFIIA) depicted similarities. ZFs were believed to be in accordance with a few of the metalloproteins, as their metal binding properties were traced back to a conserved sequence (Berg, 1988; Pellegrino and Berg, 1991). TFIIA is a zinc-finger protein (ZFP) subfamily that is distinguished by a variable zinc-finger repeat coupled to an amino (NH_2) terminal. These structural repeats tend to be conserved. On carrying out investigation, it was observed that there was the presence of zinc (Zn) finger domains in the RNA binding activity, when the procedure was carried out under *in vitro* conditions (Köster *et al.*, 1991). Kim *et al.* (1996) demonstrated that the purified enzymes could disintegrate a DNA substrate into fragments of the expected lengths, and additionally indicated that both enzymes can cleave respective DNA substrates at multiple locations. Kim *et al.* provided the research framework for the creation of recombinant man-made restriction enzymes (Kim *et al.*, 1996).

The ZFD folds independently and has a compact globular form. The zinc atom is attached to the ZF via two amino acids, that is, two cysteine (C2) and two histidine (H2) amino acids. Two β strands are layered in a hairpin orientation and packed tightly against a well-defined helix that originates as α strand and ends as 3(10) helices to form the polypeptide backbone fold (Lee *et al.*, 1989). The first structure elucidated of zinc fingers forming a complex with DNA helped understand the mechanism involved in DNA identification, which suggests that other proteins that recognize DNA may be constructed after the formation of zinc fingers (Fairall *et al.*, 1993). ZFNs must have features of a rare mix, which includes the ability to recognize long target sequences specifically and adapts to user-defined sequences. Ideally, the target sequences should be long enough to occur only once in a eukaryotic genome (Urnov *et al.*, 2010).

3.3 Modular Assembly of ZFP Structure

There are three distinct sets within the modular assembly approach. First, the Carlos Barbas set involves the α domain of zinc fingers, which recognizes nucleotide sequences in the GNN, ANN, TNN, and CNN patterns, where N represents any nucleotide (Segal *et al.*, 1999; Mandell and Barbas, 2006). Second, the Sangamo Biosciences Set, though now infrequently used due to reduced activity, produces ZFNs that recognize all possible combinations in a GNN format. The recognition occurs based on the specific position of the zinc fingers, denoted as ZF1, ZF2, or ZF3 (Scott, 2005). Finally, the ToolGen Set focuses on identifying naturally occurring ZFNs in the human system, contrasting with artificially synthesized ones. Each of these sets represents a unique strategy within the modular assembly approach, offering diverse options for designing ZFNs based on their distinctive recognition patterns and origins (Bhakta and Segal, 2010; Kim *et al.*, 2010).

These sets were formed with the basis that each of the ZFNs acts independent of the other succeeding or preceding ZFNs. But it has been brought to light that the presence of aspartate, as the second amino acid of the **α** submit has an influence on the choice made by the first finger to a subunit with starting base pairs (bps) to be either G or T. ZFNs interact with regions of DNA that are guanine rich (Pavletich and Pabo, 1991; Bhakta and Segal, 2010).

Zinc-finger nucleases, zinc-finger transcription activators (ZFAs), zinc-finger transcription repressors (ZFRs), and zinc-finger methylases (ZFMs) can all be produced from a single ZFP by combining different domains, such as the non-specific FokI cleavage domain (N), transcription activator domains (A), transcription repressor domains (R), and methylases (M) (Carroll *et al.*, 2006; Palpant and Dudzinski, 2013). Only 5′-GNN-3′ and 5′-ANN-3′ triplet-specific ZF modules have been found thus far through design and selection procedures. Recently, it has been reported that the triplets 5′-CNN-3′ exhibit ZF motif recognition preferences. In the literature, there have also been a few ZF motif designs for 5′-TNN-3′ triplets (Durai *et al.*, 2005; Gersbach *et al.*, 2014). Multiple ZFNs are joined together to produce the tools that target larger DNA segments, which involves linking individual ZFNs to create larger and more complex tools capable of targeting extended DNA segments.

The combination of multiple ZFNs to target larger DNA segments is a strategy often employed in genetic engineering. This approach involves assembling individual ZFNs to create a tool that can recognize and induce breaks at specific sequences within a larger DNA target. The modular assembly of ZFNs allows researchers to tailor these enzymes for specific applications, including targeting larger genomic regions (Kim *et al.*, 2010).

The rationale behind this approach is to enhance the specificity and effectiveness of gene editing across broader genomic regions. Each ZFN typically comprises several ZFDs, with each domain recognizing and binding to a specific triplet nucleotide sequence in the DNA. Combining multiple ZFNs, each designed to recognize distinct DNA sequences, allows for the customization of these tools to target unique and complex DNA sequences of interest (Jabalameli *et al.*, 2015). The zinc ions present as one or more tetrahedral structures present in ZFs are referred to as structural zincs, as they provide structural stability (Neuhaus, 2022).

The structure and recognizing domain of ZFNs is what makes it the most versatile ever. There are approximately 30 amino acids in each ZF, and it leads to the formation of the **ββα** structure. In the ZF, it is the α region that forms a complex with the major groove of DNA. Amid the 30 amino acid sequences, the most crucial amino acids are 1, + 1, +2, +3, +4, +5, and +6 in reference to the onset of each ZF motif's helix. These amino acids are responsible for the majority of the motif-specific interactions with the DNA site (Pavletich and Pabo, 1991). Alterations can be made to the ZF's other regions while leaving the numbered region's amino acid exactly as it is. More sequence specificity between different organisms may result from this. The presence of aspartate at position two in the recognition helix may lead to nonspecific interactions with DNA bases. Aspartate is negatively charged, and its presence at this critical position can result in reduced discrimination between different DNA sequences. This lack of specificity could lead to unintended off-target effects, where the zinc finger may bind to and cleave DNA sequences other than the intended target (Durai *et al.*, 2005; Gersbach *et al.*, 2014). The zinc is surrounded by histidine and cysteine residues creating a Cys2-His2 ZF domain, and it is among the most prevalent DNA-binding motifs. These amino acids provide stability in the folds present in the ZFN domain. It is not just this C2H2-zinc that provides stability but also the hydrophobic core that is conserved (Jacobs, 1992; Fairall *et al.*, 1993; Kim *et al.*, 1996; Davis and Stokoe, 2010; Kim and Kini, 2017).

FokI endonuclease is often used in conjunction with ZFNs for genome editing. FokI is a class II endonuclease, an enzyme that recognizes a specific DNA sequence, known as the recognition sequence or target sequence, and cleaves the DNA at or near that sequence. The N-terminal of FokI is responsible for binding to the DNA, whereas the C terminal is responsible for cleavage (Durai *et al.*, 2005; Gersbach *et al.*, 2014). When four to six ZFP domains are arranged in a stretch, then the DNA binding is accomplished, each of which can identify around 3 bps of DNA. They engage with nearby domains, increasing their binding affinity and adding a little complexity to their design (Petolino, 2015). ZFNs in their monomer form are considered inactive. For ZFNs to be catalytic, heterodimerization between ZFNs and the wild-type FokI takes place leading to nuclease activation. Reverse alignment of two ZFN monomers results in the formation of an active enzyme that can digest its target sequence (Tovkach *et al.*, 2011).

According to the studies mentioned earlier, if two FokI proteins of similar sequence come together, they form a homodimer, which has been shown to be toxic to the organism and causes nonspecific sequence cleavage (Miller *et al.*, 2007; Jabalameli *et al.*, 2015). Noticeable advancements have been reported in dimer formation, such as a 6 bp spacer is added between the FokI domains of the two ZFNs to enhance specific sequence recognition. The dimer formed thus recognizes 18 bps. Through computational methods, there have been major advancements in the redesigning of the ZFN to avoid toxicity and reduce homodimer formation. As per structural investigation, the cleavage domain's helices **α**4 and **α**5 (residues 479–490 and 528–539) primarily mediate dimerization. Combining structural and functional information, it was discovered that D483 may form a bidentate hydrogen bond to R487 in the opposite subunit (Fig. 3.1). Furthermore, the dimer may be stabilized through hydrophobic interactions between isoleucine in the inter-helical sequence (I499) and in **α**5 (I538). Variants/mutants of ZFNs also have been used to lead to single-strand annealing, and a lower level of toxicity was observed (Szczepek *et al.*, 2007).

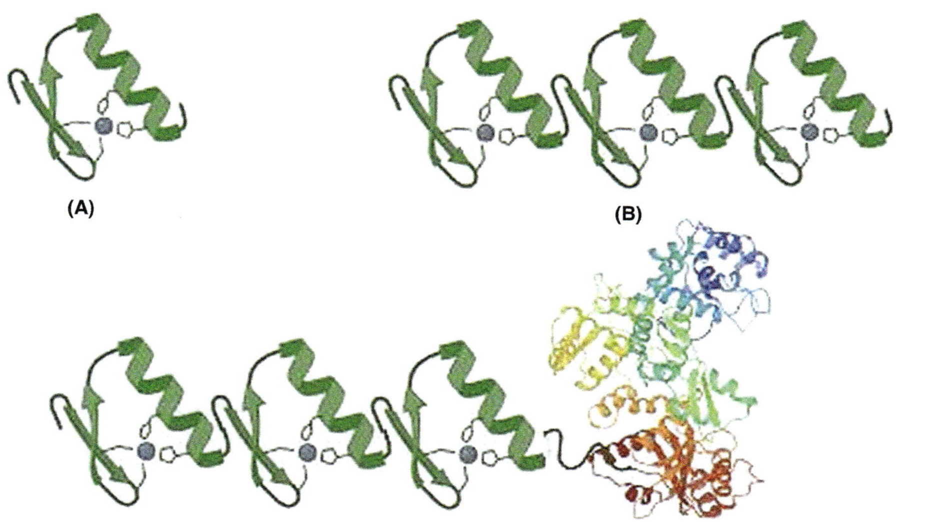

Fig. 3.1. (a) Zinc-finger protein recognizes three bps on DNA, (b) ZFPs 3–4 tethered together to increase specificity during DNA binding, and (c) ZFP couples with FokI nucleases (Created using BioRender).

From the understanding that one possesses, it is known that dimerization occurs between ***α*** units of the C-end of the protein. The NN terminal joining seems to reduce toxicity to an even higher extent, and their cleavage and recognition abilities are enhanced as well. The known gap to be present in dimers of understanding is six bps, but, in this dimerization, a range from 5 to 9 bps was maintained and studied. When the gene-modifying potential of these ZFNs was tested in K562 cells, the analysis indicates excellent average activity for NN pairs with gaps of 7, 8, and 9 bps, as well as a high hit rate (Paschon *et al.*, 2019). N6a and N7a linkers are specific types of molecular connectors used in the design of these ZFNs. The designs incorporating the N6a and N7a linkers, along with an 8 bp gap, have demonstrated the highest average insertion levels. This means that these particular ZFN configurations, with the specified linkers and gap size, are particularly effective in introducing new genetic material into the target DNA, showcasing their enhanced activity and success in the gene-editing process (Wah *et al.*, 1998; Paschon *et al.*, 2019).

The dimerization of the bacterial nuclease FokI for cleavage activation was explored by Wah *et al.* (1998). An unusual class of restriction enzymes, including FokI, can detect a specific DNA sequence and cut freely just outside of it. FokI binds to the homologous sequence 5-GGATG-3 before cutting the DNA's phosphodiester bond 13 bases distant and nine bps downstream of recognition. The three smaller subdomains (D1, D2, and D3) that make up the identification domain share a genetic history with the DNA-binding domain, which is built of a helix-turn-helix plus catabolite-gene activator protein. The nonspecific FokI cleavage region is guided to a specific DNA-target site by the DNA-binding ZF region (Wah *et al.*, 1998; Cathomen and Keith Joung, 2008).

To use ZFNs for gene targeting in research, it is required to create unique ZFNs and also thoroughly test how well they perform in cells. For this, it is crucial to show that the freshly discovered ZFNs can bind and digest target sequences *in vitro*, in addition to being able to digest target DNA *in vivo* (Tovkach *et al.*, 2009). As per studies conducted, the DNA-binding specificity of a single ZF can be altered by bringing about a change either inside or nearby one or more of the amino acid residues of the structurally conserved helix (also known as the 'recognition helix') (Cathomen and Keith Joung, 2008).

3.4 Structure of Artificially Synthesized ZFNs

Zinc-finger proteins serve as modular units in the design of ZFNs, characterized by coordinated-zinc ions and specific amino acid residues. These modular units exhibit the ability to recognize and bind to distinct DNA triplets or sequences. In the realm of artificially synthesized ZFNs, researchers custom design ZFPs to target-specific DNA sequences by selecting zinc-finger modules tailored to bind to the desired target sequence (Carroll *et al.*, 2006; Urnov *et al.*, 2010).

Linkers play a crucial role in ZFN architecture, serving as connective elements that link ZFPs to the nuclease domain. These linkers provide flexibility and enable the proper alignment of ZFPs, facilitating their binding to target sequences on the DNA (Urnov *et al.*, 2010).

The nuclease domain, a key component of ZFNs, is responsible for introducing double-stranded breaks in the DNA at or near the designated target site. Commonly utilized nuclease domains, such as those derived from the FokI endonuclease, contribute to the cleavage activity of ZFNs. Additionally, ZFNs often operate as dimers, with two nuclease domains coming together for active cleavage, a process facilitated by the binding of ZFPs to their respective target sequences (Kim *et al.*, 2010; Urnov *et al.*, 2010; Carroll, 2011).

In the context of *in vivo* applications, ZFNs are engineered to include expression elements such as promoters and terminators. These elements regulate the expression of ZFNs within the host organism, ensuring controlled and precise functionality (Tovkach *et al.*, 2009).

3.5 Mechanism of Action

Zinc-finger domains use their ability to bind to nucleotides and bind to the DNA. Once this binding is established, the FokI enzyme is placed near the DNA strand on both complementary strands (Urnov *et al.*, 2010; Carroll, 2011; Osakabe and Osakabe, 2015). When the ZFNs bind, the FokI domain dimerizes, thereby creating a nick in the target DNA strand. If the ZFNs bind at a farther distance, dimerization of Fok1 is prevented and, hence, the nick is not created. The ZFNs are designed to have left ZFNs and right ZFNs; these are designed manually to target the DNA region of interest (Smith *et al.*, 2000; Durai *et al.*, 2005; Wu *et al.*, 2007). The ZFNs are carefully programmed in such a way that they bind in close range at the target site. Once the double-stranded break is created, the nucleotides in that region are lost, causing deletion. These create mutations in the DNA segments (Smith *et al.*, 2000).

In SDM, the iterative process of cutting and repairing the target site by ZFNs persists within the designated region until a mutation becomes evident (Fig. 3.2). This cyclical activity continues until the ZFNs can no longer recognize the sequence at the target site. Experimental evidence from various studies, including those conducted in Drosophila (Bibikova *et al.*, 2002) and in plant species such as *Arabidopsis*, tobacco, and maize (Lloyd *et al.*, 2005; Krajewska *et al.*, 2015; Vítor *et al.*, 2020), has substantiated the effectiveness of ZFN-mediated SDM. The findings underscore the reliability and success of this approach in inducing specific mutations at desired target sites, validating its utility across diverse organisms.

To use ZFNs to insert a specific gene into the target sequence, the only viable way is to utilize the HR repair pathway. The gene of interest (GOI) is flanked by the DNA, which is homologous to the target DNA. When the double-stranded break occurs, this flanked gene is targeted to interact. Since the regions surrounding the GOI are homologous to host DNA, it undergoes HR repair, where the gene gets inserted into the gap of double-stranded break (Paszkowski *et al.*, 1988; Lee *et al.*, 1990; Rouet *et al.*, 1994; Bibikova *et al.*, 2003).

3.6 Applications of ZFNs

3.6.1 Herbicide resistance

ZFNs have demonstrated their effectiveness in *Arabidopsis*, tobacco, soybean, and maize (Townsend *et al.*, 2009; Zhang and Voytas, 2011; Nandula, 2019; Yadav *et al.*, 2023). A few instances of utilizing ZFNs in crop breeding involved the insertion of PAT gene cassettes, disrupting the endogenous ZmIPK1 gene in

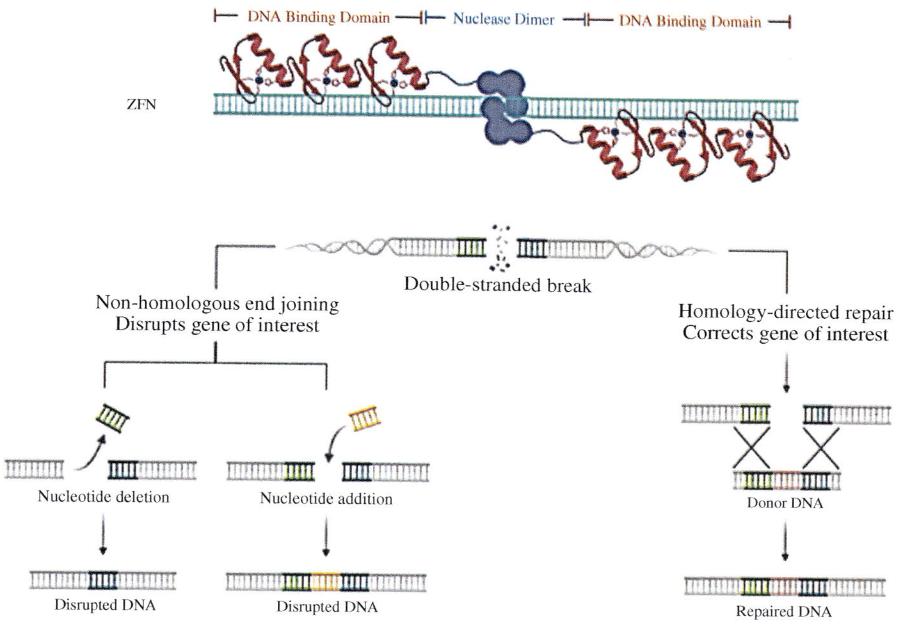

Fig. 3.2. Mechanism to create SDM using ZFNs. On the right-hand side, DSBs created using ZNFs are repaired using the NHEJ mechanism and the left image highlights the HR mechanism to repair DSBs (created using BioRender).

maize. This disruption led to changes in the inositol phosphate profile of developing maize seeds, resulting in improved herbicide resistance (Shukla *et al.*, 2009).

In another study, ZFNs were used to stimulate high-frequency gene targeting at endogenous plant genes, specifically the tobacco acetohydroxyacid synthase (SuRA and SuRB) genes. These genes are associated with known herbicide-resistance mutations. The researchers successfully introduced herbicide-resistance mutations into SuR loci using ZFN-mediated gene targeting, achieving frequencies exceeding 2% of transformed cells. Remarkably, modifications were observed up to 1.3 kb from the ZFN cleavage site. Over 40% of the resulting plants showed modifications in multiple SuR alleles, demonstrating the efficiency of targeted sequence changes in endogenous plant genes through ZFNs (Townsend *et al.*, 2009).

The creation of ZFNs employs diverse protein-engineering techniques, enabling precise targeting of virtually any unique DNA sequence. Enhanced specificity and activity in ZFNs have been achieved, facilitating the development of knockouts to deactivate gene function, as well as gain-of-function alterations (Petolino, 2015).

3.6.2 Salinity tolerance

The salt stress regulation role of the CCCH ZFP, GhC3H20, in cotton and *Arabidopsis* plants becomes evident, as it responds to salt, drought, and ABA treatments. Transgenic *Arabidopsis* lines, featuring GhC3H20 overexpression, demonstrate an increased ability to withstand salt stress, displaying longer roots and sustained leaf health. Notably, the modified plants exhibit elevated levels of catalase (CAT), indicating a reinforced stress response. Conversely, silencing GhC3H20 in cotton through virus-induced gene silencing (VIGS) experiments results in diminished salt stress tolerance. A yeast two-hybrid assay identifies two interacting proteins, GhPP2CA and GhHAB1, suggesting their involvement in the ABA signaling pathway. This suggests that within the framework of

ZFN applications, GhC3H20 collaborates with GhPP2CA and GhHAB1 to fortify salt stress tolerance in cotton through the ABA signaling pathway (Zhang *et al.*, 2023).

3.6.3 Enhanced product quality

Functional analysis of the tomato L1L4 master transcription factor, a key regulator of biosynthesis for seed storage proteins and fatty acids, has revealed its significant impact on tomato fruit quality. Mutant lines with the disrupted L1L4 gene displayed subtle morphological differences but exhibited substantial variations in moisture content, acidity, and overall metabolite profiles, including antioxidants and bioactive compounds. Specific mutant lines demonstrated enhanced levels of β-carotene, antioxidants, and other bioactive compounds, highlighting the potential of L1L4 as a molecular target for improving tomato crop quality (Gago *et al.*, 2017).

3.6.4 Nutritional value enhancement

In pursuit of improving maize (*Zea mays*), the world's most cultivated cereal crop, efforts were made to address the antinutrient effects of phytate, hindering protein and mineral absorption in animals and humans. Utilizing ZFNs, the IPK1 gene responsible for encoding inositol-1,3,4,5,6-pentakisphosphate 2-kinase was effectively blocked. This targeted genetic modification aims to reduce phytate content in maize, enhancing its nutritional value and applicability in various industries, including animal feed, biofuel, and human consumption products like maize starch and maize syrup (Shukla *et al.*, 2009).

3.7 Conclusion

Precision gene indels are possible using ZFNs. Many nonspecific recombinations caused by convectional gene targeting technologies, which lead to detrimental effects on plants or hosts, can be avoided. From the applications of ZFNs discussed earlier, it is observed they are both a boon and a bane for the scientific community. Multiple ZFNs can be coupled and used by programming them in a way to achieve higher recombinant outputs. Either HR- or NHEJ-based gene targeting using ZFNs can be used depending on the plant. SDMs are essential to study the functioning of certain gene/proteins, involved in major life processes. These SDMs can be achieved using ZFNs. Studies in Arabidopsis, Drosophila, poplar, soybean, and many higher order organisms have proved the diverse application range of ZFNs. The gene insertions using ZFNs were observed to be heritable, and no loss over generations was observed.

Indeed, despite their advantages, ZFNs come with certain drawbacks. One significant limitation is the complexity and high cost associated with constructing protein domains tailored for each specific genome locus. This process demands a considerable investment of time and resources, making it less practical for widespread application.

3.8 Future Prospects

Researchers have made several efforts to develop transgenic plants with desired characteristics, but there are still certain obstacles for utilizing this gene-editing technology, which need to be tackled. To make this method efficient, certain supporting technologies like high-resolution molecular and genome analysis and bioinformatics tools for characterizing traits are also required to be improvised. The ability of ZFNs to make a precise cut in the endogenous loci will advance the knowledge about the genome structure and the mechanism by which the gene gets expressed to give rise to the phenotype. Through this mechanism, scientists will be able to discover new genes associated with the novel trait in the near future. It will also help researchers to understand the extent of complexity of the genetic network and the interaction of genes with several environmental factors. This knowledge can be employed to control this complex plant trait. Random mutation in the plant genome increases the time and cost for screening, as a large number of transgene plants have to be screened but specific targeting using

ZFNs will reduce not only the cost and time but also the undesired characteristics. Future research will offer us the knowledge to reduce the cytotoxicity of ZFNs and also to increase the specificity. Optimization of ZFNs-protein arrays can further result in less cytotoxicity and high specificity.

There are still limitations for the transformation system because the same transformation strategy cannot be employed in all plants, as its efficiency reduced or it will not transform at all. Delivery of genetically engineered components into germ lines and shoot meristems of nontransformable species for getting transgenic offspring should also be exploited to come up with a promising strategy.

Zinc-finger nucleases, although early pioneers in the realm of gene editing, find themselves under scrutiny when compared with more recent advanced techniques like CRISPR. The emergence of CRISPR has revolutionized the field of genetic engineering, offering unparalleled simplicity, versatility, and cost-effectiveness. While ZFNs exhibit precision in targeting specific DNA sequences through custom protein domain design, their complexity and high associated costs have posed challenges in widespread adoption. In contrast, CRISPR's straightforward design using guide RNAs allows for more efficient and multiplexed genome editing, making it a preferred choice for contemporary research and therapeutic applications.

In summary, the choice between ZFNs and CRISPR-Cas9 depends on the specific requirements of the experiment or application. ZFNs may be preferred for their precision in certain cases, whereas CRISPR-Cas9 is often favored for its simplicity, cost-effectiveness, and versatility. Advances in both technologies continue, offering researchers an evolving toolkit for precise genome editing. The comparative analysis between ZFNs and CRISPR sheds light on the evolving landscape of gene-editing technologies, emphasizing the need for accessibility, precision, and adaptability in driving advancements in the field.

References

Berg, J.M. (1988) Proposed structure for the zinc-binding domains from transcription factor IIIA and related proteins. *Proceedings of the National Academy of Sciences of the United States of America* 85(1), 99–102. DOI: 10.1073/PNAS.85.1.99.

Bhakta, M.S. and Segal, D.J. (2010) The generation of zinc finger proteins by modular assembly. *Methods in Molecular Biology (Clifton, N.J)* 649, 3. DOI: 10.1007/978-1-60761-753-2_1.

Bibikova, M., Carroll, D., Segal, D.J., Trautman, J.K., Smith, J. *et al.* (2001) Stimulation of homologous recombination through targeted cleavage by chimeric nucleases. *Molecular and Cellular Biology* 21(1), 289–297. DOI: 10.1128/MCB.21.1.289-297.2001.

Bibikova, M., Golic, M., Golic, K.G. and Carroll, D. (2002) Targeted chromosomal cleavage and mutagenesis in Drosophila using zinc-finger nucleases. *Genetics* 161(3), 1169–1175. DOI: 10.1093/GENETICS/161.3.1169.

Bibikova, M., Beumer, K., Trautman, J.K. and Carroll, D. (2003) Enhancing gene targeting with designed zinc finger nucleases. *Science* 300(5620), 764. DOI: 10.1126/SCIENCE.1079512.

Bitinaite, J., Wah, D.A., Aggarwal, A.K. and Schildkraut, I. (1998) FokI dimerization is required for DNA cleavage. *Proceedings of the National Academy of Sciences of the United States of America* 95(18), 10570–10575. DOI: 10.1073/PNAS.95.18.10570.

Carroll, D. (2011) Genome engineering with zinc-finger nucleases. *Genetics* 188(4), 773–782. DOI: 10.1534/genetics.111.131433.

Carroll, D., Morton, J.J., Beumer, K.J. and Segal, D.J. (2006) Design, construction and *in vitro* testing of zinc finger nucleases. *Nature Protocols* 1(3), 1329–1341. DOI: 10.1038/nprot.2006.231.

Cathomen, T. and Keith Joung, J. (2008) Zinc-finger nucleases: the next generation emerges. *Molecular Therapy: The Journal of the American Society of Gene Therapy* 16(7), 1200–1207. DOI: 10.1038/mt.2008.114.

Chang, H.H.Y., Pannunzio, N.R., Adachi, N. and Lieber, M.R. (2017) Non-homologous DNA end joining and alternative pathways to double-strand break repair. *Nature Reviews Molecular Cell Biology* 18(8), 495–506. DOI: 10.1038/nrm.2017.48.

Davis, D. and Stokoe, D. (2010) Zinc finger nucleases as tools to understand and treat human diseases. *BMC Medicine* 8(1), 1–11. DOI: 10.1186/1741-7015-8-42/FIGURES/3.

Durai, S., Mani, M., Kandavelou, K., Wu, J., Porteus, M.H. *et al.* (2005) Zinc finger nucleases: Custom-designed molecular scissors for genome engineering of plant and mammalian cells. *Nucleic Acids Research* 33(18), 5978–5990. DOI: 10.1093/nar/gki912.

Fairall, L., Schwabe, J.W.R., Chapman, L., Finch, J.T. and Rhodes, D. (1993) The crystal structure of a two zinc-finger peptide reveals an extension to the rules for zinc-finger/DNA recognition. *Nature* 366(6454), 483–487. DOI: 10.1038/366483A0.

Gago, C., Drosou, V., Paschalidis, K., Guerreiro, A., Miguel, G. *et al.* (2017) Targeted gene disruption coupled with metabolic screen approach to uncover the LEAFY COTYLEDON1-LIKE4 (L1L4) function in tomato fruit metabolism. *Plant Cell Reports* 36(7), 1065–1082. DOI: 10.1007/S00299-017-2137-9.

Gersbach, C.A., Gaj, T. and Barbas, C.F. (2014) Synthetic zinc finger proteins: The advent of targeted gene regulation and genome modification technologies. *Accounts of Chemical Research* 47(8), 2309–2318. DOI: 10.1021/AR500039W/ASSET/IMAGES/LARGE/AR-2014-00039W_0007.JPEG.

Hershberg, R. (2015) Mutation – the engine of evolution: Studying mutation and its role in the evolution of bacteria. *Cold Spring Harbor Perspectives in Biology* 7(9). DOI: 10.1101/CSHPERSPECT.A018077.

Holliday, R. (1964) A mechanism for gene conversion in fungi. *Genetics Research* 5(2), 282–304. DOI: 10.1017/S0016672300001233.

Jabalameli, H.R., Zahednasab, H., Karimi-Moghaddam, A. and Jabalameli, M.R. (2015) Zinc finger nuclease technology: Advances and obstacles in modelling and treating genetic disorders. *Gene* 558(1), 1–5. DOI: 10.1016/J.GENE.2014.12.044.

Jacobs, G.H. (1992) Determination of the base recognition positions of zinc fingers from sequence analysis. *The EMBO Journal* 11(12), 4507. DOI: 10.1002/J.1460-2075.1992.TB05552.X.

Kim, M.S. and Kini, A.G. (2017) Engineering and application of zinc finger proteins and tales for biomedical research. *Molecules and Cells* 40(8), 533–541. DOI: 10.14348/MOLCELLS.2017.0139.

Kim, J.S., Lee, H.J. and Carroll, D. (2010) Genome editing with modularly assembled zinc-finger nucleases. *Nature Methods* 7(2), 91–91. DOI: 10.1038/nmeth0210-91a.

Kim, Y.G., Cha, J. and Chandrasegaran, S. (1996) Hybrid restriction enzymes: Zinc finger fusions to Fok I cleavage domain. *Proceedings of the National Academy of Sciences of the United States of America* 93(3), 1156. DOI: 10.1073/PNAS.93.3.1156.

Köster, M., Kühn, U., Bouwmeester, T., Nietfeld, W., el-Baradi, T. *et al.* (1991) Structure, expression and *in vitro* functional characterization of a novel RNA binding zinc finger protein from Xenopus. *The EMBO Journal* 10(10), 3087–3093. DOI: 10.1002/J.1460-2075.1991.TB07861.X.

Krajewska, M., Fehrmann, R.S.N., Vries, E.G.E. and van Vugt, M.A.T.M. (2015) Regulators of homologous recombination repair as novel targets for cancer treatment. *Frontiers in Genetics* 6, MAR. DOI: 10.3389/FGENE.2015.00096/FULL.

Lee, K.Y., Lund, P., Lowe, K. and Dunsmuir, P. (1990) Homologous recombination in plant cells after Agrobacterium-mediated transformation. *The Plant Cell* 2(5), 415–425. DOI: 10.1105/TPC.2.5.415.

Lee, M.S., Gippert, G.P., Soman, K.V., Case, D.A. and Wright, P.E. (1989) Three-dimensional solution structure of a single zinc finger DNA-binding domain. *Science* 245(4918), 635–637. DOI: 10.1126/SCIENCE.2503871.

Lieber, M.R. (2010) The mechanism of double-strand DNA break repair by the nonhomologous DNA end-joining pathway. *Annual Review of Biochemistry* 79, 181–211. DOI: 10.1146/ANNUREV.BIOCHEM.052308.093131.

Lloyd, A., Plaisier, C.L., Carroll, D. and Drews, G.N. (2005) Targeted mutagenesis using zinc-finger nucleases in *Arabidopsis*. *Proceedings of the National Academy of Sciences of the United States of America* 102(6), 2232–2237. DOI: 10.1073/PNAS.0409339102.

Mandell, J.G. and Barbas, C.F. (2006) Zinc finger tools: Custom DNA-binding domains for transcription factors and nucleases. *Nucleic Acids Research* 34 (Web Server issue). DOI: 10.1093/NAR/GKL209.

Meselson, M.S. and Radding, C.M. (1975) A general model for genetic recombination. *Proceedings of the National Academy of Sciences of the United States of America* 72(1), 358. DOI: 10.1073/PNAS.72.1.358.

Miller, J., McLachlan, A.D. and Klug, A. (1985) Repetitive zinc-binding domains in the protein transcription factor IIIA from *Xenopus oocytes*. *The EMBO Journal* 4(6), 1609–1614. DOI: 10.1002/J.1460-2075.1985.TB03825.X.

Miller, J.C., Holmes, M.C., Wang, J., Guschin, D.Y., Lee, Y.L. *et al*. (2007) An improved zinc-finger nuclease architecture for highly specific genome editing. *Nature Biotechnology* 25(7), 778–785. DOI: 10.1038/NBT1319.

Nandula, V.K. (2019) Herbicide resistance traits in maize and soybean: Current status and future outlook. *Plants* 8(9). DOI: 10.3390/PLANTS8090337.

Neuhaus, D. (2022) Zinc finger structure determination by NMR: Why zinc fingers can be a handful. *Progress in Nuclear Magnetic Resonance Spectroscopy* 130–131, 62–105. DOI: 10.1016/J.PNMRS.2022.07.001.

Osakabe, Y. and Osakabe, K. (2015) Genome editing with engineered nucleases in plants. *Plant & Cell Physiology* 56(3), 389–400. DOI: 10.1093/PCP/PCU170.

Palpant, N.J. and Dudzinski, D. (2013) Zinc finger nucleases: Looking toward translation. *Gene Therapy* 20(2), 121–127. DOI: 10.1038/gt.2012.2.

Paschon, D.E., Lussier, S., Wangzor, T., Xia, D.F., Li, P.W. *et al*. (2019) Diversifying the structure of zinc finger nucleases for high-precision genome editing. *Nature Communications* 10(1), 1–12. DOI: 10.1038/s41467-019-08867-x.

Paszkowski, J., Baur, M., Bogucki, A. and Potrykus, I. (1988) Gene targeting in plants. *The EMBO Journal* 7(13), 4021. DOI: 10.1002/J.1460-2075.1988.TB03295.X.

Pavletich, N.P. and Pabo, C.O. (1991) Zinc finger-DNA recognition: Crystal structure of a zif268-DNA complex at 2.1 å. *Science* 252(5007), 809–817. DOI: 10.1126/SCIENCE.2028256.

Pellegrino, G.R. and Berg, J.M. (1991) Identification and characterization of 'zinc-finger' domains by the polymerase chain reaction. *Proceedings of the National Academy of Sciences of the United States of America* 88(2), 671. DOI: 10.1073/PNAS.88.2.671.

Petolino, J.F. (2015) Genome editing in plants via designed zinc finger nucleases. *In Vitro Cellular & Developmental Biology Plant: Journal of the Tissue Culture Association* 51(1), 1–8. DOI: 10.1007/S11627-015-9663-3.

Poetsch, A.R. (2020) The genomics of oxidative DNA damage, repair, and resulting mutagenesis. *Computational and Structural Biotechnology Journal* 18, 207–219. DOI: 10.1016/J.CSBJ.2019.12.013.

Porteus, M.H. and Carroll, D. (2005) Gene targeting using zinc finger nucleases. *Nature Biotechnology* 23(8), 967–973. DOI: 10.1038/nbt1125.

Rouet, P., Smih, F. and Jasin, M. (1994) Expression of a site-specific endonuclease stimulates homologous recombination in mammalian cells. *Proceedings of the National Academy of Sciences of the United States of America* 91(13), 6064–6068. DOI: 10.1073/PNAS.91.13.6064.

Scott, C.T. (2005) The zinc finger nuclease monopoly. *Nature Biotechnology* 23(8), 915–918. DOI: 10.1038/NBT0805-915.

Segal, D.J., Dreier, B., Beerli, R.R. and Barbas, C.F. (1999) Toward controlling gene expression at will: Selection and design of zinc finger domains recognizing each of the 5′-GNN-3′ DNA target sequences. *Proceedings of the National Academy of Sciences of the United States of America* 96(6), 2758–2763. DOI: 10.1073/PNAS.96.6.2758/ASSET/0E731804-71D0-4E57-B979-4D727663E24B/ASSETS/GRAPHIC/PQ0595065002.JPEG.

Shukla, V.K., Doyon, Y., Miller, J.C., Dekelver, R.C., Moehle, E.A. *et al*. (2009) Precise genome modification in the crop species *Zea mays* using zinc-finger nucleases. *Nature* 459(7245), 437–441. DOI: 10.1038/nature07992.

Smith, J., Bibikova, M., Whitby, F.G., Reddy, A.R., Chandrasegaran, S. *et al*. (2000) Requirements for double-strand cleavage by chimeric restriction enzymes with zinc finger DNA-recognition domains. *Nucleic Acids Research* 28(17), 3361–3369. DOI: 10.1093/nar/28.17.3361.

Szczepek, M., Brondani, V., Büchel, J., Serrano, L., Segal, D.J. *et al*. (2007) Structure-based redesign of the dimerization interface reduces the toxicity of zinc-finger nucleases. *Nature Biotechnology* 25(7), 786–793. DOI: 10.1038/nbt1317.

Tovkach, A., Zeevi, V. and Tzfira, T. (2009) A toolbox and procedural notes for characterizing novel zinc finger nucleases for genome editing in plant cells. *The Plant Journal: For Cell and Molecular Biology* 57(4), 747–757. DOI: 10.1111/J.1365-313X.2008.03718.X.

Tovkach, A., Zeevi, V. and Tzfira, T. (2011) Expression, purification and characterization of cloning-grade zinc finger nuclease. *Journal of Biotechnology* 151(1), 1–8. DOI: 10.1016/J.JBIOTEC.2010.10.071.

Townsend, J.A., Wright, D.A., Winfrey, R.J., Fu, F., Maeder, M.L. *et al*. (2009) High-frequency modification of plant genes using engineered zinc-finger nucleases. *Nature* 459(7245), 442–445. DOI: 10.1038/NATURE07845.

Urnov, F.D., Rebar, E.J., Holmes, M.C., Zhang, H.S. and Gregory, P.D. (2010) Genome editing with engineered zinc finger nucleases. *Nature Reviews Genetics* 11(9), 636–646. DOI: 10.1038/nrg2842.

Vítor, A.C., Huertas, P., Legube, G. and de Almeida, S.F. (2020) Studying DNA double-strand break repair: An ever-growing toolbox. *Frontiers in Molecular Biosciences* 7. DOI: 10.3389/FMOLB.2020.00024/FULL.

Wah, D.A., Bitinaite, J., Schildkraut, I. and Aggarwal, A.K. (1998) Structure of FokI has implications for DNA cleavage. *Proceedings of the National Academy of Sciences of the United States of America* 95(18), 10564. DOI: 10.1073/PNAS.95.18.10564.

Wu, J., Kandavelou, K. and Chandrasegaran, S. (2007) Custom-designed zinc finger nucleases: What is next. *Cellular and Molecular Life Sciences* 64(22), 2933–2944. DOI: 10.1007/s00018-007-7206-8.

Xiao, A., Hu, Y.Y., Wang, W.Y., Yang, Z.P., Wang, Z.X. *et al.* (2011) Progress in zinc finger nuclease engineering for targeted genome modification. *Yi Chuan = Hereditas / Zhongguo Yi Chuan Xue Hui Bian Ji* 33(7), 665–683. DOI: 10.3724/SP.J.1005.2011.00665.

Yadav, R.K., Tripathi, M.K., Tiwari, S., Tripathi, N., Asati, R. *et al.* (2023) Genome editing and improvement of abiotic stress tolerance in crop plants. *Life* 13(7). DOI: 10.3390/LIFE13071456.

Zhang, F. and Voytas, D.F. (2011) Targeted mutagenesis in *Arabidopsis* using zinc-finger nucleases. *Methods in Molecular Biology (Clifton, N.J)* 701, 167–177. DOI: 10.1007/978-1-61737-957-4_9.

Zhang, Q., Zhang, J., Wei, F., Fu, X., Wei, H. *et al.* (2023) The CCCH-type zinc-finger protein ghc3h20 enhances salt stress tolerance in *Arabidopsis thaliana* and cotton through ABA signal transduction pathway. *International Journal of Molecular Sciences* 24(5). DOI: 10.3390/IJMS24055057/S1.

4 Structure and Mechanism of Transcription Activator-like Effector Nuclease-mediated Genome Editing in Plants

Sana Khalid[1], Muhammad Naveed Shahid[2], Kiran Nawaz[3*], Muddesra Naeem[1] and Khajista Jabeen[1]

[1]Department of Botany, Lahore College for Women University, Lahore, Pakistan; [2]Department of Botany, Division of Science and Technology, University of Education, Lahore, Pakistan; [3]RNAissance Ag. TechAccel, Saint Louis, Missouri, USA

Abstract

Genetic engineering tools can be used to completely disrupt/reduce normal gene expression, thereby revealing the function of the targeted gene. For over a decade, RNA interference has reigned supreme in the laboratory as the go-to method for modulating gene expression. In a wide range of organisms or cell types, the targeted DNA sequences are altered via engineered nucleases. The newly invented transcription activator-like effector nuclease (TALEN) technique encompasses the nuclease (nonspecific DNA-cleaver) that binds to a DNA-binding domain (DBD) that can be engineered easily so that TALENs can target any sequence. It has provided the minimal off-target effects as compared to other gene-editing tools and sparked the genome-editing revolution. It has the ability of targeted mutagenesis that can be applied in multiple species. Its precise positioning and flexibility are unmatched. Some research investigators become confused about which is the best genome-editing tool. This problem is addressed in this chapter by making comparison between TALENs with RNAi and clustered regularly interspaced short palindromic repeats (CRISPR) in terms of ease, workflow, setup, procedures, and validation of experiments followed by effectiveness, off-target effects, and adaptability of technologies. Herein, we discuss the essential facts, novel developments, and design guidelines of TALENs.

Keywords: Plant genome editing, Transcription activator-like effector nucleases (TALENs), RNA interface (RNAi), Clustered regularly interspaced short palindromic repeats (CRISPR), Transcription, DNA-binding domain (DBD)

4.1 Introduction

Techniques for genome (DNA) sequence manipulation, visualization, and gene expression regulation are growing speedily in the post-genomic era. Currently, some methods are highly recommended by researchers due to their maximum accessibility, safety, and

*Corresponding author: Kiran@techaccel.net

DOI: 10.1079/9781800622517.0004

efficiency for making such changes in DNA, which would result in the desired phenotype. Genome-editing revolution was ignited with the invention of transcription activator-like effector nucleases (TALENs) after the application of zinc-finger nucleases (ZFNs). They were the first highly efficient and precise methods that can target the specific genome locus because of their easy design. TALENs became the favorite genome-editing tool for livestock, crops, and model and non-model organisms, and for curing human cancer in 2015 (Qasim *et al.*, 2017) and in 2019, they were used to introduce the first genome-edited crop (Menz *et al.*, 2020). In the development of the genome-editing field, meganucleases and ZFNs laid the concept of gene editing in 1990s irrespective of their difficult use. However, in 2002, the earliest genome-edited organism of the world came into existence (Bibikova *et al.*, 2002; Carroll, 2021). In 2010, transcription activator-like effector (TALE) and RE (restriction enzyme) Fok1 were fused to develop TALENs that were more easily accessible and reachable than meganucleases and ZFNs for the scientific community.

Basically, designer nucleases are used as genome-editing tools that can create a break in double strands of DNA at a desired place, which convert into mutations of cells' own repair system. As per the DNA configurations, many repair mechanisms may be applied. For example, the NHEJ (non-homologous end joining) repair system results in (i) single-strand annealing (SSA) and (ii) microhomology-mediated end joining (MMEJ) for expected deletions. Finally, precise changes are possible if a repair template is used in a recombination-based homology-directed repair (HDR) (Jasin and Rothstein, 2013). However, NHEJ is the preferred system for knockout mutations over HDR. Of these, Cas9 and Cas12a base editors focus on targeted changes rather than on random repair processes (Anzalone *et al.*, 2020; Porto *et al.*, 2020). Moreover, Komor *et al.* (2016) and Gaudelli *et al.* (2017) introduce cytosine editors for the transition of C to T and adenine editors for the transition of A to G.

4.2 Structure and Function of TALEs and TALENs

4.2.1 DNA-binding modes of TALEs

Transcription activator-like effector nucleases are an innovative class of genome-editing tools that can alter the gene by precise deletion, insertion, or substitution of targeted genes. Currently, they have appeared as a radical genome-editing technique in many cell types and species. Basically, TALENs comprise two components: (i) a nuclease that can break dsDNA in cells and (ii) an assigned TALE domain that simulates the natural TALE proteins (Fig. 4.1). TALEs are the programmable DNA-binding proteins extracted from *Xanthomonas* (plant-pathogenic bacteria) that fundamentally modulate gene function. In order to recognize the targeted DNA motifs, transcription factors (DNA-binding domains (DBDs)) of 7–34 highly homologous direct repeats (DRs) with 33–35 amino acids are synthesized and programmed accordingly. Among 33–35 amino acids in each repeat, 12 and 13 have repeat variable diresidues (RVDs), which are the key amino acids for specificity (Joung and Sander, 2013). It is possible to design any TALE of a targeted sequence of DNA by appropriate engineering of DBDs with the cracked RVDs-binding code of DNA protein (Boch *et al.*, 2009; Moscou and Bogdanove, 2009).

Usually, TALEs balance specificity with potential off-targeting due to their specific design to spot 15–20 base pairs of DNA (Miller *et al.*, 2011; Mussolino *et al.*, 2011). So, different DBDs of approximately 500–700 amino acids are designed for each target site in the genome, accounting for a laborious process.

TALEs consist of a super-helical structure (right-handed) that covers the DNA and performs a non-rotational, one-dimensional, rapid, and unbiased search of target sequences (Cuculis *et al.*, 2015, 2016). A loop connects the two α-helices in each repeat at the RVD region (Deng *et al.*, 2012; Mak *et al.*, 2012). Other DNA-binding proteins (DBPs) detect their target sequences through a 'hopping' mechanism (re-association or deassociation of proteins

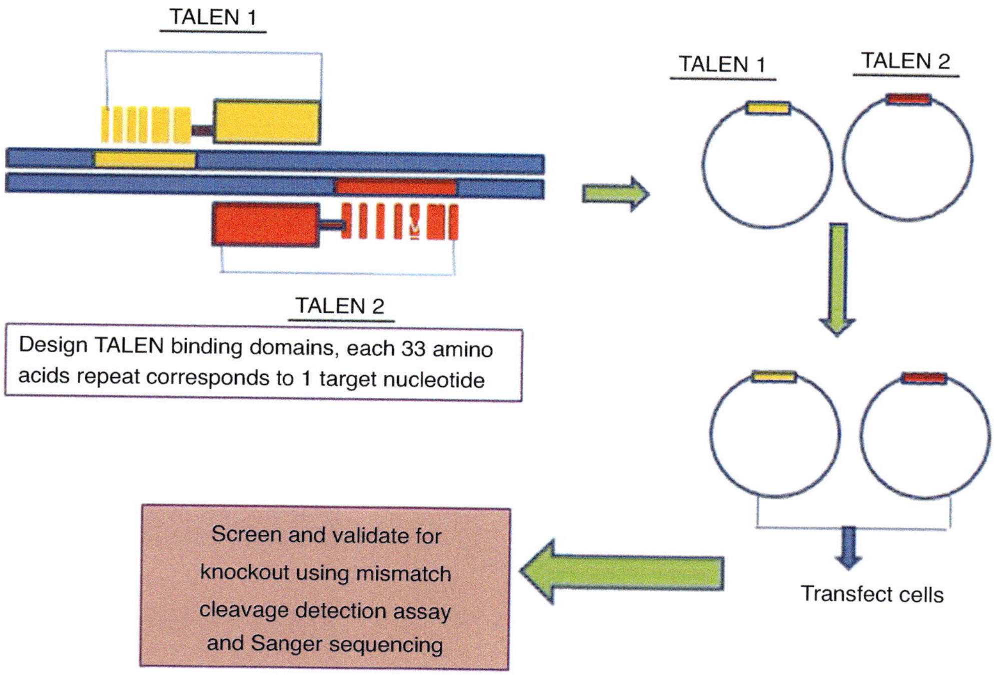

Fig. 4.1. Illustration of TALENs for achieving gene silencing. Author's own figure.

with DNA) or a 'sliding' mechanism (DBPs rotate following the major groove of DNA). Irrespective of that, TALE speed exceeds the sliding mechanism and rapidly recognizes the target sequence even in large genomes (Cuculis *et al.*, 2016). Moreover, in the N-terminal region, TALEs contain four non-canonical/degenerated repeats, that is, −3,−2, −1, and 0. Repeat −1 that preceded the TALE region is recognized by an additional thymine. Thus, in order to simplify the positioning of TALENs and TALEs, many variants of N-terminal recognize all bases (Lamb *et al.*, 2013). For best results, it is recommended to choose a minimum of two to three Cs and Gs in target sites of TALEs or TALENs that unite with strong RVDs, that is, HD and NN, respectively (Streubel *et al.*, 2012). In TALENs, repeats can be adjusted freely to precisely cut the target site nucleotide in the DNA sequence.

In addition, gene manipulation through methylation can be achieved by applying TALEs and TALENs. Two target sequences differing by only one methylated nucleotide can be easily distinguished by TALEs. Hence, two matchless aspects of the TALE that are depicted in Fig. 4.2 expand its application (Richter *et al.*, 2014; Tsuji and Imanishi, 2019).

4.2.2 TALEN design tools and off-target prediction

Nowadays, many computational tools and websites, that is, TALE-NT (Čermák *et al.*, 2011; Doyle *et al.*, 2012), PROGNOS (Fine *et al.*, 2014), idTALE (Li *et al.*, 2012), E-TALEN (Heigwer *et al.*, 2013), TALENoffer (Grau *et al.*, 2013), ChopChop (Montague *et al.*, 2014), and Mojo Hand (Neff *et al.*, 2013) are used for the specific region and proper positioning of TALENs. These tools enable scientists to identify the possible TALEN-binding sites in a given DNA sequence, recognize off-targets, and select RVDs. However, there are some basic guidelines that can be used to design functional TALENs without applying bioinformatics tools (Fig. 4.3).

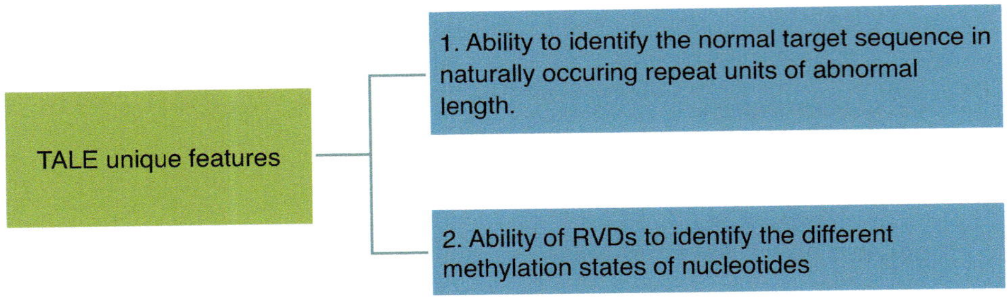

Fig. 4.2. Two unique features of TALEs. Author's own figure.

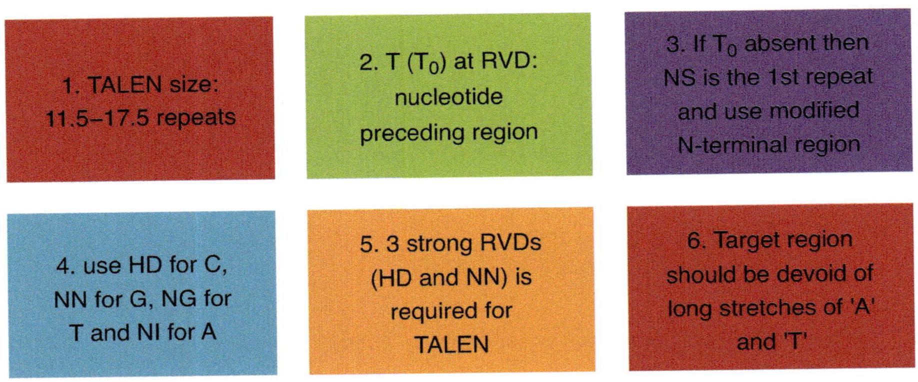

Fig. 4.3. Guidelines to design the functional TALEN. Author's own figure.

4.2.3 Assembly kits of TALEs and TALENs

Many cloning assembly kits are in use to generate TALEs and TALENs. Among them, golden gate cloning success rate is higher with maximum precision. It is cost-effective and does not require ordering new materials (Engler *et al.*, 2008). Assembly kits are of two types: (i) one-step protocol (one cloning step) and (ii) two-step (two cloning steps) protocol. Several features like N- and C-terminal points, RVD choice, and repeats number are used to design flexible TALENs. So, one-step cloning is preferred because two-step cloning is laborious, and it takes 1 week to clone TALENs (Ma *et al.*, 2016; Lee *et al.*, 2021). One-step cloning kits use preassembled multirepeat libraries to represent three to four combined repeats of DNA specifications. These libraries have 100s of plasmids that can rapidly regulate the TALEN process (Zhang *et al.*, 2020). None the less, TALEN cloning can be achieved in 1 day, thus easily matching the time required to order or clone clustered regularly interspaced short palindromic repeats (CRISPR)/Cas sgRNA.

4.3 Genome Editing with TALENs

4.3.1 Delivery of TALENs into living cells

Depending on the aim and target organism, the following methods are used to deliver TALENs into living cells (Table 4.1).

4.4 Agronomical Achievements of TALENs

Xanthomonas-resistant TALEN-edited rice is considered one of the earliest genome-edited crops. *Xanthomonas* can cause 100% yield loss. TALENs identify the natural TALE virulence factor from

Table 4.1. Different methods for TALEN delivery into living cells. Author's own table.

Physical method	*Agrobacterium*-mediated transformation	Most preferred method for model and non-model plants. It also allows delivery of Geminivirus-based DNA-TALEN replicons that show good results due to their high copy number and efficiency in genome editing.	Christian *et al*., 2013; Sun *et al*., 2013; Wendt *et al*., 2013; Liang *et al*., 2014; Lor *et al*., 2014; Baltes *et al*., 2014; Čermák *et al*., 2015; Butler *et al*., 2016
	Biolistic transformation	This particle bombardment is applicable to all plant species.	Klein *et al*., 1987; Wang *et al*., 2014; Budhagatapalli *et al*., 2015; Takahashi *et al*., 2018; Liang *et al*., 2018
	Micro-injection /electroporation	N-terminus TALEN proteins fuse/ conjugate with cell-penetrating peptides.	Ru *et al*., 2013; Liu *et al*., 2014
Bacterial-based delivery		*Type*-IV *Pseudomonas aeruginosa* bacterial secretion system	Jia *et al*., 2014; Jia *et al*., 2015
Viral-based delivery		Lentivirus (99), baculovirus (100), and adenovirus are used.	Holkers *et al*., 2013; Zhang *et al*., 2014; Chamorro *et al*., 2016
Chemical methods	Transfection reagents	TALEN DNA coupled with commercial transfection agents by using standard protocols and injected into the animal and human cell lines.	Čermák *et al*., 2011; Sun *et al*., 2012; Wang *et al*., 2013; Feng *et al*., 2014; Hu *et al*., 2015; Liao *et al*., 2015
	Liposomes or PEG	Cell protoplast of plant tissues transformed using PEG.	Nicolia *et al*., 2015; Luo *et al*., 2015; Li *et al*., 2016; Stoddard *et al*., 2016

Xanthomonas and introduce small insertions and deletions to prohibit the binding of virulence factor with the TALE target site, thus creating plant resistance against the bacterium. As these mutations are very small, so the plants were considered as non-GMO in the United States and Columbia in 2020. In some highly complex plants such as in wheat, the potential of TALENs is demonstrated by creating resistance against powdery mildew (fungal infection). Scientists create mutations in three pairs of the mlo1 gene using designed TALENs and achieve resistance without obtaining any transgene. This resistance is inherited up to two generations (Wang *et al*., 2014). Sugarcane is considered the best source for biofuel production but its lignin contents create hindrance in its efficient conversion. The size of the sugarcane genome is very large, approximately 10 Gb because of the large number of chromosomes. Only one TALEN pair mutates the COMT (caffeic acid) 107 genes/ alleles that lead to 19.7% reduced production of lignin and cell wall-bound sugar saccharification increases by 44%, so that the biofuel production is improved (Jung and Altpeter, 2016; Kannan *et al*., 2018). Soybean with reduced production of unhealthy trans-fats was also generated and commercially deployed in the market. This was the first TALEN-based commercially available crop. Desaturase genes of two fatty acids were mutated that shift the linoleic acid to oleic acid.

4.5 Additional TALE-Based Tools for Genome Editing

4.5.1 Mito-TALENs

TALENs have the ability to cut DNA not only in the nucleus but also in the mitochondria of mouse and humans besides plant cells. In this tool, MTS (mitochondrial targeting signal) is fused to the TALEN N-terminus. In mitochondria, many HJRD and HDS systems are non-functional, so mito-TALENs increase the percentage of healthy mitochondria by inducing recombination in mitochondrial chromosomes (Srivastava and Moraes, 2001; Kazama *et al.*, 2019).

4.5.2 TALE-based editors

In cellular compartments, TALE-based editors are the first genome-editing tools. Double-stranded DNA deaminase toxin A (DddA) extracted from bacteria is an antibacterial toxin and a strong cytidine deaminase that uses double-stranded DNA as substrate. Recently, DddA-based models have also been applied in animals (Mok *et al.*, 2020).

4.5.3 TALE transposases

They are used for large insertions in a targeted manner. Hyperactive variants of the piggyBac transposase are used that fuse with the TALE domain to direct the transposition of cargo DNA within the human genome (Owens *et al.*, 2013).

4.5.4 TALE recombinases

TALE DNA is fused with site-specific recombination (SSR) systems to target the genome rearrangements. A hyperactive version of the catalytic domain from the DNA invertase Gin was N-terminally fused to a TALE DBD, generating TALE recombinases (TALERs). TALERs are operated in pairs and allow for a targeted DNA recombination in bacterial and mammalian cells (Mercer *et al.*, 2012).

4.5.5 TALE transcriptome modifiers

TALEs naturally function as highly efficient gene activators in plant cells. A single protein bound to a plant promoter can induce transcription 100-fold and to levels much higher than the well-known constitutive 35S promoter (Römer *et al.*, 2007; Boch *et al.*, 2009).

4.5.6 TALE-based epigenomic modifiers

Epigenome editing has the potential to not alter the DNA sequence and allow the gene regulation/expression. TALE DBDs fuse with enzymes and generate a designer epigenome editor by involving two major proteins: (i) those that methylate/demethylate DNA nucleotides and (ii) those that modify histones directly by removing or adding ubiquitination, methylation, acetylation, and phosphorylation signals (Kungulovski and Jeltsch, 2016).

4.6 Conclusions and Future Suggestions

A wide range of different tools such as genome editing, epigenome editing, chromosome painting, and transcriptome editing have been applied using designed DNA-binding ability of the TALE. This opens up new insights due to its specification of target DNA sequences and DNA-associated proteins, thus resulting in the evolution of molecular biology, genetics, and biotechnology. TALENs offer the advantage of highly flexible positioning and pure protein nature. The constant expansion, diversification, and competition of potent genome-editing tools create a highly fruitful ground for future harvests of the genome-editing field.

References

Anzalone, A.V., Koblan, L.W. and Liu, D.R. (2020) Genome editing with CRISPR–Cas nucleases, base editors, transposases and prime editors. *Nature Biotechnology* 38(7), 824–844.

Baltes, N.J., Gil-Humanes, J., Cermak, T., Atkins, P.A. and Voytas, D.F. (2014) DNA replicons for plant genome engineering. *The Plant Cell* 26(1), 151–163.

Bibikova, M., Golic, M., Golic, K.G. and Carroll, D. (2002) Targeted chromosomal cleavage and mutagenesis in *Drosophila* using zinc-finger nucleases. *Genetics* 161(3), 1169–1175.
Boch, J., Scholze, H., Schornack, S., Landgraf, A., Hahn, S. *et al.* (2009) Breaking the code of DNA binding specificity of TAL-type III effectors. *Science* 326(5959), 1509–1512.
Budhagatapalli, N., Rutten, T., Gurushidze, M., Kumlehn, J. and Hensel, G. (2015) Targeted modification of gene function exploiting homology-directed repair of TALEN-mediated double-strand breaks in barley. *G3: Genes, Genomes, Genetics* 5(9), 1857–1863.
Butler, N.M., Baltes, N.J., Voytas, D.F. and Douches, D.S. (2016) Geminivirus-mediated genome editing in potato (*Solanum tuberosum* L.) using sequence-specific nucleases. *Frontiers in Plant Science* 7, 1045.
Carroll, D. (2021) A short, idiosyncratic history of genome editing. *Gene and Genome Editing* 1, 100002.
Čermák, T., Baltes, N.J., Čegan, R., Zhang, Y. and Voytas, D.F. (2015) High-frequency, precise modification of the tomato genome. *Genome Biology* 16, 1–15.
Čermák, T., Doyle, E.L., Christian, M., Wang, L., Zhang, Y. *et al.* (2011) Efficient design and assembly of custom TALEN and other TAL effector-based constructs for DNA targeting. *Nucleic Acids Research* 39(12), e82–e82.
Chamorro, C., Mencía, A., Almarza, D., Duarte, B., Büning, H. *et al.* (2016) Gene editing for the efficient correction of a recurrent COL7A1 mutation in recessive dystrophic epidermolysis bullosa keratinocytes. *Molecular Therapy-Nucleic Acids* 5, e307.
Christian, M., Qi, Y., Zhang, Y. and Voytas, D.F. (2013) Targeted mutagenesis of *Arabidopsis thaliana* using engineered TAL effector nucleases. *G3: Genes, Genomes, Genetics* 3(10), 1697–1705.
Cuculis, L., Abil, Z., Zhao, H. and Schroeder, C.M. (2015) Direct observation of TALE protein dynamics reveals a two-state search mechanism. *Nature Communications* 6(1), 7277.
Cuculis, L., Abil, Z., Zhao, H. and Schroeder, C.M. (2016) TALE proteins search DNA using a rotationally decoupled mechanism. *Nature Chemical Biology* 12(10), 831–837.
Deng, D., Yan, C., Pan, X., Mahfouz, M., Wang, J. *et al.* (2012) Structural basis for sequence-specific recognition of DNA by TAL effectors. *Science* 335(6069), 720–723.
Doyle, E.L., Booher, N.J., Standage, D.S., Voytas, D.F., Brendel, V.P. *et al.* (2012) TAL effector-nucleotide targeter (TALE-NT) 2.0: Tools for TAL effector design and target prediction. *Nucleic Acids Research* 40(W1), W117–W122.
Engler, C., Kandzia, R. and Marillonnet, S. (2008) A one pot, one step, precision cloning method with high throughput capability. *PloS One* 3(11), e3647.
Feng, Y., Zhang, S. and Huang, X. (2014) A robust TALENs system for highly efficient mammalian genome editing. *Scientific Reports* 4(1), 3632.
Fine, E.J., Cradick, T.J., Zhao, C.L., Lin, Y. and Bao, G. (2014) An online bioinformatics tool predicts zinc finger and TALE nuclease off-target cleavage. *Nucleic Acids Research* 42(6), e42–e42.
Gaudelli, N.M., Komor, A.C., Rees, H.A., Packer, M.S., Badran, A.H. *et al.* (2017) Programmable base editing of A• T to G• C in genomic DNA without DNA cleavage. *Nature* 551(7681), 464–471.
Grau, J., Boch, J. and Posch, S. (2013) TALENoffer: Genome-wide TALEN off-target prediction. *Bioinformatics* 29(22), 2931–2932.
Heigwer, F., Kerr, G., Walther, N., Glaeser, K., Pelz, O. *et al.* (2013) E-TALEN: A web tool to design TALENs for genome engineering. *Nucleic Acids Research* 41(20), e190–e190.
Holkers, M., Maggio, I., Liu, J., Janssen, J.M., Miselli, F. *et al.* (2013) Differential integrity of TALE nuclease genes following adenoviral and lentiviral vector gene transfer into human cells. *Nucleic Acids Research* 41(5), e63–e63.
Hu, Z., Ding, W., Zhu, D., Yu, L., Jiang, X. *et al.* (2015) TALEN-mediated targeting of HPV oncogenes ameliorates HPV-related cervical malignancy. *The Journal of Clinical Investigation* 125(1), 425–436.
Jasin, M. and Rothstein, R. (2013) Repair of strand breaks by homologous recombination. *Cold Spring Harbor Perspectives in Biology* 5(11), a012740.
Jia, J., Jin, Y., Bian, T., Wu, D., Yang, L. *et al.* (2014) Bacterial delivery of TALEN proteins for human genome editing. *PloS One* 9(3), e91547.
Jia, J., Bai, F., Jin, Y., Santostefano, K.E., Ha, U.H. *et al.* (2015) Efficient gene editing in pluripotent stem cells by bacterial injection of transcription activator-like effector nuclease proteins. *Stem Cells Translational Medicine* 4(8), 913–926.
Joung, J.K. and Sander, J.D. (2013) TALENs: A widely applicable technology for targeted genome editing. *Nature Reviews Molecular Cell Biology* 14(1), 49–55.

Jung, J.H. and Altpeter, F. (2016) TALEN-mediated targeted mutagenesis of the caffeic acid O-methyltransferase in highly polyploid sugarcane improves cell wall composition for production of bioethanol. *Plant Molecular Biology* 92, 131–142.

Kannan, B., Jung, J.H., Moxley, G.W., Lee, S.M. and Altpeter, F. (2018) TALEN-mediated targeted mutagenesis of more than 100 COMT copies/alleles in highly polyploid sugarcane improves saccharification efficiency without compromising biomass yield. *Plant Biotechnology Journal* 16(4), 856–866.

Kazama, T., Okuno, M., Watari, Y., Yanase, S., Koizuka, C. *et al.* (2019) Curing cytoplasmic male sterility via TALEN-mediated mitochondrial genome editing. *Nature Plants* 5(7), 722–730.

Klein, T.M., Wolf, E.D., Wu, R. and Sanford, J.C. (1987) High-velocity microprojectiles for delivering nucleic acids into living cells. *Nature* 327(6117), 70–73.

Komor, A.C., Kim, Y.B., Packer, M.S., Zuris, J.A. and Liu, D.R. (2016) Programmable editing of a target base in genomic DNA without double-stranded DNA cleavage. *Nature* 533(7603), 420–424.

Kungulovski, G. and Jeltsch, A. (2016) Epigenome editing: State of the art, concepts, and perspectives. *Trends in Genetics* 32(2), 101–113.

Lamb, B.M., Mercer, A.C. and Barbas III, C.F. (2013) Directed evolution of the TALEN-terminal domain for recognition of all 5′ bases. *Nucleic Acids Research* 41(21), 9779–9785.

Lee, H., Lee, S., Baek, G., Kim, A., Kang, B.C. *et al.* (2021) Mitochondrial DNA editing in mice with DddA-TALE fusion deaminases. *Nature Communications* 12(1), 1190.

Li, J., Stoddard, T.J., Demorest, Z.L., Lavoie, P.O., Luo, S. *et al.* (2016) Multiplexed, targeted gene editing in *Nicotiana benthamiana* for glyco-engineering and monoclonal antibody production. *Plant Biotechnology Journal* 14(2), 533–542.

Li, L., Piatek, M.J., Atef, A., Piatek, A., Wibowo, A. *et al.* (2012) Rapid and highly efficient construction of TALE-based transcriptional regulators and nucleases for genome modification. *Plant Molecular Biology* 78, 407–416.

Liang, Z., Zhang, K., Chen, K. and Gao, C. (2014) Targeted mutagenesis in *Zea mays* using TALENs and the CRISPR/Cas system. *Journal of Genetics and Genomics* 41(2), 63–68.

Liang, Z., Chen, K., Yan, Y., Zhang, Y. and Gao, C. (2018) Genotyping genome-edited mutations in plants using CRISPR ribonucleoprotein complexes. *Plant Biotechnology Journal* 16(12), 2053–2062.

Liao, H., Xiao, Y., Hu, Y., Xiao, Y., Yin, Z. *et al.* (2015) Suppression of cellular proliferation and invasion by HMGB1 knockdown in bladder urothelial carcinoma cells. *Oncology Research Featuring Preclinical and Clinical Cancer Therapeutics* 22(5–6), 235–245.

Liu, J., Gaj, T., Patterson, J.T., Sirk, S.J. and Barbas III, C.F. (2014) Cell-penetrating peptide-mediated delivery of TALEN proteins via bioconjugation for genome engineering. *PloS One* 9(1), e85755.

Lor, V.S., Starker, C.G., Voytas, D.F., Weiss, D. and Olszewski, N.E. (2014) Targeted mutagenesis of the tomato PROCERA gene using transcription activator-like effector nucleases. *Plant Physiology* 166(3), 1288–1291.

Luo, S., Li, J., Stoddard, T.J., Baltes, N.J., Demorest, Z.L. *et al.* (2015) Non-transgenic plant genome editing using purified sequence-specific nucleases. *Molecular Plant* 8(9), 1425–1427.

Ma, A.C., McNulty, M.S., Poshusta, T.L., Campbell, J.M., Martínez-Gálvez, G. *et al.* (2016) FusX: A rapid one-step transcription activator-like effector assembly system for genome science. *Human Gene Therapy* 27(6), 451–463.

Mak, A.N.S., Bradley, P., Cernadas, R.A., Bogdanove, A.J. and Stoddard, B.L. (2012) The crystal structure of TAL effector PthXo1 bound to its DNA target. *Science* 335(6069), 716–719.

Menz, J., Modrzejewski, D., Hartung, F., Wilhelm, R. and Sprink, T. (2020) Genome edited crops touch the market: A view on the global development and regulatory environment. *Frontiers in Plant Science* 11, 586027.

Mercer, A.C., Gaj, T., Fuller, R.P. and Barbas III, C.F. (2012) Chimeric TALE recombinases with programmable DNA sequence specificity. *Nucleic Acids Research* 40(21), 11163–11172.

Miller, J.C., Tan, S., Qiao, G., Barlow, K.A., Wang, J. *et al.* (2011) A TALE nuclease architecture for efficient genome editing. *Nature Biotechnology* 29(2), 143–148.

Mok, B.Y., Moraes, M.H., Zeng, J., Bosch, D.E., Kotrys, A.V. *et al.* (2020) A bacterial cytidine deaminase toxin enables CRISPR-free mitochondrial base editing. *Nature* 583(7817), 631–637.

Montague, T.G., Cruz, J.M., Gagnon, J.A., Church, G.M. and Valen, E. (2014) CHOPCHOP: A CRISPR/Cas9 and TALEN web tool for genome editing. *Nucleic Acids Research* 42(W1), W401–W407.

Moscou, M.J. and Bogdanove, A.J. (2009) A simple cipher governs DNA recognition by TAL effectors. *Science* 326(5959), 1501–1501.

Mussolino, C., Morbitzer, R., Lütge, F., Dannemann, N., Lahaye, T. *et al.* (2011) A novel TALE nuclease scaffold enables high genome editing activity in combination with low toxicity. *Nucleic Acids Research* 39(21), 9283–9293.

Neff, K.L., Argue, D.P., Ma, A.C., Lee, H.B., Clark, K.J. *et al.* (2013) Mojo hand, a TALEN design tool for genome editing applications. *BMC Bioinformatics* 14(1), 1–7.

Nicolia, A., Proux-Wéra, E., Åhman, I., Onkokesung, N., Andersson, M. *et al.* (2015) Targeted gene mutation in tetraploid potato through transient TALEN expression in protoplasts. *Journal of Biotechnology* 204, 17–24.

Owens, J.B., Mauro, D., Stoytchev, I., Bhakta, M.S., Kim, M.S. *et al.* (2013) Transcription activator like effector (TALE)-directed piggybac transposition in human cells. *Nucleic Acids Research* 41(19), 9197–9207.

Porto, E.M., Komor, A.C., Slaymaker, I.M. and Yeo, G.W. (2020) Base editing: Advances and therapeutic opportunities. *Nature Reviews Drug Discovery* 19(12), 839–859.

Qasim, W., Zhan, H., Samarasinghe, S., Adams, S., Amrolia, P. *et al.* (2017) Molecular remission of infant B-ALL after infusion of universal TALEN gene-edited CAR T cells. *Science Translational Medicine* 9(374), eaaj2013.

Richter, A., Streubel, J., Blücher, C., Szurek, B., Reschke, M. *et al.* (2014) A TAL effector repeat architecture for frameshift binding. *Nature Communications* 5(1), 3447.

Römer, P., Hahn, S., Jordan, T., Strauss, T., Bonas, U. *et al.* (2007) Plant pathogen recognition mediated by promoter activation of the pepper Bs3 resistance gene. *Science* 318(5850), 645–648.

Ru, R., Yao, Y., Yu, S., Yin, B., Xu, W. *et al.* (2013) Targeted genome engineering in human induced pluripotent stem cells by penetrating TALENs. *Cell Regeneration* 2(1), 1–8.

Srivastava, S. and Moraes, C.T. (2001) Manipulating mitochondrial DNA heteroplasmy by a mitochondrially targeted restriction endonuclease. *Human Molecular Genetics* 10(26), 3093–3099.

Stoddard, T.J., Clasen, B.M., Baltes, N.J., Demorest, Z.L., Voytas, D.F. *et al.* (2016) Targeted mutagenesis in plant cells through transformation of sequence-specific nuclease mRNA. *PLoS One* 11(5), e0154634.

Streubel, J., Blücher, C., Landgraf, A. and Boch, J. (2012) TAL effector RVD specificities and efficiencies. *Nature Biotechnology* 30(7), 593–595.

Sun, N., Liang, J., Abil, Z. and Zhao, H. (2012) Optimized TAL effector nucleases (TALENs) for use in treatment of sickle cell disease. *Molecular BioSystems* 8(4), 1255–1263.

Sun, Z., Li, N., Huang, G., Xu, J., Pan, Y. *et al.* (2013) Site-specific gene targeting using transcription activator-like effector (TALE)-based nuclease in *Brassica oleracea*. *Journal of Integrative Plant Biology* 55(11), 1092–1103.

Takahashi, K., Ide, Y., Hayakawa, J., Yoshimitsu, Y., Fukuhara, I. *et al.* (2018) Lipid productivity in TALEN-induced starchless mutants of the unicellular green alga *Coccomyxa* sp strain Obi. *Algal Research* 32, 300–307.

Tsuji, S. and Imanishi, M. (2019) Modified nucleobase-specific gene regulation using engineered transcription activator-like effectors. *Advanced Drug Delivery Reviews* 147, 59–65.

Wang, F., Ma, S., Xu, H., Duan, J., Wang, Y. *et al.* (2013) High-efficiency system for construction and evaluation of customized TALENs for silkworm genome editing. *Molecular Genetics and Genomics* 288, 683–690.

Wang, Y., Cheng, X., Shan, Q., Zhang, Y., Liu, J. *et al.* (2014) Simultaneous editing of three homoeoalleles in hexaploid bread wheat confers heritable resistance to powdery mildew. *Nature Biotechnology* 32(9), 947–951.

Wendt, T., Holm, P.B., Starker, C.G., Christian, M., Voytas, D.F. *et al.* (2013) TAL effector nucleases induce mutations at a pre-selected location in the genome of primary barley transformants. *Plant Molecular Biology* 83, 279–285.

Zhang, S., Li, L., Kendrick, S.L., Gerard, R.D. and Zhu, H. (2014) TALEN-mediated somatic mutagenesis in murine models of cancer. *Cancer Research* 74(18), 5311–5321.

Zhang, S., Wang, J. and Wang, J. (2020) One-day TALEN assembly protocol and a dual-tagging system for genome editing. *ACS Omega* 5(31), 19702–19714.

5 Structure and Mechanism of Clustered Regularly Interspaced Short Palindromic Repeats (CRISPR) CRISPR-associated Nuclease 9 (Cas9)-mediated Genome Editing in Plants

Lubna Rasool[1]*, Hamid Manzoor[2], Maria Rasul[3] and Abdur Rauf[1]

[1]Department of Chemistry, University of Sahiwal, Sahiwal, Pakistan; [2]Institute of Molecular Biology and Biotechnology, Bahauddin Zakariya University, Multan, Pakistan; [3]Department of Environment and Energy, Sejong University, Seoul, South Korea

Abstract

Over the past decade, scientists have employed a range of molecular and traditional breeding techniques to enhance crop yields and resilience. Genetically modified crops were developed for acquiring optimal yield and quality production, while providing protection against both biotic and abiotic stress conditions. The clustered regularly interspaced short palindromic repeats (CRISPR) and CRISPR-associated (Cas) proteins make an innate adaptive defense system in several microorganisms. Three fundamental steps comprise the CRISPR-Cas9 genome-editing mechanism: recognition, cleavage, and repair. CRISPR-Cas9 has the potential to function in eukaryotic DNA editing, which shows its tremendous application in different fields. It is used for genome editing as also to study gene regulation, epigenetics, RNA targeting, chromatin topology, chromatin imaging, and many other applications. CRISPR-Cas genome modification has been effectively demonstrated in wheat, maize, tomatoes, rice, soybeans, and barley for the improvement of crops. This chapter discusses the CRISPR-Cas system, classification, structural components, and its mechanisms of action in plants. It also brings forth the status and essentiality of CRISPR-Cas9 technology for food, feed, fiber, and fuel crops. It is a simple, fast, cheap, effective, more accurate method and exhibits a wide range of potential applications. This method facilitates molecular breeding and fast-track progress in basic research as well as in advanced research. It is the most promising tool for metabolic engineering. It is important to note that CRISPR technology has significantly accelerated the pace of plant breeding and genetic improvement by providing a more precise and effective mode to introduce genetic changes.

Keywords: Plant-genome editing, Clustered regularly interspaced short palindromic repeats (CRISPR), Genetically modified crops, Mechanisms of action, Plant breeding

*Corresponding author: lubnarasool@uosahiwal.edu.pk

DOI: 10.1079/9781800622517.0005

5.1 Introduction

The global population is projected to exceed 9 million by 2050. However, over the recent years, global agricultural production has not seen improvement, and in certain aspects, it has even declined as a result of the ongoing consequences of climate variation and the dearth of arable land. Therefore, there is an urgent need for advancements in plant research and the development of modified crops. These crops must be of high quality, high yielding, and tolerant to biotic as well as abiotic pressure to meet the growing demands of the human population. Traditional breeding programs have been highly successful in creating elite crop varieties with high yields. However, there has been limited utilization of traditional breeding to enhance resistance to abiotic stress, a critical need in the context of current climate change. Significant alleles present in traditional landraces or wild relatives of crop plants have been incorporated into superior cultivars through labor-intensive and laborious conventional breeding methods (Coyne *et al.*, 2020; Renzi *et al.*, 2022). Genome editing (also called gene editing) encompasses a set of technologies that change an organism's DNA. Numerous techniques have been developed to modify the genome through the insertion, deletion, and substitution of nucleotides at the defined loci in the genome. An innovative and novel genome-editing technique (CRISPR) was invented in 2009. This made DNA editing more feasible over the earlier techniques. It was identified in *E. coli* DNA by Ishino (Ishino *et al.*, 1987). Clustered, regularly interspaced short palindromic repeats (CRISPR) and CRISPR-associated protein 9 (CRISPR-Cas9) are employed extensively in a wide range of organisms to modify specific genes. In parallel to the transcription activator-like effector nuclease (TALEN) and zinc-finger nuclease (ZFN) methods, CRISPR-Cas9 is a more flexible and versatile gene-editing technique. This method facilitates editing of one or more genes by allowing the insertion (knock-in) or deletion (knock-out) of genes derived from the genome of the host. The crRNA's guide RNA (gRNA) sequence was changed to enable the CRISPR-Cas9 system to produce site-specific blunt-ended double-stranded breaks (DSBs). One of the two methods for repairing DSBs is error-prone nonhomologous end joining (NHEJ), which causes minor random insertions and/or deletions (indels) at the cleavage site, or precise genome modification at the DSB site using a homologous repair template. It enables the simultaneous alteration of several features or metabolic pathways by causing DSBs at several places (Mao *et al.*, 2019; Selma *et al.*, 2023). Site-directed mutagenesis in the majority of plants is additionally advantageous owing to its relatively low expense, efficiency, and effortlessness of implementation in comparison to alternative methods. Precise and rapid modification of plant genomes through targeted manipulation has enabled the addition of desired characteristics and/or the elimination of undesired ones, surpassing the speed and precision of traditional breeding methods. According to Makarova and his colleagues, CRISPR-Cas systems work in the same way as RNA interference, a process in which complexes of protein split messenger RNA (mRNA) to silence the genes. Certain Cas proteins cleave DNA, while others cleave RNA (Makarova *et al.*, 2006). Additionally, Cas9 is involved in spacer acquirement and crRNA maturation in CRISPR interference (Heler *et al.*, 2015). Furthermore, oligonucleotides and site-directed alteration have great potential to modify the plant genome at the single-nucleotide level. Nowadays, adenine-base editors (ABEs) technology has been established for A-T to G-C base pairs (bps) alternation. It requires Cas9 nickase and deoxyadenine deaminase to mutate A-T to G-C bps in the DNA string (Mohanta *et al.*, 2017). Different plant species have had their genomes modified using CRISPR-Cas9 technology, ZFNs, and TALENs. CRISPR-Cas system construction is easier than ZFNs and TALENs. It is easy, efficient, simple, low cost and allows for the simultaneous targeting of the sequence of various genes (Cong *et al.*, 2013; Mali *et al.*, 2013). Hence, the adaptation of CRISPR-Cas9 in plants has revolutionized plant biotechnology, enabling systematic, precise, and well-organized genome-editing procedures.

5.2 CRISPR-Cas System

A unique class of repeated elements found in many bacterial genomes is called CRISPR. Bacteriophages of the DNA fragment that had previously infected the prokaryote are the sources of such sequences. These nucleotides are therefore crucial to the prokaryotes' antiviral defense mechanism and offer a type of acquired immunity. In 1987, Yoshizumi Ishino and his colleagues discovered CRISPR in *E. coli* while accidentally cloning an unusual sequence of repeats interspersed with spacer sequences while examining a gene that converts alkaline phosphatase. CRISPR systems were identified for a unique DNA repair mechanism and were also involved in a defense mechanism to prevent repeated infections by the same virus. Rodolphe Barrangou and colleagues in 2007 demonstrated that changing the spacers' composition could change the cell's resistance to bacteriophages (Barrangou *et al.*, 2007). CRISPR family of DNA sequences is present in the chromosome of prokaryotic organisms, including archaea and bacteria. The building blocks of a CRISPR element are direct repeats that span approximately 28–40 bps, with a distinct sequence of approximately 25–40 bps between each copy. Tandem arrays with 4 to over 100 elements are often formed by CRISPR.

5.2.1 Structural component, mode of action, and delivery of CRISPR complex system

This system is composed of two essential components: the first is the RNA molecule and the other is the enzyme. Major classes and types of CRISPR systems have distinct features, including types of Cas protein, effector complexes, and mechanisms of identification and split of target DNA, as illustrated in Fig. 5.1.

5.2.1.1 Class I CRISPR-Cas system

Class I, CRISPR-Cas system comprises multi-protein complexes, while class II organization is composed of a single-effector protein. However, 6 main types and 29 subtypes of CRISPR have been reported and their number was promptly

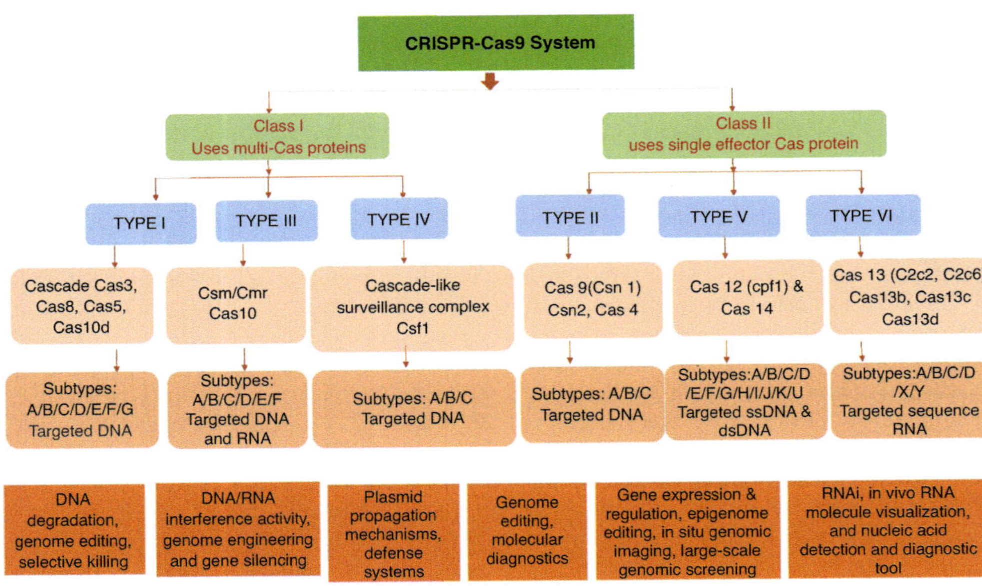

Fig. 5.1. Classification of the CRISPR-Cas system. Author's own figure.

increased (Li *et al.*, 2020a). Class I includes types I and III: type I consists of a complex with multiple Cas proteins, including Cas3. It utilized a cascade (CRISPR-associated complex for antiviral defense)-effector complex for target recognition and interference, while type III, similar to type I, also uses multiple Cas proteins, such as Cas10, and utilizes a Csm effector complex for recognition.

5.2.1.2 Class II CRISPR-Cas system

Class II, CRISPR-Cas system includes types II and V. Type II comprises the well-known CRISPR-Cas9 system, which has been extensively applied to gene modification. It involves a single Cas protein (Cas9) and utilizes a single-guide RNA (sgRNA) fragment for target recognition and cleavage. CRISPR-Cas9 systems have shown unique characteristics represented in Table 5.1. Type V includes CRISPR-Cas12 and CRISPR-Cas12a systems. It also uses a single Cas protein (Cas12 and Cas12a) and sgRNA, particularly for the target recognition of DNA sequence and cleavage. CRISPR-Cas12a has been successfully constructed in various organisms for genome editing, including human cells, animal models, and plants. The type II system is versatile, unique, and frequently used for the editing of the genome found in *Streptococcus pyogenes* Cas protein (SpCas9) (Makarova *et al.*, 2015; Li *et al.*, 2020a). Cas9 comprises the recognition domain that recognizes the complementary DNA sequence, and nuclease domains (RuvC and HNH) that cleave the DNA (Jiang and Doudna, 2017; Hassan *et al.*, 2021). It cuts on both the strings of DNA at specific sites in the genome by the addition and deletion of DNA. Cas9 has been widely used in various organisms, such as *Candida albicans, Saccharomyces cerevisiae, Danio rerio, Drosophila melanogaster, Harpegnathos saltator, Ooceraea biroi, Aedes aegypti, Caenorhabditis elegans, Mus musculus domesticus*, and *Homo sapiens* embryos. Cas13, Cas12e, Cas12a, and Cas-X are RNA-guided endonucleases. Cpf1 is also recognized as nuclease Cas12a. The CRISPR/Cpf1 system produced a staggered cut in dsDNA on a 'T rich' protospacer adjacent motif (PAM). The gRNA consists of a 20-nucleotide-long-predesigned RNA sequence and a long RNA scaffold. This scaffold RNA segment is attached to DNA and the remaining predesigned RNA Cas9 recognized and ensured the cut at the correct point in the genome. However, this system consists

Table 5.1. Distinct characteristics of the CRISPR-Cas9 system. Author's own table.

Factors	CRISPR-Cas9 system
Nuclease assembly	Simple
Choose target site	Unlimited
Gene mutation	Unlimited
Target efficiency	High with most guide RNAs
Components	Guide RNA, Cas9 protein PAM sequence
Enzymes	Cas9 protein
Time	1–3 days
Cost	Low
Recognition site	22 bps
Recognition mechanism	RNA-guided DNA endonuclease
Repairing mechanism	NHEG
Specificity	High
Sensitivity to DNA methylation	Not sensitive to CpG methylation
Off-target effects	Variable
Delivery *in vivo*	Easy
Multiple targeting	Easy

of PAMs, trans-activating RNA (tracrRNA), Cas9 nuclease, and CRISPR RNA (crRNA) (Mohanta *et al.*, 2017). PAM is a short stretch of nucleotides, 5′-NGG-3′ (N=G, C, A, or T) located downstream of the target sequence for DNA recognition. Different CRISPR arrays are typically 32–38 bps in size and a few can be as short as 23 bps and as long as 55 bps. Secondary structures of RNA hairpin and others are designed to be unstructured. Different secondary structures are available in Rfam database as CRISPR/DR2, DR5, DR6, DR8, DR9, DR19, DR41, DR52, and so on. This is an exact method of editing the genome without shuffling an organism's DNA.

5.2.2 CRISPR-mediated genome-alteration techniques

There are numerous CRISPR-mediated genome-alteration techniques, such as the usage of premade ribonucleoprotein complexes of Cas protein and gRNA, CRISPR-associated transposases, CRISPR nickases, base editors, Cas-12a, prime editing, and CRISPR endogenous systems (Hassan *et al.*, 2021). In addition, novel Cas proteins with different target requirements have been identified in several prokaryotes, and numerous Cas variants, including dCas, CRISPRa, CRISPRi, and nCas, have been created for prime and base editing that could provide unlimited gene-editing usage (Alamillo *et al.*, 2023). Transposon-associated CRISPR-Cas systems (CASTs), containing transposase, CRISPR effector, CRISPR array, *TniQ* subunit, *TnsD E. coli* homolog, and *Vibrio cholerae Tn6677* types I–F effector cascade, made a firm and stable complex that plays a major role in the DNA insertion process. The CRISPR-Cas system has been associated with the Tn7- transposon system in types I, IV, and V. Larger-size DNA fragments (10–30 kb) integrate at the specific region in CASTs, and gene expression as knock-ins and knock-outs can be achieved in which the host cell restoration system is not involved (Shmakov *et al.*, 2017; Wang *et al.*, 2020). CRISPR nickases offer a more refined and controlled approach to genome editing, offering increased specificity and minimizing unwanted genetic alterations. The Cas9 editing system was modified to generate breaks in the DNA's single strand instead of producing a DSB, as it involves the conventional CRISPR-Cas9 system. This modification has been achieved by introducing point mutations in the two nucleases' domain of Cas9 enzyme, rendering one of the nuclease domains inactive. Cas12a (CRISPR-associated protein 12a), a component of the CRISPR-Cas system, derived from the bacterium *Acidaminococcus* sp. BV3L6, is an RNA-directed endonuclease enzyme. It is similar to Cas9, a more well-known enzyme in the CRISPR toolkit, but it has some distinct properties that make it useful for genetic engineering. One key difference between Cas12a and Cas9 is the recognition site on the DNA molecule. On the other hand, Cas12a recognizes and cleaves DNA on a different PAM sequence, which is 'TTTV' (where T is thymine, and V can be any nucleotide). Another advantage of Cas12a is that it requires a shorter gRNA compared with Cas9. The gRNA is a small nucleotide strand of RNA that guides the Cas nuclease to the target DNA sequence. Shorter size of the gRNA simplifies its design and synthesis. Furthermore, Cas12a has the ability to generate staggered or sticky ends after DNA cleavage, whereas Cas9 typically creates blunt ends. These sticky ends can be useful for integrating new DNA sequences into the genome during the repair process, offering additional possibilities for genetic modifications. It holds great potential for applications in genetic research, agriculture, and potentially in therapeutic interventions for genetic diseases.

5.2.3 Mechanism of the CRISPR-Cas9 system

The CRISPR-Cas9 system exhibits three distinct phases of adaptive immunity, including adaptation, interference, and expression. In the process of adaptation, the CRISPR locus receives small fragments of cleaved genetic material (DNA) from invasive viruses or plasmids. After transcription, the CRISPR loci are changed into crRNA, which leads effector endonucleases to target infectious particles using complementarity of bases (Jiang and Doudna, 2017). A solitary Cas9 protein is necessary for DNA interference in the type II CRISPR-Cas method (Mir *et al.*, 2018; Moreira *et al.*, 2020). In Fig. 5.2, the CRISPR-Cas9 mechanism is summarized.

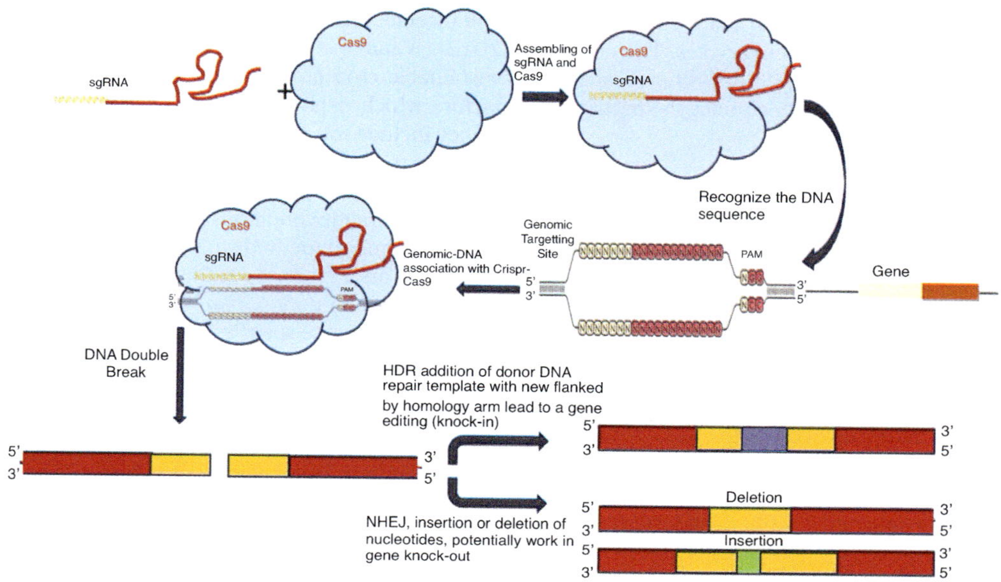

Fig. 5.2. Mechanism of CRISPR-Cas9-mediated genome editing in plants. The assembly mechanism of the CRISPR-Cas system. This system recognizes the host DNA sequences and produces nicks in the DNA strands. Author's own figure.

The Cas9 protein, being sizable, comprises multiple domains, including RuvC and HNH nuclease domains, along with tracrRNA and crRNA. RNase III, which is selective for single-stranded RNA, is engaged in the process of Cas9, which contributes to adaptation, helps convert pre-crRNA to crRNA, and introduces targeted DSBs that are directed by tracrRNA (Cánovas-Márquez *et al.*, 2021). Type III CRISPR differs from type II CRISPR in that it targets both DNA and RNA, and it interacts with the Cas10 cleavage protein. An actively transcribed promoter is one of the DNA sequences modified by the cleavage process, which is transcription dependent (Lin *et al.*, 2021). The Cas10 system gives microbes the ability to obtain viral spacer constituents under certain conditions, offering protection against foreign DNA. This defense system stops the activation of the lytic pathway, which could be deleterious to the host, when it comes to viruses or foreign DNA. Furthermore, these sequences have the power to change the organism's physical attributes, giving the host cell an edge in survival (Lin *et al.*, 2021).

The initial application of CRISPR in plants was documented by Feng *et al.* (2013) and Xie and Yang (2013). The researchers employed homologous recombination (HR) and NHEJ processes to create healthy transgenic plants by targeting different endogenous transgenes and genes through agro-infiltration and protoplast transfection. A different study achieved a mutation efficiency ranging from 3% to 8% by establishing three guide RNAs at various loci of the rice genome (Xie and Yang, 2013). Although off-target alterations were detected, their contribution to genome editing was negligible, when compared with matching sites. Numerous research projects focusing on sorghum, wheat, and maize have revealed excellent uses of CRISPR-based genome editing (Fiaz *et al.*, 2019; Zhang *et al.*, 2021; Wang *et al.*, 2023; Claeys *et al.*, 2024).

Comprehensive insights into several aspects, such as the particularity of cleavage, mutation effectiveness, wide chromosomal deletions, and the resolution of the locus structure, were provided by the preliminary research. The CRISPR-Cas9 technology has been refined throughout time to improve the precision and effectiveness of gene editing. It became critical to include nuclear localization signals at the protein ends in order to adapt the technique for genome editing in eukaryotes. The development of orthogonal CRISPR-Cas9 processes has greatly broadened the technology's spectrum of uses.

5.2.4 Dynamic working of the CRISPR system in plants

The CRISPR system's dynamics operating in plants have some common steps. In the first step, designing of gRNA is complimentary to the target's DNA sequence guided by the CRISPR system. In the second step, the CRISPR system is established by gRNA and the Cas9 protein. Once the CRISPR components are introduced into the plant cells, the gRNA guides the Cas9 protein to identify the gene of interest and produce breaks on both strands. In the third step, the plant's natural repair mechanisms become activated. The NHEJ directly ligates the dsDNA break, which often results in minor deletion/insertion at the site of the cut. These insertions or deletions can disturb the target gene functioning, leading to gene knockouts or loss of function, while in the homology-directed repair (HDR) pathway, template DNA molecules were utilized to heal the cut in a more precise manner. In the next step, the modified plant cells must be screened to detect those with the required changes. This can involve various techniques, such as DNA sequencing or phenotypic analysis, to verify the existence of the chosen modifications. In the last step, once modifications are identified in plants, they can be regenerated into whole plants. These plants can then be further bred to develop stable lines or varieties with the desired trait.

5.3 Significance of CRISPR Technology for Crop Production

According to recent data, there are 828 million starving people in the world (Zarocostas, 2022). It is predicted that this number would rise to 46 million by 2020 and by 150 million since the COVID-19 epidemic. We want to accelerate the food production and supply proficiently to avoid such condition. It can be possible by employing plant-genome integration approaches for crop enhancement (Roy, 2011; Beddington *et al.*, 2012). A major obstacle in the world today is sustainable crop production. The most serious factors which depressingly influence crop efficiency include soil quality, heat, drought, stress, flood, pests, insects, and irrigation availability. A growing world population, particularly under changing climatic conditions, increases the need of revolutionized methods to boost and develop stress-tolerance crops.

Targeted mutation can be accomplished through sequence-specific nucleases (SSNs) within the complex genome. They are classified into numerous classes: ZFNs, TALENs, meganucleases, and CRISPR. Meganucleases are naturally occurring restriction enzymes that identify and cut specific DNA sequences of 12–40 nucleotides. They can be engineered to detect and break the new nucleotide strand. Meganucleases, TALENs, and ZFNs all identify their sequence via protein/DNA interactions, while in CRISPR-Cas9, they recognize their sequence using gRNA (Cong *et al.*, 2013; Mali *et al.*, 2013). This results in an accurate and well-organized process for genome editing. This scientific skill revealed great promise in accelerating the breeding process and creating desired traits in plants, which could result in sustainable agriculture and food production in the long run.

5.4 Fate of CRISPR-Cas9

CRISPR-Cas9 is a revolutionary tool that has renovated the field of genetic engineering. It allows for precise changes in the DNA of organisms, including plants. With CRISPR-Cas9, we can focus on a particular gene and introduce changes to alter plant traits, improve crop yields, enhance disease resistance, and develop new varieties with desirable characteristics (Basu *et al.*, 2022). It can be employed to enhance important agricultural traits, like growth, grain yield, nutritional content, levels of essential vitamins or removal of allergenic compounds, and grain quality characteristics like texture, gluten content, starch composition, and environmental stress tolerance. This can lead to the development of wheat varieties with better baking properties or suitability for the specific food products (Miao *et al.*, 2018). It has great potential to generate targeted genetic modifications without introducing foreign DNA, making it different from traditional genetic modification techniques. This aspect has sparked interest in using CRISPR-edited plants in regions where genetically modified organisms (GMOs) face strict regulations. Numerous research has reported using the CRISPR-Cas system for an array of purposes, including genome editing, gene regulation, epigenome editing, base editing, RNA targeting, chromatin topological analysis, and chromatin imaging. Even though it has great potential, unintentional immune responses and toxicities, resistances to the Cas9/sgRNA system, the HRD vs NHEJ repair mechanism, delivery techniques, accurate DNA site targeting, the efficacy of the Cas9 activity, cleavage efficiency, designing of the sgRNA, and off-target effects were some of the hurdles. This system edits the genes responsible for enhancing drought tolerance in several crops like maize, rice, and wheat (Ansari *et al.*, 2020). Additionally, it can be used to make good quality crops. Stress-tolerant plants often face challenges like drought, salinity, temperature, and other environmental and abiotic stresses. Using the CRISPR-Cas9 system, stress response genes can make plants better able to withstand these harsh conditions (Xu *et al.*, 2015; Zhang *et al.*, 2019). Cultivation of herbicide-tolerant crops is an effective weed control strategy that can reduce the damage caused by herbicides. The CRISPR-Cas9 tool can be used to develop crops that are resilient to specific herbicides, allowing farmers to control weeds more effectively without damaging the crops. This approach has been applicable to crops including soybeans and maize. Using the CRISPR-Cas9-assisted gene-editing method, researchers have developed herbicide-resistant plants by focusing on endogenous genes including acetolactate synthase (*ALS*), 5-enolpyruvylshikimate-3-phosphate synthase (*EPSPS*), cellulose synthase catalytic subunit 3 (*CESA3*), and splicing factor 3 B subunit 1 (*SF3B1*) (Han and Kim, 2019). Herbicide-resistant rice plants can be developed using the CRISPR-Cpf1 system (Li *et al.*, 2020b). Plants are susceptible to numerous diseases produced by virus, bacteria, fungus, insects, and pests. Genes involved in plant defense mechanisms have been modified using CRISPR-Cas9 to increase the resistance of plants to infections (Oliva *et al.*, 2019). For instance, it has been successfully used for citrus bacterial canker disease caused by *Xanthomonas*, which has severe consequences on citrus productivity. Through the use of CRISPR-mediated genome editing, valuable results have been seen in the formation of disease-resistant cultivars of sweet oranges (Huang *et al.*, 2022). It can also be employed to modify plant traits like flower color, fruit size, and plant architecture. These modifications can be applied in horticulture, enabling the development of plants with desirable features for ornamental purposes. The CRISPR method was applied on *Ipomea nil* (blue morning glory). It has been used to understand the features of floral color, shape, and fragrance. With the potential to speed up plant breeding and develop desirable features, this scheme holds significant promise for the future of sustainable agriculture and food supply. The genome has been modified through the process of naturally occurring mutations over time. It is described that mutagens, both chemical and physical, have led to the discovery of more than 3000 genetic changes in plants that showed beneficial effects for both agriculture and humans (Songstad *et al.*, 2017). The most distinctive characteristic that sets the CRISPR-Cas9 system separate from ZFNs and TALENs is its DNA-binding site for target recognition.

5.5 CRISPR-Cas9 in Food, Feed, Fiber, and Fuel Crops

Crops were categorized into oil crops, feed crops, food crops, fiber crops, ornamental crops, industrial crops, and bioenergy-producing crops. Wheat is one of the most important cereal crops, widely cultivated and rich in protein and gluten. CRISPR technology can contribute to the formation of biofortified wheat varieties with an enhanced nutritional profile that can address various challenges faced by farmers and consumers. Genes responsible for the increasing iron or zinc content and other beneficial minerals have been targeted by using the CRISPR system. Several genes (*TaDEP1*, *TaGASR7*, *TaLOX2*, *TaNAC2*, *TaPIN1*, *TaGW2*, *TaDREB2*, and *TaERF3*) were targeted by CRISPR-Cas9 technology for height regulation, lipoxygenase activity, grain weight regulation, and drought resistance (Zhang and Gao, 2017; Kim *et al.*, 2018). Knockouts of *TaGW7*, *TaGW2-A1*, *TaGW2-D1*, and *TaGW2*-B1 genes provide improvements in grain width and weight in wheat (Zhang *et al.*, 2018; Wang *et al.*, 2019). Similarly in another report, it was applied to lower the amount of gluten in a particular type of wheat (Liu *et al.*, 2021). Worldwide, wheat production has significantly reduced, mainly due to the rust-producing fungus *Puccinia striiformis*. To reduce the impact of different pathogens, disease-resistant wheat varieties have been generated through CRISPR genome-editing. It has been reported that pathogens interact with plants through calcineurin B-like protein-interacting protein kinases that protect the plant. A knockout gene approach has been applied to the *TaClK14* gene to develop resistance to *P. striiformis* fungi. A significant change has been observed and the *TaClK14* gene can be used as a potential target for breeding broad-spectrum and highly disease-resistant wheat varieties (He *et al.*, 2023). In wheat, 100 members of the gliadin gene family were mutated using two separate consensus gRNAs. As Sánchez-León *et al.* (2018) demonstrated in 2018, this technique produced a variety of multiplex mutants, including a line with simultaneous genetic mutations in 35 different genes. The resulting transgene-free, low-gluten wheat lines will be used to introduce this characteristic into other elite wheat genotypes. Furthermore, Wang *et al.* (2018) used multiplexed genome editing to construct knockout mutants of wheat TaGW2, TaMLO, and TaLpx-one genes, resulting in increased seed size and thousand-kernel weight. More than half of the world's population relies mostly on rice as a food source. It is a type of grain plant that is utilized in genetic and molecular biology studies as a model plant. CRISPR-Cas9 was used to modify the *OsPDS* gene in rice (Banakar *et al.*, 2020). Many other genes including *OsERF922*, *OsHAK1*, *BADH2*, and *TMS5* have been mutated using the knockout strategy via the CRISPR-Cas9 procedure. A Cas-9-OsRR22-expression vector has been organized for targeting the *OsRRSS* gene to develop salinity tolerance in rice cultivars (Zhang *et al.*, 2019). Hence, it would be ideal to use this technology to more effectively and precisely achieve the goal of improving seed quality. Rice has a relatively low protein content, and the nutritional value of the rice grain has been based on the amino acids, fats, seed storage proteins (SSPs), vitamins, and other micronutrients. To accomplish a high level of beta carotene in the rice endosperm, the CRISPR-Cas9 method has been used to knock down the carotenoid catabolic genes (*OsCYP97A4*, *OsCCD4a*, *OsDSM2*, *OsCCD4b*, and *OsCCD7*) (Fiaz *et al.*, 2019). Maize is a cereal crop commonly cultivated for feed, food, fiber, and fuel worldwide. For the first time in 2014, CRISPR-Cas9-targeted genome editing was reported in the protoplast of maize (Liang *et al.*, 2014). There are distinct categories for modification that can be made using CRISPR-Cas, such as gene knockout (disrupting the function of a specific gene), gene insertion (adding a new gene), or gene editing (making precise changes to an existing gene for disease resistance, drought tolerance, nutritional value, and agronomic performance to develop maize varieties that can thrive under challenging environmental conditions). Among more than 200 volatile compounds, 2-acetyl-1-pyrroline (2-AP) is a significant flavor compound found in good fragrant rice. World-first fragrant maize was developed through CRISPR-mediated modification of the two BADH2 (betaine aldehyde dehydrogenase 2) genes. Both genes *ZmBADH2a* and *ZmBADH2b* show countless contribution in the biosynthesis of 2-AP in maize (Wang *et al.*, 2021).

Barley is an important grain crop primarily grown as animal fodder and also widely used for malt production, food, and beverages. It is cultivated in a dry environment and has shown high tolerance for abiotic stresses including drought. It is used as a model plant due to affordable, reliable, and quick plant breeding. Simple editing was carried out by RNA-guided Cas9 of the cytokinin oxidase/dehydrogenase *HvCKX1* gene, mostly showing a frameshift mutation (Gasparis *et al.*, 2019). Sorghum, commonly known as jowar, is a multi-purpose crop, widely valued as food (grains), feed (grains and biomass), fiber (paper), bioenergy fuel (making of ethanol and methane), and fertilizer (organic by-products utilization). It is especially significant in the brewing industry. For livestock feeding, its biomass is the major resource of silage. Superior quality fragrant sorghum was developed by CRISPR-Cas9 technology to achieve the high global demand and it has been revealed that high content of the 2-AP compound has been accumulated in leaves and seeds (Zhang *et al.*, 2022). Pearl millet is one of the most nutritious crops, rich in minerals, fiber, and protein. It is majorly used for food and feed and has excellent photosynthetic efficiency and biomass production potential. *Magnaporthe grisea* causes pearl millet blast disease. It is considered to have crucial impact on the destruction of millet crops worldwide. This technology has been used for editing the genome of millet for generating resistance against pearl millet blast and to improve stress resilience, as well as to improve the productivity of crops (Anil Kumar *et al.*, 2021).

5.6 Conclusion

Mutation breeding via physical radiation and chemically induced random mutagenesis was widely utilized in genome engineering. In recent decades, CRISPR-Cas-driven gene modification has emerged as one of the most powerful tools in crop biotechnology. The CRISPR system is composed of repeat and spacer sequences along with proteins guided by sgRNA. Moreover, RNA-guided Cas9 endonuclease and a sgRNA are the two key parts of it. The altered sgRNA guides the Cas9 protein to the target nucleotide sequence inside the chosen gene by using its crRNA complementary base pair component. The utilization of CRISPR-Cas-based gene editing enables efficient, precise, and targeted mutagenesis to strengthen nutritional content, yield, and disease resistance in various crops, including cereals, vegetables, fruits, and ornamental crops. These applications provide a significant advantage to ongoing genetic engineering endeavors aimed at ensuring food security and addressing nutrient deficiencies, vitamins, and nutraceuticals for the growing global population. While the CRISPR-Cas9 system stands out as a potent tool for crop betterment, genome editing in cereals and other crucial horticultural crops faces challenges such as unintended targets, DNA stability, transformation capability, and the absence of specific PAMs. It is imperative that these barriers are removed to facilitate the widespread implementation of these strategies with bright futures. Advances in providing the CRISPR-Cas9 technology, transformation method development, and editing capabilities will be critical to the growth of genome-editing approaches in major commercial crops. These advancements make CRISPR-Cas9-genome editing a promising technique for developing novel stress-tolerant cultivars, high-yielding crops, with improved resistance to pathogens, and higher nutritional value in the years to come.

References

Alamillo, J.M., López, C.M., Rivas, F.J.M., Torralbo, F., Bulut, M. *et al.* (2023) Clustered regularly interspaced short palindromic repeats/CRISPR-associated protein and hairy roots: A perfect match for gene functional analysis and crop improvement. *Current Opinion in Biotechnology* 79, 102876.

Anil Kumar, B.M., Tara Satyavathi, C., Prakash, P., Sharma, R., Hosahatti, R. *et al.* (2021) Pearl millet blast resistance: Current status and recent advancements in genomic selection and genome editing approaches. In: Nayaka, S.C., Hosahatti, R., Prakash, G., Satyavathi, C.T. and Sharma, R. (eds) *Blast Disease of Cereal Crops: Evolution and Adaptation in Context of Climate Change*. Springer International Publishing, Cham, pp. 183–200.

Ansari, W.A., Chandanshive, S.U., Bhatt, V., Nadaf, A.B., Vats, S. *et al.* (2020) Genome editing in cereals: Approaches, applications and challenges. *International Journal of Molecular Sciences* 21(11), 4040.

Banakar, R., Schubert, M., Collingwood, M., Vakulskas, C., Eggenberger, A.L. *et al.* (2020) Comparison of CRISPR-Cas9/Cas12a ribonucleoprotein complexes for genome editing efficiency in the rice phytoene desaturase (OsPDs) gene. *Rice* 13(1), 1–7.

Barrangou, R., Fremaux, C., Deveau, F., Richards, M., Boyaval, P. *et al.* (2007) CRISPR provides acquired resistance against viruses in prokaryotes. *Science* 315(5819), 1709–1712.

Basu, U., Riaz Ahmed, S., Bhat, B., Anwar, Z., Ali, A. *et al.* (2022) A CRISPR way for accelerating cereal crop improvement: Progress and challenges. *Frontiers in Genetics* 13, 866976.

Beddington, J.R., Asaduzzaman, M., Fernandez, A., Clark, M.E., Guillou, M. *et al.* (2012) *Achieving Food Security in the Face of Climate Change: Final Report from the Commission on Sustainable Agriculture and Climate Change*. CGIAR Research Program on Climate Change, Agriculture and Food Security (CCAFS), Copenhagen, Denmark. Available at: https://hdl.handle.net/10568/35589 (accessed March 2012).

Cánovas-Márquez, J.T., Falk, S., Nicolás, F.E., Padmanabhan, S., Zapata-Pérez, R. *et al.* (2021) A ribonuclease III involved in virulence of Mucorales fungi has evolved to cut exclusively single-stranded RNA. *Nucleic Acids Research* 49(9), 5294–5307.

Claeys, H., Neyrinck, E., Dumoulin, L., Pharazyn, A., Verstichele, A. *et al.* (2024) Coordinated gene upregulation in maize through CRISPR/Cas-mediated enhancer insertion. *Plant Biotechnology Journal* 22(1), 16.

Cong, L., Ran, F.A., Cox, D., Lin, S., Barretto, R. *et al.* (2013) Multiplex genome engineering using CRISPR/Cas systems. *Science* 339(6121), 819–823.

Coyne, C.J., Kumar, S., von Wettberg, E.J., Marques, E., Berger, J.D. *et al.* (2020) Potential and limits of exploitation of crop wild relatives for pea, lentil, and chickpea improvement. *Legume Science* 2(2), e36.

Feng, Z., Zhang, B., Ding, W., Liu, X., Yang, D.L. *et al.* (2013) Efficient genome editing in plants using a CRISPR/Cas system. *Cell Research* 23(10), 1229–1232.

Fiaz, S., Ahmad, S., Noor, M.A., Wang, X., Younas, A. *et al.* (2019) Applications of the CRISPR/Cas9 system for rice grain quality improvement: Perspectives and opportunities. *International Journal of Molecular Sciences* 20(4), 888.

Gasparis, S., Przyborowski, M., Kała, M. and Nadolska-Orczyk, A. (2019) Knockout of the HvCKX1 or HvCKX3 gene in barley (*Hordeum vulgare* L.) by RNA-guided Cas9 nuclease affects the regulation of cytokinin metabolism and root morphology. *Cells* 8(8), 782.

Han, Y.-J. and Kim, J.-I. (2019) Application of CRISPR/Cas9-mediated gene editing for the development of herbicide-resistant plants. *Plant Biotechnology Reports* 13(5), 447–457.

Hassan, M.M., Zhang, Y., Yuan, G., De, K., Chen, J.G. *et al.* (2021) Construct design for CRISPR/Cas-based genome editing in plants. *Trends in Plant Science* 26(11), 1133–1152.

He, F., Wang, C., Sun, H., Tian, S., Zhao, G. *et al.* (2023) Simultaneous editing of three homoeologues of TaCIPK14 confers broad-spectrum resistance to stripe rust in wheat. *Plant Biotechnology Journal* 21(2), 354–368.

Heler, R., Samai, P., Modell, J.W., Weiner, C., Goldberg, G. *et al.* (2015) Cas9 specifies functional viral targets during CRISPR–Cas adaptation. *Nature* 519(7542), 199–202.

Huang, X., Wang, Y. and Wang, N. (2022) Highly efficient generation of canker-resistant sweet orange enabled by an improved CRISPR/cas9 system. *Frontiers in Plant Science* 12, 769907.

Ishino, Y., Shinagawa, H., Makino, K., Amemura, M. and Nakata, A. (1987) Nucleotide sequence of the iap gene, responsible for alkaline phosphatase isozyme conversion in Escherichia coli, and identification of the gene product. *Journal of Bacteriology* 169(12), 5429–5433.

Jiang, F. and Doudna, J.A. (2017) CRISPR–Cas9 structures and mechanisms. *Annual Review of Biophysics* 46(1), 505–529.

Kim, D., Alptekin, B. and Budak, H. (2018) CRISPR/Cas9 genome editing in wheat. *Functional and Integrative Genomics* 18, 31–41.

Li, H., Yang, Y., Hong, W., Huang, M., Wu, M. *et al.* (2020a) Applications of genome editing technology in the targeted therapy of human diseases: Mechanisms, advances and prospects. *Signal Transduction and Targeted Therapy* 5(1), 1.

Li, S., Zhang, Y., Xia, L. and Qi, Y. (2020b) CRISPR-Cas12a enables efficient biallelic gene targeting in rice. *Plant Biotechnology Journal* 18(6), 1351.

Liang, Z., Zhang, K., Chen, K. and Gao, C. (2014) Targeted mutagenesis in *Zea mays* using TALENs and the CRISPR/Cas system. *Journal of Genetics and Genomics* 41(2), 63–68.

Lin, J., Shen, Y., Ni, J. and She, Q. (2021) A type III-A CRISPR–Cas system mediates co-transcriptional DNA cleavage at the transcriptional bubbles in close proximity to active effectors. *Nucleic Acids Research* 49(13), 7628–7643.

Liu, Q., Yang, F., Zhang, J., Liu, H., Rahman, S. *et al.* (2021) Application of CRISPR/Cas9 in crop quality improvement. *International Journal of Molecular Sciences* 22(8), 4206.

Makarova, K.S., Grishin, N.V., Shabalina, S.A., Wolf, Y.I. and Koonin, E.V. (2006) A putative RNA-interference-based immune system in prokaryotes: Computational analysis of the predicted enzymatic machinery, functional analogies with eukaryotic rnai, and hypothetical mechanisms of action. *Biol Direct* 1, 7.

Makarova, K.S., Wolf, Y.I., Alkhnbashi, O.S., Costa, F., Shah, S.A. *et al.* (2015) An updated evolutionary classification of CRISPR–Cas systems. *Nature Reviews Microbiology* 13(11), 722–736.
Mali, P., Esvelt, K.M. and Church, G.M. (2013) Cas9 as a versatile tool for engineering biology. *Nature Methods* 10(10), 957–963.
Mao, S., Ying, Y., Wu, X., Krueger, C.J. and Chen, A.K. (2019) CRISPR/dual-FRET molecular beacon for sensitive live-cell imaging of non-repetitive genomic loci. *Nucleic Acids Research* 47(20), e131–e131.
Miao, C., Xiao, L., Hua, K., Zou, C., Zhao, Y. *et al.* (2018) Mutations in a subfamily of abscisic acid receptor genes promote rice growth and productivity. *Proceedings of the National Academy of Sciences* 115(23), 6058–6063.
Mir, A., Edraki, A., Lee, J. and Sontheimer, E.J. (2018) Type II-C CRISPR-Cas9 biology, mechanism, and application. *ACS Chemical Biology* 13(2), 357–365.
Mohanta, T.K., Bashir, T., Hashem, A., Abd_Allah, E.F., Bae, H. *et al.* (2017) Genome editing tools in plants. *Genes* 8(12), 399.
Moreira, D., Pereira, A.M., Lopes, A.L. and Coimbra, S. (2020) The best CRISPR/Cas9 versus RNA interference approaches for Arabinogalactan proteins' study. *Molecular Biology Reports* 47, 2315–2325.
Oliva, R., Ji, C., Atienza-Grande, G., Huguet-Tapia, J.C., Perez-Quintero, A. *et al.* (2019) Broad-spectrum resistance to bacterial blight in rice using genome editing. *Nature Biotechnology* 37(11), 1344–1350.
Renzi, J.P., Coyne, C.J., Berger, J., Wettberg, E., Nelson, M. *et al.* (2022) How could the use of crop wild relatives in breeding increase the adaptation of crops to marginal environments. *Frontiers in Plant Science* 13, 886162.
Roy, C. (2011) Extending the Roy adaptation model to meet changing global needs. *Nursing Science Quarterly* 24(4), 345–351.
Sánchez-León, S., Gil-Humanes, J., Ozuna, C.V., Giménez, M.J., Sousa, C. *et al.* (2018) Low-gluten, nontransgenic wheat engineered with CRISPR/Cas9 16(4), 902–910.
Selma, S., Ceulemans, E., Goossens, A. and Lacchini, E. (2023) Clustered regularly interspaced short palindromic repeats tools for plant metabolic engineering: Achievements and perspectives. *Current Opinion in Biotechnology* 79, 102856.
Shmakov, S., Smargon, A., Scott, D., Cox, D., Pyzocha, N. *et al.* (2017) Diversity and evolution of class 2 CRISPR–Cas systems. *Nature Reviews Microbiology* 15(3), 169–182.
Songstad, D.D., Petolino, J.F., Voytas, D.F. and Reichert, N.A. (2017) Genome editing of plants. *Critical Reviews in Plant Sciences* 36(1), 1–23.
Wang, B., Xu, W. and Yang, H. (2020) Structural basis of a Tn7-like transposase recruitment and DNA loading to CRISPR-Cas surveillance complex. *Cell Research* 30(2), 185–187.
Wang, N., Ryan, L., Sardesai, N., Wu, E., Lenderts, B. *et al.* (2023) Leaf transformation for efficient random integration and targeted genome modification in maize and sorghum. *Nature Plants* 9(2), 255–270.
Wang, P., Zhao, F.J. and Kopittke, P.M. (2019) Engineering crops without genome integration using nanotechnology. *Trends in Plant Science* 24(7), 574–577.
Wang, W., Pan, Q., He, F., Akhunova, A., Chao, S. *et al.* (2018) Transgenerational CRISPR-Cas9 activity facilitates multiplex gene editing in allopolyploid wheat. *The CRISPR Journal* 1(1), 65–74.
Wang, Y., Liu, X., Zheng, X., Wang, W., Yin, X. *et al.* (2021) Creation of aromatic maize by CRISPR/Cas. *Journal of Integrative Plant Biology* 63(9), 1664–1670.
Xie, K. and Yang, Y. (2013) RNA-guided genome editing in plants using a CRISPR–Cas system. *Molecular Plant* 6(6), 1975–1983.
Xu, R.F., Li, H., Qin, R.Y., Li, J., Qiu, C.H. *et al.* (2015) Generation of inheritable and "transgene clean" targeted genome-modified rice in later generations using the CRISPR/Cas9 system. *Scientific Reports* 5(1), 11491.
Zarocostas, J. (2022) UN says a third of food wasted. *The Lancet* 400(10359), 1185.
Zhang, Y. and Gao, C. (2017) Recent advances in DNA-free editing and precise base editing in plants. *Emerging Topics in Life Sciences* 1(2), 161–168.
Zhang, A., Liu, Y., Wang, F., Li, T., Chen, Z. *et al.* (2019) Enhanced rice salinity tolerance via CRISPR/Cas9-targeted mutagenesis of the OsRR22 gene. *Molecular Breeding* 39, 1–10.
Zhang, D., Tang, S., Xie, P., Yang, D., Wu, Y. *et al.* (2022) Creation of fragrant sorghum by CRISPR/Cas9. *Journal of Integrative Plant Biology* 64(5), 961–964.
Zhang, S., Zhang, R., Gao, J., Song, G., Li, J. *et al.* (2021) CRISPR/Cas9-mediated genome editing for wheat grain quality improvement. *Plant Biotechnology Journal* 19(9), 1684.
Zhang, Y., Li, D., Zhang, D., Zhao, X., Cao, X. *et al.* (2018) Analysis of the functions of TaGW 2 homoeologs in wheat grain weight and protein content traits. *The Plant Journal* 94(5), 857–866.

6 Multiplex Genome Editing in Plants

Fozia Saeed[1], Hamid Manzoor[1], Sherien Bukhat[1], Farrukh Azeem[2] and Sumaira Rasul[1]*

[1]Institute of Molecular Biology and Biotechnology, Bahauddin Zakariya University, Multan, Pakistan; [2]Department of Bioinformatics and Biotechnology, Government College University Faisalabad, Faisalabad, Pakistan

Abstract

Crop yields are vulnerable to a number of a(biotic) stresses. To maintain a sufficient balance within the growing population and production of food and to prevent future food crises, crop yield reductions caused by these stresses must be addressed. To meet the rising food requirements, the land area for cultivation cannot be increased proportionately. Mitigating the damage caused by environmental stresses through biotechnological interventions seems an appropriate strategy for addressing the world's food needs. By conventional methods, a substantial amount of time and labor investment is required to produce desired genotypes. In the past few years, the emergence of a series of innovative genome engineering methods has significantly contributed to crop-improvement efforts. More recently, multiple editing events (simultaneously) have increased editing efficiency, which is named multiplexed genome editing (MGE). This method of genome editing has vastly increased the range of potential genetic modifications in a variety of hosts, including bacteria, human cells, and plants. The developmental tools for MGE comprise nucleotide-mediated as well as nuclease-mediated techniques. These techniques have produced a wide range of fruitful examples to accelerate crop improvement by modification of a crop genome sequence with specificity. In this chapter, we have explained the major innovations in the application and creation of MGE tools to speed up the production of stress-tolerant and high-yielding crop plants. Furthermore, we have addressed the obstacles, strengths, and importance of the emerging MGE technologies for fast and precise crop-improvement programs.

Keywords: Multiplexed genome editing (MGE), Stress tolerance

6.1 Introduction

The human population is expected to reach nine billion by the middle of the 21st century. To meet the food requirement of an ever-increasing population, food production needs to rise by 70% to 80%. Plants are susceptible to numerous stresses, both abiotic and biotic in nature, which result in significant loss in plant yield potential. To produce a plant with all the desired qualities using traditional breeding methods, a great deal of time and effort is required. With new advancements, the genome editing (GE) techniques offer alternatives to quickly introduce desirable traits into crops. These techniques are assisting to edit the genomes of several crops, which are challenging to improve through traditional breeding. Nonetheless, genome engineering was difficult and expensive in the early days because it was ineffective to introduce mutations at a

*Corresponding author: dr.sumaira@bzu.edu.pk

DOI: 10.1079/9781800622517.0006

single site and impossible to quickly change particular sequences easily. As a result, traditional techniques typically target one locus during one round, strictly limiting the engineering cycle turnover (Carr and Church, 2009). In this situation, tools empowering multiplexed genome editing (MGE) or high-throughput GE are urgently needed. We indicate 'multiplex' as producing modified DNA-sequences in two or more loci in a single genome during one cycle of mutant development. In this chapter, we will first discuss MGE using oligonucleotides and homologous recombination in microbes. The subsequent section will go over the nuclease-based multiplexed genome engineering with or without homologous recombination. Particularly, the double-stranded break (DSB) caused by nuclease, such as zinc-finger nucleases (ZFNs), transcriptional activator-like effector nucleases (TALENs), and the clustered regularly interspaced short palindromic repeats (CRISPR)-associated endonuclease Cas9 (CRISPR/Cas9) will be discussed. In the end, we have discussed various novel applications being developed or evolved for multiplexing, as well as their effectiveness for crop-improvement programs. We shall also provide an update on recent developments regarding MGE in plants, particularly highlighting its contribution toward creating novel varieties with increased yield, improved nutritional quality, and enhanced stress tolerance.

6.2 Background

MGE refers to the process of simultaneously introducing two or even more DSBs at target sites within the same genome in just one round of mutation (Cong *et al*., 2013a). MGE involves a rapid (desired) change in distinct quantitative trait loci (QTLs) or genes, leading to the formation of novel plant genetic variants with a number of multiple desired modifications. The ability to make such genomic variations permits scientists to better understand and investigate relationships among the members of a particular gene family with duplicative acts, as well as to examine epistatic interactions between strongly linked genetic routes (Xing *et al*., 2014). Moreover, these numerous genetic edits or alterations may provide a practical framework for editing metabolic pathways in crops. Numerous favorable crop developments can be facilitated by MGE, such as the development of high-yielding crops and improved resistance to diseases (Jiang *et al*., 2020; Mubarik *et al*., 2021), or the ability to avoid wide separation in vegetatively reproducing crops like sugarcane (Kannan *et al*., 2018). Unlike conventional random mutation, MGE technologies are adapted to precisely target particular genetic loci and generate the desired variations or mutations (Belhaj *et al*., 2015). Previously, it took years and many cycles of crossing and screening to accomplish specific mutated genes in elite germplasm. Furthermore, conventional breeding has often become more complex due to linkage pull, which hinders the transfer of genes. For example, creating high-yielding genetic makeups with superior-quality traits has been regarded as a challenging task for plant breeders who use traditional plant breeding methods. MGE methods became simpler for users, especially since the emergence as well as the modification of the CRISPR/Cas tool (Cong *et al*., 2013b). Now, researchers can construct patterned genetic circuits, which govern cellular functions or regulate metabolic processes through simultaneously modifying, activating, or down-regulating multitarget gene sequences with different Cas protein and gRNAs *in situ* (Nielsen and Voigt, 2014; Jakočiūnas *et al*., 2015; Gander *et al*., 2017).

6.3 Oligonucleotide-Based MGE

6.3.1 Multiplex automated genome engineering (MAGE)

The Church research laboratory established a high-efficiency multiplexing technique in *Escherichia coli* (*E. coli*) in 2009, which it called MAGE (Wang *et al*., 2009). The comparatively high rate of recombination in MAGE (it is possible to skip selections for modified mutants) results in a diverse mutant collection wherein individual cells may contain numerous modifications throughout the genome. The first MAGE-based example was wherein they effectively increased lycopene contents in *E. coli* by more than fivefold by optimizing the

1-deoxy-D-xylulose-5-phosphate (DXP) biosynthesis (Wang *et al.*, 2009). Multiplexing also served as a tool to recode the genetic material of *E. coli*, substituting all stop-codons with TAA-codons (Ma *et al.*, 2014). MAGE repurposed the released stop codon like a sense-codon for synthetic peptides while inserting it into the 22 *E. coli* gene sequences, resulting in recoded microbes which were fully secure to be utilized in the natural surroundings (Rovner *et al.*, 2015). Notably, conjugative assembly genome editing (CAGE) was utilized in conjunction with MAGE to combine multiple groups of modifications via horizontal transmission to allow multiple edits (usually more than ten) within one individual cell. MAGE has been used to bring more focused features and functions like promoter sequences, in addition to codon replacement and random mutations. In one study, a MAGE-modified version known as 'co-selection' MAGE (CoS-MAGE) has been utilized to simultaneously incorporate numerous different T7 promoters in 12 genetic operons, allowing researchers to study epistatic interactions and gain-of-function modifications in gene networks (Wang *et al.*, 2012). Due to MAGE's popularity, additional research focused on increasing its effectiveness by improving *in-silico* design strategy and working on cost-cutting (Bonde *et al.*, 2014, 2015).

6.3.2 MGE by yeast oligo-mediated genome engineering (YOGE)

Despite being revolutionary, the uses of MAGE are mainly limited for model bacterium *E. coli* because short oligonucleotides used in homologous recombination are ineffective in other species of microbes. Its success cannot be readily transferable to different microbial systems. For example, the MAGE-like strategy was used in *Saccharomyces cerevisiae* by the YOGE method (DiCarlo *et al.*, 2013b). Recombination rates for yeast have increased significantly as a result of strain engineering, which also included the knockout of mismatch-repair proteins, overexpression of DNA recombinases, and improvement of oligonucleotides' design and transformation. Then, using YOGE, two locations in *S. cerevisiae* genetic material were simultaneously cycled. Despite utilizing the best-performing oligonucleotides, after three rounds, the success rate of double modifications falls below 0.01%. It is possible to combine YOGE with site-directed nuclease-based phenotypic screening or negative selection to boost the effectiveness of multiplexed yeast gene editing (DiCarlo *et al.*, 2013a).

6.3.3 MGE by natural transformation (MuGENT)

MuGENT is a technique for accelerating evolution depending on the co-transformation of unrelated genetic traits in naturally competent microorganisms. The ability of various microbes to take up DNA from the environment and incorporate it into their genome via homologous recombination is referred to as natural competence (Lorenz and Wackernagel, 1994). Taking benefit of the natural competence, MuGENT was established (Dalia *et al.*, 2014). MAGE was produced in *E. coli* and has proven highly effective in 'accelerated evolution,' which was used for phenotypic as well as metabolic modifications (Sandoval *et al.*, 2012). MAGE is based on highly efficient recombination of single-stranded DNA (ssDNA) oligos. This method is based on annealing the ssDNA oligonucleotides with complementary strands throughout replication of DNA as well as to introduce insertions and deletions or point mutations into the genome with up to 20% efficiency (Ellis *et al.*, 2001). The lack of choice for genetic changes in cis is a key aspect of this method, allowing multiplexed modifications to be distributed at random in the output mutated collections, where cells belonging to this group can have any number or combination of genetic modifications. Moreover, the efficiency of this approach was indicated by combinatorically modifying the genetic sequence of *Vibrio cholerae* to improve the natural transformation characteristic through promoter replacement, gene deletions, and customizing translation initiation of five genes. MuGENT generated an intricate mutant pool in 1 week and contributed to the collection of modified strains with a 30-times' improved performance in natural transformation. Moreover, MuGENT used to alter simultaneously four redundant genes from different species lacking mismatch repair, *Streptococcus pneumoniae*, illustrating its broad functionality as a MGE tool. MAGE has shown the utility of this MGE method in microbes; however, due to the requirement for extremely efficient recombination, this method is difficult to adapt to non-model microbes.

6.4 Tools and Techniques for Nuclease-based MGE

6.4.1 ZFNs for MGE

ZFN is a gene editing technique that was developed in the 1990s and is efficient in performing precise gene mutation at specific locations. Endonucleases in MGE require a collection of 'off-the-shelf' genome detection tools that can be generally produced to create designer-specific customized endonucleases. This modularity was first discovered at the protein level in zinc-finger (ZF) structural-motifs, which can recognize unique three-nucleotide triplets. ZFNs combine the endonuclease domains of a type IIS restriction enzyme, including *Flavobacterium okeanokoites* 1 (*Fok*1), that creates a DSB from outside of a recognition pattern, with an array of zinc-finger DNA-binding modules, each with a distinct affinity for a set of three base pair sequence (Baltes and Voytas, 2015). By placing multiple zinc fingers in a cluster, a customized nuclease can be created. On each strand, to create a target with 24 base pairs, four modules are utilized, including an intervening cleavage site, that *Fok*1 cleaves non-specifically, leaving behind 4- to 5-nucleotide overhangs (Kim *et al.*, 1996). Target specificity might be developed with the help of an appropriate combination of zinc fingers. Marton *et al.* (2010) simultaneously delivered two foreign genes in petunia and tobacco plants by using ZFN-mediated targeted modification. De Pater *et al.* (2009) used a pair of ZFN-genes for site-specific modification of Arabidopsis genomes. In another study, the targeted modification of two Arabidopsis genes, *ADH1* and *TT4* ZFNs, was achieved by using ZFNs through floral-dip transformation method (Zhang *et al.*, 2010). Hilioti *et al.* (2016) applied customized engineered-ZFNs to modify the *L1L4* gene and introduce the phenotypic diversity in modified tomato plants.

This method shows excellent targeting efficacy; however, due to factors like the limited selection of ZF proteins, only a certain number of specific genetic sequences might be recognized and the high price of designing ZFNs also makes this method of genetic modification less effective, and therefore its application has gradually decreased in recent decades.

6.4.2 TALENs for MGE

Based on ZFN, the TALEN was developed in 2009 and consisted of two components: a 'cleavage domain' as well as a 'binding domain' with selective binding. The cleavage domains are the restriction-endonuclease *Fok*1, and the binding domains are made from highly repeated conserved regions (Abdelrahman *et al.*, 2021). This conserved structural-domain binds to specific regions, creating a 'dimer' with endonuclease activity that results in DSBs in TALEN's sequence spacer site, activating DNA-repair machinery and accomplishing the goal of genome modification. When MGE techniques were initially created, Wang *et al.* (2014) stated the modification of three homoalleles that encode mildew locus (MLOs) proteins in bread wheat by applying only one TALEN pair. The objective was to create a broad-spectrum and Bgt '*Blumeria graminis* f. sp. *tritici*' resistance in wheat germplasm. Multiplexing with TALENs was also employed to create high-potency glycoproteins. In order to improve their immunogenicity and constancy, Li *et al.* (2016b) successfully edited two α−1, 3-fucosyltransferase and two β−1, 2-xylosyltransferase genes in *Nicotiana benthamiana*, producing plants with an enhanced ability to generate glycoproteins totally lacking crop-specific by-products. The researchers used two pairs of TALENs to knock out the two gene family members individually and simultaneously. The N-glycan profiles of XylT and FucT mutant plants were significantly altered. Particularly, this was the first experiment in multiplex editing to target two different gene families in a single germplasm by using two separate circuits of TALEN pairs. Up to now, sugarcane crops have the top score for concurrent targeted gene editing in similar genotypes. Kannan *et al.* (2018) effectively modified 107 of 109 'Catechol-*O*-methyltransferase' (COMT) gene copies by using one TALEN pair to reconfigure the biosynthesis of lignin in sugarcane. These studies demonstrated that TALEN-mediated MGE enables editing of gene families with conserved parts effectively between multiple gene copies, which was successfully tested in polyploid genomes. Furthermore, for specific targeting applications, TALEN methods require the genetic modification of DNA-binding domains, making the process more difficult if multiple gene families are targeted at once. Besides that, TALEN techniques necessarily require the design of DNA-binding domains for each targeted application, making the process more difficult if multiple gene families are targeted at once (Lowder *et al.*, 2015).

6.4.3 The CRISPR/Cas9 toolkit for multiplexed genome engineering

The first characterized or the most commonly utilized endonuclease for gene editing is Cas9, and it is derived from the bacteria *Streptococcus pyogenes*. CRISPR/Cas9 is simpler and much more effective as compared to other GE methods, especially when editing several unrelated genes is required at once. The efficiency of MGE is greatly simplified and improved by applying several gRNAs with a single Cas protein. In CRISPR/Cas9 experiments, the simplicity of plasmid construction also makes it possible to examine the role of numerous genes that either have detrimental effects or enhance plant-specific traits. It detects the target sequences next to the 5′-NGG-3′ PAM (protospacer adjacent motif) and forms a ribonucleoprotein complex with sgRNA (single-guide RNA) to cause a double-strand break. SpCas9 contains numerous domains, the most important of which are nuclease domains and nucleic acid-binding domains. The former aids SpCas9 in binding gRNA and targeting DNA, whereas the nuclease domain is made up of two RuvC and HNH domains that cleave both the targeted as well as non-targeted DNA strands, respectively. The PAM sequence in the target organism's genome is identified by the PAM-interacting (PI) domain, which facilitates the binding of the SpCas9-complex and sgRNA toward the target. This PAM sequence requirement limits the ability of CRISPR/Cas9 systems to select target sites, which is a significant drawback. To overcome this restriction, orthologous endonuclease or engineered Cas9 variants like Cpf1 from Francisella and Prevotella, also known as Cas12a, are used. Most importantly, compared to Cas9, Cas12a identifies a variety of PAM nucleotide sequences and only acquires the RuvC domain, generating a staggered cut rather than a sharp cut. Cas12a can recognize T-rich PAM nucleotide sequences like 5′-TTV3′ or 5′-TTTV-3′, which expands the potential range of targets for genome engineering. The Cas12a system is also favorable because it merely needs 43 nt crRNA, as compared to sgRNA that contains 100 nt in the Cas9 system. StCas9 from *Streptococcus thermophilus*, BlatCas9 from *Brevibacillus laterosporus*, or SaCas9 from *Staphylococcus aureus* are three orthologues of SpCas9 that were derived from various bacterial species and utilized for CRISPR-based GE. By activating RuvC-like nuclease domains, HNH or dCas9 (dead Cas9), by only DNA-binding capability, Cas9 was further modified for nuclease domains and transformed into 'Cas nickase.' Additionally, the base editors and dCas9 are combined to modify the target genetic sequence through irreversibly changing one base pair to another without strand cleavage in DNA. Base editors like adenine base editors (ABEs), cytosine base editors (CBEs), and RNA base editors are attached with nickase or dCAS9 to perform specific alterations to produce stop-codons or SNPs.

6.5 The CRISPR/Cas9 Tool Provides Excellent MGE Ability

MGE, which is frequently used to modify numerous genes, is one of the primary benefits of CRISPR. Among two different multiplex editing approaches, the first strategy for multiplex editing utilizes one promoter for expressing several gRNA as one transcript, whereas the second strategy uses a separate promoter in each gRNA. This part presents extensive information about the approaches used during MGE.

6.5.1 Csy4 nuclease-based excision of gRNA

Csy4 is a protein that can identify its RNA substrate through specific sequence and structure contacts. Csy4 endoribonuclease-mediated genetic engineering is an additional tool for multiplexing. In this method, Csy4 from the bacterium *Pseudomonas aeruginosa* is used to release specific gRNA from numerous gRNA transcripts. This system's construct was created by using sets of Csy4 restriction sites and gRNA (Fig. 6.1A). The same vector is used to clone the bacterial origin Csy4, which is then translated as a host protein to generate single gRNA. This approach detects and attaches the stem-loop form of an RNA sequence (Qi *et al.*, 2012) and produces a nick at the 20th position after G (guanine) instead of utilizing the host RNA machines and tools to excise gRNA. When compared to individual RNA polymerase III promoters, tRNA or Csy4 schemes utilized to create dels in six genes using 12 sgRNAs in one transcript demonstrated a 100% better productivity in inducing mutations. In an early effort, Tsai *et al.* (2014) constructed a vector wherein various sgRNAs could be compiled, each surrounded by a cleavage site for a Csy4 RNase of *P. aeruginosa* to develop a polycistronic CRISPR/Cas9 mechanism for MGE. Based on this, every gRNA should be developed with a 3′ end containing a Csy4 recognition site, allowing the co-expressed Csy4 to concisely cleave the long transcript, releasing gRNAs.

6.5.2 Drosha-mediated MGE

Short hairpin RNAs (shRNA) are synthetic RNA molecules with tight hairpin turns that make it possible to use RNA interference (RNAi) to suppress the expression of any target gene. Depending mostly on the selected promoter, polymerase II as well as polymerase III can be used to transcribe shRNA. The product resembles pre-miRNA (pre-microRNA), and it is primarily processed

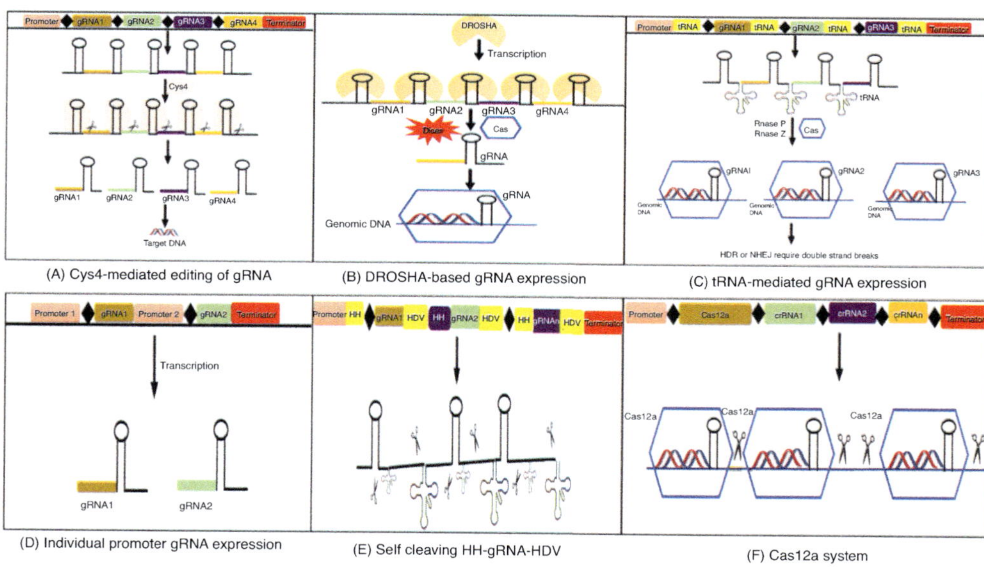

Fig. 6.1. A visual representation of various methods for expressing multiple gRNA in plants. (A) Cys4 nuclease-based multiplex genome engineering. (B) DROSHA-mediated genome editing. DROSHA separates multiple gRNAs from one another. (C) tRNA-based multiple gRNA expression. Cas nucleases cleave the tRNA after it is released from the transcript. (D) Individual promoter for each gRNA expression. (E) Self-cleaving ribozyme flanked gRNA expression. (F) Cas12a system for gRNA expression. However, this approach has a gRNA self-cleaving feature and does not need any additional posttranslational processing elements. Author's own figure.

by Drosha, a class II ribonuclease III (RNase III) enzyme15. Drosha is the fundamental nuclease that carries out the initiation step of miRNA processing in the nucleus (Yan *et al.*, 2016). A tandem array of shRNA or gRNA-micro-RNA (miRNA), expressed with a polymerase II promoter, is used in the Drosha-based MGE method (Fig. 6.1B). Because of their short lifespan, polymerase III promoters, which do not have a 3′ tail or 5′ cap, are frequently used to express sgRNA. In these circumstances, Pol II is preferred because of its flexible and tissue-specific expression. However, the 5′ cap's redundant nuclease activity raises some issues. A mi-RNA-based approach is used to deal with this situation, and Drosha (formerly known as double-stranded RNA-specific endoribonuclease) is used to excise both miRNA and gRNA.

6.5.3 tRNA-mediated expression of gRNA

The tRNA is the essential component of every living cell and is crucial for translation. RNaseP and RNaseZ are present in the tRNA that RNA processing machinery synthesizes from its primary transcript. A short and non-coding gRNA sequence, which utilizes different multiplexed gRNAs can be translated because of the smaller sequence and self-splicing ability of tRNA promoters. Utilizing endogenous RNA processing, the synthesized 'tRNA-gRNA' sequences liberate one sgRNA (Fig. 6.1C). After processing, the sgRNA is released from the transcript by the RNaseP and RNaseZ's cleavage of the tRNA's 3′ and 5′ ends (Xu *et al.*, 2017). Using this method, numerous sgRNA expression cassettes were created and successfully edited 46 target sites in rice. Dimeric RNA-guided *Fok*1 nucleases (RFNs), another method with multiple gRNAs, can edit the target without the need for a specific sequence at the 5′ end. The monomeric Cas9 nuclease, which frequently causes off-target mutations, is preferred to the dimeric RFNs nuclease. In addition, in contrast to the monomeric Cas9 nucleases, these dimeric RFNs are target-oriented because they bind to a particular target with two gRNAs. The tRNA-based method is employed for editing numerous loci or to remove brief fragments from a genome, and the Cas9 fused repressor or transcription activator may regulate the transcription of multiple genes (Ran *et al.*, 2013). Although tRNA-based methods are the most effective ones currently offered for multiplex genome engineering, this system is thought to have a limitation in that it is inefficient for editing more than six gRNAs. Furthermore, Xie *et al.* (2015) developed a polycistronic tRNA-gRNA (PTG) process for the generation of individual transcripts of gRNA. This method has led to highly efficient genome engineering, even as Xie and his colleagues noticed a decrease in PTG efficiency as the multiple sgRNAs competed for Cas9 binding (Xie *et al.*, 2015). (Any example of this method can be used in crops/plants.)

6.5.4 Multiplex expression of each gRNA with individual promoter

The most classical method for producing different sgRNAs is to express every sgRNA with its promoter as well as the terminator. This method has been employed by numerous labs, which have produced toolkits for expressing numerous sgRNAs with unique 'cassettes' in plants (Xing *et al.*, 2014; Zhang *et al.*, 2016). Based on the chosen promoter sequence, an expression cassette's size can range from 500 to 800 bp (Fig. 6.1D). Size restrictions may be an issue when expressing a number of gRNAs because the majority of technologies for plant transformation ideally use just a single vector (Li *et al.*, 2013; Ma *et al.*, 2015). Considering the tradeoffs between size of construct as well as cloning proficiency is another difficulty. The 'Golden Gate Assembly (GGA)' method was used to examine the cloning effectiveness of an increasing number of cassettes. Results showed that only up to five individual expression cassettes can be joined together in one reaction to achieve satisfactory cloning efficiency, and the authors had a lot of difficulty doing this (Li *et al.*, 2013). Accordingly, in other articles, only assemblies of two to eight expression cassettes have been reported, suggesting that this approach is unsuitable for expressing a large number of gRNAs. Surprisingly, one study discovered a process to avoid the former issue of cloning efficiency using a technique known as 'stackable arrays.' Cloning of as many as 14 gRNA expression units is possible into a single vector using stackable arrays (Peterson *et al.*, 2016). This process allows for the cloning of multiple cassettes; however, it needs part formulation by a company that might be prohibitively expensive for some investigators.

6.5.5 Self-cleaving ribozyme-mediated gRNA expression

The gRNAs maturation through self-cleaving ribozyme is a technique that initially evolved in yeast cells and is potentially applicable to multiplex genome engineering. The system makes use of the nuclease action of customized ribozyme (Gao and Zhao, 2014), which are RNA molecules and activate enzyme-like responses. Gao and Zhao (2014) created a 'Ribozyme-gRNA-Ribozyme' (RGR) gene. The gRNA in RGR is flanked over its 5′ end by 'hammerhead-type (HH)' ribozymes and on its 3′ end by a 'hepatitis delta virus (HDV)' ribozyme. This technique was initially seen to only provide one sgRNA, which is developed through self-cleaving of the ribozyme. The scientists proposed that this process can be readily converted into a multiplexed process by combining multiple RGR cassettes, which possibly express with just one promoter. RGRs have recently been used in nonplant organisms for multiplex genome 5′ and 3′ engineering. Two studies, one in *Leishmania donovani* (Zhang and Matlashewski, 2015) and the other in *Danio rerio* (Lee *et al.*, 2016), reported successful RGR-mediated processing of two gRNAs. A further study in cell lines from humans utilized RGRs for processing of four sgRNAs to mutate two genes; moreover, they noticed a decrease in sgRNA processing efficacy in comparison to just two sgRNAs, which clearly indicates system restrictions if sgRNAs expression is increased (Xu *et al.*, 2017). The RGR system is anticipated to prove useful for their effective multiple gRNAs in plants (Fig. 6.1E).

6.5.6 CRISPR/Cas12a (Cpf1) system to express multiple crRNA

One of the key barriers to the previously mentioned CRISPR/Cas9-based multiplexing is the size of a construct. Furthermore, the demand for GC-rich PAM patterns is a problem with using Cas9 (note: G stands for guanine and C stands for cytosine here). These barriers could be overcome with Cas12a, a new member to the family of Cas proteins previously recognized as 'Cpf1.' Cas12a is small in size, requires a PAM sequence to target T-rich regions (T stands for thymine here), and requires a short 42 nt crRNA to facilitate construct development and multiplexing (Zetsche *et al.*, 2015). Furthermore, the Cas12a multiplexing platform outperforms further genetic modification methods. A Cas12a system does not demand the addition of multiple pol-III promoters or a particular sequence for crRNA construct as it can produce crRNAs appropriately by using the activity of its ribonucleases (Zetsche *et al.*, 2015; Fonfara *et al.*, 2016). Cas12a is an ideal system for multiplexing because of these features (Fig. 6.1F). Furthermore, Cas12a performs DSB cleavage at the cohesive end proximal to the PAM location, allowing for constant cleaving of DNA (Zetsche *et al.*, 2015). Swarts and Jinek tried to compare the relationships between Cas9's and Cas12a's structure and function, as well as their implications for genetic modification (Swarts and Jinek, 2018). Using whole-genome sequencing, high specificity of Cas12a was found in simultaneous editing of plants (Tang *et al.*, 2018). Parallel results were obtained when using 'CIRCLE-seq' (Lee *et al.*, 2019).

6.6 MGE in Plants

The understanding of functions of individual genes in plant biology is changing dramatically due to the CRISPR/Cas system. Targeting multiple genes at once with MGE using CRISPR/Cas can now give a researcher the power to understand extremely complex biological problems.

6.6.1 MGE in crop domestication and genetic diversity

Crop domestication entailed phenotypic changes that made species reliant on humans for continued existence. With the development of GE technologies like CRISPR/Cas9, domestication of wild species can now be accomplished in a few generations, which once required thousands of years. *Solanum pimpinellifolium*, a wild tomato, was domesticated by the use of CRISPR/Cas tools. These studies demonstrated that multiplex CRISPR/Cas editing can disrupt known tomato domestication genes. In *S. pimpinellifolium*, the following genes were edited to speed up the accumulation of suitable traits for cultivation: *MULTIFLORA* (for fruit number), *OVATE* (for fruit shape), SELF-PRUNING 5G (for floral induction), *FASCIATED*, *WUSCHEL*, *CLAVATA3* and *FRUIT WEIGHT 2.2* (for fruit size), *GDP-L-GALACTOSE PHOSPHORYLASE* (for vitamin C content), and *LYCOPENE BETA CYCLASE* (for lycopene content). Surprisingly, in comparison to their wild parents, the altered plants displayed domesticated features including compact plant architecture, increased fruit quantity and size, and enhanced vitamin C and lycopene content in fruits. Similarly, Lemmon *et al.* (2018) improved domestication features in groundcherry (*Physalis pruinosa*) by interrupting three genes that control fruit size, flower production, and plant architecture. Compared with diploid rice, *Oryza alta* is an allotetraploid wild rice with a CCDD genome that has higher biomass and greater tolerance to biotic and abiotic stress (Wing *et al.*, 2018). To domesticate *O. alta*, CRISPR/Cas was used to target two genes, *OaqSH1* and *OaAN-1*, which control seed shattering and awn length, respectively. As a result, the altered plants had less seed breaking and shorter awns. In another study, genes controlling heading date (*OaGhd7* and *OaDTH7*), seed size (*OaGS3*), and plant height (*OaSD1*) were edited (Yu *et al.*, 2021). Recent research used CRISPR/Cas technology to modify domesticated features in wild tomatoes by interrupting *SpRDR6* (*RNA-DEPENDENT RNA POLYMERASE 6*), *SpSGS3* (*SUPPRESSOR OF GENE SILENCING 3*), *SpPR-1* (*PATHOGENESIS-RELATED PROTEIN-1*), *SpMlo1* (*MILDEW-RESISTANT LOCUS O*), and *SpProSys* (*PROSYSTEMIN*) genes (Lin *et al.*, 2022). Recently, another study used CRISPR/Cas mediated gene editing tool to domesticate sea barley grass by mutating postdomestication genes, for example, IPA1 and MOC (Kuang *et al.*, 2022). MGE technology could assist us in bringing wild and partially domesticated crops into mainstream farming in order to meet our present and future requirements in a sustainable manner.

6.6.2 Increasing crop quality traits and yield potential by MGE

High-yielding and crop quality traits are the most essential features for a breeding program, which drive farmers' adoption of new genotypes. Consumer interest in high-yielding cultivars with higher quality has increased, which raises the marketing value for important crop plants (Abdelrahman *et al.*, 2021). Therefore, multiple QTLs must be introduced into an elite genotype to generate a novel genetic variant in terms of plant quality, yield, and nutritive values to secure food supply. Numerous MGE studies have been conducted successfully to change the yield-attributing characteristics and nutritional properties in commercially important crops. An excellent example of MGE influencing crop development is the targeted modification of rice11 members of FT-related genes, by using one construct constituting eight gRNAs (Ma *et al.*, 2015). Xu *et al.* (2016) used CRISPR/Cas9 system for multiplex alteration of negative regulators of GRAIN WIDTH 2 (*GS3*, *GW6*, *GW5*, and *GW2*), which encode grain mass. It led to significant enhancement in grain size and weight (Xu *et al.*, 2016). An increase of 8.47% and 11.69% was seen in the *gw5tgw6* double mutant in grain length and grain width, and a total grain width which improved by 12.68%. Also, four *gw2gw5tgw6* triple mutants demonstrated an increase of 20% to 30% in the very same parameters, suggesting that genetic variations in such QTLs resulted in the expected multiplicative impacts (Xu *et al.*, 2016). In a move to generate early maturation cultivars, three genes (*Hd5*, *Hd4*, and *Hd2*) that act as floral repressor genes have been mutated in rice plants (Li *et al.*, 2017). The authors utilized five gRNAs for successful editing of single sites in *Hd5* and multiple sites in *Hd4* and *Hd2* in this experiment. It was observed that 14 of the 18 recovered lines had all the three genes modified, and the majority of these were single-nucleotide indels. According to the researcher, all the triple-mutant plants flowered up to 40 days earlier as compared to the non-transformed rice plants. Lacchini *et al.* (2020) used multiplexed tool to edit the '*OsGS3*, *OsGW2*, and *OsGn1a*' genetic factor to improve the shape of rice grains. Another recent multiplex genome engineering experiment on Nippon bare rice resulted in a combined effect of abiotic stress tolerance and yield improvement. *OsMYB30* (a gene of cold tolerance), *GS3* (a gene of grain size), and *OsPIN5b* (a gene of panicle length) were all mutated, and thus the mutated plants showed improved cold tolerance, grain size, and panicle length (Zeng *et al.*, 2019). Furthermore, the CRISPR/Cas9 platform was employed to create sweet/waxy corns (SWCs) through simultaneously engineering *ZmSH2* and *ZmWAXY1* genes (Dong *et al.*, 2019).

In wheat, multiple members (100 members) of α-gliadin gene family were mutated with two different consensus gRNAs, yielding a variety of multiplexed mutants, such as a line with genetic variations in almost 35 specific genes (Sánchez-León *et al.*, 2018). The resulting transgene-free, low-gluten wheat lines are used as a starting point for introducing this trait into other elite wheat genotypes. Knockout mutants of wheat *TaGW2*, *TaMLO*, and *TaLpx-1* genes were created via MGE and exhibited improved seed size as well as thousand kernel weight (Wang *et al.*, 2018). Different members of a *k1C* gene family, which encodes α-kafirin (a main seed-storage protein), were mutated with only one sgRNA in sorghum, which resulted in 72 mutant plants with 26 events of mutation. It included the one with concurrent genetic variations in 11 distinct genes of *k1C* and the other in *k1C3–9* group. The resulting (T1 as well as T2) seeds had lower levels of α-kafirin, and the preferred T2 events had substantially higher grain lysine content and protein digestibility (Huang *et al.*, 2018). A current study, CRISPR/Cas9-mediated multiplexed genome engineering of fruit color-related genes *SGR1*, *PSY1*, and *MYB12*, was performed to generate various transformed tomato lines with various fruit colors (Yang *et al.*, 2023). All these reports illustrating the use of CRISPR/Cas9 to obtain ideotype through targeting multiple genes simultaneously in different crops are expected to grow rapidly in the future. Table 6.1 lists the targeted genes and traits in various important crops that have been modified using different MGE techniques to construct new varieties with higher yield and nutrient content.

6.6.3 Increasing crop resistance against biotic/abiotic stresses by MGE

Numerous abiotic and biotic factors significantly harm nutrition and food security by lowering crop production. Several plant pathologists and researchers have explored the CRISPR/Cas9 mediated multiplexed gene modification to improve the defense response to various environmental stresses (Table 6.2). The primary goal of improving crop genetics is to boost the production as well as ensure food security to feed the expanding world population. Enhanced resistance to diseases brought on by fungal, bacterial, and viral pathogens is one of the traits targeted by the multiplexed CRISPR/Cas system in crops. Moreover, the CRISPR/Cas system has accelerated the development of climate-resilient varieties that can endure extreme weather conditions like floods, drought, cold, heat, or heavy metal stresses. Powdery mildew is a biotrophic fungal pathogen that is among the most prevalent diseases that severely limits wheat production (Draz *et al.*, 2019). To address the issue, Zhang *et al.* (2017) targeted three *TaEDR1* homoeologs at the same time to enhance protection against powdery mildew in wheat. By using CRISPR/Cas9 to create the sequence-specific mutations in the *Xa13*, *Pi 2*, and *TMS5* genes, the temperature-sensitive male sterile lines with resistance to both bacterial blight and rice blast fungus were created (Li *et al.*, 2019). Currently, the *PMR4* locus was subjected to large mutations by using four individual gRNAs to knock down the *PMR4* gene via CRISPR/Cas technology. The resulting mutant plants were less susceptible to powdery mildew (Santillán Martínez *et al.*, 2020). CRISPR/Cas-mediated multiplexed genome engineering was recently used to successfully improve resistance to cotton leaf curl virus (CLCuV) in cotton plants by simultaneously mutating six *CLCuV* genes (Mubarik *et al.*, 2021). The Xoo (*Xanthomonas oryzae* pv. *oryzae*) rice system has been extensively researched, and it has served as a model to study numerous

Table 6.1. A list of the MGE efforts to revolutionize crop improvement activities for important crop plants. Author's own table.

Plant species	Target genes	Gene function	Genome editing technique	Reference
Rice	FT family genes	Induce early maturation	CRISPR/ Cas9	Ma *et al.*, 2015
	OsGW2, *OsGW5*, and *OsTGW6*	Increase grain size	CRISPR/ Cas9	Xu *et al.*, 2016
	GS3, *Gn1a*, *DEP1*, and *IPA1*	Improve grain length, yield with dwarf stature	CRISPR/ Cas9	Li *et al.*, 2016b
	OsGS3, *OsGW2*, and *OsGn1a*	Increase grain size	CRISPR/ Cas9	Zhou *et al.*, 2019
	GS3 and *Gn1a*	Increase grain yield	CRISPR/ Cas9	Shen *et al.*, 2018
	Gn1a and *OsDEP1*	Improve yield	CRISPR/ Cas9	Huang *et al.*, 2018
	Group I (*PYL1–PYL6*, *PYL12*) and Group II (*PYL7–PYL11*, *PYL13*)	Improve growth and productivity	CRISPR/ Cas9	Miao *et al.*, 2018
	OsPIN5b, *GS3*, and *OsMYB30*	Increase grain size, panicle length, and cold tolerance	CRISPR/ Cas9	Zeng *et al.*, 2019
	GS3, *GW2*, and *GN1A*	Increase grain size	CRISPR/ Cas9	Lacchini *et al.*, 2020
	GS3, *GW2*, and *Gn1a*	Increase grain size, number, width, and weight	CRISPR/ Cas9	Zhou *et al.*, 2019
	FWL (FW 2.2-like) and *OsFWL4*	Increase tiller number and yield	CRISPR/ Cas9	Jiang *et al.*, 2020
	Multiple sites	Help in the domestication of wild relative (*O. alta*)	CRISPR/ Cas9	Yu *et al.*, 2021
Wheat	Two sites at *α-gliadin* gene	Produce wheat with low-gluten content	CRISPR/ Cas9	Sánchez-León *et al.*, 2018
	TaGW2, *TaLpx-1*, and *TaMLO*	Increase grain weight and seed size	CRISPR/ Cas9	Wang *et al.*, 2018
Durum wheat	*α-amylase trypsin* inhibitor genes	Decrease allergen proteins	CRISPR/ Cas9	Camerlengo *et al.*, 2020
Maize	*ZmIPK*, *ZmIPK1A*, and *ZmMRP4*	Improve the synthesis of phytic acid	CRISPR/ Cas9	Liang *et al.*, 2014
	ZmWAXY1 and *ZmSH2*	Produce sweet and waxy corn	CRISPR/ Cas9	Dong *et al.*, 2019
	48 growth-related genes	Increase yield and drought resistance	CRISPR/ Cas9	Lorenzo *et al.*, 2023

Continued

Table 6.1. Continued

Plant species	Target genes	Gene function	Genome editing technique	Reference
Barley	Two *HvPM19* paralogs	Help in hormone signaling	CRISPR/ Cas9	Lawrenson *et al.*, 2015
Sorghum	20 *K1C* paralogs	Improve digestibility and protein quality	CRISPR/ Cas9	Li *et al.*, 2018a
Tobacco	Two α–1,3-fucosyltransferase and two β–1,2-xylosyltransferase	Enhance glycoprotein production	TALENs	Li *et al.*, 2016a
Sugarcane	107 COMT gene copies	Enhance lignin production	TALENs	Kannan *et al.*, 2018
Tomato	*PSY1*, *MYB12*, and *SGR1*	Induce recoloring of tomato fruits	CRISPR/ Cas9	Yang *et al.*, 2023

susceptible (*SWEET*) genes in monocots. A mutated Kitaake line of rice was created by Oliva *et al.* (2019) using a CRISPR/Cas9-based multiplexing approach, and mutated lines of the mega varieties Ciherang-Sub1 and IR-64 with varying configurations of genetic changes in the EBEs of *OsSWEET11*, *OsSWEET13*, and *OsSWEET14* promoters conferred resistance toward a significant number of *Xoo* strains. In super basmati rice, the *SWEET14* gene's promoter region was modified with the CRISPR/Cas9 genetic editing method. This caused resistance toward the specific TALEs of *Xoo* strains (such as *TalF*, *PthXo3*, and *AvrXa7*) (Zafar *et al.*, 2020). Citrus canker develops by the *Xanthomonas citri* subsp. *citri* (*Xcc*) PthA4 effector molecule, which networks through EBEs of the *CsLOB1* gene promoter region. Wanjincheng orange resistance to *Xcc* strains was successfully conferred in homologous mutant lines by five pCas9/*CsLOB1* sgRNA constructs that were intended to genetically modify promotors of both alleles (Peng *et al.*, 2017). CRISPR/Cas9 is an extremely target-oriented, potent cellular immunity mechanism to combat various issues brought on by viruses. Recent studies have confirmed and shown how useful and effective this system is. The CRISPR/Cas9-expressing *N. benthamiana* plants were developed by agroinfiltration through tobacco rattle virus (TRV), which contains an expression cassette containing various gRNAs. The resulting plants demonstrated resistance to tomato yellow leaf curl virus, Merremia mosaic virus (MerMV), and beet curly top virus (BCTV) by degrading the viruses and introducing mutations at the virus' target sequences (Ali *et al.*, 2015). The accumulation of BSCTV (Beet severe curly top virus) was hampered when a construct of gRNA-Cas9 was incorporated into the target region in Arabidopsis and *N. benthamiana*. The incorporation of plasmid 'pHSN401-C3' in Arabidopsis and 'pHSN401-A7' in *N. benthamiana* targeting the three distinct regions C3, B7, and A7 decreased the viral concentration by 70%, 66%, and 65% correspondingly, and created virus-resistant plants (Ji *et al.*, 2015). Abdallah *et al.* (2022) used a multiplex sgRNA-CRISPR/Cas9 gene editing tool to mutate five *TaSal1* homologous genes by using three gRNAs simultaneously in hexaploid wheat. Results showed that all mutants grew better under drought-stress conditions than wild-type plants. Similarly, CRISPR/Cas9 tool was used to create deletion mutations in *DST* gene of rice plants. The resulting mutated lines were confirmed for their tolerance against drought and salinity stress (Santosh Kumar *et al.*, 2020).

Current breeders and geneticists strive to minimize heavy metal contamination in plants and prevent them from entering a food web. The rice *OsNramp5* gene is crucial in getting Cd (cadmium) into the rice grains, which then moves into the human body and causes fatal diseases. Under cadmium stress conditions, the knockout of this gene led to less cadmium deposition in roots and shoots than in the wild type (Tang *et al.*, 2017). Rice genes (*OsNramp5* and *OsLCT1*) were chosen as similar targets using

Table 6.2. A summary of MGE studies for improving defense system in important crop plants. Author's own table.

Plant species	Target genes	Gene function	Genome editing tool	Reference
Arabidopsis	*Elf(iso)4E/cds region*	Resistance to the potyvirus	CRISPR/Cas9	Pyott *et al.*, 2016
Tobacco and Arabidopsis	*IR*, *CP*, *Rep*	Produce geminivirus-resistant plants	CRISPR/Cas9	Ji *et al.*, 2015
	ORF1a, *ORFCP*, and *3′-UTR*	Decrease symptoms and virus load	CRISPR/Cas9	Zhang *et al.*, 2018
Tobacco	*LIR*, *Rep*	Decrease virus prevalence and symptoms	CRISPR/Cas9	Baltes *et al.*, 2015
	GFP, *Hc-pro*, and *CP*	Decrease virus load and symptoms	CRISPR/Cas13a	Aman *et al.*, 2018
	IR and *C1*	Resistance against CLCuMuV	CRISPR/Cas9	Yin *et al.*, 2019
Tobacco and Tomato	*CP*, *Rep*	Develop resistance to disease	CRISPR/Cas9	Tashkandi *et al.*, 2018
Cucumber	*elf4E/cds region*	Resistance against PRSMV, CVYV, and ZYMV	CRISPR/Cas9	Chandrasekaran *et al.*, 2016
Cassava	*nCBP-1* and *nCBP-2/cds region*	Decrease virus prevalence and symptoms	CRISPR/Cas9	Gomez *et al.*, 2019
Rice	*OsSWEET11* and *OsSWEET14*	Rice with broad-spectrum disease resistance	CRISPR/Cas9	Xu *et al.*, 2019
	PYL1, *PYL2*, *PYL3*, *PYL4*, *PYL5*, *PYL6*, and *PYL12*	Heat-stress tolerance	CRISPR/Cas9	Miao *et al.*, 2018
	TMS5, *Pi21*, and *Xa13*	Yield enhancement and disease resistance	CRISPR/Cas9	Li *et al.*, 2019
	OsACC-T1, *OsCDC48-T3*, *OsALS-T1*, *OsDEP1-T2*, *OsNRT1.1B-T1*, and *OsDEP1-T1*	Resistance to herbicides	CRISPR/Cas9	Li *et al.*, 2018b
	Three sgRNAs that target the *FTIP1e* and *ALS* genes	Herbicide resistance	CRISPR/Cas9	Shimatani *et al.*, 2017
	SWEET11, *SWEET13*, and *SWEET14*	Develop broad-spectrum disease resistance	CRISPR/Cas9	Oliva *et al.*, 2019
	Various loci of *OsACC gene*	Herbicide resistance	CRISPR/Cas	Liu *et al.*, 2020
	Pi21, *Bsr-d1*, and *Xa5*	Enhance resistance against bacterial blast and rice blast	CRISPR/Cas9	Tao *et al.*, 2021
	OsPIN5b, *GS3*, and *OsMYB30*	Improve yield and cold stress tolerance	CRISPR/Cas9	Zeng *et al.*, 2019

Continued

Table 6.2. Continued

Plant species	Target genes	Gene function	Genome editing tool	Reference
	Various loci at *OsALS1*	Increase herbicide resistance	CRISPR/ Cas9	Kuang *et al.*, 2020
	Two sites at *DST* gene	Improve grain yield and tolerance to drought and salinity stress	CRISPR/ Cas9	Santosh Kumar *et al.*, 2020
	OsNramp5 and *OsLCT1*	Decrease cadmium deposition	CRISPR/ Cas9	Songmei *et al.*, 2019
Wheat	Three homologs of *EDR1*	Resistance to powdery mildew	CRISPR/ Cas9	Zhang *et al.*, 2017
	TaERF3 and *TaDREB2*	Drought and heat tolerance	CRISPR/ Cas9	Kim *et al.*, 2018
	Five *TaSal1* homologs	Drought resistance	CRISPR/ Cas9	Abdallah *et al.*, 2022
	Three *MLO* homologs	Develop resistance against powdery mildew	TALENs	Wang *et al.*, 2014
Cotton	*AV2/AV1*, *AC2/AC3*, and *AC1/AC4*	Cotton leaf curl virus resistance	CRISPR/ Cas9	Mubarik *et al.*, 2021
Maize	*ZmALS1* and *ZmALS2*	Tolerance to Chlorsulfuron	CRISPR/ Cas9	Jiang *et al.*, 2020
Soybean	*Drb2a* and *Drb2b*	Enhancement of drought and salinity resistance	CRISPR/ Cas9	Curtin *et al.*, 2018
Barley	*MP*, *CP*, *Rep/Rep*, *IR*	Reduce disease symptoms and virus accumulation	CRISPR/ Cas9	Kis *et al.*, 2019
Apple	*DIPM-1*, *DIPM-2*, and *DIPM-4*	Induce resistance against fire blight disease	CRISPR/ Cas9 RNPs	Malnoy *et al.*, 2016

CRISPR/Cas9 to reduce cadmium concentration, and both genes' knockouts demonstrated a decrease in cadmium buildup (Songmei *et al.*, 2019). CRISPR/Cas9 was applied to simultaneously target the wheat protoplasm by modifying the heat-stress regulators *TaERF3* (ethylene-responsive factor 3) and *TaDREB2* (wheat dehydration-responsive element binding factor protein 2) (Kim *et al.*, 2018). The recent research showed that CRISPR/Cas9 might be effectively used to edit the wheat genome to ensure constant performance against heat stress by suppressing dehydration-related genes. To achieve long-term weed resistance, crops are frequently selected for their ability to tolerate herbicides. In this respect, Jiang *et al.* (2020) created herbicide-tolerant maize plants through prime editing (PE) that altered the *ZmALS2* as well as *ZmALS1* genes to produce 'S621I/W542L' double mutant lines that impart tolerance against numerous herbicides. Table 6.2 provides a list of the targeted genes and traits that have been altered in important crop plants with different multiplexed genome engineering technologies to enhance defense response, boost resistance to biotic/abiotic factors, and increase resistance to disease caused by pathogenic organisms.

6.7 Factors Influencing Multiplexing Efficiency

The advancement of the aforementioned methods has greatly sped up GE with multiplexing. Multiplexing still has a constrained level and is also technically difficult, while increased multiplexing is highly desirable. The core of multiplexed gene modification is a low-priced method of producing mutagenic reagents, such as oligonucleotides and nucleases. Using chemically synthesized short DNA oligos for introducing alterations in a genome, such as in YOGE and MAGE, is a cost-effective option. These approaches do not require cloning, which further lowers the cost. These methods, however, are limited to microbiological systems (Wannier *et al.*, 2021). Although TALENs have been successfully proven for multiplexed genetic modification, they are limited by high cost and complex assemblage workflows. However, specificity of Cas9 is solely dependent on the sequence of gRNA, which is extremely simple and inexpensive to produce. Therefore, Cas9 looks as a more suitable option for multiplexing compared to TALENs and ZFNs. The delivery of mutagenic agents is another factor that influences multiplexing capability (Cox *et al.*, 2015). It is crucial to make sure that multiple nucleases and oligos-encoding plasmids can be introduced into one particular cell efficiently. Furthermore, mismatched TALEN or ZFN dimers may form, exacerbating off-target effects. Cas9 still has off-target effects, but overlapping is not a problem because Cas9 operates as a monomer while its target specificity is merely in the 'cr/gRNA.' Even though multiplexed CRISPR/Cas9 techniques are flexible, there are still significant biological and technical obstacles to overcome (Gisler *et al.*, 2019). In terms of engineering, the current issues in making long arrays of gRNAs or predicting their behavior in living cells are presently the most important. The occurrence of unintended chromosomal reorganization appears to be a significant problem. Another serious issue is that as a cell's total number of gRNAs increase, it must become more competitive for a limited set of endonucleases. By reducing the repetitive sequences in sgRNA, using a powerful promoter to drive the Cas9 genes, and using a conditional gRNAs construct, these difficulties may be overcome.

6.8 Conclusion and Prospects

MGE is a gene regulatory device that allows users to simultaneously activate/suppress the activity of multiple genes. Multiplex genome engineering enables researchers to edit and rearrange specific chromosomal segments for exploring the functions of the enzymes in different metabolic pathways, interactions between pathways, and cell signaling in response to various environmental stresses. Moving forward, the advancement of these multiplexing techniques will most likely be determined by two factors. First, ongoing optimization is required to improve the efficacy whereby required genetic variants can be created with decreasing occurrences of off-target modifications. This might consist of competency enhancement to improve the uptake of mutagenic oligos as well as promoter editing to generate optimum expression profiles for core components, which include TALENs, Cas9 as well as gRNA for each organism. Second, rapid advancements in laboratory-automated systems and high-throughput processing will have a significant impact on future efforts at MGE. Efforts are also needed for increasing the efficiency for simultaneously targeting multiple loci. It is expected that developments in fundamental biology and genomics, chemistry, and computation biology will further aid in removing the bottlenecks in multiplexed genome engineering.

References

Abdallah, N.A., Elsharawy, H., Abulela, H.A., Thilmony, R., Abdelhadi, A.A. *et al.* (2022) Multiplex CRISPR/Cas9-mediated genome editing to address drought tolerance in wheat. *GM Crops & Food* 13, 1–17.

Abdelrahman, M., Wei, Z., Rohila, J.S. and Zhao, K. (2021) Multiplex genome-editing technologies for revolutionizing plant biology and crop improvement. *Frontiers in Plant Science* 12, 721203.

Ali, Z., Abulfaraj, A., Idris, A., Ali, S., Tashkandi, M. *et al.* (2015) CRISPR/Cas9-mediated viral interference in plants. *Genome Biology* 16, 1–11.

Aman, R., Ali, Z., Butt, H., Mahas, A., Aljedaani, F. *et al.* (2018) RNA virus interference via CRISPR/Cas13a system in plants. *Genome Biology* 19(1), 1–9.

Baltes, N.J. and Voytas, D.F. (2015) Enabling plant synthetic biology through genome engineering. *Trends in Biotechnology* 33(2), 120–131.

Baltes, N.J., Hummel, A.W., Konecna, E., Cegan, R., Bruns, A.N. *et al.* (2015) Conferring resistance to geminiviruses with the CRISPR-Cas prokaryotic immune system. *Nature Plants* 1, 15145.

Belhaj, K., Chaparro-Garcia, A., Kamoun, S., Patron, N.J. and Nekrasov, V. (2015) Editing plant genomes with CRISPR/Cas9. *Current Opinion in Biotechnology* 32, 76–84.

Bonde, M.T., Klausen, M.S., Anderson, M.V., Wallin, A.I.N., Wang, H.H. *et al.* (2014) MODEST: A web-based design tool for oligonucleotide-mediated genome engineering and recombineering. *Nucleic Acids Research* 42(W1), W408–W415.

Bonde, M.T., Kosuri, S., Genee, H.J., Sarup-Lytzen, K., Church, G.M. *et al.* (2015) Direct mutagenesis of thousands of genomic targets using microarray-derived oligonucleotides. *ACS Synthetic Biology* 4(1), 17–22.

Camerlengo, F., Frittelli, A., Sparks, C., Doherty, A., Martignago, D. *et al.* (2020) CRISPR-Cas9 multiplex editing of the α-amylase/trypsin inhibitor genes to reduce allergen proteins in durum wheat. *Frontiers in Sustainable Food Systems* 4, 104.

Carr, P.A. and Church, G.M. (2009) Genome engineering. *Nature Biotechnology* 27(12), 1151–1162.

Chandrasekaran, J., Brumin, M., Wolf, D., Leibman, D., Klap, C. *et al.* (2016) Development of broad virus resistance in non-transgenic cucumber using CRISPR/Cas9 technology. *Molecular Plant Pathology* 17(7), 1140–1153.

Cong, L., Ran, F., Cox, D., Lin, S., Barretto, R. *et al.* (2013a) Ingeniería de genoma multiplex utilizando sistemas CRISPR/Cas. *Ciencias* 339, 819–823.

Cong, L., Ran, F.A., Cox, D., Lin, S., Barretto, R. *et al.* (2013b) Multiplex genome engineering using CRISPR/Cas systems. *Science* 339(6121), 819–823.

Cox, D.B.T., Platt, R.J. and Zhang, F. (2015) Therapeutic genome editing: Prospects and challenges. *Nature Medicine* 21(2), 121–131.

Curtin, S.J., Xiong, Y., Michno, J.M., Campbell, B.W., Stec, A.O. *et al.* (2018) CRISPR/Cas9 and TALENs generate heritable mutations for genes involved in small RNA processing of *Glycine max* and *Medicago truncatula*. *Plant Biotechnology Journal* 16(6), 1125–1137.

Dalia, A.B., McDonough, E. and Camilli, A. (2014) Multiplex genome editing by natural transformation. *Proceedings of the National Academy of Sciences* 111(24), 8937–8942.

De Pater, S., Neuteboom, L.W., Pinas, J.E., Hooykaas, P.J.J. and Van Der Zaal, B.J. (2009) ZFN-induced mutagenesis and gene-targeting in Arabidopsis through Agrobacterium-mediated floral dip transformation. *Plant Biotechnology Journal* 7(8), 821–835.

DiCarlo, J.E., Conley, A.J., Penttilä, M., Jäntti, J., Wang, H.H. *et al.* (2013a) Yeast oligo-mediated genome engineering (YOGE). *ACS Synthetic Biology* 2(12), 741–749.

DiCarlo, J.E., Norville, J.E., Mali, P., Rios, X., Aach, J. *et al.* (2013b) Genome engineering in *Saccharomyces cerevisiae* using CRISPR-Cas systems. *Nucleic Acids Research* 41(7), 4336–4343.

Dong, L., Qi, X., Zhu, J., Liu, C., Zhang, X. *et al.* (2019) Supersweet and waxy: Meeting the diverse demands for specialty maize by genome editing. *Plant Biotechnology Journal* 17(10), 1853.

Draz, I.S., Esmail, S.M., Abou-Zeid, M.A.E.-H. and Essa, T.A.E.-M. (2019) Powdery mildew susceptibility of spring wheat cultivars as a major constraint on grain yield. *Annals of Agricultural Sciences* 64(1), 39–45.

Ellis, H.M., Yu, D., DiTizio, T. and Court, D.L. (2001) High efficiency mutagenesis, repair, and engineering of chromosomal DNA using single-stranded oligonucleotides. *Proceedings of the National Academy of Sciences* 98(12), 6742–6746.

Fonfara, I., Richter, H., Bratovič, M., Le Rhun, A. and Charpentier, E. (2016) The CRISPR-associated DNA-cleaving enzyme Cpf1 also processes precursor CRISPR RNA. *Nature* 532(7600), 517–521.

Gander, M.W., Vrana, J.D., Voje, W.E., Carothers, J.M. and Klavins, E. (2017) Digital logic circuits in yeast with CRISPR-dCas9 NOR gates. *Nature Communications* 8(1), 1–11.

Gao, Y. and Zhao, Y. (2014) Self-processing of ribozyme-flanked RNAs into guide RNAs in vitro and in vivo for CRISPR-mediated genome editing. *Journal of Integrative Plant Biology* 56(4), 343–349.

Gisler, S., Gonçalves, J.P., Akhtar, W., Jong, J., Pindyurin, A.V. *et al.* (2019) Multiplexed Cas9 targeting reveals genomic location effects and gRNA-based staggered breaks influencing mutation efficiency. *Nature Communications* 10(1), 1598.

Gomez, M.A., Lin, Z.D., Moll, T., Chauhan, R.D., Hayden, L. *et al.* (2019) Simultaneous CRISPR/Cas9-mediated editing of cassava *eif 4E* isoforms *nCBP-1* and *nCBP-2* reduces cassava brown streak disease symptom severity and incidence. *Plant Biotechnology Journal* 17(2), 421–434.

Hilioti, Z., Ganopoulos, I., Ajith, S., Bossis, I. and Tsaftaris, A. (2016) A novel arrangement of zinc finger nuclease system for in vivo targeted genome engineering: the tomato *LEC1-LIKE4* gene case. *Plant Cell Reports* 35, 2241–2255.

Huang, L., Zhang, R., Huang, G., Li, Y., Melaku, G. *et al.* (2018) Developing superior alleles of yield genes in rice by artificial mutagenesis using the CRISPR/Cas9 system. *The Crop Journal* 6(5), 475–481.

Jakočiūnas, T., Bonde, I., Herrgård, M., Harrison, S.J., Kristensen, M. *et al.* (2015) Multiplex metabolic pathway engineering using CRISPR/Cas9 in *Saccharomyces cerevisiae*. *Metabolic Engineering* 28, 213–222.

Ji, X., Zhang, H., Zhang, Y., Wang, Y. and Gao, C. (2015) Establishing a CRISPR-Cas-like immune system conferring DNA virus resistance in plants. *Nature Plants* 1, 15144.

Jiang, Y.-Y., Chai, Y.-P., Lu, M.-H., Han, X.-L., Lin, Q. *et al.* (2020) Prime editing efficiently generates W542L and S621I double mutations in two ALS genes in maize. *Genome Biology* 21(1), 257.

Kannan, B., Jung, J.H., Moxley, G.W., Lee, S.M. and Altpeter, F. (2018) TALEN-mediated targeted mutagenesis of more than 100 *COMT* copies/alleles in highly polyploid sugarcane improves saccharification efficiency without compromising biomass yield. *Plant Biotechnology Journal* 16(4), 856–866.

Kim, D., Alptekin, B. and Budak, H. (2018) CRISPR/Cas9 genome editing in wheat. *Functional & Integrative Genomics* 18(1), 31–41.

Kim, Y.-G., Cha, J. and Chandrasegaran, S. (1996) Hybrid restriction enzymes: Zinc finger fusions to Fok I cleavage domain. *Proceedings of the National Academy of Sciences* 93(3), 1156–1160.

Kis, A., Hamar, É., Tholt, G., Bán, R. and Havelda, Z. (2019) Creating highly efficient resistance against wheat dwarf virus in barley by employing CRISPR/Cas9 system. *Plant Biotechnology Journal* 17(6), 1004.

Kuang, L., Shen, Q., Chen, L., Ye, L., Yan, T. *et al.* (2022) The genome and gene editing system of sea barleygrass providea novel platform for cereal domestication and stress tolerance studies. *Plant Communications* 3(5), 100333.

Kuang, Y., Li, S., Ren, B., Yan, F., Spetz, C. *et al.* (2020) Base-editing-mediated artificial evolution of *OsALS1* in planta to develop novel herbicide-tolerant rice germplasms. *Molecular Plant* 13(4), 565–572.

Lacchini, E., Kiegle, E., Castellani, M., Adam, H., Jouannic, S. *et al.* (2020) CRISPR-mediated accelerated domestication of African rice landraces. *PLoS One* 15(3), e0229782.

Lawrenson, T., Shorinola, O., Stacey, N., Li, C., Østergaard, L. *et al.* (2015) Induction of targeted, heritable mutations in barley and *Brassica olerace*a using RNA-guided Cas9 nuclease. *Genome Biology* 16, 258.

Lee, K., Zhang, Y., Kleinstiver, B.P., Guo, J.A., Aryee, M.J. *et al.* (2019) Activities and specificities of CRISPR/Cas9 and Cas12a nucleases for targeted mutagenesis in maize. *Plant Biotechnology Journal* 17(2), 362–372.

Lee, R.T.H., Ng, A.S.M. and Ingham, P.W. (2016) Ribozyme mediated gRNA generation for *in vitro* and *in vivo* CRISPR/Cas9 mutagenesis. *PloS One* 11(11), e0166020.

Lemmon, Z.H., Reem, N.T., Dalrymple, J., Soyk, S., Swartwood, K.E. *et al.* (2018) Rapid improvement of domestication traits in an orphan crop by genome editing. *Nature Plants* 4(10), 766–770.

Li, A., Jia, S., Yobi, A., Ge, Z., Sato, S.J. *et al.* (2018a) Editing of an alpha-kafirin gene family increases, digestibility and protein quality in sorghum. *Plant Physiology* 177(4), 1425–1438.

Li, C., Zong, Y., Wang, Y., Jin, S., Zhang, D. *et al.* (2018b) Expanded base editing in rice and wheat using a Cas9-adenosine deaminase fusion. *Genome Biology* 19(1), 59.

Li, J., Stoddard, T.J., Demorest, Z.L., Lavoie, P.-O., Luo, S. *et al.* (2016a) Multiplexed, targeted gene editing in *Nicotiana benthamiana* for glyco-engineering and monoclonal antibody production. *Plant Biotechnology Journal* 14(2), 533–542.

Li, J.-F., Norville, J.E., Aach, J., McCormack, M., Zhang, D. *et al.* (2013) Multiplex and homologous recombination-mediated genome editing in *Arabidopsis* and *Nicotiana benthamiana* using guide RNA and Cas9. *Nature Biotechnology* 31(8), 688–691.

Li, M., Li, X., Zhou, Z., Wu, P., Fang, M. *et al.* (2016b) Reassessment of the four yield-related genes *Gn1a*, *DEP1*, *GS3*, and *IPA1* in rice using a CRISPR/Cas9 system. *Frontiers in Plant Science* 7, 377.

Li, S., Shen, L., Hu, P., Liu, Q., Zhu, X. *et al.* (2019) Developing disease-resistant thermosensitive male sterile rice by multiplex gene editing. *Journal of Integrative Plant Biology* 61(12), 1201–1205.

Li, X., Zhou, W., Ren, Y., Tian, X., Lv, T. *et al.* (2017) High-efficiency breeding of early-maturing rice cultivars via CRISPR/Cas9-mediated genome editing. *Journal of Genetics and Genomics* 44(3), 175–178.

Liang, Z., Zhang, K., Chen, K. and Gao, C. (2014) Targeted mutagenesis in *Zea mays* using TALENs and the CRISPR/Cas system. *Journal of Genetics and Genomics* 41(2), 63–68.
Lin, C.-S., Hsu, C.-T., Yuan, Y.-H., Zheng, P.-X., Wu, F.-H. *et al.* (2022) DNA-free CRISPR-Cas9 gene editing of wild tetraploid tomato *Solanum peruvianum* using protoplast regeneration. *Plant Physiology* 188(4), 1917–1930.
Liu, X., Qin, R., Li, J., Liao, S., Shan, T. *et al.* (2020) A CRISPR-Cas9-mediated domain-specific base-editing screen enables functional assessment of accase variants in rice. *Plant Biotechnology Journal* 18(9), 1845–1847.
Lorenz, M.G. and Wackernagel, W. (1994) Bacterial gene transfer by natural genetic transformation in the environment. *Microbiological Reviews* 58(3), 563–602.
Lorenzo, C.D., Debray, K., Herwegh, D., Develtere, W., Impens, L. *et al.* (2023) BREEDIT: A multiplex genome editing strategy to improve complex quantitative traits in maize. *The Plant Cell* 35(1), 218–238.
Lowder, L.G., Zhang, D., Baltes, N.J., Paul, J.W., Tang, X. *et al.* (2015) A CRISPR/Cas9 toolbox for multiplexed plant genome editing and transcriptional regulation. *Plant Physiology* 169(2), 971–985.
Ma, N.J., Moonan, D.W. and Isaacs, F.J. (2014) Precise manipulation of bacterial chromosomes by conjugative assembly genome engineering. *Nature Protocols* 9(10), 2285–2300.
Ma, X., Zhang, Q., Zhu, Q., Liu, W., Chen, Y. *et al.* (2015) A robust CRISPR/Cas9 system for convenient, high-efficiency multiplex genome editing in monocot and dicot plants. *Molecular Plant* 8(8), 1274–1284.
Malnoy, M., Viola, R., Jung, M.-H., Koo, O.-J., Kim, S. *et al.* (2016) DNA-free genetically edited grapevine and apple protoplast using CRISPR/Cas9 ribonucleoproteins. *Frontiers in Plant Science* 7, 1904.
Marton, I., Zuker, A., Shklarman, E., Zeevi, V., Tovkach, A. *et al.* (2010) Nontransgenic genome modification in plant cells. *Plant Physiology* 154(3), 1079–1087.
Miao, C., Xiao, L., Hua, K., Zou, C., Zhao, Y. *et al.* (2018) Mutations in a subfamily of abscisic acid receptor genes promote rice growth and productivity. *Proceedings of the National Academy of Sciences* 115(23), 6058–6063.
Mubarik, M.S., Wang, X., Khan, S.H., Ahmad, A., Khan, Z. *et al.* (2021) Engineering broad-spectrum resistance to cotton leaf curl disease by CRISPR-Cas9 based multiplex editing in plants. *GM Crops & Food* 12(2), 647–658.
Nielsen, A.A. and Voigt, C.A. (2014) Multi-input CRISPR/Cas genetic circuits that interface host regulatory networks. *Molecular Systems Biology* 10(11), 763.
Oliva, R., Ji, C., Atienza-Grande, G., Huguet-Tapia, J.C., Perez-Quintero, A. *et al.* (2019) Broad-spectrum resistance to bacterial blight in rice using genome editing. *Nature Biotechnology* 37(11), 1344–1350.
Peng, A., Chen, S., Lei, T., Xu, L., He, Y. *et al.* (2017) Engineering canker-resistant plants through CRISPR/Cas9-targeted editing of the susceptibility gene *CsLOB 1* promoter in citrus. *Plant Biotechnology Journal* 15(12), 1509–1519.
Peterson, B.A., Haak, D.C., Nishimura, M.T., Teixeira, P.J.P.L., James, S.R. *et al.* (2016) Genome-wide assessment of efficiency and specificity in CRISPR/Cas9 mediated multiple site targeting in Arabidopsis. *PloS One* 11(9), e0162169.
Pyott, D.E., Sheehan, E. and Molnar, A. (2016) Engineering of CRISPR/Cas9-mediated potyvirus resistance in transgene-free *Arabidopsis* plants. *Molecular Plant Pathology* 17(8), 1276–1288.
Qi, L., Haurwitz, R.E., Shao, W., Doudna, J.A. and Arkin, A.P. (2012) RNA processing enables predictable programming of gene expression. *Nature Biotechnology* 30(10), 1002–1006.
Ran, F.A., Hsu, P.D., Lin, C.-Y., Gootenberg, J.S., Konermann, S. *et al.* (2013) Double nicking by RNA-guided CRISPR Cas9 for enhanced genome editing specificity. *Cell* 154(6), 1380–1389.
Rovner, A.J., Haimovich, A.D., Katz, S.R., Li, Z., Grome, M.W. *et al.* (2015) Recoded organisms engineered to depend on synthetic amino acids. *Nature* 518(7537), 89–93.
Sánchez-León, S., Gil-Humanes, J., Ozuna, C.V., Giménez, M.J., Sousa, C. *et al.* (2018) Low-gluten, nontransgenic wheat engineered with CRISPR/Cas9. *Plant Biotechnology Journal* 16(4), 902–910.
Sandoval, N.R., Kim, J.Y.H., Glebes, T.Y., Reeder, P.J., Aucoin, H.R. *et al.* (2012) Strategy for directing combinatorial genome engineering in *Escherichia coli*. *Proceedings of the National Academy of Sciences* 109(26), 10540–10545.
Santillán Martínez, M.I., Bracuto, V., Koseoglou, E., Appiano, M., Jacobsen, E. *et al.* (2020) CRISPR/Cas9-targeted mutagenesis of the tomato susceptibility gene *PMR4* for resistance against powdery mildew. *BMC Plant Biology* 20(1), 1–13.

Santosh Kumar, V.V., Verma, R.K., Yadav, S.K., Yadav, P., Watts, A. *et al.* (2020) CRISPR-Cas9 mediated genome editing of drought and salt tolerance (*OsDST*) gene in indica mega rice cultivar MTU1010. *Physiology and Molecular Biology of Plants* 26, 1099–1110.

Shen, L., Wang, C., Fu, Y., Wang, J., Liu, Q. *et al.* (2018) QTL editing confers opposing yield performance in different rice varieties. *Journal of Integrative Plant Biology* 60(2), 89–93.

Shimatani, Z., Kashojiya, S., Takayama, M., Terada, R., Arazoe, T. *et al.* (2017) Targeted base editing in rice and tomato using a CRISPR-Cas9 cytidine deaminase fusion. *Nature Biotechnology* 35(5), 441–443.

Songmei, L., Jie, J., Yang, L., Jun, M., Shouling, X. *et al.* (2019) Characterization and evaluation of *OsLCT1* and *OsNramp5* mutants generated through CRISPR/Cas9-mediated mutagenesis for breeding low Cd rice. *Rice Science* 26(2), 88–97.

Swarts, D.C. and Jinek, M. (2018) Cas9 versus Cas12a/Cpf1: Structure-function comparisons and implications for genome editing. *Wiley Interdisciplinary Reviews* 9(5), e1481.

Tang, L., Mao, B., Li, Y., Lv, Q., Zhang, L. *et al.* (2017) Knockout of *OsNramp5* using the CRISPR/Cas9 system produces low Cd-accumulating indica rice without compromising yield. *Scientific Reports* 7(1), 14438.

Tang, X., Liu, G., Zhou, J., Ren, Q., You, Q. *et al.* (2018) A large-scale whole-genome sequencing analysis reveals highly specific genome editing by both Cas9 and Cpf1 (Cas12a) nucleases in rice. *Genome Biology* 19(1), 1–13.

Tao, H., Shi, X., He, F., Wang, D., Xiao, N. *et al.* (2021) Engineering broad-spectrum disease-resistant rice by editing multiple susceptibility genes. *Journal of Integrative Plant Biology* 63(9), 1639–1648.

Tashkandi, M., Ali, Z., Aljedaani, F.R., Shami, A. and Mahfouz, M.M. (2018) Engineering resistance against Tomato yellow leaf curl virus via the CRISPR/Cas9 system in tomato. *Plant Signaling & Behavior* 13, e1525996.

Tsai, S.Q., Wyvekens, N., Khayter, C., Foden, J.A., Thapar, V. *et al.* (2014) Dimeric CRISPR RNA-guided FokI nucleases for highly specific genome editing. *Nature Biotechnology* 32, 569–576.

Wang, H.H., Isaacs, F.J., Carr, P.A., Sun, Z.Z., Xu, G. *et al.* (2009) Programming cells by multiplex genome engineering and accelerated evolution. *Nature* 460(7257), 894–898.

Wang, H.H., Kim, H., Cong, L., Jeong, J., Bang, D. *et al.* (2012) Genome-scale promoter engineering by coselection MAGE. *Nature Methods* 9(6), 591–593.

Wang, W., Simmonds, J., Pan, Q., Davidson, D., He, F. *et al.* (2018) Gene editing and mutagenesis reveal inter-cultivar differences and additivity in the contribution of *TaGW2* homoeologues to grain size and weight in wheat. *Theoretical and Applied Genetics* 131(11), 2463–2475.

Wang, Y., Cheng, X., Shan, Q., Zhang, Y., Liu, J. *et al.* (2014) Simultaneous editing of three homoeoalleles in hexaploid bread wheat confers heritable resistance to powdery mildew. *Nature Biotechnology* 32(9), 947–951.

Wannier, T.M., Ciaccia, P.N., Ellington, A.D., Filsinger, G.T., Isaacs, F.J. *et al.* (2021) Recombineering and MAGE. *Nature Reviews Methods Primers* 1(1), 7.

Wing, R.A., Purugganan, M.D. and Zhang, Q. (2018) The rice genome revolution: From an ancient grain to green super rice. *Nature Reviews Genetics* 19(8), 505–517.

Xie, K., Minkenberg, B. and Yang, Y. (2015) Boosting CRISPR/Cas9 multiplex editing capability with the endogenous tRNA-processing system. *Proceedings of the National Academy of Sciences* 112(11), 3570–3575.

Xing, H.-L., Dong, L., Wang, Z.-P., Zhang, H.-Y., Han, C.-Y. *et al.* (2014) A CRISPR/Cas9 toolkit for multiplex genome editing in plants. *BMC Plant Biology* 14(1), 1–12.

Xu, L., Zhao, L., Gao, Y., Xu, J. and Han, R. (2017) Empower multiplex cell and tissue-specific CRISPR-mediated gene manipulation with self-cleaving ribozymes and tRNA. *Nucleic Acids Research* 45(5), e28–e28.

Xu, R., Yang, Y., Qin, R., Li, H., Qiu, C. *et al.* (2016) Rapid improvement of grain weight via highly efficient CRISPR/Cas9-mediated multiplex genome editing in rice. *Journal of Genetics and Genomics* 43(8), 529–532.

Xu, W., Song, W., Yang, Y., Wu, Y., Lv, X. *et al.* (2019) Multiplex nucleotide editing by high-fidelity Cas9 variants with improved efficiency in rice. *BMC Plant Biology* 19(1), 511.

Yan, Q., Xu, K., Xing, J., Zhang, T., Wang, X. *et al.* (2016) Multiplex CRISPR/Cas9-based genome engineering enhanced by Drosha-mediated sgRNA-shRNA structure. *Scientific Reports* 6(1), 1–9.

Yang, T., Ali, M., Lin, L., Li, P., He, H. *et al.* (2023) Recoloring tomato fruit by CRISPR/Cas9-mediated multiplex gene editing. *Horticulture Research* 10(1), uhac214.

Yin, K., Han, T., Xie, K., Zhao, J., Song, J. *et al.* (2019) Engineer complete resistance to cotton leaf curl multan virus by the CRISPR/Cas9 system in *Nicotiana benthamiana*. *Phytopathology Research* 1(1), 1–9.

Yu, H., Lin, T., Meng, X., Du, H., Zhang, J. *et al.* (2021) A route to de novo domestication of wild allotetraploid rice. *Cell* 184(5), 1156–1170.

Zafar, K., Khan, M.Z., Amin, I., Mukhtar, Z., Yasmin, S. *et al.* (2020) Precise CRISPR-Cas9 mediated genome editing in super basmati rice for resistance against bacterial blight by targeting the major susceptibility gene. *Frontiers in Plant Science* 11, 575.

Zeng, Y., Wen, J., Zhao, W., Wang, Q. and Huang, W. (2019) Rational improvement of rice yield and cold tolerance by editing the three genes *OsPIN5b*, *GS3*, and *OsMYB30* with the CRISPR–Cas9 system. *Frontiers in Plant Science* 10, 1663.

Zetsche, B., Gootenberg, J.S., Abudayyeh, O.O., Slaymaker, I.M., Makarova, K.S. *et al.* (2015) Cpf1 is a single RNA-guided endonuclease of a class 2 CRISPR-Cas system. *Cell* 163(3), 759–771.

Zhang, W.-W. and Matlashewski, G. (2015) CRISPR-Cas9-mediated genome editing in *Leishmania donovani*. *mBio* 6(4), e00861–00815.

Zhang, F., Maeder, M.L., Unger-Wallace, E., Hoshaw, J.P., Reyon, D. *et al.* (2010) High frequency targeted mutagenesis in *Arabidopsis thaliana* using zinc finger nucleases. *Proceedings of the National Academy of Sciences* 107(26), 12028–12033.

Zhang, T., Zheng, Q., Yi, X., An, H., Zhao, Y. *et al.* (2018) Establishing RNA virus resistance in plants by harnessing CRISPR immune system. *Plant Biotechnology Journal* 16(8), 1415–1423.

Zhang, Y., Bai, Y., Wu, G., Zou, S., Chen, Y. *et al.* (2017) Simultaneous modification of three homoeologs of *TaEDR1* by genome editing enhances powdery mildew resistance in wheat. *The Plant Journal* 91(4), 714–724.

Zhang, Z., Mao, Y., Ha, S., Liu, W., Botella, J.R. *et al.* (2016) A multiplex CRISPR/Cas9 platform for fast and efficient editing of multiple genes in Arabidopsis. *Plant Cell Reports* 35, 1519–1533.

Zhou, J., Xin, X., He, Y., Chen, H., Li, Q. *et al.* (2019) Multiplex QTL editing of grain-related genes improves yield in elite rice varieties. *Plant Cell Reports* 38(4), 475–485.

7 Delivering Genome Editing Reagents Into Plants: *Agrobacterium*- and Non-*Agrobacterium*-mediated Approaches

Mallesham Bulle[1], Kaushik Ghose[1], Hemendra Shinde[2], Altafhusain Nadaf[3], Rupesh Deshmukh[4] and Gunvant B. Patil[1]*

[1]Institute of Genomics for Crop Abiotic Stress Tolerance (IGCAST), Department of Plant and Soil Science, Texas Tech University, Lubbock, USA; [2]K.V.N. Naik Arts, Commerce and Science College, Nashik, India; [3]Department of Botany, Savitribai Phule Pune University, Pune, India; [4]Department of Biotechnology, Central University of Haryana, Mahendragarh, India

Abstract

Plant genome editing technology (CRISPR/Cas) has paved the path for the generation of transgene-free plant genomes, which has tremendously revolutionized plant breeding and crop improvement. Recent advances in this technology include the creation of more potent Cas proteins for a larger variety of targets, improved delivery, enhanced effectiveness, and greater accuracy in DNA alteration with the advent of base editing and prime editing technologies. However, frequent polyploidy and genome rearrangements, as well as rigid cell walls, make it challenging to effectively introduce genome editing reagents into plant cells. Therefore, a more thorough examination of the delivery of CRISPR components into plant cells is required. This chapter focuses on recent developments in different delivery strategies, such as the use of *Agrobacterium* as a cargo material for genome editing and non-*Agrobacterium*-mediated delivery systems constructed from a range of biological, chemical, and physical materials, such as viruses, inorganic nanoparticles, carbon nanotubes, particle bombardment techniques, and polyethylene glycol (PEG)-mediated delivery in a variety of explants across different plant species.

Keywords: *Agrobacterium*, CRISPR, Biolistic, DNA-free delivery, Ribonucleoprotein complexes, Protoplasts

7.1 Introduction

Genome editing technology presents vast opportunities for enhancing agriculture, including improving consumer health, boosting productivity, and addressing the escalating food security crisis (Vats *et al.*, 2019). The development and implementation of site-directed nucleases (SDNs), especially CRISPR/Cas9, have progressed rapidly and excitingly. Additionally, numerous alternative approaches have been concurrently developed for delivering CRISPR/Cas9 components into plant cells, ranging from *Agrobacterium* infection to nanoparticles (Chen *et al.*, 2019). However, despite these advancements in genome editing and delivery technologies, our ability to efficiently produce edited events is still hindered by the lack of robust

*Corresponding author: gunvant.patil@ttu.edu

 Genome Editing for Crop Improvement: Theory and Methodology (eds J.M. Al-Khayri *et al.*)
DOI: 10.1079/9781800622517.0007

methods for regenerating fertile plants in many crops (Altpeter *et al.*, 2016). While the recovery of transgenic plants, even at low frequency, has historically been considered a major success, genome editing applications that involve template-based modifications via homology-directed repair (HDR) often necessitate the production of hundreds of regenerated events to obtain plants with the desired edit(s). This limitation in regeneration efficiency impedes in realizing the full adoption of CRISPR/Cas9-mediated genome editing in crop plants.

The challenge of evolving efficient regeneration systems has a significant impact on various experimental options for developing a successful genome editing protocol, ultimately depending on how amenable a species is to *in vitro* manipulation (Ojha *et al.*, 2023). In light of this prerequisite, species that can be easily cultured, such as *Arabidopsis thaliana* and *Nicotiana benthamiana*, have become models for demonstrating multiple genome editing applications, using delivery to a variety of plant explants ranging from protoplasts (Cigliano *et al.*, 2013) to *in planta* (Maher *et al.*, 2020). Conversely, many important crop species (e.g. cereals, legumes, and woody perennials, among others) are difficult to manipulate *in vitro*, thus limiting the opportunities for successful application of genome editing technologies. Recalcitrant species, which typically exhibit consistently low-regeneration frequencies, are usually suitable for simple mutagenesis-based knockouts, but not for more complex (and low frequency) template-based edits. While recalcitrance for a given species can sometimes be overcome by perseverance and patience by exploring tissue culture conditions, new approaches such as the use of morphogenic genes—for example, *WUSCHEL* (*WUS*), *SHOOT MERISTEMLESS* (*STM*), and *BABY BOOM* (*BBM*)—can be an important adjunct, aiding in both accelerating progress in certain species and expanding the range of accessible plant varieties (Gordon-Kamm *et al.*, 2019; Nalapalli *et al.*, 2021; Maren *et al.*, 2022; Lee and Wang, 2023). Careful consideration must be given to selecting the optimal combination of the CRISPR/Cas delivery method and tissue culture strategy for ensuring success in any given plant species. This chapter discusses recent developments of different methods that have been deployed for genome editing reagent delivery into plant cells.

7.2 Delivery of Genome Editing Reagents Into Plant Cells

Recent reagent delivery approaches have facilitated the efficient delivery of DNA-free editing (ribonucleoproteins (RNPs)) and CRISPR/Cas-mediated gene-editing reagents into mammalian systems and plants (Graham *et al.*, 2020). The delivery can be accomplished by microinjection, transfection, viral delivery vehicles (such as adeno-associated viruses and lentiviruses), and non-viral delivery vehicles (such as nanoparticles), lipofection, and injection of cell-penetrating peptides into animal cell lines (Chavez *et al.*, 2023). However, the delivery of CRISPR reagents into plant cells has been limited to plasmid vectors delivered by *Agrobacterium* or via particle bombardment (Ghogare *et al.*, 2021). Delivery of RNPs into plant systems through particle gun or polyethylene glycol (PEG)-mediated transfection is gaining interest. However, the effectiveness of distribution remains largely reliant on the kind of tissue used (Ghogare *et al.*, 2021).

Competence and regeneration potential are essential factors to consider when choosing an explant for genome editing. Single cells (such as protoplasts or microspores), embryonic tissues (such as meristem, scutellum, or cotyledons), seedling-derived tissues (such as apical/axillary meristems, hypocotyl, or leaves), and *in planta* alternatives are transformable explant material options. In general, these procedures advance from those requiring prolonged *in vitro* culture (e.g. protoplast, embryogenic/organogenesis callus) to those that capitalize on more rapid plant growth (such as direct somatic embryo or meristem/shoot creation) to those that are conducted totally *in planta*. Following this trend is a concurrent decrease in the time needed from the introduction of DNA, RNA, and/or protein to the recovery of a fertile genome-edited plant, circumstances that have long been recognized as desirable in order to minimize exposure to tissue culture and reduce somaclonal variations (Bairu *et al.*, 2011). By using direct and indirect methods, it has been established effectively that CRISPR reagents may be delivered into plant cells both *in vivo* and *in vitro*. Delivery vectors, such as viral and non-viral vectors, can support mRNA or plasmids by inducing nuclease

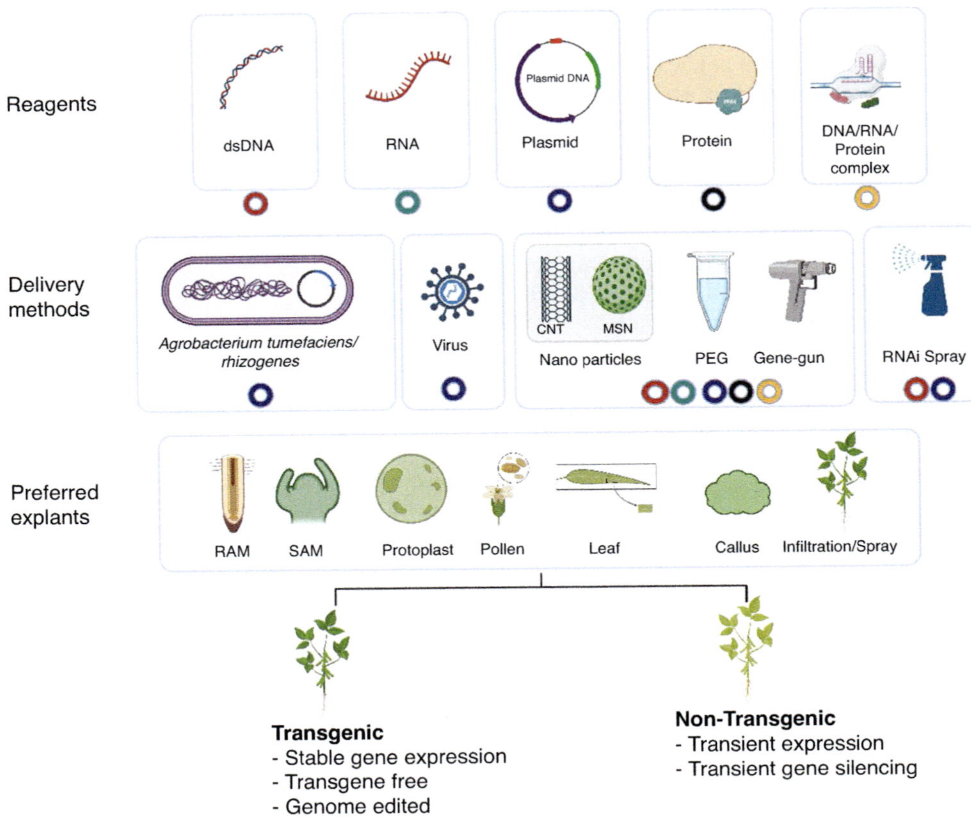

Fig. 7.1. A graphical summary of the various delivery cargos, delivery methods, and explants used in developing gene-edited plants. *Agrobacterium*- and virus-mediated transformation depends on the DNA plasmid construct, while non-*Agrobacterium*-mediated methods can deliver a variety of reagents including double-stranded DNA (dsDNA), RNA, protein, and DNA/RNA/protein complexes. PEG—Polyethylene glycol; RAM—root apical meristem; SAM—shoot apical meristem. Author's own figure.

expression in target cells. Alternatively, CRISPR reagents may also be transferred into plant cells by the use of physical methods, such as electroporation, biolistic delivery, and PEG (Fig. 7.1). In this chapter, we have summarized some of the recent developments of CRISPR reagent delivery in economically important crops.

7.2.1 *Agrobacterium*-mediated genome editing reagent delivery system

Agrobacterium-mediated delivery is a widely used technique in plant genetic transformation for introducing foreign DNA into genomes. It is based on the natural ability of *Agrobacterium tumefaciens*, a soil bacterium, to transfer a specific portion of its DNA (T-DNA, flanked with left and right borders) with the help of helper plasmids (virulence (*Vir*) genes) into host plant cells (Gelvin, 2017). The process begins with the creation of a recombinant DNA molecule that contains the desired gene of interest, which is usually inserted into a plant transformation vector. Following the contact with wounded sites in plant tissues, *Agrobacterium* secretes chemical signals to activate virulence proteins that promote the transfer of T-DNA into host plant cells (Gelvin, 2017). The T-DNA integrates into the plant genome and becomes stably inherited in subsequent generations. Once inside plant cells, the T-DNA can be expressed, leading to the production of the desired protein or the modification (via gene-editing reagents) of the plant's traits. *Agrobacterium*-mediated transformation offers several advantages over other methods, including its ability to transfer large DNA fragments and its relatively high efficiency, making it a valuable tool in plant biotechnology, enabling researchers to engineer plants for various applications in sustainable agriculture (Hwang *et al.*, 2017; Gheysen *et al.*, 2022).

7.2.2 Mechanism of T-DNA entry into the plant cell nucleus

Transformation mediated by *Agrobacterium* consists of multiple phases. All virulent *Agrobacterium* strains harbor tumor-inducing (Ti) or root-inducing (Ri) plasmids, such as *Agrobacterium tumefaciens*, *Agrobacterium vitis*, and *Agrobacterium rubi*; or *Agrobacterium rhizogenes*. Ti/Ri plasmids encode virulence (*vir*) genes that are activated by phenolics secreted by plants (Stachel *et al.*, 1985). Induction of the *vir* gene may occur before or concurrently with bacterial attachment to plant cells. T-DNA is processed from the Ti/Ri plasmid by the coordinated action of *VirD*1 and *VirD*2 after induction (De Vos and Zambryski, 1989). Transfer of a single-strand form of T-DNA (T-strand) and Virulence (Vir) effector proteins involves a series of steps: transfer of T-strands and multiple Vir effector proteins to the plant via an induced Type IV secretion system (T4SS) (Gelvin, 2017), trafficking of T-strands and Vir protein/plant protein complexes through the plant cytoplasm to the nucleus. In the 1980s, the method of *Agrobacterium*-mediated transformation was initially described by Bevan (Bevan *et al.*, 1983). Subsequently, Hamilton (Hamilton *et al.*, 1996) demonstrated the transfer of a large DNA fragment up to 150 kb to a plant cell using a binary BAC system.

7.2.3 Stable expression

Agrobacterium-mediated delivery is a widely used and well-established method for delivering genome editing reagents in various plant species, including crops like cotton, tomato, rice, maize, wheat, and soybean (Ceasar and Ignacimuthu, 2023). This approach utilizes the inherent capability of the soil bacterium *A. tumefaciens* to transfer DNA into plant cells. The process entails employing a vector, such as a binary plasmid, which contains the genome editing reagents, including the Cas nuclease and guide RNA integrated into its T-DNA. The vector is then introduced into the *Agrobacterium*, which is then used to infect the plant cells, and gets integrated into the plant genome, leading to stable genetic transformation (Zhang *et al.*, 2021). *Agrobacterium*-mediated delivery has also been shown to have a relatively high transformation efficiency, making it an attractive choice for large-scale applications. For example, in rice, multiple genes (*OsROC5*, *OsSPP*, *OsYSA*, and *Os ISA1*) were targeted using CRISPR/Cas9 to demonstrate the editing efficiency of CRISPR/Cas9 (Feng *et al.*, 2013; Shufen *et al.*, 2019). The editing efficiency of *GmIPK1* and *GmIPK2* was evaluated using two different CRISPR/Cas9 systems, a single transcriptional unit (STU) and a two-component transcriptional unit (TCTU), in the soybean genome (Carrijo *et al.*, 2021). The *Agrobacterium*-mediated genome editing system has been established in tomato and cotton for functional genetics and trait improvement (Li *et al.*, 2018, 2023; Bari *et al.*, 2019; Xia *et al.*, 2021; Tiwari *et al.*, 2023). To date, more than 25 plant species have been efficiently edited with the help of *Agrobacterium*-mediated delivery of genome editing. We have summarized the results of a few commercial crops in Table 7.1.

7.2.4 *In planta* transformation methods

Previously, diverse techniques have been utilized for the introduction of plasmids into plants. One approach involved direct application of plasmids onto the stigma surface or their combination with pollen prior to their introduction into the receptive stigma. To accomplish effective gene transfer, various parameters such as the dipping and wounding of flowers containing both male and female organs into *Agrobacterium* suspensions have been employed. Successful floral transition

Table 7.1. Different methods of CRISPR/Cas reagent delivery into major crops. Author's own table.

Crop	Explant	Target trait	References
Agrobacterium-mediated delivery of CRISPR/Cas reagents in different crops			
Soybean	Hairy roots	KO *GFP*	Jacobs *et al*., 2015
	Hairy roots	Phytic acid	Carrijo *et al*., 2021
	Half seeds	Carbohydrate content	Virdi *et al*., 2020
Rice	Callus	Starch content	Shufen *et al*., 2019
	Callus	Shoot growth	Feng *et al*., 2013
Tomato	Hypocotyls	Lycopene content	Li *et al*., 2018
	Cotyledons	Weed resistance	Bari *et al*., 2019
Cotton	Hypocotyls	Fiber	Li *et al*., 2019
Maize	Immature embryos	Anthocyanin	Char *et al*., 2017
Wheat	Immature embryos	Shoot growth	Upadhyay *et al*., 2013
Non-*Agrobacterium*-mediated delivery of CRISPR/Cas reagents in different crops			
Particle bombardment method for CRISPR/Cas reagent delivery			
Soybean	Callus	Trichome development	Campbell *et al*., 2019
Rice	Callus	Carotenoid content	Dong *et al*., 2020
	Callus	Albino phenotype	Banakar *et al*., 2020
	Callus	Albino and dwarf phenotype	Shan *et al*., 2013
	Callus	Altered growth phenotypes	Shan *et al*., 2014
Maize	Immature embryos	Herbicide resistant	Svitashev *et al*., 2015
PEG-mediated delivery of CRISPR/Cas components in different crops			
Soybean	Protoplasts	Testing gene-editing efficiency	Patil *et al*., 2022
	Protoplasts	Fatty acid biosynthesis	Kim *et al*., 2017
Rice	Protoplasts	Disease resistance	Xie and Yang, 2013
	Protoplasts	Bacterial blight resistance	Jiang *et al*., 2013
	Protoplasts	Altered growth	Shan *et al*., 2013
Tomato	Protoplasts	Flowering time	Liu *et al*., 2022
	Protoplasts	Virus resistance	Lin *et al*., 2022
Cotton	Protoplasts	Altered growth phenotypes	Chen *et al*., 2017
Maize	Protoplasts	Testing gene-editing efficiency	Liang *et al*., 2014
	Protoplasts	Herbicide resistance	Li *et al*., 2020
Plant viral vectors utilized for the delivery of CRISPR/Cas reagents in different crops			
Crop	**Explant/Virus type**	**Target trait/Target gene**	**References**
Wheat	Leaf	Testing editing efficiency	Hu *et al*., 2019
	Protoplasts	Testing editing efficiency and gene targeting	Gil-Humanes *et al*., 2017
Rice	Callus/Protoplast	Testing editing efficiency	Wang *et al*., 2017
	Callus	Plant height	Tian *et al*., 2023
	Callus	Carotenoid content	Kim *et al*., 2022

Continued

Table 7.1. Continued

Crop	Explant/Virus type	Target trait/Target gene	References
Tomato	Cotyledon	Fruit phenotype	Dahan-Meir *et al.*, 2018
Maize	Seedlings	Testing editing efficiency	Mei *et al.*, 2019
	Leaf	Testing editing efficiency	Hu *et al.*, 2019
Soybean	Hairy roots	Testing editing efficiency	Luo *et al.*, 2021
Tobacco	Tobacco rattle virus (TRV)	*NbPDS3*	Ali *et al.*, 2015
		NbPDS3, NbPCNA	Ali *et al.*, 2015
		NbPDS, NbAG	Ellison *et al.*, 2020
Petunia	TRV	Reporter genes *DsRed2, EGFP*, and *uidA*	Marton *et al.*, 2010
Tobacco	Sonchus yellow net rhabdovirus (SYNV)	*GFP*, *NbPDS*, *NbRDR6*, and *NbSGS3*	Ma *et al.*, 2020
Tobacco	Potato virus X (PVX)	*NbXT2B, NbPDS,* and *NbFT*	Uranga *et al.*, 2021
		NbTOM1	Ariga *et al.*, 2020
Wheat	Barley stripe mosaic virus (BSMV)	*TaGW2, TaUPL3, TaGW7,* and *TaQ*	Wang *et al.*, 2022
		TaHRC	Chen *et al.*, 2022
		TaPDS, TaGW2, TaGASR7	Li *et al.*, 2021
Maize		*GASR7, TMS5*	Hu *et al.*, 2019
Tobacco	Pea early browning virus (PEBV)	*NbPDS*	Ali *et al.*, 2018
Soybean	Apple latent spherical virus (ALSV)	*GmGW2*	Luo *et al.*, 2021

during this course is reliant on the developmental stage of plants (Clough, 2004; Bäurle and Dean, 2006). In addition to physical characteristics such as pH, medium composition, optical density, and temperature, major roles are played by molecular variables such as gene size, promoter preference, and vector types on achieving success in transformation (Alghuthaymi *et al.*, 2021). Constitutive promoters like *CaMV35S* and *AtUBI10* have been extensively employed to enhance editing performance. UBI10, particularly, has exhibited effectiveness in the germline of *Arabidopsis* (Verma *et al.*, 2023). This technique has several advantages. For instance, it does not require a plant-tissue culture unit, making it cost-effective and straightforward. Therefore, this method is widely employed worldwide for genome editing in model crops like *Arabidopsis*. Nevertheless, it is at present limited in its application to only a few restricted plants such as *Linum usitatissimum*, *Arabidopsis*, and *Solanum lycopersicum*. Moreover, its productivity is restricted due to limited flowering and seed production in these plants. In the field of plant transformation, strategies that harness the natural processes of growth and reproduction with minimal system disruption are regarded as critical. A recent *in planta* research targeted the axillary meristem using *Agrobacterium*-mediated delivery to demonstrate the feasibility of genome editing (Maher *et al.*, 2020; Cao *et al.*, 2023). To avoid the integration of foreign DNA, the transient expression of genome editing reagents, achieved through events created without selection, has facilitated gene editing that is free from transgenes (Yang *et al.*, 2023). This approach holds significant importance in the development of transgene-free edited plants, especially in cases where the removal of integrated transgenes through conventional breeding methods, such as crosses, is not feasible (Zhang *et al.*, 2016).

7.3 Non-*Agrobacterium*-mediated Genome Editing Reagent Delivery

Non-*Agrobacterium*-mediated delivery, on the other hand, involves the direct delivery of genome editing reagents into plant cells using methods such as electroporation, particle bombardment, PEG, and sonication. These methods do not rely on the natural ability of *Agrobacterium* to transfer DNA into plant cells. Electroporation involves the use of an electric field to create temporary pores in the plant cell membrane, allowing genome editing reagents to enter. Particle bombardment, also known as biolistics, employs high-velocity microprojectiles to deliver genome editing reagents coated on to particles such as gold and tungsten directly into the plant cell. Ultrasonic techniques create strong shear pressures, which disrupt cell walls and facilitate the transport of genome editing reagents from the surrounding solvent to the inside of plant cells. PEG is the most often utilized chemical for protoplast transformation (Kofer *et al.*, 1998). Numerous plant species, including maize, wheat, and cotton, have been successfully transformed by delivering genome editing reagents via non-*Agrobacterium*-mediated delivery techniques. However, the transformation efficiency of these approaches is lower than that of *Agrobacterium*-mediated delivery, which might pose a restriction for large-scale applications.

7.3.1 Biolistic transformation

Biolistic-mediated gene delivery into plant cells is one of the greatest agricultural inventions of the 20th century, and it is one of the most frequently used non-*Agrobacterium*-based gene delivery methods for plant transformation (Rustgi *et al.*, 2020). The main advantage of using this strategy is to circumvent the genotype and host gene interaction dependence of *Agrobacterium*-mediated gene delivery, since it is based on physical interaction rather than biological interaction for gene delivery. The technique can be used to even deliver very large transgenes, such as whole bacterial artificial chromosomes, in either linear or circular form (Rustgi *et al.*, 2022). The biolistic technique depends on the capacity of heavy-metal carriers (such as tungsten or gold) coated with cargo biomolecules to enter plant cells when accelerated by high-pressure helium in a partial vacuum. Once inside the cell, the cargo may either integrate into the genome (stable expression) or remain extrachromosomal (transient expression). Technological innovation led to the discovery of the Hepta™ adaptor for the PDS-1000/He biolistic system, which can bombard the DNA-coated particles more uniformly over a large population of cells (https://www.bio-rad.com/). This technology also minimizes the penetration force causing minimum cellular damage to explants. As part of more advanced ongoing research projects, scientists have already achieved substantial increase in transformation efficiency by replacing the conventional spermidine and calcium chloride with PEG and magnesium salt solutions (Ismagul *et al.*, 2018; Cunningham *et al.*, 2020). Gene delivery using the biolistic method gains more attention in the era of genome editing with the need to deliver genome editing cargos, that is guide RNAs complexed with Cas9 protein in a faster and genotype-independent manner (Bulle *et al.*, 2024b). Delivery of the RNP complex *via* particle bombardment can be achieved by mixing the same concentrations of gRNA and Cas9 with gold particles to generate the RNP complex, which can be subsequently delivered into the plant tissue in a high-pressure environment. This transformation mechanism can alter not just the nucleus but also the plastid and mitochondria (Sanford, 2000; Bulle *et al.*, 2023, 2024a). In addition, the biolistic method may be applied to several cell types and provides many biomolecules (Altpeter *et al.*, 2005). Multiple genome editing components may be co-delivered on distinct plasmids, which simplifies vector design and assembly; this is an additional benefit of this delivery approach. The outcome of biolistic studies relies on several variables, including particle acceleration, selection of target tissues, microparticle size, and DNA-to-particle ratio. The limitations of this technology include limited control over the distribution and penetration of microparticles, as well as the possibility of cell toxicity. In addition, chromosomal damage and complex multicopy vector DNA integration reduce the proportion of effective genome-edited events (Svitashev *et al.*, 2000, 2002; Liu *et al.*, 2019). Numerous species, including maize (Ainley *et al.*, 2013; Svitashev *et al.*, 2015, 2016; Shi *et al.*, 2017), wheat (Liang *et al.*, 2017; Hamada *et al.*, 2018), soybean (Jacobs *et al.*, 2015; Li *et al.*, 2015), chilli pepper (Bulle *et al.*, 2024b) and rice, have used biolistic delivery for genome modification application (Sun *et al.*, 2016). In a pioneering work done in maize, large 75.5 Mb chromosomal modification was achieved using RNP and guide RNAs flanking the junctions of the desired inversion (Schwartz *et al.*, 2020). RNP-mediated CRISPR-Cas9 transformation achieved 71% editing in *Arabidopsis* (Woo *et al.*, 2015). By using the totipotent nature of plant protoplasts, editing efficiencies of 46 and 68% were achieved in lettuce and potato microcalli regenerated from CRISPR-Cas9 RNPs-treated protoplasts (Woo *et al.*, 2015; Andersson *et al.*, 2018; González *et al.*, 2020). The RNP-transfected protoplast has also been used to generate 11.9% editing efficacy in petunia (Yu *et al.*, 2021) and 20% editing efficacy in cabbage (Park *et al.*, 2019). But the major drawbacks in RNP-mediated protoplast transformation are regeneration (Altpeter *et al.*, 2005, 2016; Eeckhaut *et al.*, 2013) and somaclonal variation leading to unintended mutations and morphological defects (Kawata and Oono, 1998; Tang *et al.*, 2018; Fossi *et al.*, 2019). One way scientists have tried to overcome this regeneration problem is introducing CRISPR-Cas9 RNPs using PEG, as shown by Toda *et al.* (2019) with 14–64% editing efficiency in rice zygotes.

7.3.2 Virus-mediated gene transfer in plants

Virus-mediated gene transfer in plants is widely used to introduce transgenes into plant cells; however, this method is mostly confined to transient gene expression which implies that transgenes are not integrated into the host genomes (Gil-Humanes *et al.*, 2017). Plant viruses have evolved mechanisms to infect and replicate within plant cells, and scientists have harnessed these properties to create efficient gene delivery systems (Mahmood *et al.*, 2023). The process typically involves modifying the viral genome by replacing certain viral genes with the gene of interest. The modified virus, known as a viral vector, is then introduced into plant cells through either mechanical inoculation or agro-infection (using *Agrobacterium* to deliver the viral vector) or biolistic methods (using gene guns to physically deliver the viral vector) (Mahmood *et al.*, 2023). Once inside the plant cells, the viral vector utilizes the natural infection pathways of the virus to deliver the foreign gene into the host plant genome. The viral vector typically contains regulatory elements that allow the expression of the introduced gene, enabling the production of the desired protein or the modification of the plant's traits (Oh *et al.*, 2021).

This viral system has now found application in genome editing of plants. By utilizing virus vectors, it becomes possible to introduce Cas proteins and guide RNAs, which are crucial components of the CRISPR gene-editing system, into plant cells without the need for complex experimental procedures. Remarkably, certain plant viruses have the ability to target meristematic cells, where they can express the gene-editing components, ultimately leading to the generation of mutated seeds via virus-induced genome editing (VIGE) technology in plants (Zhang *et al.*, 2022). In a recent breakthrough, (Ellison *et al.*, 2020) developed *in planta* gene-editing approach using RNA virus that expresses single-guide RNAs (sgRNAs) that augmented with sequences that promote cell-to-cell mobility. Following this, several methods have been adopted and improved for Virus-Induced Gene Silencing (VIGS) as listed in Table 7.1. VIGE has emerged as a powerful method with several advantages, including high editing efficiency and a simplified process for generating gene-edited DNA-free plants. It has been shown that tobacco rattle virus-mediated VIGE can be used in tobacco with high editing efficiency (90–100%) by simultaneous expression of sgRNAs with mobile RNA sequences. They have reported more than 65% of heritable editing (Ellison *et al.*, 2020). TRV-mediated VIGE has been used to edit PDS (Ellison *et al.*, 2020), and *PCNA* gene in tobacco (Ali *et al.*, 2015). It has also been used to edit *Arabidopsis* GL1 and TT4 genes (Ali *et al.*, 2018). Potato virus X-mediated VIGE has been successfully used in tobacco to edit *PDS*, *FT* genes (Uranga *et al.*, 2021), and *TOM1* gene (Ariga *et al.*, 2020). Barley yellow striate mosaic virus (BYSMV)-mediated VIGE has been used in tobacco to target GFP (Gao *et al.*, 2019). Sonchus yellow net rhabdovirus (SYNV) has been used to edit *GFP, NbPDS, NbRDR6,* and *NbSGS3* genes in tobacco (Ma *et al.*, 2020). Pea early browning virus (PEBV)-mediated editing has been used to edit tobacco PDS (Ali *et al.*, 2018).

7.3.3 Nanoparticles

Discovery of nanomaterial significantly advanced agricultural innovation. Nanomaterial pesticides and fertilizers are better performing than conventional application due to their physicochemical properties, and the desired nutrient can be delivered in a much more precise way which can easily enter the plant cell and they have much less cellular toxicity (Hamers, 2017; Kah *et al.*, 2018; Buriak *et al.*, 2022), although the precise entry mechanism is not fully discovered yet. Even though a lipid exchange envelope penetration (LEEP) was discovered recently for the entry of nanoparticles into the lipid bilayer, the most conventional entry mechanism is proved to be mechanically driven (Lew *et al.*, 2020) stomatal or cuticular uptake and that makes the size of the nanoparticle the most important determining factor to be used for gene delivery (Zhi *et al.*, 2022). In a recent study (Yong *et al.*, 2021), this group determined that 50 nm should be the maximum cut off value for nanomaterial to be delivered to plants. Other nanoparticle characteristics such as shape and zeta potential (desired over 30 mV) should also be taken care of for successful gene delivery (Demirer *et al.*, 2019). Use of nanomaterials for foreign gene delivery for stable and transient expression as well as successful delivery of CRISPR/Cas-based genome editing machinery into plant cells has been documented in multiple recent studies (Sharma *et al.*, 2021; Zhi *et al.*, 2022). Genes delivered into plant cells using nanoparticles have a higher self-life which is very crucial for the delivery of the RNA component (Demirer *et al.*, 2019, 2021; Wang *et al.*, 2019; Shidore *et al.*, 2021). The single-walled carbon nanotubes (SWCNTs) based on nanoparticles have been widely used to deliver electrostatically attached DNA constructs into plant cells (Demirer *et al.*, 2019). In a recent study, researchers developed a cationic and fluorescent rosette nanotube (RNT) that is permeable to wheat microspores without any external force, thus causing minimal damage to the cell (Cho *et al.*, 2020). One of the major limitations of nanoparticle-mediated delivery is their inability to move the construct systematically (Rustgi *et al.*, 2022). In summary, nanoparticles offer significant potential in advancing gene delivery and editing in plants. Nanoparticles can be precisely engineered for the targeted delivery of transgenes including gene-editing reagents, enhancing the efficiency of introducing the genetic material into plant cells. This targeted approach could improve transformation efficiency transiently and enhances the precision of genetic modifications in plants. However, the application of nanoparticles in gene delivery and editing faces notable challenges including the plant cell wall acting as a barrier, limiting efficient penetration, and achieving specific targeting (Demirer *et al.*, 2019). Cytotoxicity of certain nanoparticles may affect plant cell viability. The stability of nanoparticles, their controlled release dynamics, and potential responses triggered in plants are crucial factors that need to be overcome for the successful and responsible utilization of nanoparticles in plant genetic engineering.

7.3.4 Polyethylene glycol (PEG)-mediated delivery

Successful use of PEG and other reagents like poly-L-ornithine and polyethylenimine in transfecting genes into fungal and animal cells is well documented. The PEG, a petroleum-based polymeric compound has also been successfully used to transfect plant protoplasts (Webb and Morris, 1992; Bilang *et al.*, 1994); including *Arabidopsis* (Altmann *et al.*, 1992) as well as fungal cells (Kuwano *et al.*, 2008; Kawai *et al.*, 2010), due to its low toxicity at the cellular level. Chemical agents like PEG in the presence of divalent cations and high pH can deliver naked DNA into plant protoplasts by increasing the plasma membrane permeability (Altmann *et al.*, 1992). The simplicity and low cost of this method also makes this an alternate choice when multiple DNA constructs need to be delivered. Unlike *Agrobacterium*-mediated transformation, this method has no limitation in terms of plant species or type of explant (Bilang *et al.*, 1994). It is very well documented for transient expression and efficiency testing for genome editing (Yue *et al.*, 2021). Even though the major drawback of this technique is the regeneration of plant protoplasts into complete plants, the method of PEG-mediated protoplast transformation and successful creation of transgenic plants has been documented in multiple crops like soybean (Patil *et al.*, 2022; Nwoko *et al.*, 2023), rice (Lin *et al.*, 2020; Bes *et al.*, 2021), wheat (Brandt *et al.*, 2020), maize (Svitashev *et al.*, 2016), potato (Carlsen *et al.*, 2022), and tomato (Liu *et al.*, 2022). With the advent of genome editing technology, protoplast transfection using PEG has been frequently used in many economically important crop species to deliver Cas9 protein in the form of ribonucleoprotein particles (RNPs) and guide RNA plasmids (Subburaj *et al.*, 2016; Sant'Ana *et al.*, 2020; Liu *et al.*, 2022). A recent example of development of such technologies is the Cibus Rapid Trait Development System (RTDS™) technology that can generate precise genome-edited transgene-free lines using Oligonucleotide-Directed Mutagenesis (ODM). This simple technique is accurate and does not require to deliver the nuclease enzyme-containing vector into the plant cell and is thus transgene free and its efficiency has been demonstrated in *Arabidopsis*, canola, maize, rice, tobacco, and wheat (Gocal, 2015; Gocal *et al.*, 2015). Despite these advantages, more research needs to be done to increase the efficiency of protoplast production and regeneration of the whole plant from the protoplast to overcome the genetic bottleneck of the protoplast regeneration system (Fossi *et al.*, 2019). Efforts have been made to perform successful genome editing via PEG-mediated protoplast transformation of three soybean genes, *Glyma06g14180*, *Glyma08g02290*, and *Glyma12g37050* (Sun *et al.*, 2015) as well as *FAD2* paralogues (Kim *et al.*, 2017). In a successful study performed in cotton (Chen *et al.*, 2017), deletion mutants were created in cotton chloroplasts *alterados* 1 (*GhCLA1*) and vacuolar H+-pyrophosphatase (*GhVP*) genes with 47.6–81.8% protoplast regeneration efficiency.

7.3.5 Sonication

Permeation of the plant cell plasma membrane and cell wall can be mediated in a reversible manner by ultrasound-mediated sonoporation and can be effectively used for gene delivery in biotechnological applications (Rustgi *et al.*, 2022). Since sonication can be used on any tissue type, it gives researchers more flexibility for delivering DNA to plant cells, where the only consideration should be minimal damage to cells or tissues at the time of construct delivery (Liu *et al.*, 2006). In a recent study (Zolghadrnasab *et al.*, 2021), researchers optimized the use of positively charged poly-ethyleneimine (PEI)-coated mesoporous silica nanoparticles (MSNs) to deliver the *GUS* gene-containing vector into the tobacco cell suspension using the sonoporation method. Earlier, in a review, it was discussed that *in vivo*, the onset of cavity formation into the plant cell wall is challenging but can be optimized by microbubbles like ultrasound contrast agents. The approach has successfully been tested in mammalian cells for non-viral gene delivery (Liu *et al.*, 2006). Sonication could easily be an alternative cost-effective technique which can be used for delivering constructs into plant protoplast (Joersbo and Brunstedt, 1990) suspension-cultured cells (Liu *et al.*, 2006; Zolghadrnasab *et al.*, 2021), and even intact plant tissues such as leaf segments (Zhang *et al.*, 1991). One additional advantage sonication has is that it can be paired with other gene delivery techniques to increase the transformation efficiency, such as sonication-assisted *Agrobacterium*-mediated transformation (SAAT) (Trick and Finer, 1997; Dutta *et al.*, 2013). Sonication has also been used for direct transformation into pollens to create more effective germline-inherited transgenics (Yang *et al.*, 2017). In one of the most recent works, Nakamura *et al.* (2023) used the sonication-assisted whisker method to deliver CRISPR-Cas9 ribonucleoprotein in rice leading to the successful editing of *OsLCY* beta genes causing abnormal carotenoid accumulation (Nakamura *et al.*, 2023).

7.4 Conclusion

In conclusion, researchers' striving to enhance CRISPR reagent delivery methods without resorting to the transgenic process has been achieved in diverse crop systems. With the discovery of alternative Cas9 options like Cas Phi, and Cpf1, the adoption of novel delivery techniques becomes a feasible prospect. The present state of the field is a rapidly evolving blend of specialized solutions, each developed for a specific application. Due to the dynamic landscape and complexity associated with each species, selecting a singular, superior method is challenging, and it is difficult to predict the next significant breakthrough. There is no obvious candidate method which we could predict will become dominant in the short term. However, the development of novel delivery methods such as haploid-induction editing (HI-EDIT) (Kelliher *et al.*, 2019), inducing *de novo* meristems (Maher *et al.*, 2020), and cut–dip–budding (CDB) delivery system (Cao *et al.*, 2023) holds the potential to generate transgene-free and tissue culture-free crops with short generation times. These breakthrough technologies not only facilitate the swift development of new cultivars but also enhance existing ones.

Acknowledgement

GBP would like to thank the Governor's University Research Initiative (GURI), Texas Tech University for the financial support.

References

Ainley, W.M., Sastry-Dent, L., Welter, M.E., Murray, M.G., Zeitler, B. *et al.* (2013) Trait stacking via targeted genome editing. *Plant Biotechnology Journal* 11(9), 1126–1134.

Alghuthaymi, M.A., Ahmad, A., Khan, Z., Khan, S.H., Ahmed, F.K. *et al.* (2021) Exosome/liposome-like nanoparticles: New carriers for CRISPR genome editing in plants. *International Journal of Molecular Sciences* 22(14), 7456.

Ali, Z., Abul-Faraj, A., Li, L., Ghosh, N., Piatek, M. *et al.* (2015) Efficient virus-mediated genome editing in plants using the CRISPR/Cas9 system. *Molecular Plant* 8(8), 1288–1291.

Ali, Z., Eid, A., Ali, S. and Mahfouz, M.M. (2018) Pea early-browning virus-mediated genome editing via the CRISPR/Cas9 system in *Nicotiana benthamiana* and *Arabidopsis*. *Virus Research* 244, 333–337.

Altmann, T., Damm, B., Halfter, U., Willmitzer, L. and Morris, P.-C. (1992) Protoplast transformation and methods to create specific mutants in *Arabidopsis thaliana*. In: Koncz, C., Chua, N.-H. and Schell, J. (eds) *Methods in Arabidopsis Research*. World Scientific, Singapore, pp. 310–330.

Altpeter, F., Baisakh, N., Beachy, R., Bock, R., Capell, T. *et al.* (2005) Particle bombardment and the genetic enhancement of crops: Myths and realities. *Molecular Breeding* 15, 305–327.

Altpeter, F., Springer, N.M., Bartley, L.E., Blechl, A.E., Brutnell, T.P. *et al.* (2016) Advancing crop transformation in the era of genome editing. *The Plant Cell* 28(7), 1510–1520.

Andersson, M., Turesson, H., Olsson, N., Fält, A.S., Ohlsson, P. *et al.* (2018) Genome editing in potato via CRISPR-Cas9 ribonucleoprotein delivery. *Physiologia Plantarum* 164(4), 378–384.

Ariga, H., Toki, S. and Ishibashi, K. (2020) Potato virus X vector-mediated DNA-free genome editing in plants. *Plant and Cell Physiology* 61(11), 1946–1953.

Bairu, M.W., Aremu, A.O. and Staden, J. (2011) Somaclonal variation in plants: Causes and detection methods. *Plant Growth Regulation* 63, 147–173.

Banakar, R., Schubert, M., Collingwood, M., Vakulskas, C., Eggenberger, A.L. *et al.* (2020) Comparison of CRISPR-Cas9/Cas12a ribonucleoprotein complexes for genome editing efficiency in the rice phytoene desaturase (OsPDS) gene. *Rice* 13(1), 1–7.

Bari, V.K., Nassar, J.A., Kheredin, S.M., Gal-On, A., Ron, M. *et al.* (2019) CRISPR/Cas9-mediated Mutagenesis of *CAROTENOID CLEAVAGE DIOXYGENASE 8* in tomato provides resistance against the parasitic weed *Phelipanche aegyptiaca*. *Scientific Reports* 9(1), 11438.

Bäurle, I. and Dean, C. (2006) The timing of developmental transitions in plants. *Cell* 125(4), 655–664.

Bes, M., Herbert, L., Mounier, T., Meunier, A.-C., Durandet, F. *et al.* (2021) Efficient genome editing in rice protoplasts using CRISPR/Cas9 construct. *Rice Genome Engineering and Gene Editing: Methods and Protocols* 2238, 173–191.

Bevan, M.W., Flavell, R.B. and Chilton, M.-D. (1983) A chimaeric antibiotic resistance gene as a selectable marker for plant cell transformation. *Nature* 304(5922), 184–187.

Bilang, R., Klöti, A., Schrott, M. and Potrykus, I. (1994) PEG-mediated direct gene transfer and electroporation. In: Gelvin, S.B. and Schilperoort, R.A. (eds) *Plant Molecular Biology Manual*. Springer, Dordrecht, Netherlands, pp. 1–16.

Brandt, K.M., Gunn, H., Moretti, N. and Zemetra, R.S. (2020) A streamlined protocol for wheat (*Triticum aestivum*) protoplast isolation and transformation with CRISPR-Cas ribonucleoprotein complexes. *Frontiers in Plant Science* 11, 769.

Bulle, M., Sheri, V., Aileni, M. and Zhang, B. (2023) Chloroplast genome engineering: A plausible approach to combat chili thrips and other agronomic insect pests of crops. *Plants* 12(19), 3448.

Bulle, M., Devadasu, E., Rampuria, S., Subramanyam, R. and Kirti, P.B. (2024a) Plastid-expressed *AdDjSKI* enhances photosystem II stability, delays leaf senescence, and increases fruit yield in tomato plants under heat stress. *Physiologia Plantarum* 176(3), e14374.

Bulle, M., Venkatapuram, A.K., Abbagani, S. and Kirti, P.B. (2024b) CRISPR/Cas9 based genome editing of Phytoene desaturase (PDS) gene in chilli pepper (*Capsicum annuum* L.). *Journal of Genetic Engineering and Biotechnology* 22(2), 100380.

Buriak, J.M., Parak, W.J. and Chen, X. (2022) Nano and plants. *ACS Nano* 16(2), 1681–1684.

Campbell, B.W., Hoyle, J.W., Bucciarelli, B., Stec, A.O., Samac, D.A. *et al.* (2019) Functional analysis and development of a CRISPR/Cas9 allelic series for a CPR5 ortholog necessary for proper growth of soybean trichomes. *Scientific Reports* 9, 14757.

Cao, X., Xie, H., Song, M., Lu, J., Ma, P. *et al.* (2023) Cut–dip–budding delivery system enables genetic modifications in plants without tissue culture. *The Innovation* 4(1), 100345.

Carlsen, F.M., Johansen, I.E., Yang, Z., Liu, Y., Westberg, I.N. *et al.* (2022) Strategies for efficient gene editing in protoplasts of *Solanum tuberosum* theme: determining gRNA efficiency design by utilizing protoplast. *Frontiers in Genome Editing* 3, 795644.

Carrijo, J., Illa-Berenguer, E., LaFayette, P., Torres, N., Aragão, F.J. *et al.* (2021) Two efficient CRISPR/Cas9 systems for gene editing in soybean. *Transgenic Research* 30(3), 239–249.

Ceasar, S.A. and Ignacimuthu, S. (2023) CRISPR/Cas genome editing in plants: Dawn of *Agrobacterium* transformation for recalcitrant and transgene-free plants for future crop breeding. *Plant Physiology and Biochemistry* 196, 724–730.

Char, S.N., Neelakandan, A.K., Nahampun, H., Frame, B., Main, M. *et al.* (2017) An *Agrobacterium*-delivered CRISPR/Cas9 system for high-frequency targeted mutagenesis in maize. *Plant Biotechnology Journal* 15(2), 257–268.

Chavez, M., Chen, X., Finn, P.B. and Qi, L.S. (2023) Advances in CRISPR therapeutics. *Nature Reviews Nephrology* 19(1), 9–22.

Chen, H., Su, Z., Tian, B., Liu, Y., Pang, Y. *et al.* (2022) Development and optimization of a Barley stripe mosaic virus-mediated gene editing system to improve Fusarium head blight resistance in wheat. *Plant Biotechnology Journal* 20, 1018.

Chen, K., Wang, Y., Zhang, R., Zhang, H. and Gao, C. (2019) CRISPR/CAS genome editing and precision plant breeding in agriculture. *Annual Review of Plant Biology* 70, 667–697.

Chen, X., Lu, X., Shu, N., Wang, S., Wang, J. *et al.* (2017) Targeted mutagenesis in cotton (*Gossypium hirsutum* L.) using the CRISPR/Cas9 system. *Scientific Reports* 7(1), 1–7.

Cho, J.-Y., Bhowmik, P., Polowick, P.L., Dodard, S.G., El-Bakkari, M. *et al.* (2020) Cellular delivery of plasmid DNA into wheat microspores using rosette nanotubes. *ACS Omega* 5(38), 24422–24433.

Cigliano, R.A., Cremona, G., Paparo, R., Termolino, P., Perrella, G. *et al.* (2013) Histone deacetylase AtHDA7 is required for female gametophyte and embryo development in *Arabidopsis*. *Plant Physiology* 163(1), 431–440.

Clough, S.J. (2004) Floral dip: *Agrobacterium*-mediated germ line transformation. *Transgenic Plants: Methods and Protocols* 286, 91–101.

Cunningham, A.J., Gibson, V.P., Banquy, X., Zhu, X. and Jeanne, L.C. (2020) Cholic acid-based mixed micelles as siRNA delivery agents for gene therapy. *International Journal of Pharmaceutics* 578, 119078.

Dahan-Meir, T., Filler-Hayut, S., Melamed-Bessudo, C., Bocobza, S., Czosnek, H. *et al.* (2018) Efficient in planta gene targeting in tomato using Geminiviral replicons and the CRISPR/Cas9 system. *The Plant Journal* 95(1), 5–16.

De Vos, G. and Zambryski, P. (1989) Expression of *Agrobacterium* nopaline-specific VirD1, VirD2, and VirC1 proteins and their requirement. *Molecular Plant-Microbe Interactions* 2, 43–52.

Demirer, G.S., Zhang, H., Goh, N.S., González-Grandío, E. and Landry, M.P. (2019) Carbon nanotube–mediated DNA delivery without transgene integration in intact plants. *Nature Protocols* 14(10), 2954–2971.

Demirer, G.S., Silva, T.N., Jackson, C.T., Thomas, J.B., Ehrhardt, D.W. *et al.* (2021) Nanotechnology to advance CRISPR–CAS genetic engineering of plants. *Nature Nanotechnology* 16(3), 243–250.

Dong, O.X., Yu, S., Jain, R., Zhang, N., Duong, P.Q. *et al.* (2020) Marker-free carotenoid-enriched rice generated through targeted gene insertion using CRISPR-Cas9. *Nature Communications* 11(1), 1178.

Dutta, I., Kottackal, M., Tumimbang, E., Tajima, H., Zaid, A. *et al.* (2013) Sonication-assisted efficient *Agrobacterium*-mediated genetic transformation of the multipurpose woody desert shrub *Leptadenia pyrotechnica*. *Plant Cell, Tissue and Organ Culture (PCTOC)* 112, 289–301.

Eeckhaut, T., Lakshmanan, P.S., Deryckere, D., Bockstaele, E. and Huylenbroeck, J. (2013) Progress in plant protoplast research. *Planta* 238, 991–1003.

Ellison, E.E., Nagalakshmi, U., Gamo, M.E., Huang, P.J., Dinesh-Kumar, S. *et al.* (2020) Multiplexed heritable gene editing using RNA viruses and mobile single guide RNAs. *Nature Plants* 6(6), 620–624.

Feng, Z., Zhang, B., Ding, W., Liu, X., Yang, D.-L. *et al.* (2013) Efficient genome editing in plants using a CRISPR/CAS system. *Cell Research* 23(10), 1229–1232.

Fossi, M., Amundson, K., Kuppu, S., Britt, A. and Comai, L. (2019) Regeneration of *Solanum tuberosum* plants from protoplasts induces widespread genome instability. *Plant Physiology* 180(1), 78–86.

Gao, Q., Xu, W.Y., Yan, T., Fang, X.D., Cao, Q. *et al.* (2019) Rescue of a plant cytorhabdovirus as versatile expression platforms for planthopper and cereal genomic studies. *New Phytologist* 223(4), 2120–2133.

Gelvin, S.B. (2017) Integration of *Agrobacterium* T-DNA into the plant genome. *Annual Review of Genetics* 51, 195–217.

Gheysen, G., Angenon, G. and Montagu, M. (2022) *Agrobacterium*-mediated plant transformation: A scientifically intriguing story with significant applications. In: Lindsey, K. (ed.) *Transgenic Plant Research*. Harwood Academic Publishers, UK, pp. 1–34.

Ghogare, R., Ludwig, Y., Bueno, G.M., Slamet-Loedin, I.H. and Dhingra, A. (2021) Genome editing reagent delivery in plants. *Transgenic Research* 30, 321–335.

Gil-Humanes, J., Wang, Y., Liang, Z., Shan, Q., Ozuna, C.V. *et al.* (2017) High-efficiency gene targeting in hexaploid wheat using DNA replicons and CRISPR/Cas9. *The Plant Journal* 89(6), 1251–1262.

Gocal, G. (2015) Non-transgenic trait development in crop plants using oligo-directed mutagenesis: Cibus' rapid trait development system. In: *In NABC Report 26. New DNA-Editing Approaches: Methods, Applications and Policy for Agriculture*. North American Agricultural Biotechnology Council, Ithaca, New York, pp. 97–105.

Gocal, G.F., Schöpke, C. and Beetham, P.R. (2015) Oligo-mediated targeted gene editing. In: Zhang, F., Puchta, H. and Thompson, J.G. (eds) *Advances in New Technology for Targeted Modification of Plant Genomes*. Springer, New York, pp. 73–89.

González, M.N., Massa, G.A., Andersson, M., Turesson, H., Olsson, N. *et al.* (2020) Reduced enzymatic browning in potato tubers by specific editing of a polyphenol oxidase gene via ribonucleoprotein complexes delivery of the CRISPR/Cas9 system. *Frontiers in Plant Science* 10, 1649.

Gordon-Kamm, B., Sardesai, N., Arling, M., Lowe, K., Hoerster, G. *et al.* (2019) Using morphogenic genes to improve recovery and regeneration of transgenic plants. *Plants* 8(2), 38.

Graham, N., Patil, G.B., Bubeck, D.M., Dobert, R.C., Glenn, K.C. *et al.* (2020) Plant genome editing and the relevance of off-target changes. *Plant Physiology* 183(4), 1453–1471.

Hamada, H., Liu, Y., Nagira, Y., Miki, R., Taoka, N. *et al.* (2018) Biolistic-delivery-based transient CRISPR/Cas9 expression enables in planta genome editing in wheat. *Scientific Reports* 8(1), 1–7.

Hamers, R.J. (2017) Nanomaterials and global sustainability. *Accounts of Chemical Research* 50(3), 633–637.

Hamilton, C.M., Frary, A., Lewis, C. and Tanksley, S.D. (1996) Stable transfer of intact high molecular weight DNA into plant chromosomes. *Proceedings of the National Academy of Sciences* 93(18), 9975–9979.

Hu, J., Li, S., Li, Z., Li, H., Song, W. *et al.* (2019) A barley stripe mosaic virus-based guide RNA delivery system for targeted mutagenesis in wheat and maize. *Molecular Plant Pathology* 20(10), 1463–1474.

Hwang, H.-H., Yu, M. and Lai, E.-M. (2017) Agrobacterium-mediated plant transformation: Biology and applications. *The Arabidopsis Book* 15, e0186.

Ismagul, A., Yang, N., Maltseva, E., Iskakova, G., Mazonka, I. *et al.* (2018) A biolistic method for high-throughput production of transgenic wheat plants with single gene insertions. *BMC Plant Biology* 18(1), 1–8.

Jacobs, T.B., LaFayette, P.R., Schmitz, R.J. and Parrott, W.A. (2015) Targeted genome modifications in soybean with CRISPR/Cas9. *BMC Biotechnology* 15, 1–10.

Jiang, W., Zhou, H., Bi, H., Fromm, M., Yang, B. *et al.* (2013) Demonstration of CRISPR/Cas9/sgRNA-mediated targeted gene modification in *Arabidopsis*, tobacco, sorghum and rice. *Nucleic Acids Research* 41(20), e188.

Joersbo, M. and Brunstedt, J. (1990) Direct gene transfer to plant protoplasts by mild sonication. *Plant Cell Reports* 9, 207–210.

Kah, M., Kookana, R.S., Gogos, A. and Bucheli, T.D. (2018) A critical evaluation of nanopesticides and nanofertilizers against their conventional analogues. *Nature Nanotechnology* 13(8), 677–684.

Kawai, S., Hashimoto, W. and Murata, K. (2010) Transformation of *Saccharomyces cerevisiae* and other fungi: Methods and possible underlying mechanism. *Bioengineered Bugs* 1(6), 395–403.

Kawata, M. and Oono, K. (1998) Protoclonal variation in crop improvement. In: Jain, S.M., Brar, D.S. and Ahloowalia, B.S. (eds) *Somaclonal Variation and Induced Mutations in Crop Improvement*. Vol. 32: Current Plant Science and Biotechnology in Agriculture, Springer, Dordrecht, The Netherlands, pp. 135–148.

Kelliher, T., Starr, D., Su, X., Tang, G., Chen, Z. *et al.* (2019) One-step genome editing of elite crop germplasm during haploid induction. *Nature Biotechnology* 37(3), 287–292.

Kim, H., Kim, S.-T., Ryu, J., Kang, B.-C., Kim, J.-S. *et al.* (2017) CRISPR/Cpf1-mediated DNA-free plant genome editing. *Nature Communications* 8(1), 14406.

Kim, J.H., Yu, J., Kim, H.K., Kim, J.Y., Kim, M.-S. *et al.* (2022) Genome editing of golden SNP-carrying lycopene epsilon-cyclase (Lcye) gene using the CRSPR-Cas9/HDR and *Geminiviral* replicon system in rice. *International Journal of Molecular Sciences* 23(18), 10383.

Kofer, W., Koop, H.-U., Wanner, G. and Steinmüller, K. (1998) Mutagenesis of the genes encoding subunits A, C, H, I, J and K of the plastid NAD (P) H-plastoquinone-oxidoreductase in tobacco by polyethylene glycol-mediated plastome transformation. *Molecular and General Genetics MGG* 258, 166–173.

Kuwano, T., Shirataki, C. and Itoh, Y. (2008) Comparison between polyethylene glycol- and polyethylenimine-mediated transformation of *Aspergillus nidulans*. *Current Genetics* 54, 95–103.

Lee, K. and Wang, K. (2023) Strategies for genotype-flexible plant transformation. *Current Opinion in Biotechnology* 79, 102848.

Lew, T.T.S., Park, M., Wang, Y., Gordiichuk, P., Yeap, W.-C. *et al.* (2020) Nanocarriers for transgene expression in pollen as a plant biotechnology tool. *ACS Materials Letters* 2(9), 1057–1066.

Li, B., Rui, H., Li, Y., Wang, Q., Alariqi, M. *et al.* (2019) Robust CRISPR/Cpf1 (Cas12a)-mediated genome editing in allotetraploid cotton (*Gossypium hirsutum*). *Plant Biotechnology Journal* 17, 1862.

Li, T., Hu, J., Sun, Y., Li, B., Zhang, D. *et al.* (2021) Highly efficient heritable genome editing in wheat using an RNA virus and bypassing tissue culture. *Molecular Plant* 14, 1787–1798.

Li, X., Wang, Y., Chen, S., Tian, H., Fu, D. *et al.* (2018) Lycopene is enriched in tomato fruit by CRISPR/Cas9-mediated multiplex genome editing. *Frontiers in Plant Science* 9, 559.

Li, X., Mitchell, M., Rolland, V., Allen, S., MacMillan, C. *et al.* (2023) Pink cotton candy'—A new dye-free cotton. *Plant Biotechnology Journal* 21(4), 677.

Li, Y., Zhu, J., Wu, H., Liu, C., Huang, C. *et al.* (2020) Precise base editing of non-allelic *Acetolactate* synthase genes confers sulfonylurea herbicide resistance in maize. *The Crop Journal* 8(3), 449–456.

Li, Z., Liu, Z.-B., Xing, A., Moon, B.P., Koellhoffer, J.P. *et al.* (2015) Cas9-guide RNA directed genome editing in soybean. *Plant Physiology* 169(2), 960–970.

Liang, Z., Zhang, K., Chen, K. and Gao, C. (2014) Targeted mutagenesis in *Zea mays* using TALENs and the CRISPR/Cas system. *Journal of Genetics and Genomics* 41(2), 63–68.

Liang, Z., Chen, K., Li, T., Zhang, Y., Wang, Y. *et al.* (2017) Efficient DNA-free genome editing of bread wheat using CRISPR/Cas9 ribonucleoprotein complexes. *Nature Communications* 8(1), 14261.

Lin, C.-S., Hsu, C.-T., Yuan, Y.-H., Zheng, P.-X., Wu, F.-H. *et al.* (2022) DNA-free CRISPR-Cas9 gene editing of wild tetraploid tomato *Solanum peruvianum* using protoplast regeneration. *Plant Physiology* 188(4), 1917–1930.

Lin, Q., Zong, Y., Xue, C., Wang, S., Jin, S. *et al.* (2020) Prime genome editing in rice and wheat. *Nature Biotechnology* 38(5), 582–585.

Liu, J., Nannas, N.J., Fang-Fang, F., Shi, J., Aspinwall, B. *et al.* (2019) Genome-scale sequence disruption following biolistic transformation in rice and maize. *The Plant Cell* 31(2), 368–383.

Liu, Y., Yang, H. and Sakanishi, A. (2006) Ultrasound: Mechanical gene transfer into plant cells by sonoporation. *Biotechnology Advances* 24(1), 1–16.

Liu, Y., Andersson, M., Granell, A., Cardi, T., Hofvander, P. *et al.* (2022) Establishment of a DNA-free genome editing and protoplast regeneration method in cultivated tomato (*Solanum lycopersicum*). *Plant Cell Reports* 41(9), 1843–1852.

Luo, Y., Na, R., Nowak, J.S., Qiu, Y., Lu, Q.S. *et al.* (2021) Development of a Csy4-processed guide RNA delivery system with soybean-infecting virus ALSV for genome editing. *BMC Plant Biology* 21(1), 1–12.

Ma, X., Zhang, X., Liu, H. and Li, Z. (2020) Highly efficient DNA-free plant genome editing using virally delivered CRISPR–Cas9. *Nature Plants* 6(7), 773–779.

Maher, M.F., Nasti, R.A., Vollbrecht, M., Starker, C.G., Clark, M.D. *et al.* (2020) Plant gene editing through de novo induction of meristems. *Nature Biotechnology* 38(1), 84–89.

Mahmood, M.A., Naqvi, R.Z., Rahman, S.U., Amin, I. and Mansoor, S. (2023) Plant virus-derived vectors for plant genome engineering. *Viruses* 15(2), 531.

Maren, N.A., Duan, H., Da, K., Yencho, G.C., Ranney, T.G. *et al.* (2022) Genotype-independent plant transformation. *Horticulture Research* 9.

Marton, I., Zuker, A., Shklarman, E., Zeevi, V., Tovkach, A. *et al.* (2010) Nontransgenic genome modification in plant cells. *Plant Physiology* 154, 1079–1087.

Mei, Y., Beernink, B.M., Ellison, E.E., Konečná, E., Neelakandan, A.K. *et al.* (2019) Protein expression and gene editing in monocots using foxtail mosaic virus vectors. *Plant Direct* 3(11), e00181.

Nakamura, A., Yano, T., Mitsuda, N., Furubayashi, M., Ito, S. *et al.* (2023) The sonication-assisted whisker method enables CRISPR-Cas9 ribonucleoprotein delivery to induce genome editing in rice. *Scientific Reports* 13(1), 14205.

Nalapalli, S., Tunc-Ozdemir, M., Sun, Y., Elumalai, S. and Que, Q. (2021) Morphogenic regulators and their application in improving plant transformation. *Rice Genome Engineering and Gene Editing: Methods and Protocols* 2238, 37–61.

Nwoko, C.L., Kshetry, A.O., Devkar, V. and Patil, G.B. (2023) An improved method for protoplast isolation and gene-editing from soybean root, callus and transgenic hairy-roots. *Journal of Plant Biology and Crop Research* 6(2), 1088.

Oh, Y., Kim, H. and Kim, S.-G. (2021) Virus-induced plant genome editing. *Current Opinion in Plant Biology* 60, 101992.

Ojha, A., Zhang, F. and Patil, G.B. (2023) *Genome Editing and Chromosome Engineering in Plants*, Vol. 16. Wiley Online Library.

Park, S.-C., Park, S., Jeong, Y.J., Lee, S.B., Pyun, J.W. *et al.* (2019) DNA-free mutagenesis of GIGANTEA in *Brassica oleracea* var. *capitata* using CRISPR/cas9 ribonucleoprotein complexes. *Plant Biotechnology Reports* 13, 483–489.

Patil, G.B., Stupar, R.M. and Zhang, F. (2022) Protoplast isolation, transfection, and gene editing for soybean (*Glycine max*). In: Wang, K. and Zhang, F. (eds) *Protoplast Technology: Methods in Molecular Biology*, Vol. 2464. Springer, Humana, New York, pp. 173–186.

Rustgi, S., Kashyap, S., Ankrah, N. and Wettstein, D. (2020) Use of microspore-derived calli as explants for biolistic transformation of common wheat. In: Rustgi, S. and Luo, H. (eds) *Biolistic DNA Delivery in Plants: Methods in Molecular Biology*, Vol. 2124. Springer, Humana, New York, pp. 263–279.

Rustgi, S., Naveed, S., Windham, J., Zhang, H. and Demirer, G.S. (2022) Plant biomacromolecule delivery methods in the 21st century. *Frontiers in Genome Editing* 4, 1011934.

Sanford, J.C. (2000) The development of the biolistic process. *In Vitro Cellular and Developmental Biology—Plant* 36(5), 303–308.

Sant'Ana, R.R.A., Caprestano, C.A., Nodari, R.O. and Agapito-Tenfen, S.Z. (2020) PEG-delivered CRISPR-Cas9 ribonucleoproteins system for gene-editing screening of maize protoplasts. *Genes* 11(9), 1029.

Schwartz, S.H., Hendrix, B., Hoffer, P., Sanders, R.A. and Zheng, W. (2020) Carbon dots for efficient small interfering RNA delivery and gene silencing in plants. *Plant Physiology* 184(2), 647–657.

Shan, Q., Wang, Y., Li, J., Zhang, Y., Chen, K. *et al.* (2013) Targeted genome modification of crop plants using a CRISPR-Cas system. *Nature Biotechnology* 31(8), 686–688.

Shan, Q., Wang, Y., Li, J. and Gao, C. (2014) Genome editing in rice and wheat using the CRISPR/Cas system. *Nature Protocols* 9(10), 2395–2410.

Sharma, G., Sharma, A.R., Bhattacharya, M., Lee, S.-S. and Chakraborty, C. (2021) CRISPR-Cas9: A preclinical and clinical perspective for the treatment of human diseases. *Molecular Therapy* 29(2), 571–586.

Shi, J., Gao, H., Wang, H., Lafitte, H.R., Archibald, R.L. *et al.* (2017) ARGOS 8 variants generated by CRISPR-Cas9 improve maize grain yield under field drought stress conditions. *Plant Biotechnology Journal* 15(2), 207–216.

Shidore, T., Zuverza-Mena, N., White, J.C. and Silva, W. (2021) Nanoenabled delivery of RNA molecules for prolonged antiviral protection in crop plants: A review. *ACS Applied Nano Materials* 4(12), 12891–12904.

Shufen, C., Yicong, C., Baobing, F., Guiai, J., Zhonghua, S. *et al.* (2019) Editing of rice isoamylase gene ISA1 provides insights into its function in starch formation. *Rice Science* 26(2), 77–87.

Stachel, S.E., Messens, E., Montagu, M. and Zambryski, P. (1985) Identification of the signal molecules produced by wounded plant cells that activate T-DNA transfer in *Agrobacterium tumefaciens*. *Nature* 318(6047), 624–629.

Subburaj, S., Chung, S.J., Lee, C., Ryu, S.-M., Kim, D.H. *et al.* (2016) Site-directed mutagenesis in *Petunia*× hybrida protoplast system using direct delivery of purified recombinant Cas9 ribonucleoproteins. *Plant Cell Reports* 35, 1535–1544.

Sun, X., Hu, Z., Chen, R., Jiang, Q., Song, G. *et al.* (2015) Targeted mutagenesis in soybean using the CRISPR-Cas9 system. *Scientific Reports* 5(1), 1–10.

Sun, Y., Zhang, X., Wu, C., He, Y., Ma, Y. *et al.* (2016) Engineering herbicide-resistant rice plants through CRISPR/Cas9-mediated homologous recombination of *Acetolactate* synthase. *Molecular Plant* 9(4), 628–631.

Svitashev, S., Ananiev, E., Pawlowski, W. and Somers, D. (2000) Association of transgene integration sites with chromosome rearrangements in hexaploid oat. *Theoretical and Applied Genetics* 100, 872–880.

Svitashev, S., Young, J.K., Schwartz, C., Gao, H., Falco, S.C. *et al.* (2015) Targeted mutagenesis, precise gene editing, and site-specific gene insertion in maize using Cas9 and guide RNA. *Plant Physiology* 169(2), 931–945.

Svitashev, S., Schwartz, C., Lenderts, B., Young, J. and Cigan, M. (2016) Genome editing in maize directed by CRISPR-Cas9 ribonucleoprotein complexes. *Nature Communications* 7, 13274. DOI: 10.1038/ncomms13274.

Svitashev, S.K., Pawlowski, W.P., Makarevitch, I., Plank, D.W. and Somers, D.A. (2002) Complex transgene locus structures implicate multiple mechanisms for plant transgene rearrangement. *The Plant Journal* 32(4), 433–445.

Tang, X., Liu, G., Zhou, J., Ren, Q., You, Q. *et al.* (2018) A large-scale whole-genome sequencing analysis reveals highly specific genome editing by both Cas9 and Cpf1 (Cas12a) nucleases in rice. *Genome Biology* 19(1), 1–13.

Tian, Y., Zhong, D., Li, X., Shen, R., Han, H. *et al.* (2023) High-throughput genome editing in rice with a virus-based surrogate system. *Journal of Integrative Plant Biology* 65(3), 646–655.

Tiwari, J.K., Singh, A.K. and Behera, T.K. (2023) CRISPR/Cas genome editing in tomato improvement: Advances and applications. *Frontiers in Plant Science* 14, 1121209.

Toda, E., Koiso, N., Takebayashi, A., Ichikawa, M., Kiba, T. *et al.* (2019) An efficient DNA- and selectable-marker-free genome-editing system using zygotes in rice. *Nature Plants* 5(4), 363–368.

Trick, H.N. and Finer, J.J. (1997) SAAT: Sonication-assisted *Agrobacterium*-mediated transformation. *Transgenic Research* 6, 329–336.

Upadhyay, S.K., Kumar, J., Alok, A. and Tuli, R. (2013) RNA-guided genome editing for target gene mutations in wheat. *G3: Genes, Genomes, Genetic* 3(12), 2233–2238.

Uranga, M., Aragonés, V., Selma, S., Vázquez-Vilar, M., Orzáez, D. *et al.* (2021) Efficient Cas9 multiplex editing using unspaced sgRNA arrays engineering in a potato virus X vector. *The Plant Journal* 106(2), 555–565.

Vats, S., Kumawat, S., Kumar, V., Patil, G.B., Joshi, T. *et al.* (2019) Genome editing in plants: Exploration of technological advancements and challenges. *Cells* 8(11), 1386.

Verma, V., Kumar, A., Partap, M., Thakur, M. and Bhargava, B. (2023) CRISPR-Cas: A robust technology for enhancing consumer-preferred commercial traits in crops. *Frontiers in Plant Science* 14, 1122940.

Virdi, K.S., Spencer, M., Stec, A.O., Xiong, Y., Merry, R. *et al.* (2020) Similar seed composition phenotypes are observed from CRISPR-generated in-frame and knockout alleles of a soybean KASI ortholog. *Frontiers in Plant Science* 11, 1005.

Wang, M., Lu, Y., Botella, J.R., Mao, Y., Hua, K. *et al.* (2017) Gene targeting by homology-directed repair in rice using a Geminivirus-based CRISPR/Cas9 system. *Molecular Plant* 10(7), 1007–1010.

Wang, P., Zhao, F.-J. and Kopittke, P.M. (2019) Engineering crops without genome integration using nanotechnology. *Trends in Plant Science* 24(7), 574–577.

Wang, W., Yu, Z., He, F., Bai, G., Trick, H.N. *et al.* (2022) Multiplexed promoter and gene editing in wheat using a virus-based guide RNA delivery system. *Plant Biotechnology Journal* 20, 2332–2341.

Webb, K.J. and Morris, P. (1992) Methodologies of plant transformation. In: Gatehouse, A.M.R., Hilder, V.A. and Boulter, D. (eds) *Plant Genetic Manipulation for Crop Protection*. CAB International, Wallingford, UK, pp. 7–43.

Woo, J.W., Kim, J., Kwon, S.I., Corvalán, C., Cho, S.W. *et al.* (2015) DNA-free genome editing in plants with preassembled CRISPR-Cas9 ribonucleoproteins. *Nature Biotechnology* 33(11), 1162–1164.

Xia, X., Cheng, X., Li, R., Yao, J., Li, Z. *et al.* (2021) Advances in application of genome editing in tomato and recent development of genome editing technology. *Theoretical and Applied Genetics* 134(9), 2727–2747.

Xie, K. and Yang, Y. (2013) RNA-guided genome editing in plants using a CRISPR–Cas system. *Molecular Plant* 6(6), 1975–1983.

Yang, L., Cui, G., Wang, Y., Hao, Y., Du, J. *et al.* (2017) Expression of foreign genes demonstrates the effectiveness of pollen-mediated transformation in *Zea mays*. *Frontiers in Plant Science* 8, 383.

Yang, L., Machin, F., Wang, S., Saplaoura, E. and Kragler, F. (2023) Heritable transgene-free genome editing in plants by grafting of wild-type shoots to transgenic donor rootstocks. *Nature Biotechnology* 41(7), 958–967.

Yong, J., Zhang, R., Bi, S., Li, P., Sun, L. *et al.* (2021) Sheet-like clay nanoparticles deliver RNA into developing pollen to efficiently silence a target gene. *Plant Physiology* 187(2), 886–899.

Yu, J., Tu, L., Subburaj, S., Bae, S. and Lee, G.-J. (2021) Simultaneous targeting of duplicated genes in *Petunia* protoplasts for flower color modification via CRISPR-Cas9 ribonucleoproteins. *Plant Cell Reports* 40, 1037–1045.

Yue, J.-J., Yuan, J.-L., Wu, F.-H., Yuan, Y.-H., Cheng, Q.-W. *et al.* (2021) Protoplasts: from isolation to CRISPR/Cas genome editing application. *Frontiers in Genome Editing* 3, 717017.

Zhang, C., Liu, S., Li, X., Zhang, R. and Li, J. (2022) Virus-induced gene editing and its applications in plants. *International Journal of Molecular Sciences* 23(18), 10202.

Zhang, L.-J., Cheng, L.-M., Xu, N., Zhao, N.-M., Li, C.-G. *et al.* (1991) Efficient transformation of tobacco by ultrasonication. *Bio/Technology* 9(10), 996–997.

Zhang, Y., Liang, Z., Zong, Y., Wang, Y., Liu, J. *et al.* (2016) Efficient and transgene-free genome editing in wheat through transient expression of CRISPR/Cas9 DNA or RNA. *Nature Communications* 7(1), 12617.

Zhang, Y., Iaffaldano, B. and Qi, Y. (2021) CRISPR ribonucleoprotein-mediated genetic engineering in plants. *Plant Communications* 2(2), 100168.

Zhi, H., Chen, H., Yu, M., Wang, C., Cui, B. *et al.* (2022) Layered double hydroxide nanosheets improve the adhesion of fungicides to leaves and the antifungal performance. *ACS Applied Nano Materials* 5(4), 5316–5325.

Zolghadrnasab, M., Mousavi, A., Farmany, A. and Arpanaei, A. (2021) Ultrasound-mediated gene delivery into suspended plant cells using polyethyleneimine-coated mesoporous silica nanoparticles. *Ultrasonics Sonochemistry* 73, 105507.

8 Recent Developments to Accelerate On-target and Minimize Off-target Mutations during Genome Editing in Plants

Alekya Allu[1], Phanikanth Jogam[2], Renuka Suravajhala[3] and Anu Bansal[1*]
[1]School of Bioengineering and Biosciences, Lovely Professional University, Jalandhar, India; [2]Department of Biotechnology, Kakatiya University, Warangal, India; [3]Amrita School of Biotechnology, Amrita Vishwa Vidyapeetham, Kerala, India

Abstract

The third-generation revolutionary genome editing technology, which emerged in recent years, creating high-throughput results, is the CRISPR/Cas system. The CRISPR/Cas system functions with a single gRNA and Cas nucleases, where the gRNA finds and binds to the target DNA strand and activates the Cas nucleases to cleave the DNA, resulting in double-strand breaks, which are repaired by DNA repair mechanisms such as non-homologous end joining (NHEJ) or homology-directed repair (HDR). The CRISPR/Cas systems, being accurate genome editing tools to date, still have the risk of producing off-target mutations, thereby requiring room for improvement. The off-target mutations are caused due to the gRNA's inefficiency in binding at the target site and binding to a non-target sequence, creating double-strand breaks in DNA and leading to mutations. The other possibility of these off-target mutations can be during the DNA repair mechanism, as it can be the insertion, deletion, or substitution of base pairs even after binding at the specified target site. This chapter discusses various approaches to minimize off-target modifications and, at the same time, enhance site-specific genome editing by designing the highly efficient gRNA, predicting the efficiency, stability, and possible off-target mutations by analyzing the gRNA sequence along with analysis of the secondary structure of gRNA to predict its binding and cleaving stability.

Keywords: CRISPR/Cas, sgRNA, Genome editing, Off-target analysis

8.1 Introduction

Genome editing plays an important role in studying the function of a desired gene by introducing or suppressing its transcription into a protein and consequently obtaining tolerance to drought, bacterial infection, and other biotic and abiotic stress conditions (Parmar *et al.*, 2017). With the unparalleled ability to manipulate the plant genome precisely, CRISPR has emerged as a powerful genome editing tool. The clustered regularly interspaced palindromic repeats (CRISPR) and the associated nuclease proteins are known as Cas proteins, like CRISPR-Cas9. CRISPR-Cas system is a complex of single-guide RNA (sgRNA) and a nuclease protein, where the gRNA identifies the target DNA sequence with protospacer adjacent motif (PAM), which is specific to every Cas nuclease. The sgRNA contains 100 bps, while the gRNA is the 20 nucleotides' long seed sequence called

*Corresponding author: anu.21971@lpu.co.in

DOI: 10.1079/9781800622517.0008

CRISPR RNA or CrRNA; the remaining 80 bps length is known as trans-activating CRISPR RNA (tracrRNA). The gRNA activates the Cas nuclease once it locates the target sequence, and the nuclease cleaves the DNA, leaving double-strand breaks (DBS) (Jinek *et al.*, 2012; Jiang *et al.*, 2013; Gupta and Musunuru, 2014; Barrangou and Doudna, 2016; Bogdanove *et al.*, 2018; Foss *et al.*, 2019). This tool comes under the new breeding technologies used in the field of biotechnology, relying on site-dependent nucleases for genome manipulations. Apart from CRISPR-Cas systems, other genome editing techniques like zinc-finger nucleases (ZFNs) and transcription activator-like effector nucleases (TALENs) are used for genome alterations. ZFNs are restriction endonucleases having two separable domains – DNA-binding and DNA-cleavage domains. The DNA-cleavage domain, *Fok*I, is a natural type IIS restriction endonuclease with cleavage activity and physically separable binding. ZFNs approximately recognize three base pairs of DNA, with this the DNA-binding specificity can be changed due to its small number of residue recognition, as the DNA has huge residue repetition. Longer DNA sequences can be recognized by joining the engineered ZFNs together as extended arrays (Carroll, 2011). An alternative technique to ZFNs, TALENs has emerged to introduce DBS comprising a non-specific nuclease domain, *Fok*I, which is fused with the DNA-binding domain. The transcription activator-like effectors (TALEs) are the Xanthomonas bacteria–secreted proteins for altering the gene transcription in host plant cells. These are highly conserved repeats, which make up the composition of the DNA-binding domain (Joung and Sander, 2013). All these genome engineering techniques trigger the same DNA repair mechanism but differ in their domain composition, mechanism to recognize and induce DBS and methods of delivery along with their specificity. Once these nucleases induce the DBS, they trigger the DNA repair mechanisms to further repair these double-strand breaks through NHEJ, HDR, or homologous recombination. The NHEJ type of DNA repair occurs at the sticky ends of the DNA strands to produce indels. In contrast, the HDR repair occurs in the blunt ends of DNA strands, which is helpful to induce point mutation with the provided DNA template. The last homologous recombination helps in integrating the DNA construct with homologous ends, resulting in the insertion of DNA fragments (Jiang *et al.*, 2013; Shah *et al.*, 2018; Molla *et al.*, 2021). The bacterial adaptive immune system protecting the integrity of the genome from viral invasion is the native CRISPR/Cas9 system. The high specificity of this system in bacteria is due to its small genome compared to the plant or animal genome, which is 1000-fold greater than the bacterial genome. Hence, genome alteration might result in high chances of off-target effects when CRISPR/Cas9 is used in plants or animals (Jinek *et al.*, 2012; Pattanayak *et al.*, 2013).

Off-target mutations are the unintended insertion or deletion of bases at a site other than the target site. These off-target mutations result from the mismatches between the 20-nucleotide sequence gRNA and target DNA, where more than three mismatches at the distal end of the PAM sequence affect the gRNA efficiency and specificity. The off-target mutations occur due to the sequence homology of target loci; the binding of gRNA in the genome other than at the target site. These off-target mutations can be lethal as they may have severe effects at genomic levels. Loss of gene function, undesired traits in plants, and carcinogenicity in animals are some complicated effects of these off-target mutations (Fu *et al.*, 2013; Guilinger *et al.*, 2014; Singh *et al.*, 2016; Alkan *et al.*, 2018; Cai *et al.*, 2018). The potential off-target genome manipulations can be identified through analyzed strategic biased and unbiased methods. The small DNA modifications or the DSB are screened whole genome-wide in the unbiased method, while the biased method involves *in-silico*-based analysis of the target genomic site for DNA mutations to predict the off-target sites (Bae *et al.*, 2014; Liu *et al.*, 2017).

Scientists generate vast amounts of data through various CRISPR systems on off-target sites, and these data are used to develop algorithms that predict off-target sites and mutations. These algorithms are designed in two types of tools: alignment-based and scoring-based. These bioinformatics tools are categorized under biased methods. The unbiased analysis can be categorized into *in vitro* genome-wide detection and *in vivo* detection, that is, cell-based genome-wide detection. Thus, the precise designing of gRNA is essential as the genome editing outcomes solely depend on the gRNA.

The GC content in the gRNA is crucial to maintain stability and specificity (Cong *et al.*, 2013; Cho *et al.*, 2014; Liao *et al.*, 2015; Wilson *et al.*, 2018; Naeem *et al.*, 2020).

8.2 Guide RNA Designing

The critical factor that reflects the target specificity and enhances the cleavage efficiency of the gRNA is the 20-nucleotide length, that is, the sgRNA at the 5′ end of the gRNA. As the genome editing tool, CRISPR/Cas9 systems mainly depend on the specificity, that is, the length of gRNA bound to the target site, and efficiency, that is, the number of DSBs induced at the target sites. The perfect gRNA can minimize the off-target mutations and maximize the on-target mutagenesis; hence, CRISPR/Cas9 technology has emerged as the application to enhance or reduce the targeted gene expression, gene knock-out, and knock-in genes. Therefore, it is essential to carefully design and modify the gRNA, where the GC content, length of gRNA, chemical modification, and truncated gRNA are intelligently developed to enhance the specificity and minimize the off-targets (Hsu *et al.*, 2013; Endo *et al.*, 2015; Ren *et al.*, 2016; Hiranniramol *et al.*, 2020).

The CRISPR/Cas9 genome editing on-target activity is also affected by the gRNA structural sequence. Therefore, structural analysis is needed to test its performance. The studies discovered that 40–60% of the GC content in gRNA is ideal for enhancing its on-target efficacy. The higher the GC content, the more stable the DNA: RNA duplex, and this destabilizes the off-target bindings. The efficiency increases with the purine residue position at the end of four nucleotides, and the gRNA position sequence affects the editing. Cytosine is preferred at the 16th position, and guanine is preferred at the 20th position. At the same time, the positive correlation between the GC percentage of the proximal part of sgRNA and PAM is also one of the parameters that increase the editing efficiency in designing a better gRNA (Doench *et al.*, 2014; Ren *et al.*, 2014; Wang *et al.*, 2014; Wong *et al.*, 2015). The gRNA length may affect the editing efficiency, as it can bind to many other sites due to its homology with the DNA. To minimize the off-target effects, the length of the gRNA should be up to 17 nucleotides for higher efficiency in genome editing. In some studies, the 20 bp gRNA did not show off-target mutations; in contrast, the 18–20 bp gRNA was observed to have low efficiency. In addition to this, the strategy of partial DNA replacement with chimeric gRNA was observed to have fewer off-target mutations (Fu *et al.*, 2014; Lin *et al.*, 2014). The other strategy to minimize the off-target mutations is (i) within the 7–10 bps range to the PAM, more than three mismatches should be avoided, and (ii) the sgRNA bulges inside the 12 bp of the PAM should be avoided in order to reduce off-target effects (Fu *et al.*, 2014; Lin *et al.*, 2014; Sugano *et al.*, 2018; Yin *et al.*, 2018).

Recent studies evaluated that the truncated gRNA, that is, shortening gRNA from 20 to 17/18 nucleotides, decreased off-targets by 500-fold (Ran *et al.*, 2013; Fu *et al.*, 2014; Naeem *et al.*, 2020). The latest strategy developed to minimize the undesired effects without DSBs is the dead RNA off-target suppression (dOTS), in which gRNA directs the Cas9 binding but suppresses the cleavage, resulting in no DSB increasing the on-target effect by 40-fold (Ran *et al.*, 2013; Fu *et al.*, 2014). Alongside this, the strategy of mimicking the gRNA can also result in the least undesirable effects of genome editing. The 2′-O-methyl-3′- phosphonoacetate is incorporated in the ribose phosphate backbone of gRNA. This leads to the site-specific alteration preserving the on-target activity and decreasing the off-target modification by 40- to 120-fold (Rose *et al.*, 2020). The specificity of Cas9 and Cas12 is improved by modifying the hairpin structure of gRNA 5′ upstream, which reduces the off-target effects (Ryan *et al.*, 2018; Kocak *et al.*, 2019).

8.3 Off-target Detection Tools

The prediction for off-target effects became a mandatory process, as these off-target mutations can cause severe risks in genome editing, and sometimes, these changes can be lethal. This led to the development of off-target detection methods. Algorithm-based computational methods and *in vitro/vivo* biochemical assays are the approaches the researchers discovered for detecting and quantifying the undesired effects and increasing the on-target activity. These tools are discussed below in detail.

8.3.1 CRISPR-PLANT

This tool works on the basis of an alignment-based method. The spacer sequence of gRNA specificity is predicted through genome-wide comparison of the sequence with other spacers. The specificity calculations are performed based on the number of mismatches and their position in their alignment with other spacers. Two PAM sequences – 5′-NGG-3′ and 5′-NAG-3′ – are recognized by Cas9, but less affinity is shown with NAG-PAM (Fig. 8.1A). Hence, it is recommended that the NGG-PAM sequence be used. Each gRNA spacer is aligned, checked for mismatches, and categorized based on the number of mismatches into Class 0, Class 1, and Class 2 (Fig. 8.1B).

1. **Class 0:** gRNA spacer has no sequence identity with the target genome.
2. **Class 1:** Match between gRNA and other NGG-PAM sites with ≥4 mismatches or three mismatches adjacent to PAM.
3. **Class 2:** Match with gRNA and other NGG-PAM sites with ≤3 mismatches or three mismatches distant to PAM.

The developers also introduced the minMM value, which means the minimum mismatch number after the genome-wide sequence pairing

(A)

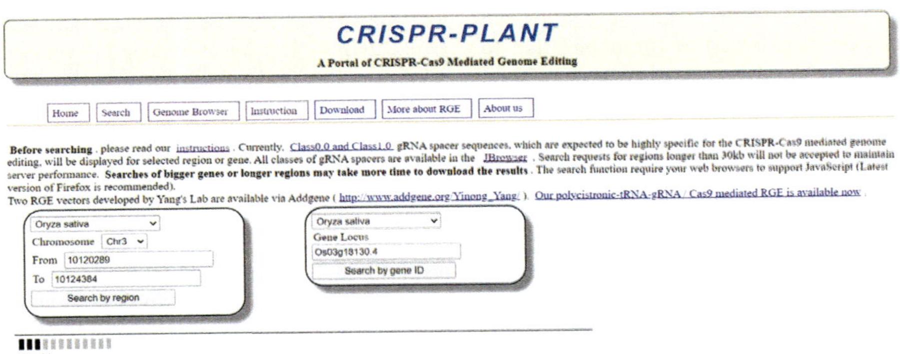

(B)

Class0.0 gRNA

SeqID	minMM_GG	minMM_AG	Spacer seq (5'->3')	PAM (5'->3')	strand	location
Chr3:10120603-10120623:c	NA	3	GGAAGCATAACGCTCGAAGC	TGGCCGGCCG	-	exon
Chr3:10123306-10123326	NA	5	TTGATTACGTTCCGCAGGGG	CGGATTTCGG	+	intron

Total of 2 class0.0 gRNA seeds were found in this region

Class1.0 gRNA

SeqID	minMM_GG	minMM_AG	Spacer seq (5'->3')	PAM (5'->3')	strand	location
Chr3:10120324-10120344	3	3	TATCGTTTCAAGGATGAGTT	CGGATGAGTG	+	exon
Chr3:10120355-10120375	3	3	TCGATCACTCTCCACATACT	TGGATCGGTC	+	exon
Chr3:10120360-10120380	4	4	CACTCTCCACATACTTGGAT	CGGTCTCTCT	+	exon
Chr3:10120388-10120408	3	3	CTCGCACACGTACAACGACA	AGGGGAGACG	+	exon
Chr3:10120390-10120410	3	3	CGCACACGTACAACGACAAG	GGGAGACGGT	+	exon

Fig. 8.1. (A,B) A screenshot of the CRISPR-PLANT web server interface and the output page of the CRISPR-PLANT web server showing the predicted gRNAs (Xie *et al.*, 2014).

in all alignments, for the evaluation of spacer sequence specificity to design the gRNAs. Here, the tools consider both substitution and gap as a mismatch in the sequence alignment to calculate the minMM.

To access the tool CRISPR-PLANT data (http://omap.org/crispr/CRISPRsearch.html), first select the plant species, the gene of interest or chromosome region, gene locus, or input a transcript unit to design the gRNA spacer. The search results show the gRNA spacer information like seqID, minMM_GG, minMM_AG, seq (gRNA sequence), PAM is the NGG, and seven additional bp of the genomic sequence, the location. The last column in the results provides a button that analyses the spacer sequence. The analysis report gives information about the best gRNA spacer for further use (Xie *et al.*, 2014).

8.3.2 CHOPCHOP

CHOPCHOP is an online intuitive domain tool for designing a gRNA targeting the specific DNA sequence, optimizing efficiency and specificity. It calculates the mismatch positions in 20 bp and the weight matrix to determine the off-target effects for each and every gRNA. The CHOPCHOP program accepts multiple input formats like fasta sequence, genomic coordinates, gene identity, RefSeq, and ENSEMBL ID available for a wide range of organisms (Fig. 8.2A). Various advanced options are also available as CRISPR modes, such as gene knock-out, knock-in, activation, repression, and nanopore enrichment. This tool also gives the option of using different Cas variants like CRISPR/Cas9, CRISPR/Cas9 nickases, CRISPR-Cpf1, and CRISPR/Cas13. The results of the tool provide the gRNA sequence with PAM, genome location, strand, GC content, self-complementarity, mismatches (MM), and efficiency (Fig. 8.2B). The result of gRNAs is given in the ranking order (Montague *et al.*, 2014; Labun *et al.*, 2016, 2019).

8.3.3 CCTop

CRISPR/Cas9 target online predictor (https://cctop.cos.uni-heidelberg.de:8043/), CCTop, is an efficient, reliable, and rapid identification tool for high-quality target sites. The friendly interface provides adjustable parameters to select the target sequence for any given DNA

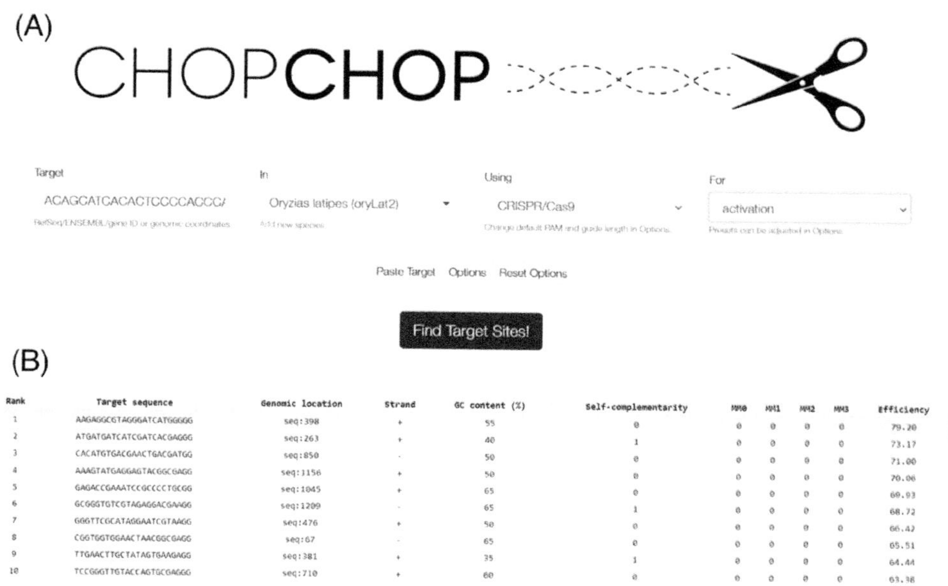

Rank	Target sequence	Genomic location	Strand	GC content (%)	Self-complementarity	MM0	MM1	MM2	MM3	Efficiency
1	AAGAGGCGTAGGGATCATGGGGG	seq:398	+	55	0	0	0	0	0	79.20
2	ATGATGATCATCGATCACGAGGG	seq:263	+	40	1	0	0	0	0	73.17
3	CACATGTGACGAACTGACGATGG	seq:850	-	50	0	0	0	0	0	71.00
4	AAAGTATGAGGAGTACGGCGAGG	seq:1156	+	50	0	0	0	0	0	70.06
5	GAGACCGAAATCCGCCCCTGCGG	seq:1045	+	65	0	0	0	0	0	69.93
6	GCGGGTGTCGTAGAGGACGAAGG	seq:1209	-	65	1	0	0	0	0	68.72
7	GGGTTCGCATAGGAATCGTAAGG	seq:476	+	50	0	0	0	0	0	66.42
8	CGGTGGTGGAACTAACGGCGAGG	seq:67	-	65	0	0	0	0	0	65.51
9	TTGAACTTGCTATAGTGAAGAGG	seq:381	+	35	1	0	0	0	0	64.44
10	TCCGGGTTGTACCAGTGCGAGGG	seq:710	+	60	0	0	0	0	0	63.38

Fig. 8.2. (A,B) A screenshot of the CHOPCHOP webserver interface and the CHOPCHOP gRNA result list (Montague *et al.*, 2014; Labun *et al.*, 2016; Labun *et al.*, 2019).

sequence. The access to the advanced parameters allows one to choose the type of PAM (NGG, NRG, NAG, etc.) and also identify the two most 5′ nucleotides (NN, GG, GN) along with 3′ two most nucleotides (NN, GG). It was easy to select the type of PAM separately to predict the off-target sites. The nuclease activity of Cas9 is experimentally evident in that it correlates strongly with the sgRNA/DNA heteroduplex and the mismatch positions (Fig. 8.3A). The mismatches distal to PAM are tolerated, while the proximal ones are taken into consideration, as they can abolish the DSB introduction. The nucleotides adjacent to PAM are the core sequence, that is, 12 bp length by default, and two mismatches in the core region can prohibit the introduction of DSB.

The interface accepts the input in FASTA sequence in single mode and uploads the sequence file in batch mode in which multiple

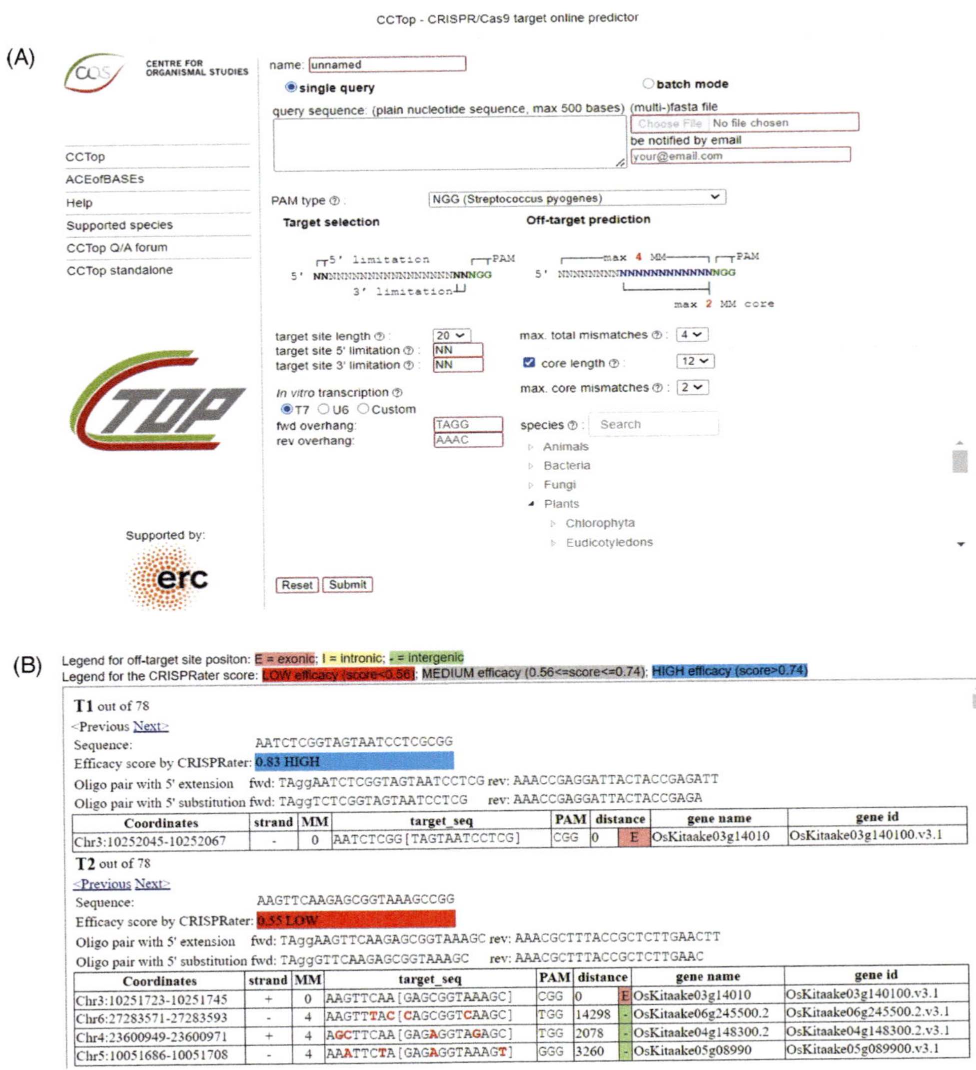

Fig. 8.3. (A,B) A screenshot of CCTop webserver interface and the gRNA list from the CCTop web server (Stemmer *et al.*, 2017; Labuhn *et al.*, 2018; Yamamoto *et al.*, 2019).

gene locus sequences can be targeted. It selects the PAM type, target sequence length, number of mismatches, core length, and plant and animal species. The resultant page that appears after the search submission contains copious information like legends for off-target site positions such as exonic, intronic, and intergenic, legends for the CRISPRater score, that is, low efficacy (score <0.56); medium efficacy (0.56 ≤ score ≤ 0.74); and high efficacy (score >0.74). Each gRNA sequence has genomic coordinates, strand, MM (mismatches), target seq, PAM, distance, gene name, and gene ID (Fig. 8.3B). The best gRNA can be selected based on the high efficacy score (Stemmer *et al.*, 2017; Labuhn *et al.*, 2018; Yamamoto *et al.*, 2019).

8.3.4 Cas-Designer and Cas-OFFinder

Cas-Designer and Cas-OFFinder is a rapid gRNA designing tool for different type II Cas9 nucleases, listing all the gRNAs possible from the provided DNA sequence. The Cas-Designer potentially calculates the off-target sites in a genome of interest through the DNA or RNA bulges and provides the out-of-frame score to select the appropriate gRNA for further experiments. With the provided DNA sequence, the Cas-Designer looks for possible target sites with a user-defined PAM sequence like 5′NGG3′, 5′NAG3′, 5′NRG3′ for SpCas9, 5′NNAGAAW3′ for StCas9, 5′NNGRRT3′ for SaCas9. The tool also allows the user to select the plant and animal species. The calculation of microhomology-associated out-of-frame score correlates positively with the frame-shift mutations (Fig. 8.4A). The Cas-Designer output page has cumulative information on cleavage position, GC content (%), out-of-frame score, and mismatches (Fig. 8.4B). The tool Cas-OFFinder identifies the potential off-target sites in the user-defined genome (Fig. 8.4C). It recognizes the off-target sites with DNA or RNA bulges using a simple brute-force approach in which the program inserts three 'N' nucleotides or deletes three nucleotides in the target sequence to recognize the off-target sites having either DNA or RNA bulges (Fig. 8.4D) (Bae *et al.*, 2014; Park *et al.*, 2015).

8.3.5 CRISPOR

CRISPOR is a tool for designing and selecting the gRNA, which has two modes, default, and batch. In default mode, the user is prompted to provide the target DNA sequence and select the organism species along with the PAM, length of gRNA, and Cas9 nucleases. Meanwhile, in batch mode, one can use the predesigned gRNAs which accept one or multiple gene identifiers from the various genome-wide libraries. The results provide broad information on gRNA, such as position/strand and gRNA sequence with PAM, along with various scores like MIT specificity score, CFD specificity score, predicted efficiency score, outcome score and mismatches, and genome browser links (Fig. 8.5A).

Specificity score: The scores (MIT, CFD) give the predicted values about the gRNA and how much it can lead to off-target sites other than the target sites. The score ranges from 0 to 100, with 100 being the best score, meaning it has not found a single sequence differing from the target at up to four positions.

Efficiency score: This score gives us information about the gRNA and its efficiency in introducing the DSB in the target DNA sequence, ranging from 0 to 100, with 100 being the best score. There are two scores displayed calculated from the two scoring methods; one by Doench *et al.* (2016) and the other by Moreno-Mateos *et al.* (2015). To calculate the efficiency score for the gRNA expressed in the cells from the U6 promoter, the Doench *et al.* (2016) scoring method is the best, while the Moreno-Mateos *et al.* (2015) method provides the best score for the gRNA expressed from the T7 promoter.

Outcomes (Out-of-frame, Lindels): This score gives information about the out-of-frame deletions led by the gRNA, generally ranging from 0 to 100. The DNA repair outcomes can be checked by clicking on the score. The Lindel score is the probability of insertions or deletions that can happen during the cleavage and repair mechanism while using the provided gRNA (Fig. 8.5B).

Off-target mismatch score: This score gives us the information for each mismatch, the number of possible off-targets in the genome. The mismatch is shown as 0-1-2-3-4, and the score represents the number of off-targets with that particular mismatch. If the score is

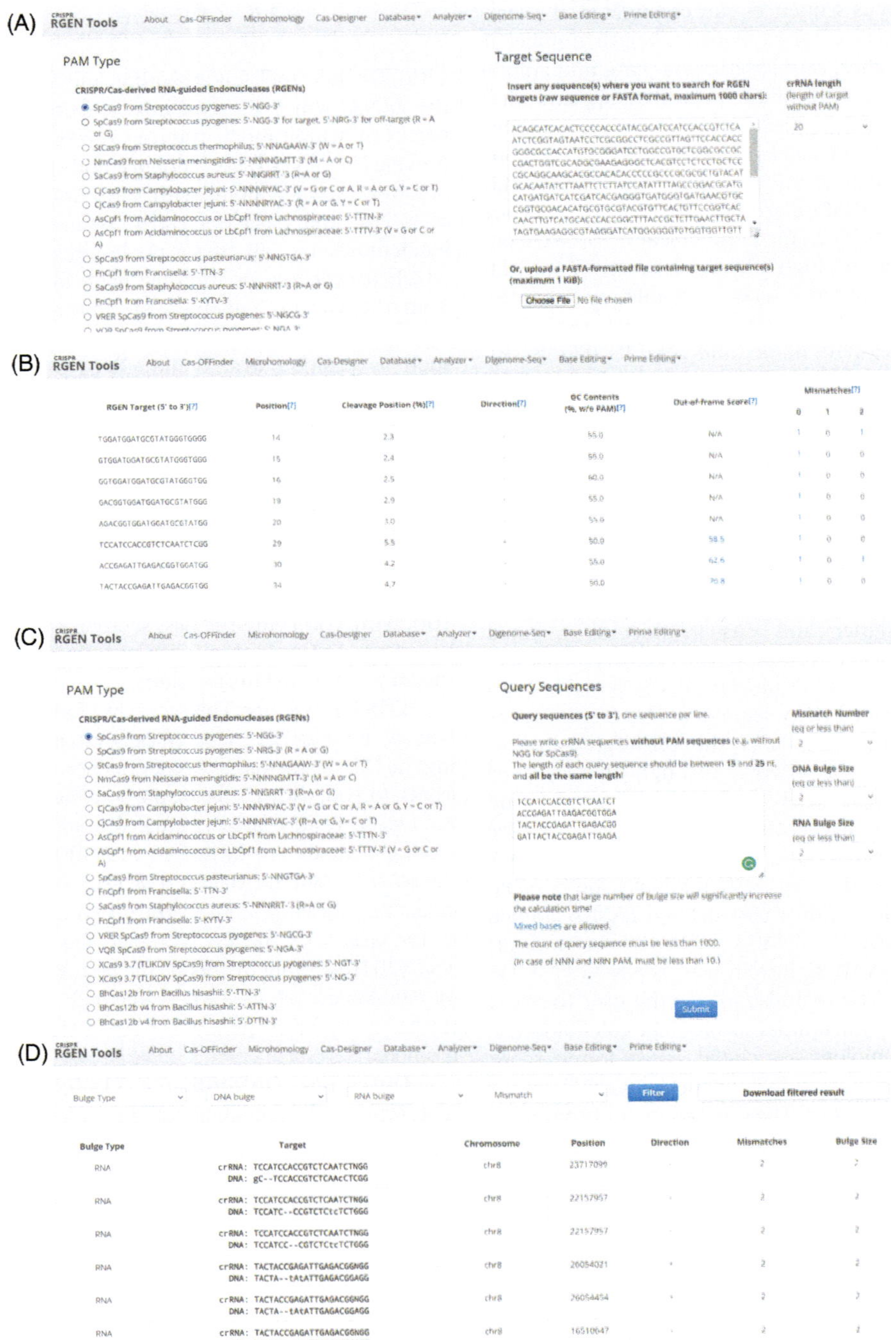

Fig. 8.4. A screenshot of Cas-Designer web server tool (A), gRNA list from Cas-designer web server (B), Cas-OFFinder webserver tool (C), and Cas-OFFinder output page showing gRNA characteristics list (D) Bae *et al*., 2014; Park *et al*., 2015.

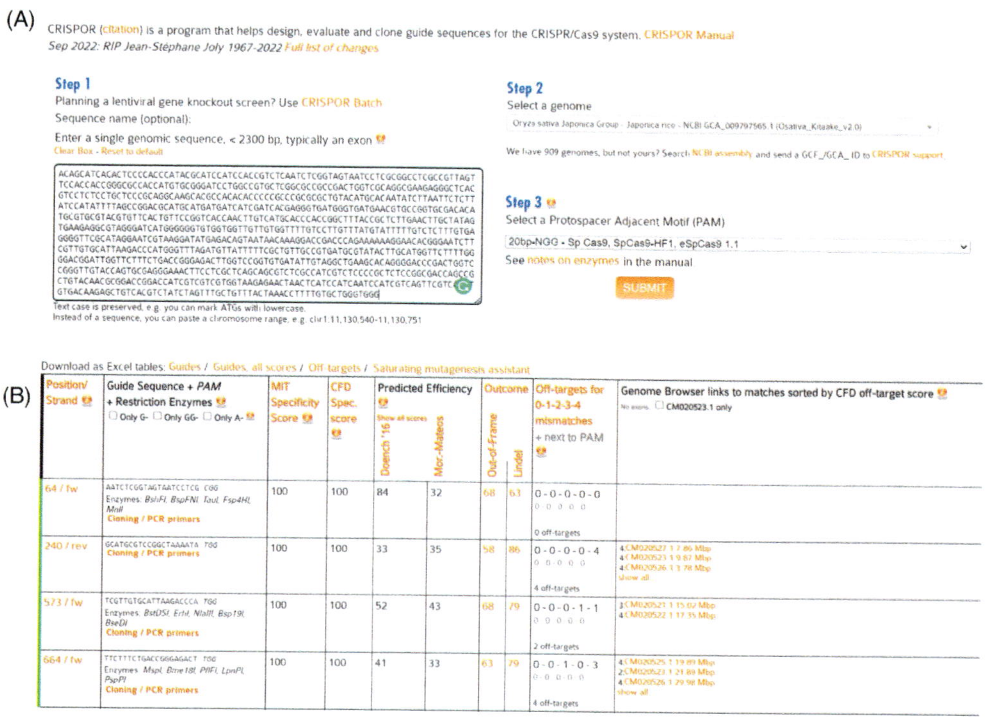

Fig. 8.5. (A,B) A screenshot of CRISPOR web server tool and CRISPOR gRNA results list.

0-3-10-8-3, there are 0 locations in the genome with 0 mismatches, 3 off-targets with one mismatch, 10 off-targets with two mismatches, 8 off-targets with three mismatches, and 3 off-targets with four mismatches adjacent to PAM (Concordet and Haeussler, 2018).

8.4 Analyzing Secondary Structure of gRNA

The sequence features of the gRNA primarily determine the specificity and efficacy of the CRISPR-Cas9-associated genome editing; in addition, the success rate is also determined by the secondary structure of sgRNA/gRNA. The nucleotides of the functionally active gRNA positioned at 18–20 at the 3′ end are more easily accessible when compared to the functionally inactive gRNA. Along with the seed sequence (18–20 nucleotides 3′ end), the availability of nucleotides at position 51–53 of sgRNA is also distinguishable in functionally active and inactive sgRNA. Thus, secondary structure analysis is required to differentiate the active and inactive sgRNAs (Nishimasu *et al.*, 2014; Ma *et al.*, 2015; Wong *et al.*, 2015; Xu *et al.*, 2017; Bruegmann *et al.*, 2019; Kocak *et al.*, 2019; Matson *et al.*, 2019; Zhang *et al.*, 2019). There are four stem-loop structures in an effective sgRNA for its efficient processing. They are RAR stem-loop, stem-loop 1, stem-loop 2, and stem-loop 3. The stem-loop has GAAA as the sequence, which is unpaired (Fig. 8.6). The gRNA (crRNA) processing from its precursor crRNA (pre-crRNA) is made available to bind with Cas9 protein by the RAR stem-loop, and the functional Cas9-sgRNA-DNA complex formation requires the stem-loop 1. The stem-loops 2 and 3 promote the stability of the complex, improving the *in vivo* activity of Cas9. It was demonstrated that the stem-loop 1 structure is lost in plants, which implies that it does not contribute to the efficacy. Hence, in plants, stem-loop 1 is neglected. The functionally active sgRNA shows, when analyzing its predictive structure, a stable stem-loop structure at the nucleotides positioned between 21 and 50, indicating the loop formation in the tracrRNA region. In contrast, the functionally inactive sgRNA predictive structure analysis shows that the 50–53 positioned nucleotides are paired with the 18–20 positioned nucleotides at 3′ end, which is the seed sequence region, forming an extended loop. As studies show, the seed region (18–20 bps) helps recognize the PAM sequence, and the extended loop makes these nucleotides inaccessible, losing their activity and resulting in functionally inactive sgRNA. Thus, this shows the importance of intramolecular structures in the efficiency of gRNA (Nishimasu *et al.*, 2014; Liang *et al.*, 2016).

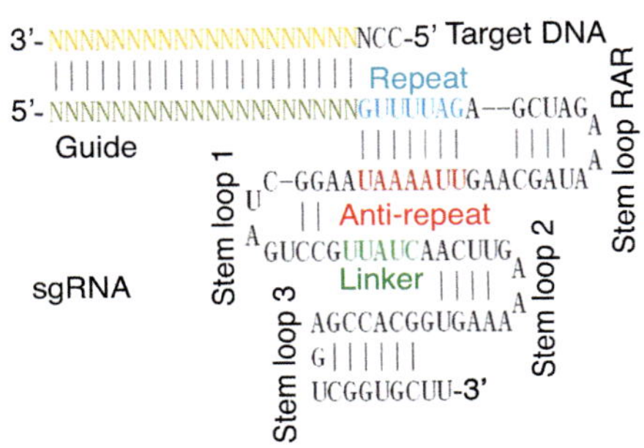

Fig. 8.6. Schematic diagram of the secondary structure of sgRNA. Black lines indicate the Watson-Crick and non-Watson-Crick base pairs, respectively. The dashed line box indicates disordered nucleotides, the sequence is obtained from Liang *et al.*, 2016.

Along with these aspects, the free energy (ΔG) of the self-folding potential of the gRNA also influences the functionality. The self-folding potential is the formation of secondary structures within the gRNA, a part of sgRNA. Therefore, the more negative ΔG value, the more possibility of the formation of secondary structures within the gRNA, making it inaccessible to bind with the target. Hence, the secondary structure analysis of the gRNA is mandatory for which the software RNAfold is used to determine the self-folding potential of sgRNA (Wong *et al.*, 2015; Jensen *et al.*, 2017). The RNAfold (http://rna.tbi.univie.ac.at/cgi-bin/RNAWebSuite/RNAfold.cgi) interface provides three sections: the first one allows the user to input the sequence information; the second allows the user to select the appropriate algorithm; and, finally, the third one allows the output options (Fig. 8.7) (Gruber *et al.*, 2008; Lorenz *et al.*, 2011). To analyze the secondary structure of sgRNA, it is better to select the minimum free energy (MFE) and partition fold option in the fold algorithm and basic options section, keeping the advanced option default. Check all three options for output. Once the user inserts the sequence of RNA and selects the appropriate option to proceed with the job, the software is redirected to the output page. The graphical

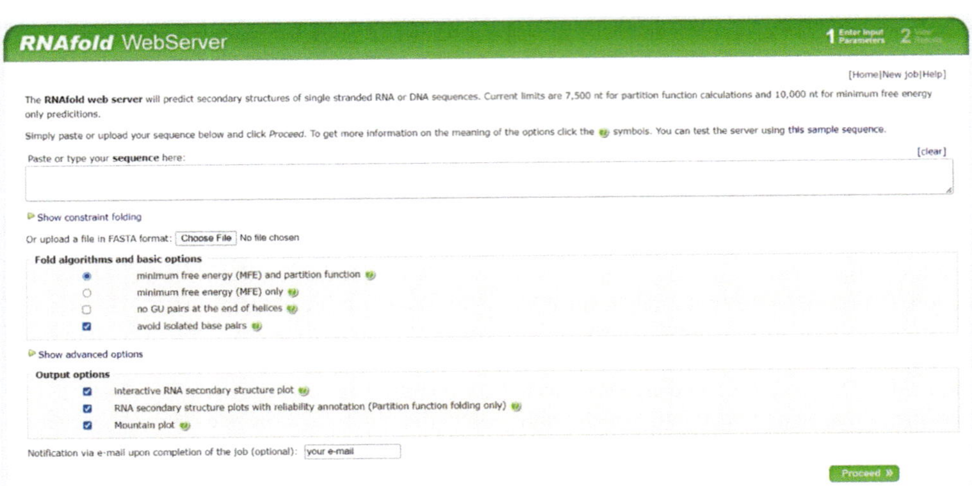

Fig. 8.7. A screenshot of the RNAfold web server (Gruber *et al.*, 2008; Lorenz *et al.*, 2011).

representation section shows two output images: MFE secondary structure and centroid secondary structure. The high similarities in these structures give a more reliable prediction.

One also has to look for the presence of key structural motifs like RAR loop, stem-loop 2, and stem-loop 3 to interpret gRNA secondary structure. In the absence of any of these stem-loops, the gRNA is not a good candidate for genome editing, as it lacks the structural properties for processing and binding of Cas9. The gRNA part of sgRNA is analyzed using the same procedure, and the self-folding free energy is calculated for the predictive secondary structure of the gRNA (Fig. 8.8A, B). In the first section of the output page, the self-folding free energy is provided. The studies reported that the value of self-folding free energy should be between 0 and 2.0 kcal/mol, and then the efficiency of the gRNA will be the highest. The secondary structure of gRNA will be investigated, whether it is paired or unpaired, at the seed sequence region. The paired seed

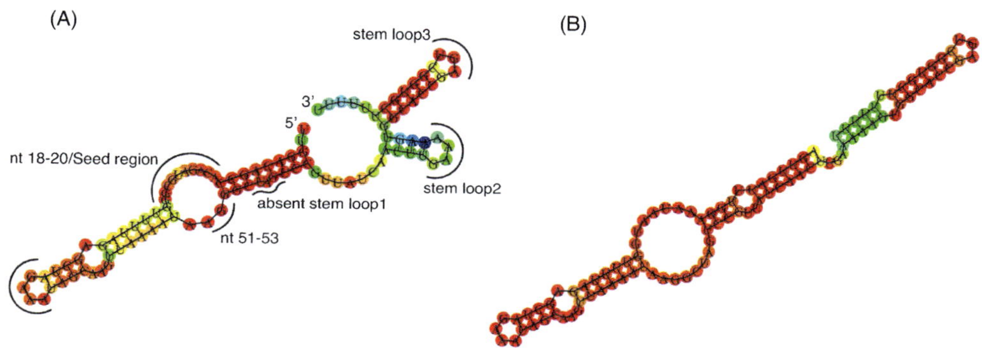

Fig. 8.8. The secondary structure of sgRNA's schematic diagram. (A) The high-performing sgRNA's predictive secondary structure. (B) The low-performing sgRNA's predictive secondary structure. In this case, the essential stem-loop structure required for gRNA processing and binding to the Cas9 is absent in the low-performing sgRNA. The sequence of sgRNA for this analysis was obtained from Yamamoto *et al.*, 2019.

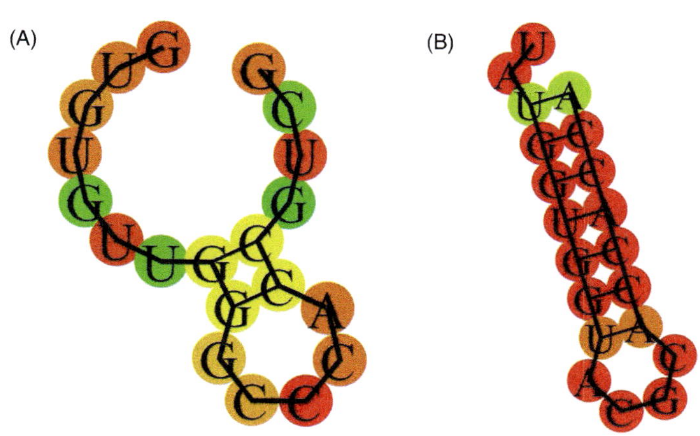

Fig. 8.9. Schematic representation of the predictive secondary structure of gRNA. (A) Represents the high-performing gRNA's predictive secondary structure; the free energy of self-folding potential of high-performing gRNA is ΔG = 1.30 kcal/mole. (B) Represents the low-performing gRNA's predictive secondary structure and low-performing gRNA is ΔG = −2.40 kcal/mole, respectively. gRNA sequence for this analysis was obtained from Yamamoto *et al.*, 2019.

region of the gRNA will not be able to access the target site (Fig. 8.9A, B). In the case of paired seed sequence region, selecting another gRNA that is unpaired is suggested (Hassan *et al.*, 2021).

8.5 Conclusion and Future Perspectives

The cutting-edge and high-throughput genome editing tool that has emerged in recent years is the CRISPR/Cas technology. By exploiting the bacterial immune system, this advanced technology is widely used in editing the genome of almost all plant and animal species. The CRISPR/Cas technology is an important part of the single-guide RNA (sgRNA), which helps identify, unwind, and bind, thus activating the nuclease for DNA cleavage. The CRISPR/Cas-mediated genome editing outcomes depend solely on the gRNA used, ultimately leading to off-target mutations, affecting the organism's normal mechanism, and sometimes can be lethal. Hence, minimizing the off-target effects and increasing the on-target engineering is necessary. Therefore, the careful designing of gRNA is very important because the sequence of gRNA information plays a vital role. Along with sequence information, the formation of secondary structures within gRNA also alters the outcomes of genome editing. In this book chapter, we presented some details of the tools that help the user design and evaluate the efficiency and specificity of the gRNA. These tools take into account various aspects in calculating their efficiency, depending on the sequence information of gRNA. The RNAfold tool helps identify the secondary structures like loops and free energy potential of gRNA, which lead to it forming structures within itself, making it unavailable to bind with the target. Designing the strategy using these tools helps minimize off-target mutations and helps in genome editing with high-throughput results. All the tools we discussed in this book chapter are *in silico* tools that help in designing the best gRNA to proceed with genome editing. Once these predicted best gRNAs are shortlisted, one can evaluate their efficiency through *in vivo* techniques. It is clear that the procedure of continuing to experiment with genome editing of the organism can be followed with success with the best-chosen gRNA.

References

Alkan, F., Wenzel, A., Anthon, C., Havgaard, J.H. and Gorodkin, J. (2018) CRISPR-Cas9 off-targeting assessment with nucleic acid duplex energy parameters. *Genome Biology* 19, 1–13.

Bae, S., Park, J. and Kim, J.S. (2014) Cas-OFFinder: A fast and versatile algorithm that searches for potential off-target sites of Cas9 RNA-guided endonucleases. *Bioinformatics* 30(10), 1473–1475.

Barrangou, R. and Doudna, J.A. (2016) Applications of CRISPR technologies in research and beyond. *Nature Biotechnology* 34, 933–941.

Bogdanove, A.J., Bohm, A., Miller, J.C., Morgan, R.D. and Stoddard, B.L. (2018) Engineering altered protein–DNA recognition specificity. *Nucleic Acids Research* 46, 4845–4871.

Bruegmann, T., Deecke, K. and Fladung, M. (2019) Evaluating the efficiency of grnas in CRISPR/Cas9 mediated genome editing in poplars. *International Journal of Molecular Sciences* 20(15), 3623.

Cai, Y., Chen, L., Sun, S., Wu, C., Yao, W. *et al.* (2018) CRISPR/Cas9-mediated deletion of large genomic fragments in soybean. *International Journal of Molecular Sciences* 19, 3835.

Carroll, D. (2011) Genome engineering with zinc-finger nucleases. *Genetics* 188(4), 773–782. DOI: 10.1534/genetics.111.131433.

Cho, S.W., Kim, S., Kim, Y., Kweon, J., Kim, H.S. *et al.* (2014) Analysis of off-target effects of CRISPR/cas-derived RNA-guided endonucleases and nickases. *Genome Research* 24, 132–141.

Concordet, J.P. and Haeussler, M. (2018) CRISPOR: Intuitive guide selection for CRISPR/Cas9 genome editing experiments and screens. *Nucleic Acids Research* 46(W1), W242–W245. DOI: 10.1093/nar/gky354.

Cong, L., Ran, F.A., Cox, D., Lin, S., Barretto, R. *et al.* (2013) Multiplex genome engineering using CRISPR/Cas systems. *Science* 339, 819–823.

Doench, J.G., Hartenian, E., Graham, D.B., Tothova, Z., Hegde, M. *et al.* (2014) Rational design of highly active sgRNAs for CRISPR-Cas9–mediated gene inactivation. *Nature Biotechnology* 32, 1262–1267.

Doench, J.G., Fusi, N., Sullender, M., Hegde, M., Vaimberg, E.W. *et al.* (2016) Optimized sgRNA design to maximize activity and minimize off-target effects of CRISPR-cas9. *Nature Biotechnology* 34(2), 184–191. DOI: 10.1038/nbt.3437.

Endo, M., Mikami, M. and Toki, S. (2015) Multigene knockout utilizing off-target mutations of the CRISPR/Cas9 system in rice. *Plant Cell & Physiology* 56, 41–47.

Foss, D.V., Hochstrasser, M.L. and Wilson, R.C. (2019) Clinical applications of CRISPR-based genome editing and diagnostics. *Transfusion* 59, 1389–1399.

Fu, Y., Foden, J.A., Khayter, C., Maeder, M.L., Reyon, D. *et al.* (2013) High-frequency off-target mutagenesis induced by CRISPR-Cas nucleases in human cells. *Nature Biotechnology* 31, 822–826.

Fu, Y., Sander, J.D., Reyon, D., Cascio, V.M. and Joung, J.K. (2014) Improving CRISPR-Cas nuclease specificity using truncated guide RNAs. *Nature Biotechnology* 32, 279–284.

Gruber, A.R., Lorenz, R., Bernhart, S.H., Neuböck, R. and Hofacker, I.L. (2008) The vienna RNA website. *Nucleic Acids Research* 36(Suppl-2), W70–W74.

Guilinger, J.P., Pattanayak, V., Reyon, D., Tsai, S.Q., Sander, J.D. *et al.* (2014) Broad specificity profiling of TALENs results in engineered nucleases with improved DNA-cleavage specificity. *Nature Methods* 11, 429–435.

Gupta, R.M. and Musunuru, K. (2014) Expanding the genetic editing tool kit: ZFNs, TALENs, and CRISPR-Cas9. *Journal of Clinical Investigation* 124, 4154–4161.

Hassan, M.M., Chowdhury, A.K. and Islam, T. (2021) *In silico* analysis of gRNA secondary structure to predict its efficacy for plant genome editing. *CRISPR-Cas Methods* 2, 15–22.

Hiranniramol, K., Chen, Y., Liu, W. and Wang, X. (2020) Generalizable sgRNA design for improved CRISPR/Cas9 editing efficiency. *Bioinformatics* 36, 2684–2689.

Hsu, P.D., Scott, D.A., Weinstein, J.A., Ran, F.A., Konermann, S. *et al.* (2013) DNA targeting specificity of RNA-guided Cas9 nucleases. *Nature Biotechnology* 31, 827–832.

Jensen, K.T., Fløe, L., Petersen, T.S., Huang, J., Xu, F. *et al.* (2017) Chromatin accessibility and guide sequence secondary structure affect CRISPR-Cas9 gene editing efficiency. *FEBS Letters* 591(13), 1892–1901.

Jiang, W., Bikard, D., Cox, D., Zhang, F. and Marraffini, L.A. (2013) RNA-guided editing of bacterial genomes using CRISPR-CAS systems. *Nature Biotechnology* 31, 233–239.

Jinek, M., Chylinski, K., Fonfara, I., Hauer, M., Doudna, J.A. *et al.* (2012) A programmable dual-RNA–guided DNA endonuclease in adaptive bacterial immunity. *Science* 337, 816–821.

Joung, J.K. and Sander, J.D. (2013) TALENs: A widely applicable technology for targeted genome editing. *Nature Reviews Molecular Cell Biology* 14(1), 49–55.

Kocak, D.D., Josephs, E.A., Bhandarkar, V., Adkar, S.S., Kwon, J.B. *et al.* (2019) Increasing the specificity of CRISPR systems with engineered RNA secondary structures. *Nature Biotechnology* 37, 657–666.

Labuhn, M., Adams, F.F., Ng, M., Knoess, S., Schambach, A. *et al.* (2018) Refined sgRNA efficacy prediction improves large-and small-scale CRISPR–Cas9 applications. *Nucleic Acids Research* 46(3), 1375–1385.

Labun, K., Montague, T.G., Gagnon, J.A., Thyme, S.B. and Valen, E. (2016) CHOPCHOP v2: A web tool for the next generation of CRISPR genome engineering. *Nucleic Acids Research* 44(W1), W272–W276.

Labun, K., Montague, T.G., Krause, M., Torres Cleuren, Y.N., Tjeldnes, H. *et al.* (2019) CHOPCHOP v3: Expanding the CRISPR web toolbox beyond genome editing. *Nucleic Acids Research* 47(W1), W171–W174.

Liang, G., Zhang, H., Lou, D. and Yu, D. (2016) Selection of highly efficient sgRNAs for CRISPR/Cas9-based plant genome editing. *Scientific Reports* 6(1), 21451.

Liao, H.K., Gu, Y., Diaz, A., Marlett, J., Takahashi, Y. *et al.* (2015) Use of the CRISPR/Cas9 system as an intracellular defense against HIV-1 infection in human cells. *Nature Communication* 6, 1–10.

Lin, Y., Cradick, T.J., Brown, M.T., Deshmukh, H., Ranjan, P. *et al.* (2014) CRISPR/Cas9 systems have off-target activity with insertions or deletions between target DNA and guide RNA sequences. *Nucleic Acids Research* 42, 7473–7485.

Liu, H., Ding, Y., Zhou, Y., Jin, W., Xie, K. *et al.* (2017) CRISPR-P 2.0: an improved CRISPR-Cas9 tool for genome editing in plants. *Molecular Plant* 10(3), 530–532.

Lorenz, R., Bernhart, S.H., Höner Zu Siederdissen, C., Tafer, H., Flamm, C. *et al.* (2011) ViennaRNA package 2.0. *Algorithms for Molecular Biology: AMB* 6, 26. DOI: 10.1186/1748-7188-6-26.

Ma, X., Zhang, Q., Zhu, Q., Liu, W., Chen, Y. *et al.* (2015) A robust CRISPR/Cas9 system for convenient, high-efficiency multiplex genome editing in monocot and dicot plants. *Molecular Plant* 8(8), 1274–1284.

Matson, A.W., Hosny, N., Swanson, Z.A., Hering, B.J. and Burlak, C. (2019) Optimizing sgRNA length to improve target specificity and efficiency for the GGTA1 gene using the CRISPR/Cas9 gene editing system. *PLoS One* 14(12), e0226107.

Molla, K.A., Sretenovic, S., Bansal, K.C. and Qi, Y. (2021) Precise plant genome editing using base editors and prime editors. *Nature Plants* 7(9), 1166–1187.

Montague, T.G., Cruz, J.M., Gagnon, J.A., Church, G.M. and Valen, E. (2014) CHOPCHOP: A CRISPR/Cas9 and TALEN web tool for genome editing. *Nucleic Acids Research* 42(Web Server issue), W401–W407.

Moreno-Mateos, M.A., Vejnar, C.E., Beaudoin, J.D., Fernandez, J.P., Mis, E.K. *et al.* (2015) CRISPRscan: Designing highly efficient sgRNAs for CRISPR-Cas9 targeting *in vivo*. *Nature Methods* 12(10), 982–988. Available at: https://doi.org/10.1038/nmeth.3543

Naeem, M., Majeed, S., Hoque, M.Z. and Ahmad, I. (2020) Latest developed strategies to minimize the off-target effects in CRISPR-Cas-mediated genome editing. *Cells* 9(7), 1608.

Nishimasu, H., Ran, F.A., Hsu, P.D., Konermann, S., Shehata, S.I. *et al.* (2014) Crystal structure of Cas9 in complex with guide RNA and target DNA. *Cell* 156(5), 935–949.

Park, J., Bae, S. and Kim, J.S. (2015) Cas-designer: A web-based tool for choice of CRISPR-Cas9 target sites. *Bioinformatics* 31(24), 4014–4016.

Parmar, N., Singh, K.H., Sharma, D., Singh, L., Kumar, P. *et al.* (2017) Genetic engineering strategies for biotic and abiotic stress tolerance and quality enhancement in horticultural crops: A comprehensive review. *3 Biotech* 7, 239.

Pattanayak, V., Lin, S., Guilinger, J.P., Ma, E., Doudna, J.A. *et al.* (2013) High-throughput profiling of off-target DNA cleavage reveals RNA-programmed Cas9 nuclease specificity. *Nature Biotechnology* 31, 839.

Ran, F.A., Hsu, P.D., Lin, C.Y., Gootenberg, J.S., Konermann, S. *et al.* (2013) Double nicking by RNA-guided CRISPR Cas9 for enhanced genome editing specificity. *Cell* 154, 1380–1389.

Ren, C., Liu, X., Zhang, Z., Wang, Y., Duan, W. *et al.* (2016) CRISPR/Cas9-mediated efficient targeted mutagenesis in chardonnay (*Vitis vinifera* L.). *Scientific Reports* 6, 32289.

Ren, X., Yang, Z., Xu, J., Sun, J., Mao, D. *et al.* (2014) Enhanced specificity and efficiency of the CRISPR/Cas9 system with optimized sgRNA parameters in *Drosophila*. *Cell Reports* 9, 1151–1162.

Rose, J.C., Popp, N.A., Richardson, C.D., Stephany, J.J., Mathieu, J. *et al.* (2020) Suppression of unwanted CRISPR-Cas9 editing by co-administration of catalytically inactivating truncated guide RNAs. *Nature Communication* 11, 1–11.

Ryan, D.E., Taussig, D., Steinfeld, I., Phadnis, S.M., Lunstad, B.D. *et al.* (2018) Improving CRISPR–Cas specificity with chemical modifications in single-guide RNAs. *Nucleic Acids Research* 46, 792–803.

Shah, T., Andleeb, T., Lateef, S. and Noor, M.A. (2018) Genome editing in plants: Advancing crop transformation and overview of tools. *Plant Physiology and Biochemistry* 131, 12–21.

Singh, D., Sternberg, S.H., Fei, J., Doudna, J.A. and Ha, T. (2016) Real-time observation of DNA recognition and rejection by the RNA-guided endonuclease Cas9. *Nature Communication* 7, 1–8.

Stemmer, M., Thumberger, T., Sol Keyer, M., Wittbrodt, J. and Mateo, J.L. (2017) Correction: CCTop: An intuitive, flexible and reliable CRISPR/Cas9 target prediction tool. *PLoS ONE* 12(4), e0176619.

Sugano, S.S., Nishihama, R., Shirakawa, M., Takagi, J., Matsuda, Y. *et al.* (2018) Efficient CRISPR/Cas9-based genome editing and its application to conditional genetic analysis in *Marchantia polymorpha*. *PLoS ONE* 13, e0205117.

Wang, Y., Cheng, X., Shan, Q., Zhang, Y., Liu, J. *et al.* (2014) Simultaneous editing of three homoeoalleles in hexaploid bread wheat confers heritable resistance to powdery mildew. *Nature Biotechnology* 32, 947–951.

Wilson, L.O., O'Brien, A.R. and Bauer, D.C. (2018) The current state and future of CRISPR-Cas9 gRNA design tools. *Frontiers in Pharmacology* 9, 749.

Wong, N., Liu, W. and Wang, X. (2015) WU-CRISPR: Characteristics of functional guide RNAs for the CRISPR/Cas9 system. *Genome Biology* 16(1), 1–8.

Xie, K., Zhang, J. and Yang, Y. (2014) Genome-wide prediction of highly specific guide RNA spacers for CRISPR–Cas9-mediated genome editing in model plants and major crops. *Molecular Plant* 7(5), 923–926.

Xu, J., Lian, W., Jia, Y., Li, L. and Huang, Z. (2017) Optimized guide RNA structure for genome editing via Cas9. *Oncotarget* 8(55).

Yamamoto, A., Ishida, T., Yoshimura, M., Kimura, Y. and Sawa, S. (2019) Developing heritable mutations in *Arabidopsis thaliana* using a modified CRISPR/Cas9 toolkit comprising PAM-altered Cas9 variants and gRNAs. *Plant and Cell Physiology* 60(10), 2255–2262.

Yin, H., Song, C.Q., Suresh, S., Kwan, S.Y., Wu, Q. *et al*. (2018) Partial DNA-guided Cas9 enables genome editing with reduced off-target activity. *Nature Chemical Biology* 14, 311–316.

Zhang, D., Hurst, T., Duan, D. and Chen, S.J. (2019) Unified energetics analysis unravels SpCas9 cleavage activity for optimal gRNA design. *Proceedings of the National Academy of Sciences* 116(18), 8693–8698.

9 Estimation and Quantification Techniques to Analyze Targeted Mutations during Genome Editing

Asadullah[1] and S.R.U. Shah[2*]

[1]Department of Plant Breeding and Genetics, Directorate of Agriculture Research Wayaro Farm, Lasbela, Pakistan; [2]Department of Soil and Environment, Swedish University of Agricultural Sciences, Uppsala, Sweden

Abstract

In the last few decades, conventional breeding, combined with transgenic methods and mutagenesis, has played a crucial role in accelerating the pace of genetic research and the application of genome editing in various fields, including agriculture and medicine. Genome editing technologies have revolutionized the field of genetics, offering precise regulation over the genetic makeup of organisms. A rapid assay for estimating the frequencies of various induced mutations during genome editing is of key importance. Understanding and quantifying the mutations introduced during genome editing processes are crucial for assessing the efficiency and specificity of these technologies. Therefore, many techniques are used to quantify genome editing-derived mutations, depending on their type and the level of sensitivity required. This chapter provides an in-depth exploration of estimation and quantification techniques used to analyze targeted mutations, with a focus on high-resolution melting (HRM) analysis, targeting induced local lesions in genomes (TILLING), CRISPR ribonucleoprotein (RNP) complexes, and mutation tracking tools like tracking of indels by decomposition (TIDE) and tracking of insertion, deletions and recombination events (TIDER). In addition, next-generation sequencing (NGS), including Nanopore, Ion Torrent, Illumina, Pacific Biosciences (PacBio), sequencing by oligonucleotide ligation and detection (SOLiD), BGI sequencer (BGISEQ), and 454 pyrosequencing, is briefly discussed to evaluate the sensitivity of existing bioinformatics tools and proposing strategies to minimize the error to identify unintended mutations.

Keywords: Plant genome editing, Targeted mutations, High-resolution melting analysis (HRM), Targeting Induced Local Lesions in Genomes (TILLING), CRISPR ribonucleoprotein (RNP) complexes, Mutation tracking, Next-generation sequencing (NGS)

9.1 Introduction

The regeneration of plants with enhanced or desirable traits has traditionally depended on time-consuming and labor-intensive breeding techniques. Recently, modern technologies, such as genome editing in plants, have led to a new era of genetic engineering, facilitating rapid, effective as well as precise modifications within plant genomes. It serves as a crucial tool, with the potential to unlock significant benefits for the consumers, the environment, and farmers by helping them cultivate much healthier food, more sustainable fuel, and fiber. In recent years,

*Corresponding author: syed.shah@slu.se

 Genome Editing for Crop Improvement: Theory and Methodology (eds J.M. Al-Khayri *et al.*)
DOI: 10.1079/9781800622517.0009

this technology has been utilized to perform site-directed genome mutagenesis, which provides an effective strategy to create mutations in DNA sequences for targeted genes.

In general, the process of inducing mutations within the targeted DNA through genetic engineering, typically involves three stages. Primarily, an exogenous engineered nuclease, comprising a nuclease domain and a recognition module, identifies the designated sequence of DNA. Subsequently, this nuclease attaches to the targeted sequence, inducing double-strand breaks (DSBs) at or near the designated site, which is repaired by homology-directed repair (HDR) or endogenous non-homologous end joining (NHEJ) pathways. However, HDR provides a precise restoration of DSBs, whereas, NHEJ, which is an error-prone repair process, frequently leads to the introduction of mutations in the form of small insertions and deletions (Indels) (Osakabe and Osakabe, 2015; Wang *et al.*, 2016).

Three major approaches have recently been developed as major genome editing technologies (Jaganathan *et al.*, 2018). Initially, zinc-finger nuclease (ZFN) and transcription activator-like effector nuclease (TALEN) emerged as a much more flexible engineered nuclease. Subsequently, the new discovery of clustered regularly interspaced short palindromic repeat (CRISPR) enables more precise and efficient targeted point mutations and is broadly adopted for the regeneration of novel germplasm along with genetic variations that exhibit unique morphological features (Molla *et al.*, 2021). Hence, a primary challenge lies in developing cost-effective and high-throughput strategies to assess the target mutations induced by these engineered nucleases as well as the non-targeted mutations or the downstream altered responses of the targeted alterations.

The following are some of the molecular techniques used for the quantification of mutations derived from genome editing, depending on their type and the required level of sensitivity.

9.2 High-resolution Melting

High-resolution melting (HRM) is a rapid, homogeneous, and extremely effective genotyping technique for the detection of mutations and genotyping across the multiple platforms of genome editing (ZFNs, TALENs, and CRISPRs). In general, HRM is a post-PCR technique based on fluorescence that assesses nucleotide variations in twisted or double-stranded DNA at various temperatures (Table 9.1). This allows for the determination of the melting temperature (T_m) of a specific PCR product, as it relies on PCR melting curve techniques (Reed *et al.*, 2007). It characterizes nucleic acids by evaluating their dissociation behavior and identifies small sequential variations in PCR-amplified sequences through direct melt curve analysis. Additionally, samples are discriminated based on factors such as GC content, sequence length as well as strand complementarity. By employing sophisticated analysis software, advanced instruments, and specific DNA dyes, genetic differences can be accurately estimated using HRM analysis.

In addition, the HRM technique was successfully implemented to quantify a single base pair deleted mutant within a sample of apple scab (fungus) with 5% mutant genome only, signifying high sensitivity for the screening of mutation (Rocafort *et al.*, 2022). Similarly, another study reported the PCR-HRM-based estimation of mutations in a genome of coffee (*Coffea arabica* L.) through chemical mutagenesis, where this protocol permits to quantify variants in pooled DNA samples of up to four M_2 mutant plants (Gatica-Arias *et al.*, 2023). Moreover, Raizada and Souframanien (2021) reported that 19 genic-SNP markers produced distinct polymorphic melting curves and were able to discriminate intraspecific accessions of black gram germplasm through HRM analysis.

9.2.1 Mechanism

The working principle for HRM analysis relies on monitoring changes in fluorescence during the melting behavior of double-stranded DNA. This is achieved using a highly precise optical detection system and saturated DNA-binding dye, eliminating the need for expensive labeled probes. In cases where a homoduplex product is derived from homozygous DNA, it exhibits a specific melting temperature (T_m). Conversely, a heteroduplex product originating from a heterozygous individual carries an additional T_m, typically representing a significantly lower T_m sign, which expedites the proof of identity for mutant alleles as shown in Fig. 9.1 (A, B). This protocol necessitates no further manipulations,

Table 9.1. Mutation detection techniques to identify and study various traits in crops and model plants. Author's own table.

Crop species (ploidy level)	Techniques to detect mutation	Size (M_2)	Frequency of mutation (1 mutation per kb)	Characteristics	Bibliography
Rice (2×)	TILLING	1860	–	–	Casella *et al*. (2013)
	NGS	2048	1/293	Phytic acid metabolism	Kim and Tai (2014)
Tobacco (2×)		3072	1/1423	Leaf yield	Reddy *et al*. (2012)
Groundnut (4×)		768	1/1066	Stress resistance	Guo *et al*. (2015)
Tomato (2×)	TILLING	3052	–	Ascorbate biosynthesis	Baldet *et al*. (2013)
		–	–	Fruit set mechanisms	Mazzucato *et al*. (2015)
	HRM	8225	1/737	Proline biosynthesis	Gady *et al*. (2009)
Brassica napus (Canola) (4×)	NGS	3158	1/109	Mutation in genes of interest	Gilchrist *et al*. (2013)
Sorghum (2×)		1024	<0.1%	–	Nida *et al*. (2016)
Wheat (4×)	HRM	3992	–	Starch metabolism	Bovina *et al*. (2014), Sestili *et al*. (2015)
Wheat (6×)		4500	30,682	Starch quality	Sestili *et al*. (2010), Botticella *et al*. (2011)
		2020	44,952	Powdery mildew disease	Acevedo-Garcia *et al*. (2016)
Crambe abyssinica		2015		CaFAD2 genes	Cheng *et al*. (2015)

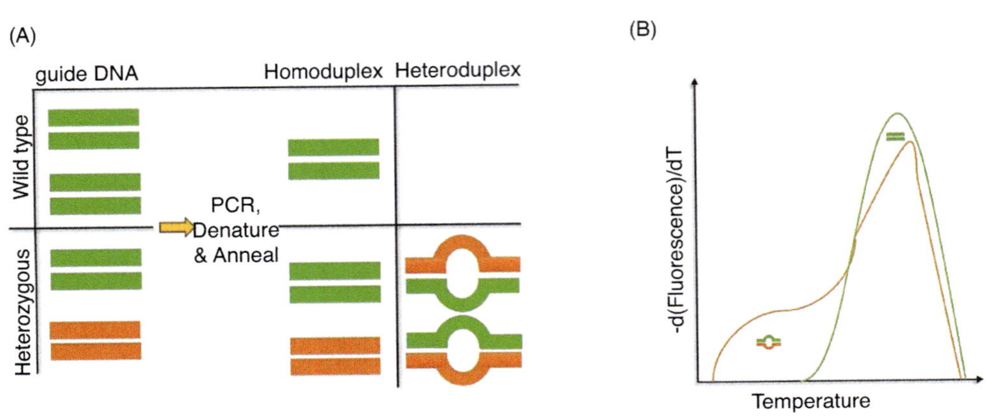

Fig. 9.1. Heteroduplex signature expedites identification of mutant alleles. Author's own figure.

such as agarose gels or digestions, following the PCR. Instead, it involves a simple 5 min HRM scan, followed by *in silico* analysis (Thomas *et al.*, 2014).

In the case of a homozygous wild-type sample (depicted by double green lines in Fig. 9.1A), only one species is generated after PCR, resulting in a distinct HRM curve (illustrated by the green curve in Fig. 9.1B). However, in the case of a heterozygous sample with mutation (double orange lines), two homoduplex species are generated after PCR application (green and orange double lines). Each homoduplex species holds a distinct HRM curve that can be distinguished if the temperature differences (T_m) are substantial enough. Significantly, for the detection of novel mutations, if the samples from PCR are subjected to a high temperature of 95°C followed by rapid cooling, heteroduplex products (indicated by orange over green and green over orange in A) can be formed. This is attributed to reduced complementarity, symbolized by bubbles in the lines in A, leading to a substantially lower (T_m) and, consequently, a curve profile appearing to the left of the homoduplex (see green over orange in B).

Advantages

- It is a cost-effective and time-saving tool.
- It depicts high sensitivity.
- It involves reduced enzymatic screening steps.

Disadvantages

- The detection sensitivity is restricted to amplicons of less than 450 base pairs.
- It requires multidimensional pooling.
- The detection of small indels might be hard as compared to substitutions.
- It strongly depends on high-quality dyes and PCR instruments.
- It may have limitations in detecting small insertions and deletions, particularly in larger amplicons than 0.5 kbs.

9.3 Targeting Induced Local Lesions in Genomes (TILLING)

McCallum and his research colleagues developed TILLING for the first time in *Arabidopsis* (McCallum *et al.*, 2000). This is referred to as a high-throughput reverse genetic technique designed to measure single-nucleotide mutations within a precise locus of the targeted genes (Table 9.1). It enables the discovery of novel alleles and the functional analysis of genes in various organisms, including plants. In TILLING, various platforms have been employed for the detection of mutations in target genes inside genomic samples. Such techniques, in majority, involve PCR amplification of the wild type and targeted mutant sites altogether to identify divergences through the analysis of heteroduplex amplicons, either through HRM or by using CEL-I followed by electrophoresis. Moreover, TILLING is applicable to any species, irrespective of the ploidy level and genome size, and has demonstrated success in numerous crop plants (Table 9.1). Furthermore, a similar strategy was implemented by Fanelli *et al.* (2021) to screen 21 candidate sunflower genes, involved in the auxin transport, chromosome segregation, circadian clock, and control of flowering period, via high-throughput sequencing, to quantify mutations within a sunflower TILLING population (sunTILL) previously developed by Sabetta *et al.* (2011).

9.3.1 Mechanism

The TILLING procedure always starts by seed mutagenesis with a chemical mutagen such as ethyl methanesulfonate (EMS), which induces point mutations by alkylating guanine bases, leading to base pair mismatches during DNA replication. Mutagenized seeds are grown to get M_1 generation, which are then permitted for self-pollination to produce M_2 generation. After that, these M_2 seeds, carrying the heritable mutation, are collected for banking, and the DNA extracted from these plants is pooled into groups to reduce the number of samples to be screened, making the process more efficient for mutation discovery. Following PCR amplification of designated genes with gene-specific primers, during subsequent steps of variable temperature, heteroduplexes are generated. These heteroduplexes are then subjected to digestion during the incubation step with a mismatch endonuclease, such as CEL-I, to facilitate cleavage of heteroduplex mismatches.

After the separation through denaturing agarose or polyacrylamide gel electrophoresis, these cut stands are visualized through fluorescence detection (Fig. 9.2).

Advantages

- It is appropriate for polyploids with a high frequency of mutations.
- TILLING-mutagenized populations are valued resources for various research purposes.
- It is highly sensitive with a high throughput.
- It provides insights into gene function and facilitates the study of diverse plant species.

Disadvantages

- It requires costly and time-consuming enzymatic reactions.
- EMS-mutagenized populations contain <5% truncations.
- It requires a multidimensional pooling to reduce the cost.

9.4 Next-Generation Sequencing (NGS) Technologies

DNA sequencing of plant genomes has profoundly transformed our understanding and interpretation of evolution and genome organization. Remarkable advancements in NGS technologies have opened avenues for the identification and utilization of genetic mutations within coding and non-coding regions of the targeted DNA. The integration of TILLING to detect mutation along with sequencing technology showcases significant potential by incorporating NGS into it. An illustration of the application of Illumina sequencing in TILLING involves the implementation of a versatile and efficient combination of TILLING-NGS methods to estimate genetic variations in designated loci in cereals such as wheat and rice (Granier *et al.*, 2015; Tsai *et al.*, 2015). For the mutant population, the targeted amplicon sequencing (TAS) or targeted resequencing approach is being used, which involves selectively sequencing specific genomic regions of interest rather than the entire genome.

9.4.1 Mechanism

This approach is particularly valuable when studying a particular locus or a set of loci, such as those associated with mutations generated by genome editing. The workflow typically involves the target enrichment of the genomic regions of

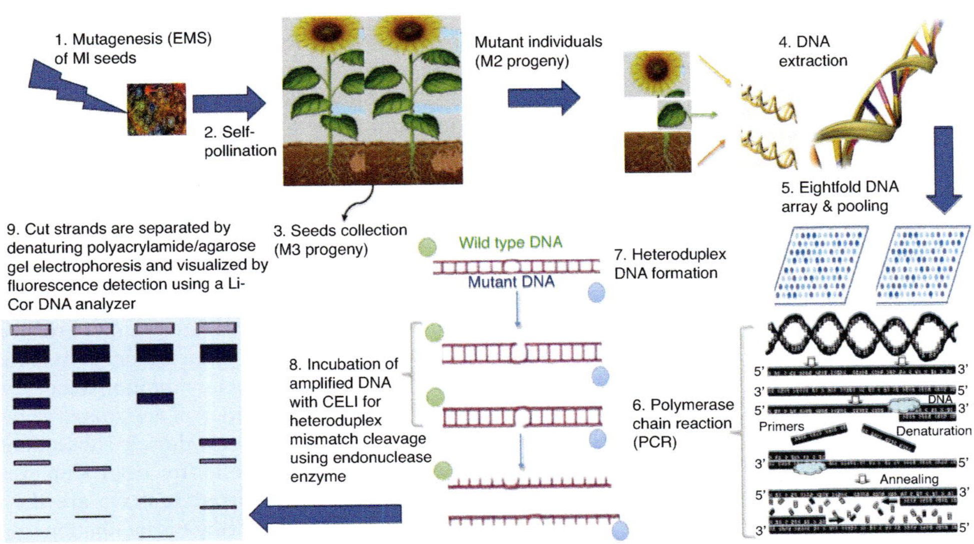

Fig. 9.2. Diagram of the TILLING strategy. Author's own figure.

interest by PCR amplification of specific targets, capturing hybridization using biotinylated probes, or other target-enrichment strategies. After target enrichment, libraries are prepared from the enriched DNA. Adapters with unique barcodes are often added to each sample to allow for multiplexing, especially when dealing with a pooled sample of mutants. The prepared libraries are then subjected to NGS on platforms such as Illumina, Ion Torrent, or others. The sequencing reads obtained will correspond to the specific regions of interest. Different bioinformatics tools are used to align the sequencing reads to a reference genome to identify mutations or variations within the targeted regions (Fig. 9.3).

This targeted approach is advantageous in terms of cost and data analysis, as it allows for deep sequencing of specific regions, providing high coverage and sensitivity for detecting mutations within those regions.

Several NGS technologies are commonly used for mutation detection (Table 9.2). Each technology has its strengths, limitations, and applications. The choice of NGS technology depends on factors such as the type of mutations of interest, the desired read length, throughput, and cost considerations. Often, a combination of technologies or a targeted approach is used to achieve comprehensive mutation detection in research and clinical applications.

Similar research experiments for estimation of mutations through NGS have been reported, where Illumina (WGS) is combined with traditional linkage mapping to quantify the occurred mutations in four mutants (loss of function angulate), which marked deficits in pigmentation and leaf structure (Mateo-Bonmatí *et al.*, 2014). Likewise, NGS for backcrossed bulk segregants was employed to estimate abscisic acid-tolerant root elongation mutants (ABA-AR) in Arabidopsis (Thole *et al.*, 2014; Thole and Strader, 2015). In another study carried out by Cheng *et al.* (2015), they achieved success in the application of NGS for marked mutation estimation in a hexaploidy crop (*Crambe abyssinica*). This approach was utilized to identify causative mutations in the *CaFAD2* gene, resulting in alterations in the oil composition of this crop. Moreover, similar research used NGS in sorghum plants (EMS-induced mutants), which was defected in hydrogen cyanide (Krothapalli *et al.*, 2013). They reported a point mutation that led to a premature stop codon for coding within the DNA sequence of *dhurrinase2*. This gene translates a protein, which is involved in the dhurrin catabolic pathway (causes mutation in acyanogenic sorghum).

Advantages

- It detects mutation in gene pools much deeper than any eight individuals.

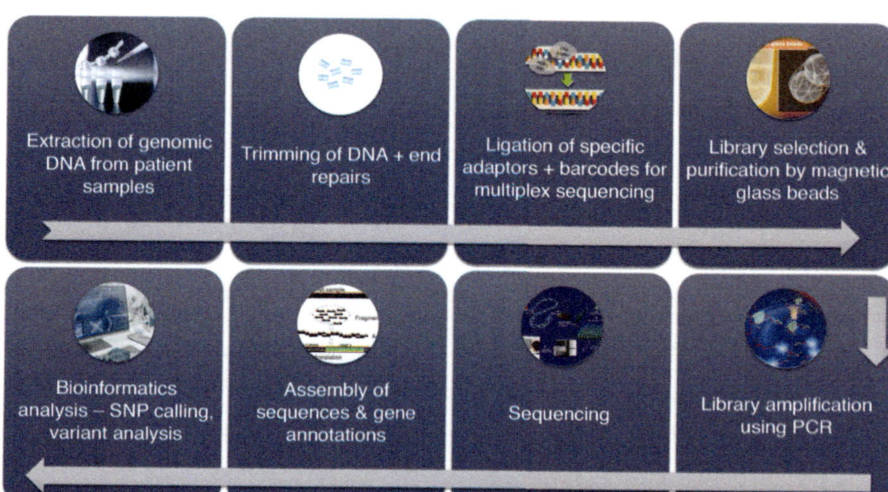

Fig. 9.3. Next-Generation Sequencing workflow. Author's own figure.

Table 9.2. Some of the key NGS platforms are used for mutation detection. Author's own table.

Principles	Mutation Detections	Applications
Illumina Sequencing		
Sequencing by synthesis	CNVs, small indels, SNVs	Target-amplicon, WES and WGS
Ion Torrent Sequencing		
Sequencing by synthesis using semi-conductor technology	Indels and SNVs	Targeted sequencing and amplicon sequencing
454 Pyrosequencing		
Pyrosequencing based on the detection of released pyrophosphate	SNVs, indels, and haplotype phasing	Targeted sequencing and amplicon sequencing
PacBio Sequencing Single-Molecule Real-Time (SMRT) Sequencing		
Sequencing by synthesis in real-time using circular consensus sequencing	Structural variations, long indels, and complex genomic rearrangements	Long-read sequencing for *de novo* genome assembly and targeted sequencing
Nanopore Sequencing		
Sequencing by passing DNA through nanopores and measuring changes in electrical current	SNVs, indels, and structural variations	Long-read sequencing for structural variation analysis and targeted sequencing
SOLiD Sequencing		
Sequencing by ligation of labeled oligonucleotides	SNVs and Indels	WES and targeted sequencing
BGISEQ (BGI Sequencer)		
Sequencing by synthesis	SNVs and Indels	WGS and targeted sequencing

Abbreviations: Copy number variations (CNVs), single-nucleotide variations (SNVs), whole exome sequencing (WES), and whole genome sequencing (WGS).

- It is a time-saving and high-throughput technique.
- It identifies mutations by means of targeted DNA sequencing.
- It is non-enzymatic and more effective precisely in polyploids.

Disadvantages

- The storage and processing of a large amount of data requires specialized skills in bioinformatics to obtain information accurately.
- Trillions of base cells result in billions of mistakes due to an abundant amount of inaccurately identified nucleotide bases in the DNA sequences, which poses a challenge that is difficult to diagnose.
- It requires multidimensional pooling with highly expensive tools.

9.5 CRISPR Ribonucleoprotein (RNP) Complexes

An effective detection method becomes especially crucial in situations where screening a substantial population is required or when the frequency of mutation is quite low. It has been investigated that pre-assembled CRISPR RNP complexes, including CRISPR/Cpf1 and CRISPR/Cas9, can assess such mutations during gene editing. The present technique relies on the gRNA and enzymes such as FnCpf1 or

SpCas9 to identify indels produced by CRISPR-mediated Cas9. It enables the differentiation of heterozygous mutants from bi-allelic and wild type from homozygous mutations. In addition, it is suggested that indels, whether induced by precise nucleases or naturally occurring, can be readily identified using the PCR/RNP technique. The existence of TT or GG (AA or CC on the complementary strands) is the only requirement, within protospacer adjacent motif (PAM) sequences, for gRNA enzymes (FnCpf1 or SpCas9). An additional benefit of identifying mutation through this technique is that it remains unaffected by external noise, such as SNPs, in the vicinity of the targeted region, which is particularly significant for polyploid plants.

In addition, Liang *et al.* (2018) described an alternative gene editing (DNA-free) method in wheat crop, for the first time, utilizing decontaminated TALEN proteins where they demonstrated that the resulting mutations can also be quantified using the PCR/RNP technique. Even though their findings provided influential tools for precisely detecting genetic mutations, they also helped with high-throughput mutation estimation requirements in various genomes of plants, although in complex backgrounds. Similarly, Nakamura *et al.* (2023) presented a novel RNP delivery approach in rice where they successfully isolated *OsLCYβ* genome-edited lines generated by RNPs in combination with marker gene delivery and detected, although at a low frequency, genome editing events in rice calli cells by RNP delivery using the sonication-assisted whisker method.

9.6 TIDE and TIDER

Recently, scientists have devised two rapid, cost-effective, and novel methods capable of accurately quantifying the proficiency of either template or indels-directed genetic variations within a pool of cells. First, the method TIDE quantifies and detects indels, as it has the requirements of only a pair of Sanger sequencing traces from PCR-based products followed by analysis through an easily accessed web tool (Brinkman *et al.*, 2014). However, it is to be noted that this technique only identifies frequencies of total indels, excluding precisely designed indels or base substitution within nucleotides.

Furthermore, a similar method, namely tracking of insertions, deletions, and recombination events (TIDER), was developed by Brinkman and Steensel (2019). This technique can assess the integration frequency of template-directed mutations, encompassing mutations (base substitution), and differentiate them from a background of extra indels originating from challenging inaccurate repair pathways. Although TIDER can detect indels alone, TIDE is slightly easier to be implemented, making it more appropriate for evaluating non-templated editing experimentations. Internet web tools for both of these techniques are easily reachable online at http://tide.nki.nl (accessed 25 November 2024).

9.6.1 Mechanism

As a result of defective repair, including repair through HDR with a donor template, after targeted nuclease cleavage, the DNA in the gene pool comprises a combination of insertions or deletions, along with the designed mutations as presented in Fig. 9.4. Several induced genetic mutations are disentangled in the gene pool either through TIDER or TIDE, which are briefly defined as under:

a. TIDE comprises two sequences, namely, composite mixed pool and wild-type control, and a gRNA sequence string as input.
b. Aberrant sequence signal is pictured in both the treated sample (in blue color) and control (depicted as black) to ensure quality control. This visualization includes the anticipated break site (dotted line drawn vertically), a decomposition region (depicted by a light brown poll), and an alignment region (represented by a green bar).
c. The spectrum of insertions and deletions, including their frequencies, is yielded through trace decomposition.
d. The composition of base is assessed in the occurrence of +1 insertions.
e. TIDER comprises three sequences, namely, mixed pool, wild type (control), and donor template (with designed mutations), along with gRNA sequence string as input.

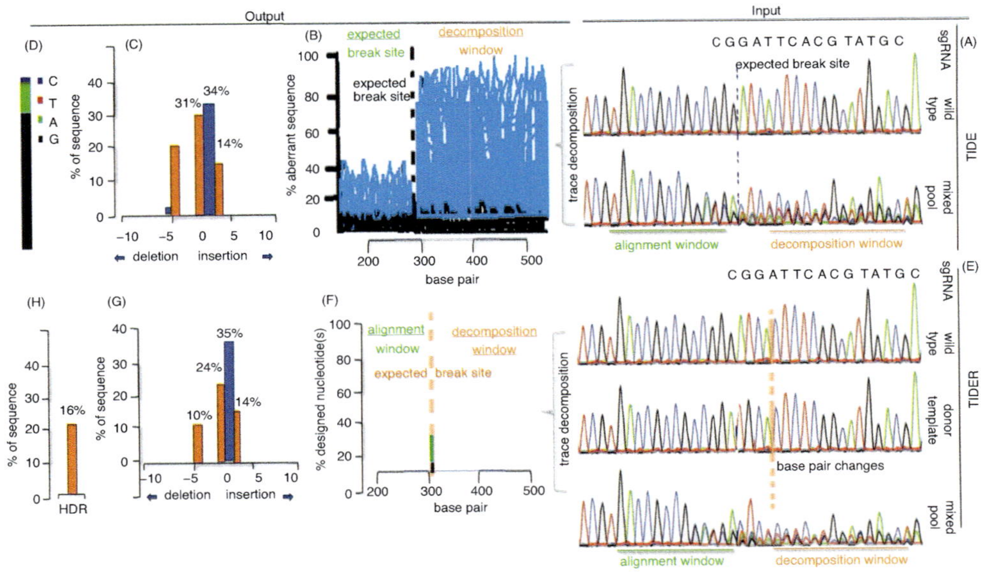

Fig. 9.4. Overview of TIDE and TIDER algorithm. Author's own figure.

f. The quality plot illustrates only the proportion of the intended mutated nucleotides present in both the treated sample (green in color) and control (black colored), as considered in the donor template. It comprises a decomposition region (as represented by a light brown color) and an alignment region (shown as a green bar), as employed in TIDER.
g. Decomposition gives the proportion of insertions and deletions (indels) and events of HDR.
h. Moreover, this also provides the spectrum of indels and frequencies of HDR in terms of sequence percentage.

9.7 Conclusion

Manipulation of the targeted genome in cells and model organisms has become mainstream in agricultural research and has empowered researchers to dissect and understand genomic alterations with unprecedented precision. A rapid assay for estimating the frequencies of the various induced mutations during genome editing is of key significance. Therefore, many techniques are used for the quantification of genome editing-derived mutations depending on their type and the level of sensitivity required. These quantification methods not only provide quantitative insights into the frequency and types of mutations but also offer a deeper understanding of off-target effects, enhancing the safety and reliability of genetic modifications. Moreover, these techniques are highly efficient and appropriate for any species, irrespective of the ploidy level and their genome size, which have successfully been employed to a wide range of germplasms to estimate and quantify induced mutations while genome editing.

References

Acevedo-Garcia, J., Spencer, D., Thieron, H., Reinstädler, A., Hammond-Kosack, K. *et al.* (2016) Mlo-based powdery mildew resistance in hexaploid bread wheat generated by a non-transgenic TILLING approach. *Plant Biotechnology Journal* 15, 1–12.

Baldet, P., Bres, C., Okabe, Y., Mauxion, J.-P., Just, D. *et al.* (2013) Investigating the role of vitamin C in tomato through TILLING identification of ascorbate-deficient tomato mutants. *Plant Biotechnology* 30, 309–314.

Botticella, E., Sestili, F., Hernandez-Lopez, A., Phillips, A. and Lafiandra, D. (2011) High resolution melting analysis for the detection of EMS induced mutations in wheat sbella genes. *BMC Plant Biology* 11, 156.

Bovina, R., Brunazzi, A., Gasparini, G., Sestili, F., Palombieri, S. *et al.* (2014) Development of a TILLING resource in durum wheat for reverse-and forward-genetic analyses. *Crop and Pasture Science* 65, 112–124.

Brinkman, E.K. and Steensel, B. (2019) Rapid quantitative evaluation of CRISPR genome editing by TIDE and TIDER. *CRISPR Gene Editing: Methods and Protocols* 29–44.

Brinkman, E.K., Chen, T., Amendola, M. and Steensel, B. (2014) Easy quantitative assessment of genome editing by sequence trace decomposition. *Nucleic Acids Research* 42, e168.

Casella, L., Greco, R., Bruschi, G., Wozniak, B., Dreni, L. *et al.* (2013) TILLING in European rice: Hunting mutations for crop improvement. *Crop Science* 53, 2550–2562.

Cheng, J., Salentijn, E.M., Huang, B., Denneboom, C., Qi, W. *et al.* (2015) Detection of induced mutations in CaFAD2 genes by next-generation sequencing leading to the production of improved oil composition in *Crambe abyssinica*. *Plant Biotechnology Journal* 13, 471–481.

Fanelli, V., Ngo, K.J., Thompson, V.L., Silva, B.R., Tsai, H. *et al.* (2021) A TILLING by sequencing approach to identify induced mutations in sunflower genes. *Scientific Reports* 11(1), 9885.

Gady, A.L., Hermans, F.W., Wal, M.H., Loo, E.N., Visser, R.G. *et al.* (2009) Implementation of two high through-put techniques in a novel application: Detecting point mutations in large EMS mutated plant populations. *Plant Methods* 5, 1–14.

Gatica-Arias, A., Bolívar-González, A., Sánchez-Barrantes, E., Araya-Valverde, E. and Molina-Bravo, R. (2023) High resolution melt (HRM) genotyping for detection of induced mutations in coffee (*Coffea arabica* L. var. Catuaí). In: Ingelbrecht, I.L.W., Céu Lavado da Silva, M. and Jankowicz-Cieslak, J. (eds) *Mutation Breeding in Coffee with Special Reference to Leaf Rust: Protocols*. Springer, Berlin, Heidelberg, pp. 275–290.

Gilchrist, E.J., Sidebottom, C.H., Koh, C.S., MacInnes, T., Sharpe, A.G. *et al.* (2013) A mutant *Brassica napus* (canola) population for the identification of new genetic diversity via TILLING and next generation sequencing. *PLoS One* 8, e84303.

Granier, F., Lemaire, A., Wang, Y., LeBris, P., Antelme, S. *et al.* (2015) Chemical and radiation mutagenesis: Induction and detection by whole genome sequencing. In: Vogel, J.P. (ed.) *Genetics and Genomics of Brachypodium*. Springer, Berlin, pp. 155–170.

Guo, Y., Abernathy, B., Zeng, Y. and Ozias-Akins, P. (2015) TILLING by sequencing to identify induced mutations in stress resistance genes of peanut (*Arachis hypogaea*). *BMC Genomics* 16, 1.

Jaganathan, D., Ramasamy, K., Sellamuthu, G., Jayabalan, S. and Venkataraman, G. (2018) CRISPR for crop improvement: An updated review. *Frontiers in Plant Science* 9, 985.

Kim, S.I. and Tai, T.H. (2014) Identification of novel rice low phytic acid mutations via TILLING by sequencing. *Molecular Breeding* 34, 1717–1729.

Krothapalli, K., Buescher, E.M., Li, X., Brown, E., Chapple, C. *et al.* (2013) Forward genetics by genome sequencing reveals that rapid cyanide release deters insect herbivory of *Sorghum bicolor*. *Genetics* 195, 309–318.

Liang, Z., Chen, K., Yan, Y., Zhang, Y. and Gao, C. (2018) Genotyping genome-edited mutations in plants using CRISPR ribonucleoprotein complexes. *Plant Biotechnology Journal* 16(12), 2053–2062.

Mateo-Bonmatí, E., Casanova-Sáez, R., Candela, H. and Micol, J.L. (2014) Rapid identification of *Angulata* leaf mutations using next-generation sequencing. *Planta* 240, 1113–1122.

Mazzucato, A., Cellini, F., Bouzayen, M., Zouine, M., Mila, I. *et al.* (2015) A TILLING allele of the tomato Aux/IAA9 gene offers new insights into fruit set mechanisms and perspectives for breeding seedless tomatoes. *Molecular Breeding* 35, 1–15.

McCallum, C.M., Comai, L., Greene, E.A. and Henikoff, S. (2000) Targeting induced local lesions in genomes (TILLING) for plant functional genomics. *Plant Physiology* 123(2), 439–442.

Molla, K.A., Sretenovic, S., Bansal, K.C. and Qi, Y. (2021) Precise plant genome editing using base editors and prime editors. *Nature Plants* 7(9), 1166–1187.

Nakamura, A., Yano, T., Mitsuda, N., Furubayashi, M., Ito, S. *et al.* (2023) The sonication-assisted whisker method enables CRISPR-Cas9 ribonucleoprotein delivery to induce genome editing in rice. *Scientific Reports* 13(1), 14205.

Nida, H., Blum, S., Zielinski, D., Srivastava, D.A., Elbaum, R. *et al.* (2016) Highly efficient de novo mutant identification in a *Sorghum bicolor* TILLING population using the ComSeq approach. *The Plant Journal* 86, 349–359.

Osakabe, Y. and Osakabe, K. (2015) Genome editing with engineered nucleases in plants. *Plant and Cell Physiology* 56(3), 389–400.

Raizada, A. and Souframanien, J. (2021) SNP genotyping and diversity analysis based on genic-SNPs through High Resolution Melting (HRM) analysis in black gram [*Vigna mungo* (L.) Hepper]. *Genetic Resources and Crop Evolution* 68, 1331–1343.

Reddy, T.V., Dwivedi, S. and Sharma, N.K. (2012) Development of TILLING by sequencing platform towards enhanced leaf yield in tobacco. *Industrial Crops and Products* 40, 324–335.

Reed, G.H., Kent, J.O. and Wittwer, C.T. (2007) High-resolution DNA melting analysis for simple and efficient molecular diagnostics. *Pharmacogenomics* 8, 597–608.

Rocafort, M., Arshed, S., Hudson, D., Sidhu, J.S., Bowen, J.K. *et al.* (2022) CRISPR-Cas9 gene editing and rapid detection of gene-edited mutants using high-resolution melting in the apple scab fungus, *Venturia inaequalis*. *Fungal Biology* 126(1), 35–46.

Sabetta, W., Alba, V., Blanco, A. and Montemurro, C. (2011) SunTILL: A TILLING resource for gene function analysis in sunflower. *Plant Methods* 7(1), 1–13.

Sestili, F., Botticella, E., Bedo, Z., Phillips, A. and Lafiandra, D. (2010) Production of novel allelic variation for genes involved in starch biosynthesis through mutagenesis. *Molecular Breeding* 25, 145–154.

Sestili, F., Palombieri, S., Botticella, E., Mantovani, P., Bovina, R. *et al.* (2015) TILLING mutants of durum wheat result in a high amylose phenotype and provide information on alternative splicing mechanisms. *Plant Science* 233, 127–133.

Thole, J. and Strader, L. (2015) Next-generation sequencing as a tool to quickly identify causative EMS-generated mutations. *Plant Signaling and Behavior* 10, 1–4.

Thole, J.M., Beisner, E.R., Liu, J., Venkova, S.V. and Strader, L.C. (2014) Abscisic acid regulates root elongation through the activities of auxin and ethylene in *Arabidopsis thaliana*. *G3 (bethesda)* 4, 1259–1274.

Thomas, H.R., Percival, S.M., Yoder, B.K. and Parant, J.M. (2014) High-throughput genome editing and phenotyping facilitated by high resolution melting curve analysis. *PloS One* 9(12), e114632.

Tsai, H., Ngo, K., Lieberman, M., Missirian, V. and Comai, L. (2015) Tilling by sequencing. In: Alonso, J.M. and Stepanova, A.N. (eds) *Plant Functional Genomics: Methods and Protocols*, Vol. 3. Springer, Berlin, pp. 359–380.

Wang, H., Russa, M. and Qi, L.S. (2016) CRISPR/Cas9 in genome editing and beyond. *Annual Review of Biochemistry* 85, 227–264.

10 Application of CRISPR-Cas9 in Agriculture and Forestry

Muhamed Adem[1]* and Oumer Abdie[2]

[1]Department of Forestry, College of Agriculture and Natural Resources, Madda Walabu University, Bale Robe, Ethiopia; [2]Department of Biotechnology, College of Natural and Computational Science, Wolkite University, Wolkite, Ethiopia

Abstract

For the exact modification of endogenous genomes, dependable and adaptable genetic techniques have been developed over the preceding 10 years. The majority of research techniques used in the 1990s and early 2000s, such as gene silencing, overexpression, and the introduction of foreign genes causing unique phenotypes, have rapidly and completely changed. Presently, one of the most popular methods for altering the genome relies on CRISPR (clustered regularly spaced short palindromic repeats)-Cas (CRISPR-associated protein). The vast applications of CRISPR-Cas systems in agriculture and medicine are ready to investigate how these tools might enhance the quality of human life. To provide enough food and feed for people and animals, as well as protect the environment from additional ecosystem malfunctions, there is a need to adopt novel methods and make desirable interventions. In particular, the CRISPR-Cas system has emerged as a game-changing tool for three different major uses (gene function research, genome-wide regulatory network construction, and regulation of sequence-specific gene expression). The fact that DNA, RNA, and protein are combined in a programmed and predictable sequence is the finest characteristic of the CRISPR technique. This enables CRISPR to draw enhancers and repressors to a particular DNA region. Numerous Cas proteins, including Cas9, Cas12, Cas13, and CasX, have so far been created and have a wide range of uses, including base editing, multiplex genome editing, primer editing, RNA editing, and directed evolution. It is believed that the CRISPR/CAS system can be used to explore the potential of nature to improve life. This chapter presents the recent research findings on the use of CRISPR-Cas systems, specifically in the broad fields of agriculture and forestry.

Keywords: Agriculture, CRISPR-Cas, Forestry, Applications

10.1 Introduction

According to the United Nations, by 2050, the world population is projected to reach 9.8 billion people. In 2015, the UN had projected a zero-hunger goal among the sustainable development goals to achieve global food security by 2030 (Ahmad *et al.*, 2021). The Green Revolution's increased plant production is steadily falling, and the agriculture and forestry sectors must meet the enormous demands of such a large population in the face of adverse problems including climate change, decreased land productivity, biotic and abiotic stress, and others (Bhambhani *et al.*, 2021; Fladung and Kersten, 2022). Agriculture and forestry are

*Corresponding author: muhamed.adem@aau.edu.et; muhamedadem@gmail.com

 Genome Editing for Crop Improvement: Theory and Methodology (eds J.M. Al-Khayri *et al.*)
DOI: 10.1079/9781800622517.0010

no longer guaranteed to be sustainable due to the rising threat of climate change. Despite the vital importance of both agriculture and forests, human demands have overshadowed their co-existence, complicating the future of life on Earth (Wang *et al.*, 2019a; Camerlengo *et al.*, 2022; Cao *et al.*, 2022; Sharma *et al.*, 2022).

The basic objective of sustainable agriculture is to provide inclusive economic growth while preserving future generations' capacity to meet their own needs. To address the inevitable environmental threats and meet the growing demand for agriculture, we must employ and advance non-conventional technologies, especially in biotechnology (Wang *et al.*, 2018a; Kumar *et al.*, 2019; Camerlengo *et al.*, 2022). Current plant breeding methods require longer time and significant labor to produce superior traits, which has led researchers to seek faster and more efficient breeding techniques (Sharma *et al.*, 2022). A lack of biodiversity, ineffective strategies now in use, and the negative effects of climate change render agriculture incapable of supplying the future world's food needs (Wang *et al.*, 2019b). It has been discovered that the CRISPR-Cas system is a game-changing technology that can precisely modify the genetic make-up of plant features. It has been discovered to be effective in improving agricultural germplasm resources and creating superior varieties to address the global food and feed challenges (Wang *et al.*, 2018b; Razzaq *et al.*, 2019: Wang *et al.*, 2019b; Liu *et al.*, 2022; Van *et al.*, 2022).

Because of the extended vegetative phases of trees in reaching maturity and their longevity, genetic improvements in tree species by conventional breeding are challenging. Recent advancements in molecular and biotechnological methods are expected to offer numerous possibilities for tree improvement (Fladung and Kersten, 2022). Most importantly, the advancement of various biotechnological methods is expected to deepen our understanding of the genetic basis of forest tree traits relevant to climate change and the biological functions of their underlying genes.

Genome editing methods used before the development of CRISPR-Cas9 include zinc-finger nucleases (ZFNs) that employ roughly 30 amino acid fingers that fold around a zinc ion to create a compact structure that can identify a triple-base pair of DNA. A large portion of the target DNA can be recognized and targeted by successive finger repeats (Segal and Meckler, 2003). Given that ZFNs are designed to recognize 18–24 base pairs from the target site, whereas transcription activator-like effector nucleases (TALENs) are intended to recognize 30–36 base pairs, TALENs are more adept at targeting their chromosomal site, have a wider range of sequences that can be targeted, and have a lower chance of going off-target (Ran *et al.*, 2013). While many species can have their genomes edited, ZFNs and TALENs have limited applications due to their high labor and time costs, their ability to only access specific regions of the genome (the chromatin site), their inability to perform multiple genome editing, and the fact that they both require the synthesis of new proteins for each DNA sequence they target (Christian *et al.*, 2013; Gaj *et al.*, 2013). Because plants have a low homologous recombination rate, random mutagenesis is a challenging task. However, strong and adaptable genome editing methods became increasingly sought after due to the increased demand to increase plant productivity. As more plant genomes become available, it is becoming possible to use CRIPSR-Cas9 in a variety of plant species by taking advantage of the potent method of genome editing (Liang *et al.*, 2014). The technique enables opening of the genome like a book and proceed to words; in this case, the DNA sequences then engineer the sequences to end up with the desired product (Adem *et al.*, 2017).

Since the invention of the CRISPR-Cas-based methodology as a genome editing method, two decades ago, the science of genome editing has made amazing strides. The CRISPR-Cas system has been used in almost all areas of the biological sciences, including agriculture and medicine. New methods and Cas proteins with diverse functions are rapidly expanding the CRISPR-Cas system capabilities. Several Cas proteins, including Cas9, Cas12, Cas13, and CasX, have been introduced recently. Multiplex genome editing, primer editing, RNA editing, base editing, and directed evolution are just a few of the uses for such proteins (Wang *et al.*, 2019b; Ahmad *et al.*, 2022). For genome editing in a variety of taxa, including bacteria, mammals, plants, and model organisms, numerous CRISPR-Cas tools have been created and made accessible (Hess *et al.*, 2017;

Wang *et al.*, 2018c; Razzaq *et al.*, 2019). Such expanded uses of the CRISPR-Cas system and its use to rewrite genetic codes offer important benefits that should attract the attention of top-tier genome editing experts (Razzaq *et al.*, 2019; Ahmad *et al.*, 2022). The goal of this book chapter is to summarize the most recent scientific findings regarding the use of CRISPR-Cas systems in enhancing vital traits in the broad fields of agriculture and forestry.

10.2 Application of CRISPR-Cas9 in Agriculture

CRISPR-Cas applications have immense potential in the field of biological science. In plant sciences, it offers an unprecedented opportunity for the rapid development of economically valuable traits. This ranges from enhancing crop yields to fortifying resistance against various stressors, a prospect that was unimaginable in previous eras of plant breeding (Fig. 10.1). Most crucially, accurate genome editing makes it possible to add mutations that code for highly desirable features or to remove harmful genetic components that cause undesirable traits (Zhu *et al.*, 2020; Neelakandan *et al.*, 2022; Sharma *et al.*, 2022).

10.2.1 CRISPR-Cas9 in major cereals

The effective editing capabilities of the CRISPR-Cas9 system were demonstrated by Kumagai *et al.* (2022). In addition to introducing the CRISPR-Cas9 system for editing recalcitrant wheat cultivars, Kumagai and colleagues also established a non-culture method for recalcitrant wheat cultivar transformation and created a direct delivery system for Cas9/gRNA RNP into shoot apical meristems (SAMs). In another experiment, Zhang *et al.* (2019) created high-yield phenotypes in wheat by deleting the gene that encodes the enzyme cytokinin oxidase/dehydrogenase (CKX), which catalyzes the breakdown of cytokinins. The grain weight of *Triticum aestivum* has increased in the GW2

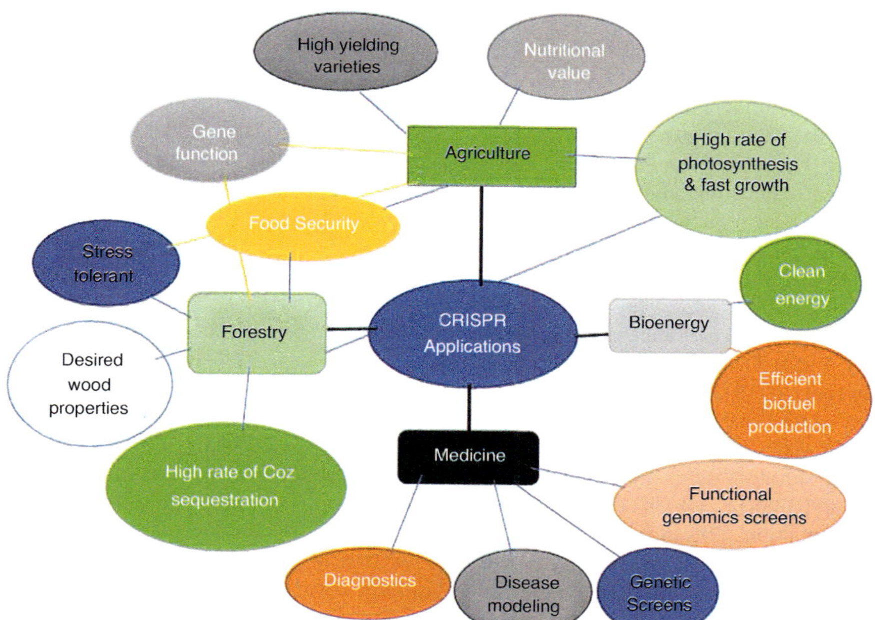

Fig. 10.1. Application areas of CRISPR-Cas9. Author's own figure.

edited lines (Zhang *et al.*, 2018). The nutritional quality of food has also been improved using the CRISPR-Cas9 technology. According to Sánchez-León *et al.* (2018), wheat lines with low gluten content have been developed. Three mildew-resistant loci O (MLO) homologs have been simultaneously mutated to create a wheat variety with broad-spectrum resistance to powdery mildew (Wang *et al.*, 2018b). Similar to this, CRISPR-Cas simultaneous mutation of the three wheat EDR1 homologs resulted in plants with increased resistance to *Blumeria graminis* f. sp. *tritici* (Zhang *et al.*, 2017b).

The editing technique has also had a significant impact on rice. A rice gene called *Oryza sativa* LOGL5 encodes an enzyme that activates cytokinin by editing the encoding gene which affects the C-terminus of this protein and increased grain production under various environmental circumstances (Wang *et al.*, 2020). Using CRISPR-Cas technology, rice yields and stress tolerance could be increased by manipulating these: *GW2* and *GW5* which regulate grain weight; *PIN5b* which regulates panicle size; *GS3* which regulates grain size, and *OsMYB30* which improves cold tolerance (Liu *et al.*, 2017; Zeng *et al.*, 2019; Zhou *et al.*, 2019). Editing a gene that encodes amino acid permease 3, which is essential in nutrient partitioning, allowed rice types to be bred with enhanced tiller numbers and increased yields in addition to maintaining grain quality (Lu *et al.*, 2018). Rice lines with low amylose contents have been produced by changing the CBE's GBSS1 amino acid sequence (Xu *et al.*, 2021). By using CRISPR-Cas9 to modify the promoter region of *O. sativa's SWEET11, SWEET13*, and *SWEET14*, researchers have created rice lines with broad-spectrum resistance to *Xanthomonas oryzae* pv. *oryzae* bacterial blight (Oliva *et al.*, 2019; Xu *et al.*, 2019).

Waxy maize variants were produced in 12 elite inbred lines as a result of the CRISPR-Cas9-driven mutation on granule-bound starch synthase 1 (GBSS1), which is essential for amylose production. The introduction of such a waxy gene into 12 elite inbred maize lines was reportedly done without linkage drag (Gao *et al.*, 2020b). In maize, on-target editing occurs often (up to 90% of the time), while there is no evidence of off-target cleavage activity (Young *et al.*, 2019; Gao *et al.*, 2020a). To enable trait stacking in maize, a complex trait locus (CTL) was created using CRISPR-Cas9. The extensive use of CRISPR in maize enhances many features, including stress tolerance, seed oil content, and a gift of herbicide strength reviewed by Gillani *et al.* (2021). Numerous researchers have focused on specific genes to design diverse features in maize. For instance, Char *et al.* (2017) used ubiquitin as the promoter driving Cas9 expression and U6 as the promoter driving strand expression to target dihydroflavonol 4-reductase (dfr), Argonaute protein Ago18, anthocyaninless 1(anl1), and homolog (A4) genes.

In barley, CRISPR-Cas9 gene editing technology was used to target the PM19 multi-copy gene, which encodes a plasma membrane protein that is affected by ABA and affects grain dormancy. Independent of the T-DNA construct, two distinct constructions (*PM19-1* and *PM19-3*) were transformed by *Agrobacterium* into the cultivar Golden Promise, resulting in 10–23% alterations in the primary transformants (T0) (Lawrenson *et al.*, 2015). Using CRISPR-mediated genome editing, researchers effectively transformed hulled barley grains into bare grains, increasing grain production and improving brewing quality (Gasparis *et al.*, 2018). With the targeting of the barley *PAPhy*, a promoter, a pro-nutritional phytase, its level of expression has been increased (Holme *et al.*, 2017). Similar to this, CRISPR-Cas9 was used to investigate the function of the cytokinin dehydrogenase enzyme (*CKX1*) by silencing the *HcCKK1* gene, which led to transgenic lines producing more grains (Holubová *et al.*, 2018).

10.2.2 CRISPR-Cas9 in fruits and vegetables

Based on transitory expression in protoplasts, Zhang *et al.* (2022) claimed that CRISPR-Cas9-mediated genome editing has successfully modified banana. The study showed that in banana protoplasts, the MaU6c promoter was around four times more active than the OsU6a promoter. In comparison to the prior CRISPR-Cas9 cassette for bananas, the application of this promoter and banana codon-optimized Cas9 resulted in a fourfold improvement in mutation efficiency. The study showed that an improved CRISPR-Cas9 system was successfully used in altering specific genes in bananas and in enhancing applications for fundamental functional genomics (Zhang *et al.*, 2022).

The editing of *ENO* genes (Yuste-Lisbona *et al.*, 2020) and *CLV* (Rodríguez-Leal *et al.*, 2017), which control meristem size, increased the yield of fruit crops. The editing of *CREs* in promoter regions could generate a continuum of phenotypes. The CRISPR-Cas introduced mutation has conferred powdery mildew tolerances in tomato *Solanum lycopersicum MLO1* (Nekrasov *et al.*, 2017). CRISPR-Cas has also facilitated the breeding of high-quality crops with enriched carotenoids (Li X. *et al.*, 2018) and γ-aminobutyric acid (Li R. *et al.*, 2018).

Using CRISPR-Cas9, Charrier *et al.* (2019) knocked out the *TFL1* gene of pear and apple. The study resulted in early flowering of 93% and 9% of the apple and pear mutant lines, respectively. The knocking out of the *AcCEN4* and *AcCEN* genes of kiwi fruit *Actinidia chinensis* by CRISPR-Cas9 produced an early blooming mutant with more compact fruit (Varkonyi-Gasic *et al.*, 2019). The knocking out of the *IdnDH*, a key tartaric acid (TA) synthesis gene in grapes using CRISPR-Cas9, resulted in the reduction of the TA content in the mutant grapes and confirmed the feasibility of applying the technology to grape genome transformation (Ren *et al.*, 2016).

10.2.3 CRISPR-Cas9 in legumes and oil seed crops

For model legumes, including *Medicago truncatula* and *Lotus japonicus*, as well as soybean, cowpea, chickpea, and others, gene editing techniques have been effectively established (Bhowmik *et al.*, 2021; Baloglu *et al.*, 2022). *Agrobacterium rhizogenes* strain K599 was utilized to mediate a temporary transformation system of hairy roots to test the effectiveness of a CRISPR-Cas9 system in peas. An effective vector, PsU6.3-tRNA-PsPDS3-en35S-PsCas9, was developed following specific optimization. This CRISPR-Cas9 system was successfully used to edit the pea phytoene desaturase (*PsPDS*) gene (Li *et al.*, 2022).

A single-guided RNA (sgRNA) sequence was created to target various regions of the FAD2-2 locus in soybeans using a variety of bioinformatics methods. *Agrobacterium tumefaciens* was used to successfully transform the binary vector (pCas9-AtU6-sgRNA) into soybean cotyledon. The target gene showed evidence of intended mutation in soybean transgenic transformants (Al Amin *et al.*, 2019). In the targeted *GmFAD2* genes of the soybean study, CRISPR-Cas9 resulted in a variety of alterations, including big and minor DNA deletions, insertions, and inversions. Oleic acid content significantly increased to over 80% in T1 seeds from CRISPR-edited plants which were homozygous for both *GmFAD2* genes, while linoleic acid content dramatically decreased to 1.3–1.7% (Do *et al.*, 2019).

CRISPR-Cas9 system along with hairy root transformation was used to induce targeted mutagenesis in sesame. Two sgRNAs were designed to target two sesame cytochrome P450 genes (*CYP81Q1* and *CYP92B14*), which are the key biosynthetic genes of sesamin and sesamolin, respectively. Sequencing data illustrated the expected indel mutations at the target sites, with 90.63% and 93.33% mutation frequency in *CYP81Q1* and *CYP92B14*, respectively (You *et al.*, 2022).

To induce insertion/deletion mutations in the targeted gene for a loss of function research, two distinct constructs were used to investigate the editing effectiveness of the CRISPR-Cas9 system in peanuts. The reported editing efficiency with hairy root transformation ranged between 24% and 32%. The expanded scaffold had a greater editing efficiency of 32% compared to the conventional scaffold's 24%. Additionally, the extended scaffold performed better at modifying individual targets (*AhFAD2A* and *AhFAD2B*) than the standard scaffold did (Neelakandan *et al.*, 2022).

10.3 Application of Genome Editing Approaches in Forest Trees

The first successful genetic modification of forest trees was reported in 1987 (Fillatti *et al.*, 1987). Since then, there has been a remarkable acceleration in the genomics and genome editing of numerous forest trees (Fan *et al.*, 2015) (Table 10.1). Today, a sizable genetic database has been developed and is available for use in

Table 10.1. CRISPR-Cas-edited plant species along with tools used, target genes, and altered traits. Author's own table.

Plant name	Tools used and targeted genes	Target trait	Reference
Apple	*Agrobacterium*-based Cas9 targeting *MdTFL1.1*	Early flowering	Charrier *et al*., 2019
Apple	*Agrobacterium*-based Cas9 targeting *PDS*	Clear and partial albino phenotypes	Nishitani *et al*., 2016
Cassava	*Agrobacterium*-based Cas9 targeting *ncbp-1*, *ncbp-2*	Cassava Brown Spot Virus resistance	Gomez *et al*., 2019
Cassava	*Agrobacterium*-based Cas9 targeting *MePDS*	Visible albino phenotypes	Odipio *et al*., 2017
Castanea sativa	RNP editing Cas9 targeting *PDS*	Transfection capacity of protoplasts	Pavese *et al*., 2022
Citrus	*Agrobacterium*-based Cas9 targeting *PDS*	Albino and mosaic phenotypes	Zhang *et al*., 2017a
Coffea canephora	*Agrobacterium*-based Cas9 targeting *CcPDS*	Albino phenotype	Breitler *et al*., 2018
Cotton	*Agrobacterium*-based Cas9 targeting *GhCLA1*	Red fluorescence and albino phenotype	Wang *et al*., 2018a
Cotton	*Agrobacterium*-based Cas9 targeting *GhARG*	Proanthocyanidin accumulation	Wan *et al*., 2017
Cryptomeria japonica	*Agrobacterium*-based Cas9 targeting *GFP* and *CjCHLI*	Green, pale green, and albino phenotypes	Nanasato *et al*., 2021
Duncan grape	*Agrobacterium*-based Cas9 targeting *CsLOB1*	Citrus canker resistance	Jia *et al*., 2016; Jia *et al*., 2019
Eucalyptus grandis	*Agrobacterium*-based Cas9 targeting *CCR1*	Wood properties (lignin biosynthesis and auxin-dependent transcription)	Dai *et al*., 2020
Grape (*Vitis vinifera*)	*Agrobacterium*-based Cas9 targeting *IdnDH*	Reduced tartaric acid content	Ren *et al*., 2016
Vitis vinifera	*Agrobacterium*-based Cas9 targeting *VvWRKY52*	*Botrytis cinerea* resistant	Wang *et al*., 2018b
Green bamboo	PEG-based Cas9 targeting *PDS*	Albino regenerates	Lin *et al*., 2018
Jatropha curcas	*Agrobacterium*-based Cas9 targeting *JcCYP735A*	Plant height reduction	Cai *et al*., 2018
Juglans regia	*Agrobacterium*-based Cas9 targeting *PDS*	Albino phenotype	Walawage *et al*., 2019
Kiwifruit (*Actinidia chinensis*)	*Agrobacterium*-based Cas9 targeting *AcCEN4*, *AcCEN*	Early flowering	Varkonyi-Gasic *et al*., 2019
Ma bamboo (*Dendrocalamus latiflorus*)	*Agrobacterium*-based Cas9 targeting *GRG1*	Plant height	Ye *et al*., 2020

Continued

Table 10.1. Continued

Plant name	Tools used and targeted genes	Target trait	Reference
Parasponia andersonii	*Agrobacterium*-based Cas9 targeting *PanHK4*, *PanEIN2*, *PanNSP1*, and *PanNSP2*	Nodulation, bisexual flowers	Van Zeijl *et al*., 2018
Pear	*Agrobacterium*-based Cas9 targeting *PcTFL1.1*	Early flowering	Charrier *et al*., 2019
Pear and apple	*Agrobacterium*-based Cas9 targeting *ALS*	Herbicide-resistant	Malabarba *et al*., 2020
Picea glauca (*Picea engelmannii*)	*Agrobacterium*-based Cas9 targeting *DXS1*	Chimerism and mosaicism	Cui *et al*., 2021
Pinus radiata	RNP editing Cas9 targeting *GUX1*	Increased productivity	Poovaiah *et al*., 2021
Plasmopara viticola	*Agrobacterium*-based Cas9 targeting *VvPR4b*	Downy mildew resistance	Li *et al*., 2020
Pomegranate	*Agrobacterium*-based Cas9 targeting *gUGT84A23*, *PgUGT84A24*	Accumulation of gallic acid	Chang *et al*., 2019
Populus	*Agrobacterium*-based Cas9 targeting *PdNF-YB21*	Root growth and drought resistance	Zhou *et al*., 2020
Populus	*Agrobacterium*-based Cas9 targeting *PtoPDS*	Albino phenotype	Fan *et al*., 2015
Populus	*Agrobacterium*-based Cas9 targeting *PDS*	Albino and pale green	An *et al*., 2020
Populus	*Agrobacterium*-based Cas9 targeting *JMJ25*	Anthocyanin biosynthesis and content	Fan *et al*., 2018
Populus	*Agrobacterium*-based Cas9 targeting *UGT71L1*	Partial reduction in salicylin content	Fellenberg *et al*., 2020
Populus	*Agrobacterium*-based Cas9 targeting *BRANCHED1*, *BRANCHED2*	Enhance the growth of shoots	Muhr *et al*., 2018
Populus alba var. *pyramidalis*	*Agrobacterium*-based Cas9 targeting *PalWRKY77*	Salt resistant	Jiang *et al*., 2021
Populus tomentosa	*Agrobacterium*-based Cas9 targeting *PtoMYB156*	Phenylpropanoid metabolism and secondary cell wall biosynthesis	Yang *et al*., 2017
Populus tomentosa	*Agrobacterium*-based Cas9 targeting *PtoDWF4*	Reduced xylem development	Shen *et al*., 2018
Populus tomentosa	*Agrobacterium*-based Cas9 targeting *Populus tomentosa*	Increased anthocyanins and procyanidins	Wan *et al*., 2017
Populus tremula × *P. alba*	*Agrobacterium*-based Cas9 targeting *CSE1*, *CSE2*	Reduced lignin and increased cellulose	de Vries *et al*., 2021
Populus tremula × *P. alba* (*P.* × *canescens*)	*Agrobacterium*-based Cas9 targeting *4CL*	Decreased lignin content, discoloration of stems	Zhou *et al*., 2015

Continued

Table 10.1. Continued

Plant name	Tools used and targeted genes	Target trait	Reference
Populus tremula × Populus tremuloides	*Agrobacterium*-based Cas9 targeting *VNS*	Secondary cell wall thinning	Takata *et al.*, 2019
Populus trichocarpa	*Agrobacterium*-based Cas9 targeting *PtrHSFB3-1, PtrMYB092*	Reduced lignin and increased cellulose	Liu *et al.*, 2021
Populus trichocarpa	*Agrobacterium*-based Cas9 targeting *PtrWRKY18, PtrWRKY3*	Melampsora resistance	Jiang *et al.*, 2017
Populus trichocarpa	*Agrobacterium*-based Cas9 targeting *PtrADA2b-3*	Drought resistance	Li *et al.*, 2019
Sweet orange	*Agrobacterium*-based Cas9 targeting *CsPDS*	Citrus canker resistance	Jia *et al.*, 2019
Theobroma cacao	*Agrobacterium*-based Cas9 targeting *TcNPR3*	Phytophthora resistance	Fister *et al.*, 2018
Vitis vinifera	*Agrobacterium*-based Cas9 targeting *VvCCD7, VvCCD8*	Increased stem branching	Ren *et al.*, 2020
Vitis vinifera cv. Neo Muscat	*Agrobacterium*-based Cas9 targeting *VvPDS*	Albinism	Nakajima *et al.*, 2017
Walnut	*Agrobacterium*-based Cas9 targeting *JrPDS*	Albinism	Walawage *et al.*, 2019
Wanjincheng orange	*Agrobacterium*-based Cas9 targeting *CsLOB1*	Citrus canker resistance	Peng *et al.*, 2017
Wanjincheng orange	*Agrobacterium*-based Cas9 targeting *CsWRKY22*	Citrus canker resistance	Wang *et al.*, 2019a

projects to improve a variety of tree species. The quantity and types of genetic engineering and genome editing techniques used for numerous tree species have increased as a result of this fantastic advancement. Most crucially, it has shown considerable promise in regulating lignin biosynthesis and shortening the breeding cycle of forest trees. Its application has produced excellent results in improving key economic features in forest trees, including drought and disease stress resistance (Yingnan and Jing, 2020; Goralogia *et al.*, 2021; Fladung and Kersten, 2022; Min *et al.*, 2022).

10.3.1 Growth and development

Several variables influence plant growth and development. Among other physiological processes, plant hormones have an impact on germination, roots, growth, and fruiting. Plant hormones play a significant role in controlling the growth and development of plants and can impact physiological processes like germination, roots, and fruiting. Numerous genes encoding for the hormones that control plant growth and branching have been silenced using CRISPR-Cas9. Strigolactone (SL) *CCD7* and *CCD8* are newly identified plant hormones that prevent woody plants from branching. The SL biosynthesis genes in *Vitis vinifera*, *VvCCD7* and *VvCCD8*, were effectively altered. The results highlighted the crucial function of *VvCCD8* in controlling grape stem branches because the *ccd8* mutant had more branches than the matching wild-type plant (Ren *et al.*, 2016). Using CRISPR-Cas9 technology, the transcription factors *PcBRC1-1* and *PcBRC2-1* of Populus were deleted. The mutant plant's branches greatly expanded, and the *pcbrc2-1* plant showed a faster rate of bud formation (Muhr *et al.*, 2018). Similarly, CRISPR-Cas9 was used to eliminate the gibberellin response gene *GRG1* in hexaploid Ma bamboo (*Dendrocalamus latiflorus*), which led to a notable increase in plant height. Studying the mechanism of Gramineae bamboo's quick elongation and growth is extremely important (Ye *et al.*, 2020).

Tropical tree *Parasponia andersonii*, capable of nodulation and nitrogen fixation, has *PanHK*, *PanEIN2*, *PanNSP1*, and *PanNSP2* genes, which regulate cytokinin, ethylene, or strigolactone hormone networks, respectively. Through the production of knocked-out mutants using the CRISPR-Cas9 system, it was discovered that the *PanNSP1* and *PanNSP2* genes are essential for nodule formation, but the *PanHK4* mutation reduces the stem's cambium activity and the *PanEIN2* mutation causes differentiation (Zeijl *et al.*, 2018).

10.3.2 Timber properties

Examining the relationship between the functional characteristics of genes and those of tree wood has been done using the CRISPR-Cas system (Dai *et al.*, 2020). The CRISPR-Cas9 method successfully eliminated the *4CL1* and *4CL2* genes in Populus with up to 100% mutation efficiency (Zhou *et al.*, 2015). The amount of lignin monophenol decreased along with a 23% reduction in the total lignin content of the mutant plants. By employing CRISPR-Cas9 to eliminate the *PtrHSFB3-1* and *PtrMYB092* genes, the resultant mutant lignin and high cellulose content were compromised (Liu *et al.*, 2021).

The deletion of *PtoMYB156* in *Populus tomentosa* demonstrated the ectopic deposition of lignin, xylan, and cellulose during the establishment of the secondary cell wall (Yang *et al.*, 2017). Due to the *PtoDET2* gene being edited by CRISPR-Cas9, the mutant's secondary cell wall grew thinner and the ratio of cellulose crystallinity index (CRI), degree of polymerization (DP), and hemicellulose content increased (Fan *et al.*, 2020). Similar outcomes, such as growth restriction and a 35% reduction in lignin content, were observed in poplar when the *CSE1* and *CSE2* genes were knocked out. The finding showed that the *CSE* gene was an important target for improving the wood properties of woody plants (de Vries *et al.*, 2021).

10.3.3 Secondary metabolism

Secondary metabolites that control stress tolerance, nutrient availability, regeneration, and growth and development are abundantly produced by plants. The CRISPR-Cas9 system has been used to successfully modify genes encoding for several enzymes. Salicylate content was partially decreased as a result of the CRISPR-Cas9-mediated knockdown of the *UGT71L1* gene, which is responsible for the synthesis of salicylate in poplar (Fellenberg *et al.*, 2020). The CRISPR-Cas9-mediated deletion of the poplar *MYB115* gene reduced procyanidin accumulation. The findings showed that procyanidins' production was positively regulated by the *MYB115* gene (Wang *et al.*, 2017).

Angoranin, a phenolic compound found in pomegranate, which is rich in phenolic metabolites, has been dramatically reduced in content by the CRISPR-Cas9-mediated knockout of the two UDP-glycosyltransferase genes *PgUGT84A23* and *PgUGT84A24* (Chang *et al.*, 2019). The CRISPR-Cas9 system was used to knock out the *PtrMYB57* gene in *P. tomentosa*, which increased the plant's production of proanthocyanidins and anthocyanins (Wan *et al.*, 2017). The *JMJ25* gene, which encodes histone *H3K9* demethylase, was also knocked out in Populus, and this increased the accumulation of ectopic anthocyanin (Fan *et al.*, 2018).

The *PDS* gene plays a crucial part in the carotenoid production pathway, as shown by the mutant's albino phenotype produced by the CRISPR-Cas9 and Cas12 systems' knockout of the *PDS* gene. The first rate-limiting enzyme in the carotenoid production pathway is phytoene dehydrogenase (PDS). According to An *et al.* (2020), poplar's *PDS* gene was knocked out using CRISPR-Cas12, yielding plants with an albino phenotype. Numerous studies have found that using CRISPR-Cas9 technology to knock out the *PDS* gene results in albino phenotypes in a variety of plants, including green bamboo (Lin *et al.*, 2018), poplar (Fan *et al.*, 2015), cassava (Odipio *et al.*, 2017), grape (Nakajima *et al.*, 2017), and sweet orange (Jia and Wang, 2014).

10.3.4 Abiotic and biotic stress tolerances

As sessile organisms, the growth, development, and yield of forest trees are affected by drought, salt, extreme temperatures, and diseases. Knocking out of many genes using CRISPR-Cas9 has resulted in both decreased and improved stress tolerance in forest trees. The drought tolerance of mutant *Populus trichocarpa* plants was significantly decreased in the knockout mutant of the PtrADA*2b-3* gene (Li *et al.*, 2019). The knocking out of the *PdNF-YB21* gene in Populus simultaneously decreased the root growth and drought tolerance of the mutant plants (Zhou *et al.*, 2020). The knockout of the *PalWRKY77* gene in *Populus alba* var. *pyramidalis*, on the other hand, increased the salt tolerance of poplar plants (Jiang *et al.*, 2021). Similar to how *PdNF-YB21* was overexpressed, Populus showed better drought resistance and root growth, showing that the NF-Y transcription factor is crucial for poplar's drought resistance (Zhou *et al.*, 2020). An increased resistance to *Phytophthora tropicalis* infection was seen in the CRISPR-Cas9-mediated knockout mutant of the *TcNPR3* gene of the *Theobroma cacao* (Fister *et al.*, 2018). The cassava eIF4E isoforms, *ncbp-1* and *ncbp-2* genes, were edited using CRISPR-Cas9, and single and double mutants of *ncbp-1* and *ncbp-2* were produced. Cassava brown streak disease incidence decreased on inoculation of these mutants (Gomez *et al.*, 2019).

10.4 Conclusions

It is anticipated that the effects of stress (both biotic and abiotic) on agriculture and forestry will be more nuanced. Because of the negative effects on biodiversity, soil fertility, microbial activity, and carbon sequestration, this will have a severe effect on agriculture and forest tree production and quality. Creating stress-resistant crops and forest trees under unfavorable climates is important to see a world without hunger and malnutrition. One of the major concerns for our world is how to increase plant yield without expanding the area of arable land. To do this, it is thought that applying cutting-edge breeding techniques like CRISPR will revolutionize current systems and create a variety of stress-resistant crops and forest trees. Both basic and practical plant science studies have been transformed by CRISPR technology. This technology has made it possible to investigate gene function, build genome-wide regulatory networks, and control the expression of genes according to specified sequences. CRISPR is the ideal intervention at the most critical period of our planet's history because of its ability to improve economically valuable agricultural and forestry species as well as its wider uses across the entire field of life sciences. The development of crops that are resilient to stress, high-yielding, nutrient-dense, photosynthesis-efficient, and quick-growing is predicted to benefit from the use of CRISPR technology. The growth of high biomass, good wood quality, high carbon sequestration, and stress-resistant forest trees is made possible by CRISPR, which is well positioned to help the forestry sector achieve its environmental and forestry goals. Thus, the novel and wide-ranging applications of CRISPR-Cas technology will transform the study of agricultural and forestry biotechnology and herald the beginning of a new green revolution.

References

Adem, M., Feyissa, T. and Beyene, D. (2017) Application of precise genome editing in plants. *SDRP Journal of Plant Science. Sift Desk*. DOI: 10.25177/JPS.2.1.2.

Ahmad, S., Tang, L., Shahzad, R., Mawia, A.M., Rao, G.S. *et al.* (2021) CRISPR-based crop improvements: A way forward to achieve zero hunger. *Journal of Agricultural and Food Chemistry* 69(30), 8307–8323. DOI: 10.1021/acs.jafc.1c02653.

Ahmad, A., Khan, S.H. and Khan, Z. (eds) (2022) *The CRISPR/Cas Tool Kit for Genome Editing*. Springer, Berlin. DOI: 10.1007/978-981-16-6305-5.

Al Amin, N., Ahmad, N., Wu, N., Pu, X., Ma, T. *et al*. (2019) CRISPR-Cas9 mediated targeted disruption of FAD2-2 microsomal omega-6 desaturase in soybean (*Glycine max.L*). *BMC Biotechnology* 19, 9. DOI: 10.1186/s12896-019-0501-2.

An, Y., Geng, Y., Yao, J., Fu, C., Lu, M. *et al*. (2020) Efficient genome editing in Populus using CRISPR/Cas12a. *Frontiers in Plant Science* 11, 593938.

Baloglu, M.C., Celik Altunoglu, Y., Baloglu, P., Yildiz, A.B., Türkölmez, N. *et al*. (2022) Gene-editing technologies and applications in legumes: Progress, evolution, and future prospects. *Frontiers in Genetics* 13, 859437. DOI: 10.3389/fgene.859437.

Bhambhani, S., Kondhare, K.R. and Giri, A.P. (2021) Advanced genome editing strategies for manipulation of plant specialized metabolites pertaining to biofortification. *Phytochemistry Reviews* 1, 282.

Bhowmik, P., Konkin, D., Polowick, P., Hodgins, C.L., Subedi, M. *et al*. (2021) CRISPR/Cas9 gene editing in legume crops: Opportunities and challenges. *Legume Science* 3, e96. DOI: 10.1002/leg3.96.

Breitler, J.C., Dechamp, E., Campa, C., Zebral Rodrigues, L.A., Guyot, R. *et al*. (2018) CRISPR/cas9-mediated efficient targeted mutagenesis has the potential to accelerate the domestication of *Coffea canephora*. *Plant Cell, Tissue and Organ Culture* 134, 383–394.

Cai, L., Zhang, L., Fu, Q. and Xu, Z.F. (2018) Identification and expression analysis of cytokinin metabolic genes IPTs, CYP735A and CKXs in the biofuel plant *Jatropha curcas*. *PeerJ* 6, e4812.

Camerlengo, F., Frittelli, A. and Pagliarello, R. (2022) CRISPR towards a sustainable agriculture. *Encyclopedia* 2, 538–558. DOI: 10.3390/encyclopedia2010036.

Cao, H.X., Vu, G.T.H. and Gailing, O. (2022) From genome sequencing to CRISPR-based genome editing for climate-resilient forest trees. *International Journal of Molecular Sciences* 23(966). DOI: 10.3390/%20ijms23020966.

Chang, L., Wu, S. and Tian, L. (2019) Effective genome editing and identification of a regiospecific gallic acid 4-O-glycosyltransferase in pomegranate (*Punica granatum* L.). *Horticulture Research* 6, 1–15.

Char, S.N., Neelakandan, A.K., Nahampun, H., Frame, B., Main, M. *et al*. (2017) An *Agrobacterium*-delivered CRISPR/Cas9 system for high-frequency targeted mutagenesis in maize. *Plant Biotechnology Journal* 15(2), 257–268. DOI: 10.1111/pbi.126111.

Charrier, A., Vergne, E., Dousset, N., Richer, A., Petiteau, A. *et al*. (2019) Efficient targeted mutagenesis in apple and first time edition of pear using the CRISPR-Cas9 system. *Frontiers in Plant Science* 10, 40.

Christian, M., Qi, Y., Zhang, Y. and Voytas, D.F. (2013) Targeted mutagenesis of *Arabidopsis thaliana* using engineered TAL effector nucleases. *G3: Bethesda* 3, 1697–1705.

Cui, Y., Zhao, J., Gao, Y., Zhao, R., Zhang, J. *et al*. (2021) Efficient multi-sites genome editing and plant regeneration via somatic embryogenesis in *Picea glauca*. *Frontiers in Plant Science* 12, 751891. DOI: 10.3389/fpls.2021.751891.

Dai, Y., Hu, G., Dupas, A., Medina, L., Blandels, N. *et al*. (2020) Implementing the CRISPR/Cas9 technology in *Eucalyptus* hairy roots using wood-related genes. *International Journal of Molecular Sciences* 21, 3408.

de Vries, L., Brouckaert, M., Chanoca, A., Kim, H., Regner, M.R. *et al*. (2021) CRISPR-Cas9 editing of CAFFEOYL SHIKIMATE ESTERASE 1 and 2 shows their importance and partial redundancy in lignification in *Populus tremula* × *P. alba*. *Plant Biotechnology Journal* 19, 2221–2234.

Do, P.T., Nguyen, C.X., Bui, H.T., Tran, L.T.N., Stacey, G. *et al*. (2019) Demonstration of highly efficient dual grna CRISPR/Cas9 editing of the homeologous GmFAD2-1A and GmFAD2-1B genes to yield a high oleic, low linoleic and α-linolenic acid phenotype in soybean. *BMC Plant Biology* 19, 311.

Fan, C., Yu, H., Qin, S., Li, Y., Alam, A. *et al*. (2020) Brassinosteroid overproduction improves lignocellulose quantity and quality to maximize bioethanol yield under green-like biomass process in transgenic poplar. *Biotechnology for Biofuels* 13, 9.

Fan, D., Liu, T., Li, C., Jiao, B., Li, S. *et al*. (2015) Efficient CRISPR/Cas9-mediated targeted mutagenesis in Populus in the first generation. *Scientific Reports* 5, 12217.

Fan, D., Wang, X., Tang, X., Ye, X., Ren, S. *et al*. (2018) Histone H3K9 demethylase JMJ25 epigenetically modulates anthocyanin biosynthesis in poplar. *The Plant Journal* 96, 1121–1136.

Fellenberg, C., Corea, O., Yan, L.H., Archinuk, F., Piirtola, E.M. *et al*. (2020) Discovery of salicyl benzoate UDP-glycosyltransferase, a central enzyme in poplar salicinoid phenolic glycoside biosynthesis. *The Plant Journal* 102, 99–115.

Fillatti, J.J., Sellmer, J., McCown, B., Haissig, B. and Comai, L. (1987) Agrobacterium mediated transformation and regeneration of Populus. *Molecular and General Genetics MGG* 206, 192–199.

Fister, A.S., Landherr, L., Maximova, S.N. and Guiltinan, M.J. (2018) Transient expression of CRISPR/Cas9 machinery targeting *TcNPR3* enhances defense response in *Theobroma cacao*. *Frontiers in Plant Science* 9, 268.

Fladung, M. and Kersten, B. (2022) Tree genetic engineering, genome editing and genomics. *International Journal of Molecular Sciences* 23, 13980. DOI: 10.3390/ijms232213980.

Gaj, T., Gersbach, C.A. and Barbas, C.F. (2013) ZFN, TALEN, and CRISPR/CAS-based methods for genome engineering. *Trends in Biotechnology* 31, 397–405. DOI: 10.1016/j.tibtech.2013.04.004.

Gao, H., Gadlage, M.J., Lafitte, H.R., Lenderts, B., Yang, M. *et al*. (2020a) Superior field performance of waxy corn engineered using CRISPR-Cas9. *Nature Biotechnology* 38, 579–581.

Gao, H., Mutti, J., Young, J.K., Yang, M., Schroder, M. *et al*. (2020b) Complex trait loci in maize enabled by CRISPR-Cas9 mediated gene insertion. *Frontiers in Plant Science* 11, 535. DOI: 10.3389/fpls.2020.00535.

Gasparis, S., Kała, M., Przyborowski, M., Łyżnik, L.A., Orczyk, W. *et al*. (2018) A simple and efficient CRISPR/Cas9 platform for induction of single and multiple, heritable mutations in barley (*Hordeum vulgare* L.). *Plant Methods* 14, 111. DOI: 10.1186/s13007-018-0382-8.

Gillani, S.F.A., Rasheed, A., Majeed, Y., Tariq, H. and Yunling, P. (2021) Recent advancements on use of CRISPR /Cas9 in maize yield and quality improvement. *Notulae Botanicae Horti Agrobotanici Cluj-Napoca* 49, 12459. DOI: 10.15835/nbha49312459.

Gomez, M.A., Lin, Z.D., Moll, T., Chauhan, R.D., Hayden, L. *et al*. (2019) Simultaneous CRISPR/Cas9-mediated editing of cassava EIF4E isoforms NCBP-1 and NCBP-2 reduces cassava brown streak disease symptom severity and incidence. *Plant Biotechnology Journal* 17, 421–434.

Goralogia, G.S., Redick, T.P. and Strauss, S.H. (2021) Gene editing in tree and clonal crops: Progress and challenges. *In Vitro Cellular & Developmental Biology - Plant*. DOI: 10.1007/s11627-021-10197-x.

Hess, G.T., Tycko, J., Yao, D. and Bassik, M.C. (2017) Methods and applications of CRISPR-mediated base editing in eukaryotic genomes. *Molecular Cell* 68(1), 26–43. DOI: 10.1016/j.molcel.2017.09.029.

Holme, I.B., Wendt, T., Gil-Humanes, J., Deleuran, L.C., Starker, C.G.V.DF. *et al*. (2017) Evaluation of the mature grain phytase candidate HvPAPhy_a gene in barley (*Hordeum vulgare* L.) using CRISPR/Cas9 and TALENs. *Plant Molecular Biology* 95, 111–121. DOI: 10.1007/s11103-017-0640-6.

Holubová, K., Hensel, G., Vojta, P., Tarkowski, P., Bergougnoux, V. *et al*. (2018) Modification of barley plant productivity through regulation of cytokinin content by reverse-genetics approaches. *Frontiers in Plant Science* 9, 1676. DOI: 10.3389/fpls.2018.01676.

Jia, H. and Wang, N. (2014) Targeted genome editing of sweet orange using cas9/sgRNA. *PLoS ONE* 9(4), e93806. DOI: 10.1371/journal.pone.0093806.

Jia, H., Orbovic, V., Jones, J.B. and Wang, N. (2016) Modification of the PthA4 effector binding elements in type I CsLOB1 promoter using Cas9/sgRNA to produce transgenic Duncan grapefruit alleviating XccDptha4:DCsLOB1.3 infection. *Plant Biotechnology Journal* 14, 1291–1301.

Jia, H., Orbović, V. and Wang, N. (2019) CRISPR-LBCas12a-mediated modification of citrus. *Plant Biotechnology Journal* 17, 1928–1937.

Jiang, Y., Guo, L., Ma, X., Zhao, X., Jiao, B. *et al*. (2017) The WRKY transcription factors PtrWRKY18 and PtrWRKY35 promote *Melampsora* resistance in Populus. *Tree Physiology* 37, 665–675.

Jiang, Y., Tong, S., Chen, N., Liu, B., Bai, Q. *et al*. (2021) The PalWRKY77 transcription factor negatively regulates salt tolerance and abscisic acid signaling in Populus. *The Plant Journal* 105, 1258–1273.

Kumagai, Y., Liu, Y., Hamada, H., Luo, W., Zhu, J. *et al*. (2022) Introduction of a second "Green Revolution" mutation into wheat via in planta CRISPR/Cas9 delivery. *Plant Physiology* 188, 1838–1842. DOI: 10.1093/plphys/kiab570.

Kumar, R., Kaur, A., Pandey, A., Mamrutha, H.M. and Singh, G.P. (2019) CRISPR-based genome editing in wheat: A comprehensive review and future prospects. *Molecular Biology Reports* 46(3), 3557–3569. DOI: 10.1007/s11033-019-04761-3.

Lawrenson, T., Shorinola, O., Stacey, N., Li, C., Østergaard, L. *et al*. (2015) Induction of targeted, heritable mutations in barley and *Brassica oleracea* using RNA-guided Cas9 nuclease. *Genome Biology* 16, 258. DOI: 10.1186/s13059-015-0826-7.

Li, G., Liu, R., Xu, R., Varshney, R.K., Ding, H. *et al*. (2022) Development of an *Agrobacterium*-mediated CRISPR/Cas9 system in pea (*Pisum sativum* L.). *The Crop Journal* 11(1), 132–139.

Li, M.Y., Jiao, Y.T., Wang, Y.T., Zhang, N., Wang, B.B. *et al*. (2020) CRISPR/Cas9-mediated *Vvpr4b* editing decreases downy mildew resistance in grapevine (*Vitis vinifera* L.). *Horticulture Research* 7, 1–11.

Li, R., Li, R., Li, X., Fu, D., Zhu, B. *et al*. (2018) Multiplexed CRISPR/Cas9-mediated metabolic engineering of γ-aminobutyric acid levels in *Solanum lycopersicum*. *Plant Biotechnology Journal* 16, 415–427.

Li, S., Lin, Y.C.J., Wang, P., Zhang, B., Li, M. *et al.* (2019) The AREB1 transcription factor influences histone acetylation to regulate drought responses and tolerance in *Populus trichocarpa*. *The Plant Cell* 31, 663–686.

Li, X., Wang, Y., Chen, S., Tian, H., Fu, D. *et al.* (2018) Lycopene is enriched in tomato fruit by CRISPR/Cas9-mediated multiplex genome editing. *Frontiers in Plant Science* 9, 559.

Liang, Z., Zhang, K., Chen, K.L. and Gao, C.X. (2014) Targeted mutagenesis in *Zea mays* using TALENs and the CRISPR/Cas system. *Journal of Genetics and Genomics* 41, 63–68. DOI: 10.1016/j.jgg.2013.12.001.

Lin, C.S., Hsu, C.T., Yang, L.H., Lee, L.Y., Fu, J.Y. *et al.* (2018) Application of protoplast technology to CRISPR/Cas9 mutagenesis: From single-cell mutation detection to mutant plant regeneration. *Plant Biotechnology Journal* 16, 1295–1310.

Liu, B., Liu, J., Yu, J., Wang, Z., Sun, Y. *et al.* (2021) Transcriptional reprogramming of xylem cell wall biosynthesis in tension wood. *Plant Physiology* 186, 250–269.

Liu, H., Chen, W., Li, Y., Sun, L., Chai, Y. *et al.* (2022) CRISPR/Cas9 technology and its utility for crop improvement. *International Journal of Molecular Sciences* 23(10442). DOI: 10.3390/%20ijms231810442.

Liu, J., Chen, J., Zheng, X., Wu, F., Lin, Q. *et al.* (2017) GW5 acts in the brassinosteroid signalling pathway to regulate grain width and weight in rice. *Nature Plants* 3, 17043.

Lu, K., Wu, B., Wang, J., Zhu, W., Nie, H. *et al.* (2018) Blocking amino acid transporter OsAAP3 improves grain yield by promoting outgrowth buds and increasing tiller number in rice. *Plant Biotechnology Journal* 16, 1710–1722.

Malabarba, J., Chevreau, E., Dousset, N., Veillet, F., Moizan, J. *et al.* (2020) New strategies to overcome present CRISPR/Cas9 limitations in apple and pear: Efficient dechimerization and base editing. *International Journal of Molecular Sciences* 22, 319.

Min, T., Hwarari, D., Li, D., Movahedi, A. and Yang, L. (2022) CRISPR-based genome editing and its applications in woody plants. *International Journal of Molecular Sciences* 23, 10175. DOI: 10.3390/ijms231710175.

Muhr, M., Paulat, M., Awwanah, M., Brinkkötter, M. and Teichmann, T. (2018) CRISPR/Cas9-mediated knockout of populus BRANCHED1 and BRANCHED2 orthologs reveals a major function in bud outgrowth control. *Tree Physiology* 38(10), 1588–1597.

Nakajima, I., Ban, Y., Azuma, A., Onoue, N., Moriguchi, T. *et al.* (2017) CRISPR/Cas9-mediated targeted mutagenesis in grape. *PLoS One* 12, e0177966.

Nanasato, Y., Mikami, M., Futamura, N., Endo, M., Nishiguchi, M. *et al.* (2021) CRISPR/Cas9-mediated targeted mutagenesis in Japanese cedar (*Cryptomeria japonica* D. Don). *Scientific Reports* 11(1), 16186. DOI: 10.1038/s41598-021-95547-w.

Neelakandan, A.K., Wright, D.A., Traore, S.M., Ma, X., Subedi, B. *et al.* (2022) Application of CRISPR/Cas9 system for efficient gene editing in peanut. *Plants* 11, 1361. DOI: 10.3390/plants11101361.

Nekrasov, V., Wang, C., Win, J., Lanz, C., Weigel, D. *et al.* (2017) Rapid generation of a transgene-free powdery mildew resistant tomato by genome deletion. *Scientific Reports* 7, 482.

Nishitani, C., Hirai, N., Komori, S., Wada, M., Okada, K. *et al.* (2016) Efficient genome editing in apple using a CRISPR/Cas9 system. *Scientific Reports* 6, 31481.

Odipio, J., Alicai, T., Ingelbrecht, I., Nusinow, D.A., Bart, R. *et al.* (2017) Efficient CRISPR/Cas9 genome editing of *Phytoene desaturase* in Cassava. *Frontiers in Plant Science* 8, 1780.

Oliva, R., Ji, C., Atienza-Grande, G., Huguet-Tapia, J.C., Perez-Quintero, A. *et al.* (2019) Broad-spectrum resistance to bacterial blight in rice using genome editing. *Nature Biotechnology* 37, 1344–1350.

Pavese, V., Moglia, A., Abbà, S., Milani, A.M., Torello Marinoni, D. *et al.* (2022) First report on genome editing via ribonucleoprotein (RNP) in *Castanea sativa* Mill. *International Journal of Molecular Sciences* 23(10), 5762. DOI: 10.3390/ijms23105762.

Peng, A., Chen, S., Lei, T., Xu, L., He, Y. *et al.* (2017) Engineering canker-resistant plants through CRISPR/Cas9-targeted editing of the susceptibility gene CsLOB1 promoter in citrus. *Plant Biotechnology Journal* 15, 1509–1519.

Poovaiah, C., Phillips, L., Geddes, B., Reeves, C., Sorieul, M. *et al.* (2021) Genome editing with CRISPR/Cas9 in *Pinus radiata* (D. Don). *BMC Plant Biology* 21(1), 363. DOI: 10.1186/s12870-021-03143-x.

Ran, F.A., Hsu, P.D., Lin, C.Y., Gootenberg, J.S., Konermann, S. *et al.* (2013) Double nicking by RNA-guided CRISPR Cas9 for enhanced genome editing specificity. *Cell* 154, 1380–1389. DOI: 10.1016/%20j.cell.2013.08.021.

Razzaq, A., Saleem, F., Kanwal, M., Mustafa, G., Yousaf, S. *et al.* (2019) Modern trends in plant genome editing: An inclusive review of the CRISPR/Cas9 toolbox. *International Journal of Molecular Sciences* 20, 4045. DOI: 10.3390/ijms20164045.

Ren, C., Liu, X., Zhang, Z., Wang, Y., Duan, W. *et al.* (2016) CRISPR/Cas9-mediated efficient targeted mutagenesis in chardonnay (*Vitis vinifera* L.). *Scientific Reports* 6, 32289.

Ren, C., Guo, Y., Kong, J., Lecourieux, F., Dai, Z. *et al.* (2020) Knockout of VvCCD8 gene in grapevine affects shoot branching. *BMC Plant Biology* 20, 47.

Rodríguez-Leal, D., Lemmon, Z.H., Man, J., Bartlett, M.E. and Lippman, Z.B. (2017) Engineering quantitative trait variation for crop improvement by genome editing. *Cell* 171, 470–480.

Sánchez-León, S., Gil-Humanes, J., Ozuna, C.V., Giménez, M.J., Sousa, C. *et al.* (2018) Low-gluten, nontransgenic wheat engineered with CRISPR/Cas9. *Plant Biotechnology Journal* 16, 902–910.

Segal, D. and Meckler, J. (2003) Genome engineering at the dawn of the golden age. DOI: 10.1146/annurevgenom-091212-153435.

Sharma, P., Singh, S.P., Iqbal, H.M.N., Parra-Saldivar, R., Varjani, S. *et al.* (2022) Genetic modifications associated with sustainability aspects for sustainable developments. *Bioengineered* 13(4), 9509–9521. DOI: 10.1080/21655979.2022.2061146.

Shen, Y., Li, Y., Xu, D., Yang, C., Li, C. *et al.* (2018) Molecular cloning and characterization of a brassinosteriod biosynthesis-related gene *PtoDWF4* from *Populus tomentosa*. *Tree Physiology* 38, 1424–1436.

Takata, N., Awano, T., Nakata, M.T., Sano, Y., Sakamoto, S. *et al.* (2019) Populus NST/SND orthologs are key regulators of secondary cell wall formation in wood fibers, phloem fibers and xylem ray parenchyma cells. *Tree Physiology* 39, 514–525.

Van, V.T., Das, S., Hensel, G. and Kim, J.Y. (2022) Genome editing and beyond: What does it mean for the future of plant breeding? *Planta* 25, 130.

Van Zeijl, A., Wardhani, T.A.K., Seifi Kalhor, M., Rutten, L., Bu, F. *et al.* (2018) CRISPR/Cas9-mediated mutagenesis of four putative symbiosis genes of the tropical tree *Parasponia andersonii* reveals novel phenotypes. *Frontiers in Plant Science* 9, 284.

Varkonyi-Gasic, E., Wang, T., Voogd, C., Jeon, S., Drummond, R.S.M. *et al.* (2019) Mutagenesis of kiwifruit CENTRORADIALIS-like genes transforms a climbing woody perennial with long juvenility and axillary flowering into a compact plant with rapid terminal flowering. *Plant Biotechnology Journal* 17, 869–880.

Walawage, S.L., Zaini, P.A., Mubarik, M.S., Martinelli, F., Balan, B. *et al.* (2019) Deploying genome editing tools for dissecting the biology of nut trees. *Frontiers in Sustainable Food Systems* 3, 100.

Wan, S., Li, C., Ma, X., Luo, K. and Repressor, P.Á. (2017) PtrMYB57 contributes to the negative regulation of anthocyanin and proanthocyanidin biosynthesis in poplar. *Plant Cell Reports* 36, 1263–1276.

Wang, C., Wang, G., Gao, Y., Lu, G., Habben, J.E. *et al.* (2020) A cytokinin-activation enzyme-like gene improves grain yield under various field conditions in rice. *Plant Molecular Biology* 102, 373–388.

Wang, L., Ran, L., Hou, Y., Tian, Q., Li, C. *et al.* (2017) The transcription factor MYB115 contributes to the regulation of proanthocyanidin biosynthesis and enhances fungal resistance in poplar. *The New Phytologist* 215, 351–367.

Wang, L., Chen, S., Peng, A., Xie, Z., He, Y. *et al.* (2019a) CRISPR/Cas9-mediated editing of CsWRKY22 reduces susceptibility to *Xanthomonas citri* subsp. *citri* in Wanjincheng orange (*Citrus sinensis* (L.) Osbeck). *Plant Biotechnology Reports* 13, 501–510.

Wang, P., Zhang, J., Sun, L., Ma, Y., Xu, J. *et al.* (2018a) High efficient multisites genome editing in allotetraploid cotton (*Gossypium hirsutum*) using CRISPR/Cas9 system. *Plant Biotechnology Journal* 16, 137–150.

Wang, T., Zhang, H. and Zhu, H. (2019b) CRISPR technology is revolutionizing the improvement of tomato and other fruit crops. *Horticulture Research* 6(1), 1–13.

Wang, W., Simmonds, J., Pan, Q., Davidson, D., He, F. *et al.* (2018b) Gene editing and mutagenesis reveal inter-cultivar differences and additivity in the contribution of TaGW2 homoeologues to grain size and weight in wheat. *Theoretical and Applied Genetics* 131(11), 2463–2475.

Wang, X., Tu, M., Wang, D., Liu, J., Li, Y. *et al.* (2018c) CRISPR/Cas9-mediated efficient targeted mutagenesis in grape in the first generation. *Plant Biotechnology Journal* 16, 844–855.

Xu, Y., Lin, Q., Li, X., Wang, F., Chen, Z. *et al.* (2021) Fine-tuning the amylose content of rice by precise base editing of the WX gene. *Plant Biotechnology Journal*. DOI: 10.1111/pbi.13433.

Xu, Z., Xu, X., Gong, Q., Li, Z., Li, Y. *et al.* (2019) Engineering broad-spectrum bacterial blight resistance by simultaneously disrupting variable TALE-binding elements of multiple susceptibility genes in rice. *Molecular Plant* 12, 1434–1446.

Yang, L., Zhao, X., Ran, L., Li, C., Fan, D. *et al.* (2017) PtoMYB156 is involved in negative regulation of phenylpropanoid metabolism and secondary cell wall biosynthesis during wood formation in poplar. *Scientific Reports* 7, 41209.

Ye, S., Chen, G., Kohnen, M.V., Wang, W., Cai, C. *et al.* (2020) Robust CRISPR/Cas9 mediated genome editing and its application in manipulating plant height in the first generation of hexaploid ma bamboo (*Dendrocalamus latiflorus Munro*). *Plant Biotechnology Journal* 18, 1501–1503.

Yingnan, C. and Jing, L. (2020) Application of CRISPR/Cas9 mediated gene editing in trees. *Hereditas* 42(7), 657–668. DOI: 10.16288/j.yczz.20-092.

You, J., Li, D., Yang, L., Dossou, S.S.K., Zhou, R. *et al.* (2022) CRISPR/Cas9-mediated efficient targeted mutagenesis in sesame (*Sesamum indicum* L.). *Frontiers in Plant Science* 13, 935825. DOI: 10.3389/fpls.2022.935825.

Young, J., Zastrow-Hayes, G., Deschamps, S., Svitashev, S., Zaremba, M. *et al.* (2019) CRISPR-Cas9 editing in maize: Systematic evaluation of off-target activity and its relevance in crop improvement. *Scientific Reports* 9, 6729. DOI: 10.1038/s41598-019-43141-6.

Yuste-Lisbona, F.J., Fernández-Lozano, A., Pineda, B., Bretones, S., Ortíz-Atienza, A. *et al.* (2020) ENO regulates tomato fruit size through the floral meristem development network. *Proceedings of the National Academy of Sciences of the United States of America* 117, 8187–8195.

Zeijl, V.A., Wardhani, T.A.K., Kalhorm, R.L., Bu, F., Hartog, M. *et al.* (2018) CRISPR/Cas9-mediated mutagenesis of four putative symbiosis genes of the tropical tree *Parasponia andersonii* reveals novel phenotypes. *Frontiers in Plant Science* 9, 284.

Zeng, Y., Wen, J., Zhao, W., Wang, Q. and Huang, W. (2019) Rational improvement of rice yield and cold tolerance by editing the three genes OsPIN5b, GS3, and OsMYB30 with the CRISPR-Cas9 system. *Frontiers in Plant Science* 10, 1663.

Zhang, F., LeBlanc, C., Irish, V.F. and Jacob, Y. (2017a) Rapid and efficient CRISPR/Cas9 gene editing in citrus using the YAO promoter. *Plant Cell Reports* 36, 1883–1887.

Zhang, S., Wu, S., Hu, C., Yang, Q., Dong, T. *et al.* (2022) Increased mutation efficiency of CRISPR/Cas9 genome editing in banana by optimized construct. *PeerJ* 10, e12664. DOI: 10.7717/peerj.12664.

Zhang, Y., Bai, Y., Wu, G., Zou, S., Chen, Y. *et al.* (2017b) Simultaneous modification of three homoeologs of TaEDR1 by genome editing enhances powdery mildew resistance in wheat. *The Plant Journal* 91, 714–724.

Zhang, Y., Li, D., Zhang, D., Zhao, X., Cao, X. *et al.* (2018) Analysis of the functions of TaGW2 homoeologs in wheat grain weight and protein content traits. *The Plant Journal* 94, 857–866.

Zhang, Z., Hua, L., Gupta, A., Tricoli, D., Edwards, K.J. *et al.* (2019) Development of an *Agrobacterium*-delivered CRISPR/Cas9 system for wheat genome editing. *Plant Biotechnology Journal* 17, 1623–1635.

Zhou, J., Xin, X., He, Y., Chen, H., Li, Q. *et al.* (2019) Multiplex QTL editing of grain-related genes improves yield in elite rice varieties. *Plant Cell Reports* 38, 475–485.

Zhou, X., Jacobs, T.B., Xue, L.J., Harding, S.A. and Tsai, C.J. (2015) Exploiting SNPs for biallelic CRISPR mutations in the outcrossing woody perennial populus reveals 4-coumarate: CoA ligase specificity and redundancy. *The New Phytologist* 208, 298–301.

Zhou, Y., Zhang, Y., Wang, X., Han, X., An, Y. *et al.* (2020) Root-specific NF-Y family transcription factor, PdNF-YB21, positively regulates root growth and drought resistance by abscisic acid-mediated indoylacetic acid transport in Populus. *The New Phytologist* 227, 407–426.

Zhu, H., Li, C. and Gao, C. (2020) Applications of CRISPR–Cas in agriculture and plant biotechnology. *Nature Reviews Molecular Cell Biology* 21, 661–677.

11 Application and Prospects of DNA-free Genome Modifications in Crop Plants through Genome Editing

Ambreen Gul[1] and Muhammad Naeem Sattar[2]*

[1]CEMB, University of the Punjab, Lahore, Pakistan; [2]Central Laboratories, King Faisal University, Al-Ahsa, Saudi Arabia

Abstract

CRISPR (an acronym for clustered regularly interspaced short palindromic repeats) technology is transforming plant breeding by enabling precise genome editing without introducing foreign DNA. This chapter explores achieving transgene-free editing, where the final edited plant lacks any permanent insertion of foreign genetic material. The past decade has seen significant advancements in genome editing, particularly through nucleotide sequence alteration, leading to remarkable progress in site-directed mutagenesis. However, research has increasingly focused on enhancing precision, targeting efficacy, and minimizing off-target mutations, especially in DNA-editing applications. This has driven the current emphasis on harnessing precise technology while avoiding transgenic outcomes. This chapter delves into various technological advancements that underscore the importance of transgene-free genome editing. We critically examine the advantages and limitations of each technology to provide a transparent perspective. A systematic analysis of published research reveals a global research effort to address the challenges posed by transgenic regulations. This approach has significant implications for mitigating concerns regarding acceptance, reliability, and ethical considerations surrounding gene editing technology, while offering sustainable solutions toward ensuring global food security. Overall, the chapter highlights the substantial progress made in achieving transgene-free CRISPR-based genome editing in plants, opening exciting possibilities for crop improvement without the concerns associated with traditional transgenic modifications.

Keywords: Transgene-free editing, Plant genome editing, Plant breeding, Transgenic regulations, CRISPR-based genome editing

11.1 Introduction

The CRISPR-Cas system has emerged as a powerful tool in genomics, offering immense potential for gene function elucidation and crop improvement. This innovative technology leverages CRISPRs alongside flanking, non-repetitive spacers (Ishino *et al.*, 1987). These spacers demonstrate striking similarity to exogenous DNA sequences originating from viruses and plasmids. Indeed, bacteria and archaea use this system to defend against foreign genetic elements. Programmable nucleases for precise genome editing often rely on the CRISPR-Cas system. Within the CRISPR locus, a short CRISPR RNA (crRNA) guides target specificity (Prajapati and Tyagi, 2024). Additionally, a key motif called the protospacer-adjacent motif

*Corresponding author: Naeem.sattar1177@gmail.com; mnsattar@kfu.edu.sa

 Genome Editing for Crop Improvement: Theory and Methodology (eds J.M. Al-Khayri *et al.*)
DOI: 10.1079/9781800622517.0011

(PAM) is crucial for accurate recognition of the target site (Jinek *et al.*, 2012). By combining crRNA with a trans-activating short CRISPR RNA (tracrRNA), a single-guide RNA (sgRNA) forms a stable complex with Cas9, ultimately facilitating targeted editing of the nucleic acid region. Precision genome editing in plant cells relies on engineered nucleases, or site-directed nucleases (SDNs), to introduce targeted double-stranded breaks (DSBs) at specific DNA locations (Mamrutha *et al.*, 2023). These specialized enzymes combine a non-specific cleavage domain coupled with the DNA-binding domain sequence specifically, enabling accurate identification and cutting of the desired DNA duplex (Hilscher *et al.*, 2017). SDN-mediated editing falls into three main categories:

I. SDN1: It generates random point mutations at the DSB site without utilizing a donor template for repair (non-homologous end joining, [NHEJ]).
II. SDN2: It induces precise, defined mutations at the DSB site via homology-directed repair (HDR), incorporating a provided repair template with homologous sequences.
III. SDN3: It facilitates the site-specific insertion of foreign DNA sequences through two potential mechanisms: (a) HDR-mediated integration using a repair template flanked by identical sequences, or (b) non-homologous integration of a donor sequence without homology-dependent repair.

CRISPR technology involves targeting a specific DNA sequence through sgRNA designed to recognize the PAM sequence (Pausch *et al.*, 2020). The sgRNA is typically delivered using a vector engineered with suitable promoters, such as AtU6 for Arabidopsis, TaU6 for wheat, or OsU6/OsU3 for rice, to drive the expression of both Cas9 and sgRNA genes. This vector is then introduced into the host cell, followed by screening and validation of editing events (Sattar *et al.*, 2017). Besides its wide applications in various crop plants (Iqbal *et al.*, 2016; Sattar *et al.*, 2017), this technology faces various challenges, including issues with target specificity, efficiency, non-transgenic applications, adaptability to perennial crops, and the development of an ideal delivery system (Li *et al.*, 2023). These challenges pose significant obstacles to realizing the full potential of the technology. The application of transgene-free editing is imperative for its use in agricultural crops, presenting difficulties in maintaining proper gene containment. Furthermore, plants generated through traditional CRISPR-based gene editing methods are subject to stringent regulations as transgenic organisms and are classified as genetically modified organisms (GMOs). This categorization, based on recombinant DNA constructs involved in the process (Adeel and Jones, 2024), imposes restrictions on the utilization of editing technology for crop enhancement and overall agricultural practices.

Plant breeders worldwide are currently directing their efforts toward using this robust technology through various modifications aimed at producing agricultural products free of foreign DNA. This chapter provides a comprehensive discussion on technological interventions, aiming to facilitate readers' understanding of the necessity and potential of CRISPR as a non-genetically modified (non-GM) technology in agricultural biotechnology.

11.2 Unlocking New Possibilities: A Glimpse Into DNA-free Genome Editing

The past few years (2019–2022) have witnessed a surge in innovative ideas for achieving DNA-free genome editing (Najafi *et al.*, 2023). Initially, a dominant focus during this period was on ribonucleoprotein (RNP)-mediated editing, a promising approach for removing foreign DNA in edited organisms. However, since 2020, there has been a diversification of research efforts, with exploration of novel and modified technologies gaining traction (Norfaezah *et al.*, 2024). This includes investigations into RNA-based editing, base editing, and the application of nanoparticles as delivery vehicles (Tao *et al.*, 2023). Despite these advancements, the use of these techniques remains limited to a small number of crop species. This highlights the crucial need for wider dissemination of these cutting-edge tools among researchers across the globe. Broader adoption and application of these novel approaches can significantly advance the field of DNA-free genome editing. Unlike traditional methods, DNA-free genome editing modifies an organism's genomic sequence without leaving behind any traces of foreign DNA. However, it is important to clarify that being 'transgene free' does not equate to the complete absence of foreign DNA throughout the entire process (Tao *et al.*, 2023). Rather, it signifies that the final edited organism lacks any permanent insertion of foreign nucleic acids. Additionally, for sexually propagated crops, the ability of traits to segregate naturally across generations offers a clear advantage, allowing both desirable and undesirable modifications to be introduced or removed over time.

In contrast, asexually propagated crops often face challenges where the introduction of a trait may result in the incorporation of undesirable DNA fragments (Sattar *et al.*, 2021; Huang *et al.*, 2023). Moreover, in plant transformation, undesired mutations arising from illegitimate recombination events can increase the likelihood of product rejection, despite resembling natural mutation events. Addressing these challenges becomes crucial to meet the global demand for a stable food supply, utilizing the robust CRISPR technology in conjunction with modern tools to generate abundant products labeled as 'transgene free.' Numerous complementary tools and technologies have been applied to traditional CRISPR-Cas9 systems, enhancing their versatility and effectiveness.

11.3 Minimizing Transgene Interference in CRISPR-edited Plants for Faster Breeding

The CRISPR gene editing system (Cas9 protein, sgRNA cassette, and an appropriate antibiotic-resistant selection marker like the kanamycin-resistant gene) can be delivered in a single vector. This delivery vector introduces the system into plant cells through either *Agrobacterium*-mediated or biolistic transformation (gene gun) (Bhattacharjee *et al.*, 2023). Once integrated into the plant's chromosomes, the transgenes express the CRISPR enzymes and gRNAs, enabling targeted modifications of specific genes. While most T0 plants inherit the CRISPR elements, only a portion exhibit actual editing at the target loci (He and Zhao, 2020). Typically, T0 plants show heterozygosity at the transgene location. During subsequent generations, the transgene and edited loci follow Mendelian inheritance patterns. With a single transgene insertion, approximately 25% of T1 plants are expected to be transgene free. However, multiple transgene insertions are common with techniques like gene gun and *Agrobacterium*-mediated transformation, potentially affecting segregation patterns (Figs 11.1 and 11.2A).

The simplicity and cost-effectiveness of this approach have made it the go-to approach for CRISPR-Cas delivery. Studies across various plant species, including Arabidopsis, rice, tomato, maize, and grapevine, have demonstrated successful integration and high-editing efficiencies using this technique (Uranga and Daròs, 2023). However, if the transgene and edited locus are closely linked, eliminating the transgenes via Mendelian segregation becomes difficult (Schaart *et al.*, 2021). Achieving transgene-free, edited plants then requires an additional generation, either through selfing T0 plants or backcrossing them to wild-type individuals. Hence, isolating plants free of both transgenes and edits from T0 progeny or backcrossed populations can be a laborious and time-intensive process due to genetic segregation limitations. Notably, alternative methods like recombinase systems can also be used for transgene removal (Rahman *et al.*, 2023). While this strategy substantially minimizes transgene integration, it does not entirely eliminate it. Degraded DNA fragments, though diminished, might still get incorporated into the host genome.

11.4 Repurposing Plant Viruses for Non-transgenic CRISPR Delivery and Genome Editing

The latest innovations in next-generation sequencing (NGS) and genome editing technologies have spurred the use of plant viruses as expression vectors for their ability to deliver heterologous genetic elements into plant cells (Shahid *et al.*, 2021; Uranga and Daròs, 2023). Building upon existing knowledge, researchers have developed innovative *Agrobacterium*-based viral vectors, enabling efficient delivery into

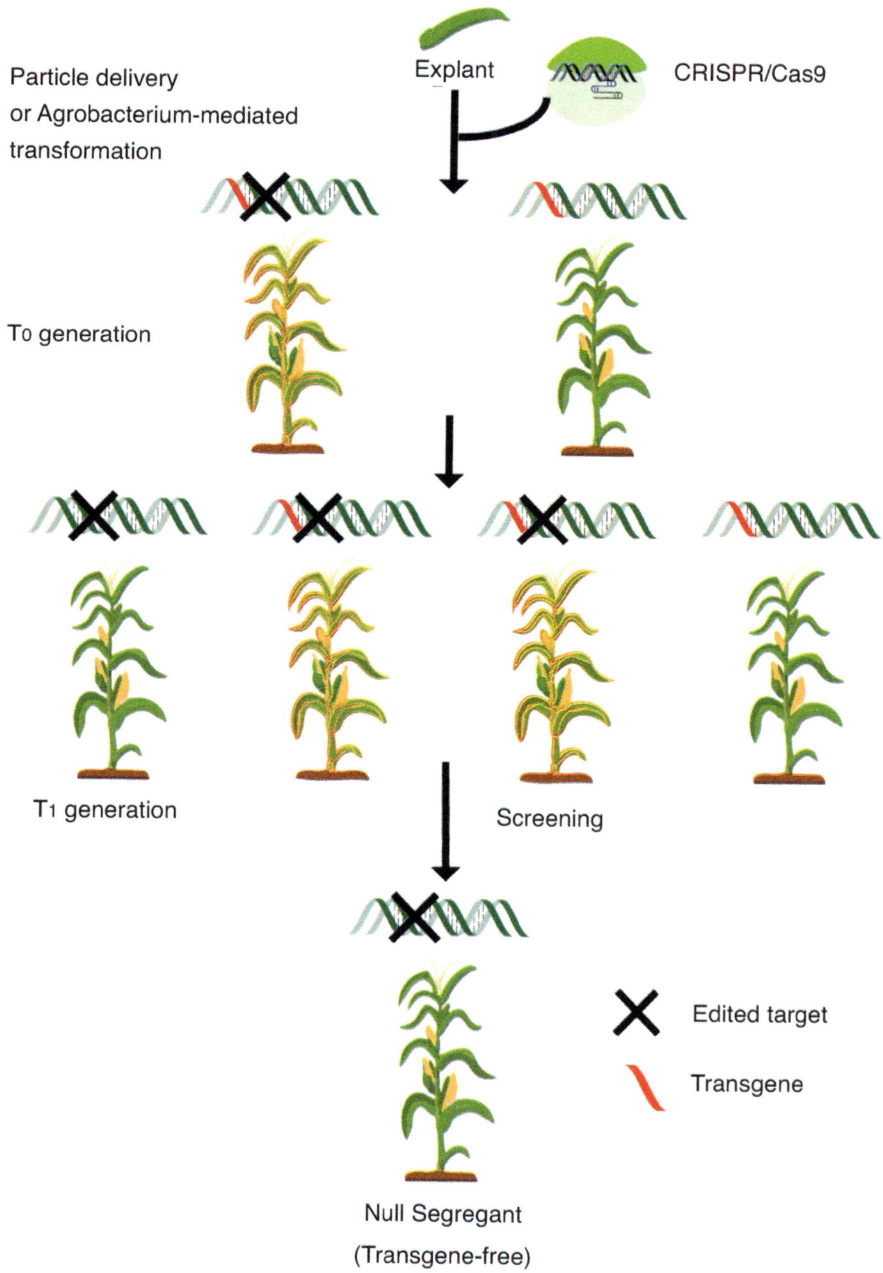

Fig. 11.1. Traditional transgene-free approach to produce DNA-free genome-edited plants. Author's own figure.

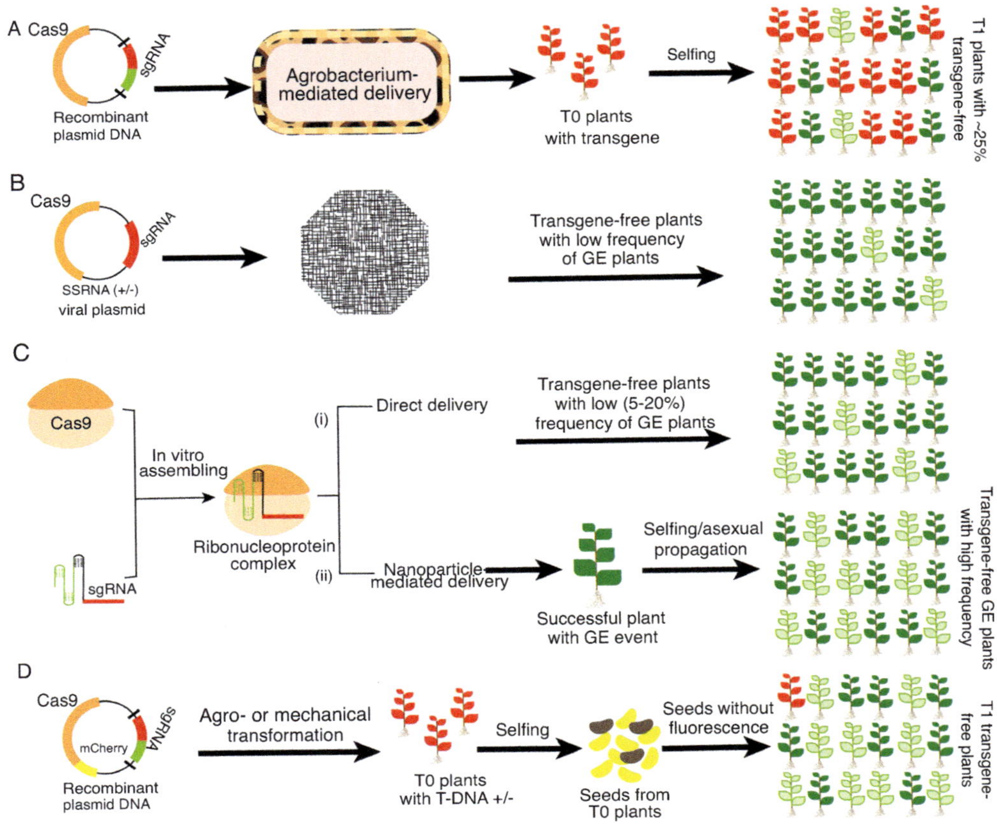

Fig. 11.2. Fast and efficient strategies for isolating transgene-free and CRISPR/Cas9-edited plants. (A) Genetic segregation can eliminate the transgene cassette in the CRISPR-edited plants. When the transgene locus is heterozygous in the first generation (T0), it segregates following Mendelian inheritance in subsequent generations (T1). PCR-based genotyping can then identify T1 plants that lack the transgene and carry the desired edit. (B) Utilizing viral vectors offers a promising avenue for delivering CRISPR-Cas components to engineer plants without leaving behind transgenes; limitations exist due to the inherently restricted cargo capacity and target specificity of these vectors. (C) The RNP complex uses *in vitro* hybridization of expressed Cas9 and sgRNA. This complex is then delivered into plant cells, either through (i) biolistic methods (e.g. particle bombardment) or (ii) via nanoparticles. The novel approach using nanoparticles coated with DNA, RNA or RNP offers a promising strategy for delivering CRISPR components to meristematic cells. The resulting cell population, containing both transformed and untransformed cells, is subsequently used to generate transgene-free seedlings. These seedlings can be distinguished after undergoing a rigorous screening process. In the illustration, wild-type plants are shown in dark green, genome-edited plants carrying the transgene are depicted in red, and genome-edited plants free of transgenes are indicated in light green with shaded leaves. (D) Fluorescence marker-assisted transgene elimination leverages the *mCherry* gene as a visual indicator linked to the gene editing machinery within the same plasmid. This visual screening method simplifies the identification and selection of transgene-free seeds, considerably reducing the time and effort required for growing plants and performing genotyping analyses. Author's own figure.

plant cells using straightforward inoculation methods. Recent research underscores the promise of these vectors as transient carriers for CRISPR-Cas components in diverse biological contexts, particularly emphasizing their value in plant applications (Xu *et al.*, 2019). Recent advancements have harnessed the potential of viral vectors in a revolutionary technique called virus-induced genome editing (VIGE). While adapting diverse plant viruses into VIGE vectors holds immense promise, translating this success to crops with recalcitrance in tissue culture and regeneration remains a significant hurdle (Wu *et al.*, 2024). High-frequency mutagenesis and chromosome deletions (up to 90%) can be achieved utilizing a vector based on a plant negative-strand RNA virus enabling DNA-free delivery of the CRISPR-Cas9 system (Lee *et al.*, 2024).

Early VIGE reports focused on utilizing geminiviruses to create plant knock-out lines via their compact single-stranded, circular DNA genomes (Lozano-Duran, 2016). While their ability to manipulate genes is promising, limited cargo capacity for inserting foreign DNA posed a challenge. Bipartite geminiviruses offered some hope, allowing replacement of the coat protein (CP) with heterologous DNA (up to 800–1000 bp) while maintaining viral functions. However, this approach could not accommodate larger pieces like Cas nucleases. To overcome this limitation, scientists explored the elimination of coding sequences for CP and movement proteins, resulting in the development of non-infectious replicons. This strategy, however, requires direct delivery into plant cells through *Agrobacterium*-mediated transformation, unlike techniques like agroinfiltration or mechanical inoculation that use infectious replicons. Despite this limitation, bean yellow dwarf virus (BeYDV) proved its worth by successfully expressing ZFNs, TALENs, Cas9, and sgRNA in plants, achieving higher editing efficiency compared to traditional T-DNA delivery (Butler *et al.*, 2016). Geminiviruses in VIGE also show promise for gene knock-in. Their high-copy replication within the host nucleus generates abundant repair templates, boosting HDR frequency. This phenomenon has been observed in various geminiviruses, including BeYDV (Dahan-Meir *et al.*, 2018), cabbage leaf curly virus (CaLCuV) (Yin *et al.*, 2015), and wheat dwarf virus (WDV) (Gil-Humanes *et al.*, 2017).

While DNA viruses offer promise for gene editing, their potential integration into the host genome raises concerns. In contrast, RNA viruses complete their infectious cycles within the cytoplasm, offering a safer alternative that avoids foreign DNA integration and associated regulatory hurdles. This opens doors for exciting applications in plant biotechnology. Several plant RNA-virus-based vectors, like tobacco rattle virus (TRV), have emerged for delivering sgRNAs for targeted editing (Varanda *et al.*, 2021). TRV boasts a broad host range and efficient movement within growing tissues, making it ideal for VIGE. Initial studies successfully delivered ZFNs and TALENs into tobacco and petunia, enabling targeted modifications and their transmission to offspring. Similarly, TRV-based vectors equipped with sgRNAs have demonstrated efficient editing in Arabidopsis and tobacco genomes using CRISPR-Cas technology (Ali *et al.*, 2015). Tobacco mosaic virus (TMV), an RNA virus, has been repurposed for sgRNA delivery. While its CP deletion hinders systemic movement, high sgRNA concentrations can be delivered, resulting in efficient editing in tobacco plants (Cody *et al.*, 2017). Transgenic plants expressing Cas9 constitutively offer a valuable platform for exploring VIGE with plant using plant viral vectors. Notably, beet necrotic yellow vein virus (BNYVV) replicons engineered to co-express foreign proteins have exhibited promising results in *Nicotiana benthamiana* (Jiang *et al.*, 2019). This strategy has been extended to non-model plant species like wheat and maize using barley stripe mosaic virus (BSMV) and foxtail mosaic virus (FoMV), demonstrating VIGE potential across diverse plant taxa. These findings showcase the versatility and immense potential of RNA-virus-based vectors in revolutionizing plant gene editing technologies.

11.5 Transgene-free Editing Through Particle Bombardment and Protoplasts

Preassembled Cas9-sgRNA complexes delivered via RNPs offer a promising strategy for DNA-free plant genome editing, as demonstrated in protoplasts (Fig. 11.2C). This approach aligns with previous studies showcasing the effectiveness of RNP-mediated editing in plants (Woo *et al.*, 2015; Liang *et al.*, 2019). However, protoplast-based delivery, where RNPs are introduced into isolated plant cells followed by regeneration into whole plants, allows for more precise targeting and avoids potential integration-related issues. Plant protoplasts, obtained by enzymatically removing their cell walls, share some similarities with animal cells. This lack of a cell wall facilitates their efficient transfection using techniques like electroporation or polyethylene glycol treatment. Unlike conventional DNA-based CRISPR-Cas expression methods, the RNP approach allows the pre-formed gene editing complex to directly cleave target DNA upon delivery, negating the need for cellular transcription and translation machinery. This promotes rapid degradation of the RNP after cleavage, thereby significantly reducing the potential for off-target edits compared to DNA-based methods. By utilizing RNPs for gene knock-out, DNA-free edited plants can be generated, potentially alleviating regulatory concerns related to GMOs. Furthermore, the RNP-editing toolbox for plants is expanding beyond Cas9. Kim *et al.* (2017) successfully delivered both LbCas12a- and AsCas12a-crRNA RNPs into protoplasts of soybean (*Glycine max* (L.) Merr.) and tobacco (*N. tabacum* L.), demonstrating the potential of Cas12a for RNP-based editing in plants. However, it is essential to acknowledge that regenerating whole plants from single-celled protoplasts invariably involves tissue culture techniques, which can occasionally introduce undesired somaclonal mutations. Or other limitations in full-plant regeneration using protoplasts or biolistics can restrict the applicability of this method in certain species (Bhattacharjee *et al.*, 2023).

11.6 Nano-CRISPR for Transgene-free Edited Plants

Several alternative techniques have been investigated for genome editing in plants besides gene gun and pathogen-based vectors, but none have been widely employed for plant cells due to the barrier presented by the cell wall. Removal of the cell wall creates protoplasts, simplifying delivery to the cell membrane. Several methods, including lipofection, lipid nanoparticles (LNPs), electroporation, and PEG/Ca^{2+}-mediated transfection, have been shown to effectively deliver proteins, RNA, or DNA to plant protoplasts and mammalian cells (Woo *et al.*, 2015; Andersson *et al.*, 2018; Murovec *et al.*, 2018). Notably, for species where protoplast regeneration is feasible, these methods offer a potentially advantageous approach for achieving transgene-free gene editing due to their convenience and high efficiency. However, regenerating plants from protoplasts remains challenging in many species. Targeting zygotes shortly after fertilization presents another possibility for circumventing the cell wall barrier. These early-stage embryos can be transfected using methods similar to those employed for protoplasts (Toda *et al.*, 2019). However, working with explants containing mature cell walls is often more practical for researchers due to various limitations.

Various types of NPs, including inorganic (gold, silver, iron oxide), carbon-based, silicon-based, and polymeric NPs, have been identified for overcoming these hurdles by effectively harnessing tissue culture-free delivery of biomolecules for genome editing (Kocsisova and Coneva, 2023). Notably, inorganic NPs can transport only DNA as their genetic cargo into plants. Recent innovations have enabled the conjugation of CRISPR/Cas9 components with diverse nanoparticles, encompassing nanocapsules, gold nanoparticles, hydrogels, peptide-based nanoparticles, DNA nanoclews, polymeric nanoparticles, unilamellar and multilamellar liposomes, and magnetic nanoparticles. This fusion between bioengineering and gene editing holds promise for delivering CRISPR/Cas9 cargo to various plant cell types, including meristematic cells (Aydin, 2024). Delivering gene editing reagents through NPs to meristematic cells holds promise for generating chimeric edits in plants. Regeneration of transgene-free and edited plants can be achieved through tissue culture or propagation of cuttings.

Previous research has shown the utility of single-wall carbon nanotubes as carriers for delivering plasmids, including gene editing reagents and siRNA, with a capacity of up to ~10 kb (Zhang *et al.*, 2022). A comprehensive overview of nanotube and nanoparticle methodologies can be found in the work by Demirer *et al.* (2021). Studies have shown that both clay nanoparticles and carbon nanodots can be effective carriers for delivering genetic material into intact plant cells. For example, Yong *et al.* (2022) demonstrated that clay nanoparticles can deliver RNA into plant cells, while Doyle *et al.* (2019) reported successful delivery of plasmid DNA to wheat leaves using carbon nanodots. This method holds potential for broader applicability, offering a straightforward, rapid, and cost-effective approach to plant genome editing. In a recent development, maize pollen transformation has been achieved using DNA-coated magnetic nanoparticles (Wang *et al.*, 2022). Zhao *et al.* (2017) demonstrated the potential of magnetic nanoparticles (MNPs) for exogenous DNA delivery in pollen, paving the way for genetic modification. This pollen magnetofection technique involves positively charged MNPs forming complexes with negatively charged DNA, guided by a magnetic field into pollen apertures before pollination. This method preserves DNA functionality and pollen viability, efficiently generating insect-resistant transgenic cotton seeds. Pollen magnetofection emerges as a promising technology for plant genetic modification due to its unique advantages. It bypasses the need for tissue culture, making it field-friendly and genotype-independent. Additionally, the process is simple, rapid, and capable of delivering multiple genes simultaneously. While the use of magnetic nanoparticles for delivering RNP complexes has not yet been demonstrated in intact plant cells, similar approaches have shown success in mammalian cells. For a broader understanding of nanocarrier delivery methods, including carbon nanotubes, carbon dots, and mesoporous silicon nanoparticles, please refer to the review by Ahmar *et al.* (2021). Furthermore, cell-penetrating peptides (CPPs) offer an alternative method for delivering DNA and proteins into plant cells through a mechanism involving endosomal entry and escape. Recent studies suggest combining CPPs might enhance delivery, possibly through micropinocytosis-like mechanisms (Miyamoto *et al.*, 2022). Direct assimilation of specific NPs, including quantum dots, carbon nanotubes, metal oxides, and silica-based NPs, has been demonstrated offering the possibility of transgene-free editing in diversified crop species. Gold nanoparticles (AuNPs) exhibit exceptional stability and biocompatibility compared to others. Gold-based nanomaterials, especially AuNPs and AuNCs, have been used for delivering CRISPR-Cas9 modules (Chen *et al.*, 2019; Vats *et al.*, 2022).

Identifying universal strengths and weaknesses of these methods is complex, and broad implementation of this approach faces ongoing challenges (Rustgi *et al.*, 2023). Optimizing the dose/concentration of NPs or functionalized materials for coating is crucial due to their high reactivity and limited stability. Additionally, a clear understanding of the biosafety profile of NPs used for gene delivery in specific plant/tissue types is essential. Furthermore, the binding affinity between cargos and nanoparticles plays a critical role in determining transformation efficiency. Finally, studying the delivery mechanism and fate of nanoparticles and/or functionalized materials is necessary to comprehend their biocompatibility, biosafety, and potential toxicity. Further research is needed to optimize delivery methods for specific applications and overcome remaining challenges.

11.7 Ribozymes and Fluorescent Markers Simplify Transgene-free Plant Generation

Like antibiotic- and herbicide-resistant genes, fluorescent proteins have become established tools for identifying stably transformed plants in a rapid and non-destructive manner. Their ease of detection based on the presence or absence of fluorescence allows efficient screening for potential genome-edited plants free of integrated CRISPR-Cas9 editing components. Visual identification of transgene-free genome-edited seeds/plants can be achieved by incorporating fluorescent marker proteins such as mCherry into the genome editing machinery (Fig. 11.2D). This approach results in transgenic seeds with a distinct color, making them easily distinguishable from non-transgenic, genome-edited seeds (He *et al.*, 2017; Aliaga-Franco *et al.*, 2019). This method has been successfully implemented in various crops like rice, Arabidopsis, tomato, and maize, utilizing different marker proteins like DsRED and GFPs for further diversification (Yan *et al.*, 2021). These studies have shown that mutations introduced in transgene-free edited plants are stably inherited by subsequent generations. This opens up a powerful new approach for efficiently identifying desired edited plants and distinguishing them from those still containing the CRISPR-Cas9 machinery. While fluorescence-based screening offers a valuable initial assessment, it is crucial to combine it with genetic analysis for definitive confirmation of transgene absence. This combined approach provides a reliable and efficient means to isolate transgene-free CRISPR-Cas9-edited plants.

Unlike protein-based strategies, ribozymes offer a unique approach for generating sgRNAs. These RNA molecules possess self-cleaving capabilities, enabling them to precisely cleave themselves at specific sites. Ribozyme-mediated sgRNA production offers a unique approach for conducting advanced gene editing experiments beyond the limitations of RNA Pol III promoter-based technology (He *et al.*, 2017). This strategy uses a Ribozyme1-gRNA-Ribozyme2 cassette, where ribozymes flank the sgRNA sequence at both ends. These ribozymes act as self-cleaving enzymes, automatically releasing the mature sgRNA during transcription. This design allows for the use of any promoter for sgRNA production, including cell/tissue-specific promoters for targeted editing. However, ribozyme cleavage requires specific features near the cleavage site, such as conserved sequences or secondary structures. Commonly used ribozymes for constructing the 'ribozyme-sgRNA-ribozyme' (RGR) unit include hammerhead and hepatitis delta virus (HDV) ribozymes (Gao and Zhao, 2014). The compact hammerhead ribozymes (often <50 bp) form characteristic stem-loop structures and are often positioned at the 5′ end of the sgRNA in the RGR unit. This ribozyme-based strategy has demonstrated successful gene editing in plants (Fang *et al.*, 2023) and forms the foundation for multiplexing technologies.

11.8 Beyond Cas9: Unleashing the Potential of Cpf1 for DNA-free Genome Editing

CRISPR-Cpf1 (Cpf1 for short) stands out as a powerful, DNA-free tool for genome editing, offering several advantages over the well-known Cas9 system. Instead of requiring a G-rich PAM sequence like Cas9, Cpf1 efficiently targets sites flanked by a T-rich motif (5′-TTTN-3′). Additionally, Cpf1 works with a single-guide RNA (sgRNA), streamlining the targeting process compared to Cas9's dependence on both CRISPR RNA (crRNA) and trans-activating crRNA (tracrRNA) (Gong *et al.*, 2021). Notable members of the Cpf1 family include LbCpf1 from *Lachnospiraceae bacterium* ND 2006 and the highly effective AsCpf1 from *Acidaminococcus* sp. BV3L6. CrRNAs for Cpf1 are significantly shorter than their Cas9 counterparts, offering the unique advantage of chemical synthesis, further simplifying the editing process. Furthermore, Cpf1 induces larger deletions compared to SpCas9, which can facilitate the insertion of desired sequences via the NHEJ pathway, leading to precise edits known as 'scarless deletions' (SDN1) and 'scarless insertions' (SDN2).

Utilizing RNPs instead of plasmids holds considerable promise for minimizing off-target effects and cell toxicity associated with traditional DNA transfection. Recent studies demonstrate successful protoplast transfection with pre-formed RNPs-containing Cpf1 (LbCpf1 or AsCpf1) and crRNA at a roughly 1:6 molar ratio, facilitated by PEG solution. Studies have shown that Cpf1-induced mutations predominantly involve deletions of several nucleotides, with no detected off-target mutations or foreign DNA traces (Kim *et al.*, 2017). This positions Cpf1 as an effective, DNA-free genome editing tool for agricultural applications. While Cpf1 holds immense potential, one limitation to consider is its specificity. While tolerating up to one or two base mismatches, Cpf1 demonstrates no tolerance for four or more mismatches (Alok *et al.*, 2020). Therefore, careful crRNA design is crucial to minimize off-target effects. This involves selecting crRNAs with no three-nucleotide mismatches within the target sequence, identified through comprehensive homology searches in the relevant reference genome, excluding the actual target site.

11.9 Base Editors Offer Transgene-free Mutations

Precise and predictable introduction of point mutations in genome editing is paramount for deciphering the intricacies of natural genomic variations and accelerating crop improvement.

However, current techniques often grapple with error-prone repair pathways (as with NHEJ) or the requirement for additional donor DNA (as with HDR). This poses particular challenges in ensuring heritability and eliminating transgene remnants. The development of base editors offers a compelling alternative (Molla *et al.*, 2021). These tools utilize Cas9 nickase (nCas9), a modified Cas9 protein lacking full nuclease activity, fused to either adenine deaminase or cytidine deaminase. This fusion allows for precise C-to-T or A-to-G substitutions without creating DSBs. Guided by sgRNA, the nCas9/deaminase complex deaminates a specific base on the non-complementary DNA strand, triggering subsequent repair mechanisms that incorporate the desired edited base.

Target-AID (target activation-induced cytidine deaminase) has emerged as a powerful cytidine base editor (CBE) capable of precise sequence editing in plant genomes. Its successful application in various crops exemplifies its potential to produce DNA-free genome edits. Meyer *et al.* (2022) demonstrated a high mutation frequency of approximately 70% by transiently expressing a CBE and sgRNA in carrot protoplasts. Subsequent regeneration of these protoplasts into somatic embryos yielded edited carrot plants. Crucially, the plasmid design ensured transient expression of the gene editing machinery. This prevented permanent integration into the host genome, guaranteeing targeted editing within a defined timeframe. While this system offers advantages, such as utility for heterozygous and polyploid crops like potatoes, its only demerit is involving plant transformation through the labor-intensive *Agrobacterium*-mediated vector system (Molla *et al.*, 2021). This process requires the cloning of crRNA and recombinant nCas9 before insertion. Additionally, the system relies on Mendelian segregation, necessitating time and selection pressure to confirm the non-transgenic status.

Unlike CBEs, which primarily generate C-to-T and C-to-G conversions, adenine base editors (ABEs) offer a cleaner editing profile with less frequent C-to-A mutations (Veillet *et al.*, 2019). This characteristic makes them advantageous for targeted genome modifications where precise control is crucial. Additionally, the inclusion of selectable markers, like herbicide tolerance in plants, allows for efficient selection of edited cells after multiplexed base substitutions, a technique vital for achieving complex genetic traits (Shimatani *et al.*, 2018). Furthermore, the potential for simultaneous editing of multiple bases opens exciting avenues for diverse applications of this versatile technology.

11.10 Transgene-free Gene Editing Through TKC and the CASE Toolkit

Transgene killer CRISPR (TKC) technology leverages spatially and temporally controlled 'suicide cassettes' to eliminate transgene-carrying sperm and embryos, facilitating the identification of transgene-free plants (He *et al.*, 2018). This strategy employs two crucial components:

I. Barnase cassette
This cassette incorporates the Barnase (a portmanteau of 'BActerial RiboNucleASE') gene from bacteria, which encodes a ribonuclease with toxic properties. The activity of this gene is controlled by the rice REG2 promoter, which is specifically active during the early stages of embryo development. This targeted expression selectively eliminates embryos that contain the transgene.

II. CMS2 cassette
This cassette encodes the CMS2 protein (ORFH79), which disrupts mitochondrial function in male gametophyte development, leading to pollen sterility. It targets and eliminates male gametophytes carrying transgenes.

State-of-the-art developments enhance TKC technology's efficiency. Notably, the customized assembly and simplified editing (CASE) toolkit integrates TKC with multiplex gene editing in rice (Liu *et al.*, 2022). This toolkit enables the efficient creation of plants with targeted edits in multiple genes without introducing foreign DNA, achieving results in the first generation (T1).

The CASE toolkit offers four separate vectors, each equipped with distinct snoRNA promoters (U3, U6a, U6b or U6c) for individual gRNA cloning. They converge on a central TKC-MCS-U3 backbone vector. This modular design facilitates the straightforward construction of tailored gRNA combinations through compatible restriction sites. Once constructed, these combinations can be seamlessly integrated into the backbone vector using the MCS. As an alternative to existing methods, this approach explores the direct assembly of gRNA cassettes incorporating self-splicing tRNAs onto the TKC-MCS-U3 backbone. This novel strategy leverages the unique properties of self-splicing tRNAs to facilitate multiplex editing within the TKC system. Further improvement in rice came with the use of alternative promoters (Yubing *et al.*, 2019). Replacing the weak CaMV35S promoter with the stronger rice ACTIN1 promoter specifically in microspore cells (pollen precursors) enhanced transgene expression and editing efficiency. Additionally, a pollen-specific promoter (OsGEX2) for CMS2 expression further improved the isolation of transgene-free, edited rice plants within a single generation. The potential of TKC technology to self-eliminate non-transgenic elements holds immense potential for transgene-free genome editing, especially in complex, multiplex editing scenarios with high frequencies. However, its broader applicability to other crops beyond rice requires further investigation.

11.11 Editing Promoters for Disease Resistance

In a 2020 study, Chinese researchers demonstrated the successful application of the CRISPR-Cas9 system to delete a specific region within the *xa13* gene promoter in rice. This approach employed two targeted guide RNAs (gRNAs) to address the challenges associated with directly editing the xa13 coding sequence (CDS) for disease resistance. The *xa13* gene is critical for both susceptibility to bacterial blight and pollen fertility. By precisely removing a portion of the promoter, the researchers effectively disabled the activation of *xa13* gene by bacterial blight, thereby conferring disease resistance without compromising pollen development (Li *et al.*, 2020). This strategic promoter modification achieved several advantages. Editing in this region avoided the introduction of transgene fragments while simultaneously modulating gene expression without triggering the production of unwanted mRNA or protein, often a pitfall of coding region edits. Additionally, the double-sgRNA design facilitated mutation detection using PCR, simplifying the identification of homozygous, heterozygous, or unaltered individuals and enhancing mutant selection accuracy. Furthermore, the two-target deletion strategy exhibited greater stability than single-target approaches, ensuring more predictable outcomes.

Another noteworthy development emerged in 2019, this time in maize. Kelliher *et al.* (2019) introduced Hi-Edit technology, a transgene-free editing strategy utilizing haploid induction. They first introduced the CRISPR-Cas9 construct into a standard transformation line, NP2222. To identify individuals carrying both the Cas9 insertion and the haploid-inducing gene, the Cas9-positive regenerated plants and the native haploid-inducer line (RWKS) were crossed together. Elite inbred lines were then pollinated with the resulting F2 individuals, and their haploid offspring were screened to isolate the desired transgene-free mutant.

11.12 Harmonizing Regulations for DNA-free Plant Editing

Genome editing and natural mutations both involve changes in an organism's DNA sequence, but they differ in their source and precision. Natural mutations typically arise spontaneously during DNA replication or due to environmental factors, while genome editing allows scientists to introduce targeted changes with greater control. While both processes can involve substitutions or insertions/deletions (indels) in DNA bases, genome editing offers the ability to make specific modifications at chosen locations, which is not possible with natural mutations (Ahmad *et al.*, 2024). Certain genome editing techniques utilize temporary endonucleases to create breaks at specific DNA sites. The subsequent repair process can lead to desired mutations. Studies have shown that the activity of these endonucleases degrades rapidly within the target cell, minimizing the risk of unintended off-target effects and ensuring that no transgenes are permanently incorporated into the plant genome (Liang *et al.*, 2017). This transgene-free approach holds promise for generating improved plant lines without raising concerns about transgenic modifications.

The international regulatory landscape for GMOs and genome-edited organisms exhibits significant variation across different countries (Buchholzer and Frommer, 2023). In the United States, Canada, and four countries in South America, the regulatory focus is on the final product rather than the specific techniques used to produce it implementing regulations that treat genome-edited plants through SDN-1 and SDN-2 largely the same as conventionally bred lines. Following the lead of several South American nations, the governments of Guatemala, Honduras, Ecuador, and Paraguay implemented a case-by-case evaluation approach for genome-edited crops in 2019 (Whelan and Lema, 2019). Plants with stably integrated transgenes are considered GMOs and fall under the purview of the Department of Agriculture, the Food and Drug Administration, and the Environmental Protection Agency. In contrast, the European Union takes a more stringent approach, subjecting all products derived from genome editing processes to the same regulatory framework as GMOs (Laaninen, 2021). This decision by the European Court of Justice in 2019 has raised concerns about potential trade barriers and stifled innovation, as argued by Voigt and Münichsdorfer (2019). Some countries, however, have adopted more flexible regulatory frameworks for genome-edited organisms. For example, United Kingdom is anticipated to release guidelines, initially exempting genome-edited crops from GMO field trial regulations and, potentially, from the GMO definition itself. This would streamline the approval process for such crops, allowing them to undergo field trials and potentially gain commercial approval without the need for individual case-by-case assessments (Stokstad, 2021). India has also taken steps to streamline regulations for specific CRISPR-based applications, exempting certain categories from biosafety assessments (Gauchan and Joshi, 2019).

However, not all countries are moving in the same direction. While Australia has opted to deregulate certain genome-edited products that meet specific criteria, New Zealand remains cautious and categorizes all gene-edited products as GMOs (Jones *et al.*, 2024). The diverse regulatory landscape highlights the ongoing debate about how to balance the potential benefits of genome editing with concerns about food safety, environmental impact, and ethical considerations.

Advancements in DNA-free plant genome editing aim to enhance transformation efficiency, reduce off-target effects, and improve explant regeneration. While numerous crops have undergone trials using this approach to enhance various traits (Table 11.1), a robust protocol for transfection yielding effective regeneration remains a challenge. Researchers must prioritize efforts to enhance the efficacy of this technology. Addressing the increasing global population (expected to reach approximately 10 billion by 2050) and the ongoing challenges of climate change and biotic stresses, a 60–100% increase in global food production is needed with limited genetic resources. International harmonization of regulations underscores the critical role of DNA-free genome editing in understanding and safeguarding the innovative potential of genome-edited plants (Buchholzer and Frommer, 2023).

11.13 Conclusion

CRISPR/Cas has revolutionized genomics, offering immense potential for advancing plant breeding and agriculture. However, concerns around regulatory hurdles and public perception of transgenic modifications have driven the development of novel DNA-free genome editing strategies. This approach promises significant progress without triggering regulatory roadblocks. RNP delivery, nano-CRISPR, base editors, and TKC technology are actively explored to prevent foreign DNA integration during editing. These methods create crops with desired traits while mitigating regulatory concerns. Advancements in sgRNA design, Cas enzyme engineering, and delivery methods enhance precision, efficiency, and enable the development of complex genetic traits with minimal off-target effects. While some techniques are

Table 11.1. The effectiveness of DNA-free genome editing across diverse crop species, achieved through varied transfection methods applied to specific explants. Author's own table.

Plant species	Targeted explant	Target gene	Transformation method	Study outcomes	References
Arabidopsis thaliana	Protoplast	*PHYB*	PEG-assisted fusion	• RGENs cleave target sites immediately after transfection. • Mutations are induced even before a complete cell division cycle. • No indels were detected.	Woo *et al*., 2015
Nicotiana attenuata		*AOC*		• Monoallelic mutations occurred at a frequency of 5.7%.	
Oryza sativa		*P450*, *DWD1*		• Biallelic mutations were observed in 40% of cases.	
Triticum aestivum	*Developing embryos*	*TaGASR7, TaDEP1, TaNAC2, TaPIN1, TaLOX2, TaGW2*	Transient expression through particle bombardment	• The occurrence rates of mutants lacking detectable transgenes were: • 53.8% for TaDEP1 • 75.0% for TaNAC2 • 62.5% for TaPIN1 • 86.8% for TaLOX2	Zhang *et al*., 2016
Triticum durum		*TdGASR7*			
Malus domestica	Protoplast	*MdMLO19*	PEG-assisted fusion	• The editing frequency ranged from 0.1% to 6.9%.	Malnoy *et al*., 2016
Vitis vinifera					
Zea mays	Developing embryos	*MS26* and *MS45* male fertility genes *LIG*, *ALS2*	Particle bombardment	• Plants exhibiting mutations were successfully isolated, with occurrence rates ranging from 2.4% to 9.7%, and no mutations detected at off-target sites.	Svitashev *et al*., 2016

Petunia × atkinsiana	Protoplast	*PhNR*	PEG-assisted fusion	• Mutations occurred at specific sites within the PhNR (photopic negative response) gene locus, with frequencies ranging from 2.4% to 21% at four distinct sites. • The average mutation efficiency was determined to be 14.9 ± 2.2%.	Subburaj *et al.*, 2016
Oryza sativa	Protoplast	*OsCDC48, OsNRT1.1B, OsSPL1*	Agrobacterium-mediated T-DNA-free base editing	• Precise conversion of cytosine to thymine took place.	Zong *et al.*, 2017
Zea mays		*ZmCENH3*			
Triticum aestivum		*TaLOX2*			
Glycine max	Protoplast	*DAD2*	PEG-assisted fusion	• Cpf1 ribonuclease exhibits intolerance to mismatches exceeding three nucleotides and is incapable of cleaving at homologous sequences within the genome.	Kim *et al.*, 2017
Nicotiana attenuata		*AOC*			
Triticum aestivum	Developing embryos	*TaGW*	Particle bombardment	• The entire procedure is completed within a brief time frame, spanning only 7–9 weeks. • It lowered the incidence of off-target mutations.	Liang *et al.*, 2017
Solanum tuberosum	Protoplast	*GBSS*	PEG-assisted fusion	A mutagenesis rate of 25% without any off-target effects was observed.	Andersson *et al.*, 2018
Brassica oleracea	Protoplast	*FRI* and *PDS*	PEG-assisted fusion	Mutation frequency ranged 0.09% to 2.25%.	Murovec *et al.*, 2018
Brassica napus				Mutation frequency ranged 1.15% to 24.51%.	
Brassica rapa					

Continued

Table 11.1. Continued

Plant species	Targeted explant	Target gene	Transformation method	Study outcomes	References
Oryza sativa	Zygote	*DROOPING, DL, GW7, GCS1*	PEG–Ca^{2+}-assisted fusion	Highly specific mutations with a mutation rate of 4% to 64% was observed.	Toda *et al*., 2019
Solanum lycopersicum	Cotyledonal segments	*ALS*	Agrobacterium-mediated transient expression of cytidine deaminase	Transgene-free mutation rate of 12.9% was observed.	Veillet *et al*., 2019
Solanum tuberosum				Transgene-free mutation rate of 10% was observed.	
Triticum aestivum	Shoot apical meristems	*SD1, TaOr, TaQsd1, TaHRGPL1*	RNP complex through particle bombardment	Mutation frequency ranged 1.9% to 8.3%.	Kumagai *et al*., 2022
Zea mays		*ZmEPSPS*	Particle bombardment	Transgene-free progeny exhibiting triple amino acid substitutions resulted.	Kaul *et al*., 2024
Solanum lycopersicum	Plant cotyledons	*PDS*	Virus-induced genome editing	Mutation rate ranged 40.3% and 36.5% for TRV and PVX.	Lee *et al*., 2024

broadly applicable, others require further optimization. Continuous research is crucial for widespread accessibility and application. Simultaneous editing of multiple genes offers exciting possibilities for generating crops with combined desirable traits like improved yield, disease resistance, and enhanced nutritional value. Optimizing multiplex editing strategies across diverse crops is crucial. Tailored genome editing approaches hold the key to overcoming challenges related to climate change, pest resistance, and resource scarcity. Developing resilient crops adapted to specific environments using these tools is essential for sustainable agriculture. Open dialogue and collaboration between scientists, policymakers, and the public are necessary to combat these challenges. Establishing trust and creating clear regulatory frameworks for the responsible use of DNA-free genome editing in agriculture is paramount. In conclusion, the field of DNA-free genome editing is rapidly evolving, presenting thrilling prospects for transforming agriculture. By harnessing the power of these innovative tools responsibly and collaboratively, we can envision a future where sustainable and resilient crops contribute to global food security and a healthier planet.

References

Adeel, M. and Jones, M.G. (2024) Challenges and prospects in the regulation of CRISPR-edited crops. In: Abd-Elsalam, K.A. and Ahmad, A. (eds) *Global Regulatory Outlook for CRISPRized Plants*. Academic Press, London, pp. 447–459.

Ahmad, A., Munir, A., Zafar, H., Zahoor, M.K., Hassan, S. *et al*. (2024) Tracking footprints of CRISPR-based genome editing. In: Abd-Elsalam, K.A. and Ahmad, A. (eds) *Global Regulatory Outlook for CRISPRized Plants*. Academic Press, London, pp. 113–145.

Ahmar, S., Mahmood, T., Fiaz, S., Mora-Poblete, F., Shafique, M.S. *et al*. (2021) Advantage of nanotechnology-based genome editing system and its application in crop improvement. *Frontiers in Plant Science* 12, 663849.

Ali, Z., Abul-Faraj, A., Li, L.X., Ghosh, N., Piatek, M. *et al*. (2015) Efficient virus-mediated genome editing in plants using the CRISPR/Cas9 system. *Molecular Plant* 8(8), 1288–1291.

Aliaga-Franco, N., Zhang, C., Presa, S., Srivastava, A.K., Granell, A. *et al*. (2019) Identification of transgene-free CRISPR-edited plants of rice, tomato, and arabidopsis by monitoring dsred fluorescence in dry seeds. *Frontiers in Plant Science* 10, 1150.

Alok, A., Sandhya, D., Jogam, P., Rodrigues, V., Bhati, K.K. *et al*. (2020) The rise of the CRISPR/Cpf1 system for efficient genome editing in plants. *Frontiers in Plant Science* 11, 264.

Andersson, M., Turesson, H., Olsson, N., Falt, A.S., Ohlsson, P. *et al*. (2018) Genome editing in potato via CRISPR-Cas9 ribonucleoprotein delivery. *Physiologia Plantarum* 164(4), 378–384. DOI: 10.1111/ppl.12731.

Aydin, M. (2024) Role of nanotechnology in the advancement in genome editing in plants. In: Upadhyay, S.K. (ed.) *Applications of Genome Engineering in Plants*. Wiley, West Sussex, UK, pp. 78–90.

Bhattacharjee, S., Bhowmick, R., Kant, L. and Paul, K. (2023) Strategic transgene-free approaches of CRISPR-based genome editing in plants. *Molecular Genetics and Genomics* 298(3), 507–520.

Buchholzer, M. and Frommer, W.B. (2023) An increasing number of countries regulate genome editing in crops. *New Phytologist* 237(1), 12–15.

Butler, N.M., Baltes, N.J., Voytas, D.F. and Douches, D.S. (2016) Geminivirus-mediated genome editing in potato (*Solanum tuberosum* L.) using sequence-specific nucleases. *Frontiers in Plant Science* 7, 1045. DOI: 10.3389/fpls.2016.01045.

Chen, G., Abdeen, A.A., Wang, Y., Shahi, P.K., Robertson, S. *et al*. (2019) A biodegradable nanocapsule delivers a Cas9 ribonucleoprotein complex for *in vivo* genome editing. *Nature Nanotechnology* 14(10), 974–980.

Cody, W.B., Scholthof, H.B. and Mirkov, T.E. (2017) Multiplexed gene editing and protein overexpression using a *Tobacco mosaic virus* viral vector. *Plant Physiology* 175(1), 23–35. DOI: 10.1104/pp.17.00411.

Dahan-Meir, T., Filler-Hayut, S., Melamed-Bessudo, C., Bocobza, S., Czosnek, H. *et al.* (2018) Efficient in planta gene targeting in tomato using geminiviral replicons and the CRISPR/Cas9 system. *The Plant Journal* 95(1), 5–16. DOI: 10.1111/tpj.13932.

Demirer, G.S., Silva, T.N., Jackson, C.T., Thomas, J.B., Ehrhardt, D.W. *et al.* (2021) Nanotechnology to advance CRISPR–Cas genetic engineering of plants. *Nature Nanotechnology* 16(3), 243–250.

Doyle, C., Higginbottom, K., Swift, T.A., Winfield, M., Bellas, C. *et al.* (2019) A simple method for spray-on gene editing in planta. *BioRxiv*. DOI: 10.1101/805036.

Fang, R., Chen, X., Shen, J. and Wang, B. (2023) Targeted mrna demethylation in *Arabidopsis* using plant M6A editor. *Plant Methods* 19(1), 81.

Gao, Y. and Zhao, Y. (2014) Self-processing of ribozyme-flanked RNAs into guide RNAs *in vitro* and *in vivo* for CRISPR-mediated genome editing. *Journal of Integrative Plant Biology* 56(4), 343–349. DOI: 10.1111/jipb.12152.

Gauchan, D. and Joshi, B.K. (2019) Seeds without borders initiative for enhanced food and nutrition security in South Asia. In: Pandey, P.R. and Bhandari, H. (eds) *Seeds Without Borders Initiative in South Asia Current Status and Future Directions*. SAARC Agricultural Centre (SAC) and International Rice Research Institute, Dhaka, Bangladesh, p. 203.

Gil-Humanes, J., Wang, Y.P., Liang, Z., Shan, Q.W., Ozuna, C.V. *et al.* (2017) High-efficiency gene targeting in hexaploid wheat using DNA replicons and CRISPR/Cas9. *Plant Journal* 89(6), 1251–1262. DOI: 10.1111/tpj.13446.

Gong, C., Huang, S., Song, R. and Qi, W. (2021) Comparative study between the CRISPR/Cpf1 (Cas12a) and CRISPR/Cas9 systems for multiplex gene editing in maize. *Agriculture* 11(5), 429.

He, Y. and Zhao, Y. (2020) Technological breakthroughs in generating transgene-free and genetically stable CRISPR-edited plants. *Abiotech* 1(1), 88–96.

He, Y., Wang, R., Dai, X. and Zhao, Y. (2017) On improving CRISPR for editing plant genes: Ribozyme-mediated guide RNA production and fluorescence-based technology for isolating transgene-free mutants generated by CRISPR. *Progress in Molecular Biology and Translational Science* 149, 151–166. DOI: 10.1016/bs.pmbts.2017.03.012.

He, Y., Zhu, M., Wang, L., Wu, J., Wang, Q. *et al.* (2018) Programmed self-elimination of the CRISPR/Cas9 construct greatly accelerates the isolation of edited and transgene-free rice plants. *Molecular Plant* 11(9), 1210–1213.

Hilscher, J., Bürstmayr, H. and Stoger, E. (2017) Targeted modification of plant genomes for precision crop breeding. *Biotechnology Journal* 12(1), 1600173.

Huang, X., Jia, H., Xu, J., Wang, Y., Wen, J. *et al.* (2023) Transgene-free genome editing of vegetatively propagated and perennial plant species in the T0 generation via a co-editing strategy. *Nature Plants* 9(10), 1591–1597.

Iqbal, Z., Sattar, M.N. and Shafiq, M. (2016) CRISPR/Cas9: A tool to circumscribe cotton leaf curl disease. *Frontiers in Plant Science* 7, 475.

Ishino, Y., Shinagawa, H., Makino, K., Amemura, M. and Nakata, A. (1987) Nucleotide sequence of the iap gene, responsible for alkaline phosphatase isozyme conversion in *Escherichia coli*, and identification of the gene product. *Journal of Bacteriology* 169(12), 5429–5433.

Jiang, N., Zhang, C., Liu, J.Y., Guo, Z.H., Zhang, Z.Y. *et al.* (2019) Development of beet necrotic yellow vein virus-based vectors for multiple-gene expression and guide RNA delivery in plant genome editing. *Plant Biotechnology Journal* 17(7), 1302–1315.

Jinek, M., Chylinski, K., Fonfara, I., Hauer, M., Doudna, J.A. *et al.* (2012) A programmable dual-RNA–guided DNA endonuclease in adaptive bacterial immunity. *Science* 337(6096), 816–821.

Jones, M.G., Iqbal, S. and Fosu-Nyarko, J. (2024) Regulation of CRISPR-edited plants in Australia and New Zealand. In: Abd-Elsalam, K.A. and Ahmad, A. (eds) *Global Regulatory Outlook for CRISPRized Plants*. Academic Press, London, pp. 281–292.

Kaul, T., Thangaraj, A., Jain, R., Bharti, J., Kaul, R. *et al.* (2024) CRISPR/Cas9-mediated homology donor repair base editing system to confer herbicide resistance in maize (*Zea mays* L.). *Plant Physiology and Biochemistry* 207, 108374.

Kelliher, T., Starr, D., Su, X., Tang, G., Chen, Z. *et al.* (2019) One-step genome editing of elite crop germplasm during haploid induction. *Nature Biotechnology* 37(3), 287–292.

Kim, H., Kim, S.T., Ryu, J., Kang, B.C., Kim, J.S. *et al.* (2017) CRISPR/Cpf1-mediated DNA-free plant genome editing. *Nature Communications* 8(1), 14406.

Kocsisova, Z. and Coneva, V. (2023) Strategies for delivery of CRISPR/CAS-mediated genome editing to obtain edited plants directly without transgene integration. *Frontiers in Genome Editing* 5, 1209586.

Kumagai, Y., Liu, Y., Hamada, H., Luo, W., Zhu, J. *et al.* (2022) Introduction of a second "Green revolution" mutation into wheat via in planta CRISPR/Cas9 delivery. *Plant Physiology* 188(4), 1838–1842.

Laaninen, T. (2021) New genomic techniques: European Commission study and first reactions. EPRS: European Parliamentary Research Service, Brussels, Belgium. Available at: https://coilink.org/20.500.12592/zwrj4x (accessed 17 June 2024).

Lee, S.Y., Kang, B., Venkatesh, J., Lee, J.H., Lee, S. *et al.* (2024) Development of virus-induced genome editing methods in Solanaceous crops. *Horticulture Research* 11(1), uhad233.

Li, C., Li, W., Zhou, Z., Chen, H., Xie, C. *et al.* (2020) A new rice breeding method: CRISPR/Cas9 system editing of the Xa13 promoter to cultivate transgene-free bacterial blight-resistant rice. *Plant Biotechnology Journal* 18(2), 313.

Li, H., Xie, Y., Chen, F., Bai, H., Xiu, L. *et al.* (2023) Amplification-free CRISPR/Cas detection technology: Challenges, strategies, and perspectives. *Chemical Society Reviews* 52(1), 361–382.

Liang, Z., Chen, K.L., Li, T.D., Zhang, Y., Wang, Y.P. *et al.* (2017) Efficient DNA-free genome editing of bread wheat using CRISPR/Cas9 ribonucleoprotein complexes. *Nature Communications* 8, 14261. DOI: 10.1038/ncomms14261.

Liang, Z., Chen, K. and Gao, C. (2019) Biolistic delivery of CRISPR/Cas9 with ribonucleoprotein complex in wheat. *Plant Genome Editing with CRISPR Systems: Methods and Protocols* 1917, 327–335.

Liu, J.L., Chen, M.M., Chen, W.Q., Liu, C.M., He, Y. *et al.* (2022) A CASE toolkit for easy and efficient multiplex transgene-free gene editing. *Plant Physiology* 188(4), 1843–1847.

Lozano-Duran, R. (2016) Geminiviruses for biotechnology: The art of parasite taming. *New Phytologist* 210(1), 58–64. DOI: 10.1111/nph.13564.

Malnoy, M., Viola, R., Jung, M.H., Koo, O.J., Kim, S. *et al.* (2016) DNA-free genetically edited grapevine and apple protoplast using CRISPR/Cas9 ribonucleoproteins. *Frontiers in Plant Science* 7, 1904. DOI: 10.3389/fpls.2016.01904.

Mamrutha, H., Zeenat, W., Kapil, D., Budhagatapalli, N., Tikaniya, D. *et al.* (2023) Evidence and opportunities for developing non-transgenic genome edited crops using site-directed nuclease 1 approach. *Critical Reviews in Biotechnology* 44(6), 1140–1150.

Meyer, C.M., Goldman, I.L., Grzebelus, E. and Krysan, P.J. (2022) Efficient production of transgene-free, gene-edited carrot plants via protoplast transformation. *Plant Cell Reports* 41(4), 947–960.

Miyamoto, T., Toyooka, K., Chuah, J.A., Odahara, M., Higchi-Takeuchi, M. *et al.* (2022) A synthetic multidomain peptide that drives a macropinocytosis-like mechanism for cytosolic transport of exogenous proteins into plants. *Journal of the American Chemical Society Au* 2(1), 223–233.

Molla, K.A., Sretenovic, S., Bansal, K.C. and Qi, Y. (2021) Precise plant genome editing using base editors and prime editors. *Nature Plants* 7(9), 1166–1187.

Murovec, J., Guček, K., Bohanec, B., Avbelj, M. and Jerala, R. (2018) DNA-free genome editing of *Brassica oleracea* and *B. rapa* protoplasts using CRISPR-Cas9 ribonucleoprotein complexes. *Frontiers in Plant Science* 9, 1594.

Najafi, S., Bertini, E., D'incà, E., Fasoli, M. and Zenoni, S. (2023) DNA-free genome editing in grapevine using CRISPR/Cas9 ribonucleoprotein complexes followed by protoplast regeneration. *Horticulture Research* 10(1), uhac240.

Norfaezah, J., Masani, M.Y.A., Fizree, M.P.M.A.A., Bahariah, B., Shaharuddin, N.A. *et al.* (2024) DNA-free CRISPR/Cas9 genome editing system for oil palm protoplasts using multiple ribonucleoproteins (RNPs) complexes. *Industrial Crops and Products* 208, 117795.

Pausch, P., Al-Shayeb, B., Bisom-Rapp, E., Tsuchida, C.A., Li, Z. *et al.* (2020) CRISPR-casφ from huge phages is a hypercompact genome editor. *Science* 369(6501), 333–337.

Prajapati, R. and Tyagi, K. (2024) *CRISPR/Cas-Mediated Multiplex Genome Editing in Plants and Applications*. Applications of Genome Engineering in Plants.

Rahman, S.U., Khan, M.O., Ullah, R., Ahmad, F. and Raza, G. (2023) Agrobacterium-mediated transformation for the development of transgenic crops: Present and future prospects. *Molecular Biotechnology* 66(8), 1836–1852.

Rustgi, S., Zhang, H. and Oz, T.M. (2023) Overcoming genome editing challenges in plants: New tools and nanotechnologies. *Frontiers in Genome Editing* 5, 1230424.

Sattar, M.N., Iqbal, Z., Tahir, M.N., Shahid, M.S., Khurshid, M. *et al.* (2017) CRISPR/Cas9: A practical approach in date palm genome editing. *Frontiers in Plant Science* 8, 1469. DOI: 10.3389/fpls.2017.01469.

Sattar, M.N., Iqbal, Z., Al-Khayri, J.M. and Jain, S.M. (2021) Induced genetic variations in fruit trees using new breeding tools: Food security and climate resilience. *Plants* 10(7), 1347.

Schaart, J.G., Van De Weil, C.C. and Smulders, M.J. (2021) Genome editing of polyploid crops: Prospects, achievements and bottlenecks. *Transgenic Research* 30(4), 337–351.

Shahid, M.S., Sattar, M.N., Iqbal, Z., Raza, A. and Al-Sadi, A.M. (2021) Next-generation sequencing and the CRISPR-CAS nexus: A molecular plant virology perspective. *Frontiers in Microbiology* 11, 609376.

Shimatani, Z., Fujikura, U., Ishii, H., Matsui, Y., Suzuki, M. *et al.* (2018) Inheritance of co-edited genes by CRISPR-based targeted nucleotide substitutions in rice. *Plant Physiology and Biochemistry* 131, 78–83.

Stokstad, E. (2021) UK set to loosen rules for gene-edited crops and animals. *Science*. DOI: 10.1126/science.abj6955.

Subburaj, S., Chung, S.J., Lee, C., Ryu, S.M., Kim, D.H. *et al.* (2016) Site-directed mutagenesis in Petunia X hybrida protoplast system using direct delivery of purified recombinant Cas9 ribonucleoproteins. *Plant Cell Reports* 35(7), 1535–1544. DOI: 10.1007/s00299-016-1937-7.

Svitashev, S., Schwartz, C., Lenderts, B., Young, J.K. and Mark Cigan, A. (2016) Genome editing in maize directed by CRISPR-Cas9 ribonucleoprotein complexes. *Nature Communications* 7, 13274. DOI: 10.1038/ncomms13274.

Tao, J., Bauer, D.E. and Chiarle, R. (2023) Assessing and advancing the safety of CRISPR-Cas tools: From DNA to RNA editing. *Nature Communications* 14(1), 212.

Toda, E., Koiso, N., Takebayashi, A., Ichikawa, M., Kiba, T. *et al.* (2019) An efficient DNA-and selectable-marker-free genome-editing system using zygotes in rice. *Nature Plants* 5(4), 363–368.

Uranga, M. and Daròs, J.A. (2023) Tools and targets: The dual role of plant viruses in CRISPR–CAS genome editing. *The Plant Genome* 16(2), e20220.

Varanda, C.M., Félix, M.D.R., Campos, M.D., Patanita, M. and Materatski, P. (2021) Plant viruses: From targets to tools for CRISPR. *Viruses* 13(1), 141.

Vats, S., Kumawat, S., Brar, J., Kaur, S., Yadav, K. *et al.* (2022) Opportunity and challenges for nanotechnology application for genome editing in plants. *Plant Nano Biology* 1, 100001.

Veillet, F., Perrot, L., Chauvin, L., Kermarrec, M.P., Guyon-Debast, A. *et al.* (2019) Transgene-free genome editing in tomato and potato plants using Agrobacterium-mediated delivery of a CRISPR/Cas9 cytidine base editor. *International Journal of Molecular Sciences* 20(2), 402.

Voigt, B. and Münichsdorfer, A. (2019) Regulation of genome editing in plant biotechnology: European Union. In: Dederer, H.G. and Hamburger, D. (eds) *Regulation of Genome Editing in Plant Biotechnology: A Comparative Analysis of Regulatory Frameworks of Selected Countries and the EU*. Springer, Cham, Switzerland, pp. 137–238.

Wang, Z.P., Zhang, Z.B., Zheng, D.Y., Zhang, T.T., Li, X.L. *et al.* (2022) Efficient and genotype independent maize transformation using pollen transfected by DNA-coated magnetic nanoparticles. *Journal of Integrative Plant Biology* 64(6), 1145–1156.

Whelan, A.I. and Lema, M.A. (2019) Regulation of genome editing in plant biotechnology: Argentina. In: Dederer, H.G. and Hamburger, D. (eds) *Regulation of Genome Editing in Plant Biotechnology: A Comparative Analysis of Regulatory Frameworks of Selected Countries and the EU*. Springer, Cham, Switzerland, pp. 19–62.

Woo, J.W., Kim, J., Kwon, S.I., Corvalán, C., Cho, S.W. *et al.* (2015) DNA-free genome editing in plants with preassembled CRISPR-Cas9 ribonucleoproteins. *Nature Biotechnology* 33(11), 1162–1164.

Wu, X., Zhang, Y., Jiang, X., Ma, T., Guo, Y. *et al.* (2024) Considerations in engineering viral vectors for genome editing in plants. *Virology* 589, 109922.

Xu, C.L., Ruan, M.Z., Mahajan, V.B. and Tsang, S.H. (2019) Viral delivery systems for CRISPR. *Viruses* 11(1), 28.

Yan, Y., Zhu, J., Qi, X., Cheng, B., Liu, C. *et al.* (2021) Establishment of an efficient seed fluorescence reporter-assisted CRISPR/Cas9 gene editing in maize. *Journal of Integrative Plant Biology* 63(9), 1671–1680.

Yin, K., Han, T., Liu, G., Chen, T., Wang, Y. *et al.* (2015) A geminivirus-based guide RNA delivery system for CRISPR/Cas9 mediated plant genome editing. *Scientific Reports* 5, 14926. DOI: 10.1038/srep14926.

Yong, J., Wu, M., Zhang, R., Bi, S., Mann, C.W. *et al.* (2022) Clay nanoparticles efficiently deliver small interfering RNA to intact plant leaf cells. *Plant Physiology* 190(4), 2187–2202.

Yubing, H., Min, Z., Lihao, W., Junhua, W., Qiaoyan, W. *et al.* (2019) Improvements of TKC technology accelerate isolation of transgene-free CRISPR/Cas9-edited rice plants. *Rice Science* 26(2), 109–117.

Zhang, H., Goh, N.S., Wang, J.W., Pinals, R.L., González-Grandío, E. *et al.* (2022) Nanoparticle cellular internalization is not required for RNA delivery to mature plant leaves. *Nature Nanotechnology* 17(2), 197–205.

Zhang, Y., Liang, Z., Zong, Y., Wang, Y.P., Liu, J.X. *et al.* (2016) Efficient and transgene-free genome editing in wheat through transient expression of CRISPR/Cas9 DNA or RNA. *Nature Communications* 7, 12617. DOI: 10.1038/ncomms12617.

Zhao, X., Meng, Z., Wang, Y., Chen, W., Sun, C. *et al.* (2017) Pollen magnetofection for genetic modification with magnetic nanoparticles as gene carriers. *Nature Plants* 3(12), 956–964.

Zong, Y., Wang, Y., Li, C., Zhang, R., Chen, K. *et al.* (2017) Precise base editing in rice, wheat and maize with a Cas9-cytidine deaminase fusion. *Nature Biotechnology* 35(5), 438–440.

12 Plant Genome Editing for Nutritional Improvement

Sana Khalid*, Shabnum Shaheen, Sumera Iqbal, Nimra Naeem and Tooba Wajid

Department of Botany, Lahore College for Women University, Lahore, Pakistan

Abstract

In comparison to conventional breeding, nutrient efficiency of some staple, oilseed, and horticultural crops has been improved efficiently by adopting genome-editing tools, namely meganucleases, zinc-finger nucleases, transcription activator-like effector nucleases, and clustered regularly interspaced short palindromic repeats (CRISPR) system. Recently, CRISPR has shown great potential for the improvement of crops and is proved to be the most cost-effective, feasible, and handy technique for precisely editing plant genomes. Many global and local companies have developed commercially valuable genome-edited crops. Genome-edited crops have the advantage over genetically modified crops due to their easy commercial release in the market. Mutation created by genome editing is very much similar to natural spontaneous mutation. In recent times, two genome-edited crops, that is, high gamma-aminobutyric acid (GABA) tomato and high oleic acid soybean have been introduced in the market. In this chapter, we discuss the enhancement of crop improvement nutritionally and functionally by employing genome-editing tools. Moreover, we elaborate on the effectiveness of genome-editing tools by targeting the suitable traits that are responsible for nutritional improvement in crops by controlling and regulating the gene expression. In future, the combination of conventional breeding with genome-edited crops can lead to its high yield and commercialization.

Keywords: Genome editing, CRISPR, Agricultural crops, Nutritional enhancement, High GABA tomatoes, High oleic acid soybean

12.1 Introduction

Efforts in crop improvement have consistently aimed at achieving higher yields, bolstering resilience against biotic/abiotic stresses, and enhancing nutritional excellence. To attain these goals, different breeding techniques have been broadly operated to selectively breed desirable traits across different species through controlled crossbreeding. The intentional induction of mutations, whether through chemical or radioactive means, alongside naturally occurring mutations, has substantially enriched the pool of selectable traits (Pacher and Puchta, 2017). Subsequently, desired traits carried by genetically modified crops (developed by the insertion of transgenes) have been developed in 1996. However, concerns persist regarding the environmental security and potential health of genetically modified (GM) crops stemming from the introduction of foreign genes. Considerable

*Corresponding author: sanakhalidpu@gmail.com

 Genome Editing for Crop Improvement: Theory and Methodology (eds J.M. Al-Khayri *et al.*)
DOI: 10.1079/9781800622517.0012

financial investment and time are required for the exhaustive and thorough safety assessments needed to address these concerns (Prado *et al.*, 2014).

Genome-editing implementations have arisen as a promising breeding technology, offering a potential solution to the concerns surrounding foreign gene introduction. These tools have been effectively utilized to modify crop genomes in a wide range of plants, without the incorporation of foreign genetic material. Within a relatively short period, they have revolutionized crop improvement efforts, offering superior efficiency and accuracy compared with traditional breeding methods (Chen *et al.*, 2019). The advent of genome editing (GE) began with the deployment of various endonucleases for introducing precise DNA double-stranded breaks (DSBs). These include zinc-finger nucleases (ZFNs) (Kim *et al.*, 1996), transcription activator-like effector nucleases (TALENs) (Christian *et al.*, 2010), and CRISPR-associated 9 (Cas9) endonucleases (Jinek *et al.*, 2012).

With global population growth driving up the demand for food, projections indicate a 1.7-fold and 1.8-fold increase in cereals and livestock products, respectively, will be needed by 2050. Consequently, it is imperative to intensify efforts to boost food production through every available means. Moreover, shifts in global dietary preferences, transitioning from starch-centric diets to those rich in fats and proteins, alongside the aging population, have contributed to a surge in lifestyle-related diseases, amplifying health concerns worldwide.

In recent times, a significant flow has been observed in the application of genome-editing technologies for crop improvement. Of the various traits targeted for enhancement, nutritional and functional qualities stand out as a key feature in the production of raw material, food, and animal feed. This chapter undertakes an overview of the latest advancements in staple, oilseed, and horticultural crops with the aim of augmenting diverse nutritional–functional quality characters with respect to their commercial value. In addition, we provide a summary of their current status and explore the prospects of genome-editing technology for further enhancing the quality of these crops.

In this study, we review the recent developments of staple, oilseed, and horticultural crops aiming to enhance diverse nutritional and functional quality traits with a focus primarily on commercial value. Besides, nutritional traits of crops can also be enhanced by controlling and regulating the gene expression. Additionally, we summarize their current status and discuss the perspectives of genome-editing technology for further quality improvement in these crops.

12.2 Genome Editing (GE) in Agricultural Crops

With respect to commercial value and improvement of the varied nutritional/functional characters, GE technologies are used in staple, oilseed, and horticultural crops (Fig. 12.1).

12.2.1 Staple crops

Ensuring food security is vital to sustain human life, meeting the needs of the rapidly expanding global population and addressing the growing demand for meat consumption. Staple crops play a pivotal role in meeting such high supply and enhancing crop quality, particularly in terms of nutritional cost. As a result, GE is actively

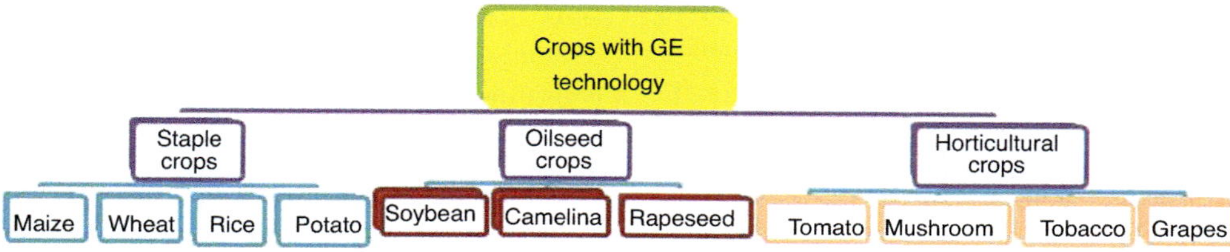

Fig. 12.1. Crops with genome editing technology. Author's own figure.

employed in staple crop agriculture, including corn, wheat, rice, and potato.

12.2.1.1 Corn (Zea mays)

Globally, the most extensively cultivated cereal crop is corn (*Zea mays*) for its application from animal feed to biofuel. Many corn-based nutritional items – for example, corn syrup and corn starch – are used for human consumption. However, the main storage protein of corn 'zeins' is deficient in some essential amino acids, such as lysine and tryptophan. RNAi and CRISPR are the most widely used techniques that target the ZmMADS47 gene (encode MADS-box protein) to activate the zein gene promoter by interacting with O_2 (Qi *et al.*, 2016). MADS-box gene name stands for the yeast minichromosome maintenance 1 (**M**CM1), Arabidopsis agamous (**A**G), Antirrhinum deficiens (**D**EF) and human serum response factor (**S**RF) (Surabhi and Badajena, 2020). Furthermore, inositol 1,2,3,4,5,6-hexakisphosphate (commonly called phytate) forms insoluble complexes by hampering the proteins/minerals absorption in humans and animals and is considered an antinutrient compound. Shukla *et al.* (2009) reported that ZFNs can inhibit the target genes IPK1A, IPK, and MRP4 encoding the phytate biosynthetic pathway (Liang *et al.*, 2014).

12.2.1.2 Wheat (Triticum aestivum)

Wheat (*Triticum aestivum*) is the second most cultivated cereal crop and one of the most ancient crops. Celiac, non-celiac, and gluten ataxia diseases are caused by gluten that is the wheat storage protein (Sapone *et al.*, 2011). GE efforts have been made for the introduction of low-gluten elite varieties of wheat by creating knockouts of α-gliadin genes using CRISPR/Cas9 technology. About 85% immunoreactivity reductions have been achieved. Moreover, CRISPR can increase the grain weight and boost up the overall protein content by targeting GW2 gene that encode RING-type E3 ubiquitin ligase, which is involved in regulation of spikelet hull cell numbers (Su *et al.*, 2011; Zhang *et al.*, 2018).

12.2.1.3 Rice (Oryza sativa)

Rice (*Oryza sativa*) provides 20% dietary need for 3 billion people worldwide. Thai jasmine and Indian Basmati rice are considered the most favorite and demanding. GE efforts have been made on improving its special aromatic property that is due to 2-AP and TALEN has been employed to target the $BADH_2$ gene that encodes BADH (betaine aldehyde dehydrogenase) in order to minimize the GABA (competing compound) and elevate the synthesis of 2-AP from the same precursor, γ-aminobutyraldehyde (Shan *et al.*, 2015).

12.2.1.4 Potato (Solanum tuberosum)

Fourth-largest demanding crop of the world is potato (*Solanum tuberosum*). In developing countries, it is a rich source of nutrition. Its postharvest shelf life is extended in cold storage by restricting sprouting. This refrigeration leads to the production of fructose and glucose that react with free amino acids on high temperatures causing acrylamide accumulation that is carcinogenic (Tareke *et al.*, 2002). As a solution, Cellectis Plant Sciences (now Calyxt Inc.), produced potatoes with negligible sugar using TALEN technology for targeting vacuolar invertase enzyme encoding gene VInv to hinder the fructose/glucose breakdown (Clasen *et al.*, 2016). Furthermore, CRISPR/Cas9 and TALEN technologies have been used to disrupt the functioning of genes related to steroidal glycoalkaloids (SGA) pathways and prevent its accumulation in tubers. SGA has a bitter taste and is lethal to organisms (Nakayasu *et al.*, 2018; Yasumoto *et al.*, 2019).

12.2.2 Oilseed crops

Some crops are used for the extraction of essential oils used for both 'consumption' and 'industrial' purposes. Transgenic approaches (RNAi) have been employed to produce seed oils with altered saturation levels of fatty acids by targeting the desaturase gene. However, GE is a good substitute for RNAi in order to avoid pleiotropic effects caused by gene knockout in the entire plant.

12.2.2.1 Soybean (*Glycine max*)

One of the most cultivated edible oil crops is soybean. Globally, 57 mt were produced during 2019–2020, fulfilling 28% of the world's oil demand (USDA-FAS, 2020). TALEN has been used in recent years to create the mutation in two FAD2 genes that increase the oleic acid production tremendously from 20 % to 80 % and decrease the production of linoleic acid from 50 % to 4.7 % (Haun *et al.*, 2014). Linoleic acid level can be further reduced by targeting FAD3 gene using TALEN (Demorest *et al.*, 2016). Calyxt (2019) reported the commercial deployment of high-oleic soybean oil in US markets.

12.2.2.2 Camelina (*Camelina sativa*)

Camelina (*Camelina sativa*) has unique feature of having abundant amount of tocopherols and omega-3 lipids, making it most suitable for GE technologies. A unique oil trait of *C. sativa* makes it distinct from other oilseed crops (Haun *et al.*, 2014; Demorest *et al.*, 2016; Faure and Tepfer, 2016). Morineau *et al.* (2017) and Jiang *et al.* (2017) used CRISPR/Cas9 to alter the three FAD2 genes in order to reduce the polyunsaturated fatty acid (PUFA) level and increase the oleic acid level.

12.2.3 Horticultural crops

Horticultural crops are a rich source of macro- and micronutrients, but they are threatened by postharvest early ripening. So, nutritional enhancement, improved fruit quality, and prolonged shelf life are the main goals of breeders to boost their market demand. Tomato (*Solanum lycopersicum*) has been subjected to many successful GE applications (e.g. CRISPR/Cas9 and TALEN), which has raised its demand notably in the market as compared with other crops (Cermák *et al.*, 2015). Emergences of purple tomatoes have been achieved by constructing CaMV 35S promoter upstream of ANT1, Myb transcription factor to boost anthocyanin content. Besides long shelf-life, tomatoes, for example Golden Bell, Ailsa Craig, and Mamirio varieties have been developed by CRISPR that target RIN gene (ripening inhibitor gene; Ito *et al.*, 2015; Jung *et al.*, 2018).

Garden tomatoes (*Solanum pimpinellifolium*) namely currant tomatoes are the wild species and have incredible disease resistance and stress tolerance ability making them favourable future cultivars. Table 12.1 illustrates the application of CRISPR technology targets and its outcomes in currant tomato.

Table 12.1. Application of CRISPR technology targets and its outcomes in currant tomato. Author's own table.

GE Technology	Target Genes	Outcomes	References
CRISPR	• Ovate (O-fruit shape) multiflora (MULT-fruit number) • Lycopene betacyclase (CycB-functional quality) • Self-pruning (SP-general plant growth habit) • Fruit weight 2.2 (FW2.2-fruit size) • Fasciated (FAS)	• Fruit morphology altered • Fruit size enhancement from threefold • Yield increase (tenfold)	Zsögön *et al.*, 2018
	• Clavata3 • Self-pruning • Wuschel • Self-pruning 5G • Vitamin C biosynthetic enzyme (SlGGP1) encoding gene	• Nutritional value enhancement, especially lycopene	Li *et al.*, 2018
	• uORF upstream open reading frame of SlGGP1	• Ascorbic acid level elevation	Ku and Ha, 2020

GE technologies have gained much importance in horticultural crops, namely mushroom, grapes, and tobacco. Postharvest browning of mushrooms due to melanin production by phenol oxidation that is initiated by polyphenol oxidase (PPO) greatly reduced its commercial value. To address this issue, six PPO genes of *Agaricus bisporus* have been knocked out by deleting some base pairs using CRISPR/Cas9 technique (Waltz, 2016). In grape wine, tartaric acid (TA) plays an important role in flavor profile. CRISPR/Cas9 is an important tool to reduce TA content by 36% by targeting IdnDH gene that is responsible for TA biosynthesis and L-idonate dehydrogenase regulation (Ren *et al.*, 2016). Similarly, custom-designed meganucleases have led to the production of low-nicotine tobacco by deleting and inserting some base pairs in berberine bridge enzyme-like genes (USDA-APHIS, 2020).

12.3 Controlling Gene Expression for the Improvement of Nutrient Composition

Mainly there are two purposes for improvement of nutrient composition by controlling the gene expression. These purposes are to modify the far upstream region of (1) UTGs (modification of untranslated regions of genes/proteins) and (2) CDS (synthesis of incomplete protein due to premature termination codon as an outcome of frameshift mutation).

12.3.1 High-oleic soybean oil development

Soybean oil is a pool of high and low amounts of polyunsaturated and monounsaturated fatty acids, respectively. Unfortunately, transunsaturated fatty acid is produced in an effort to enhance fatty acid composition due to hydrogenation process (Ascherio *et al.*, 1999). Calyxt in 2019 introduced soybean named CalynoTM using TALEN technique to target FAD2-1A, FAD2-1B, and FAD3-A genes to harbor 80% oleic acid and below 3% linolenic acid.

12.3.2 Regulation of gene expression for the improvement of nutrient composition

GE lessens the protein activities encoded by genes linked with undesirable and specific nutrients metabolism. They target the enzyme activity regulatory enzyme active site. Structural and physiological properties of target genes are very important for the selection of specific GE technology.

12.3.3 Development of Sicilian Rouge High GABA tomatoes

A high GABA genome-edited tomato variety named 'Sicilian Rouge High GABA' was produced in 2021 by a Japanese company, Sanatech Seed Co., Ltd., in Tsukuba. For this purpose, GABA inhibitors (GAD3) were excised by the introduction of a premature translation codon due to frameshift mutation employing CRISPR/Cas9 technology. This is a much-revolutionized contribution for the introduction of 1.5 times rich GABA tomatoes in the market. Conspicuously, GAD3 expression level remains unaltered during the process (Nonaka *et al.*, 2017).

12.4 Genome Editing-based Quality Improvement: Current Status

Crop GE is advancing more rapidly than animal GE due to practical advantages and fewer ethical concerns. Many nutritional and functional traits have been successfully modified in various crops using genome-editing technologies. However, some edited crops, such as CRISPR-wx1 corn, TALEN-ppo potato, and TALEN-fad2 soybean, have not yet been commercialized. To expedite the commercialization of edited crops, technical barriers related to intellectual and policy rights must be addressed. Country-specific regulations vary, with some countries exempting genome-edited products from genetically modified organism regulation if no foreign DNA is present. For instance, the United States employs a program called 'Am I Regulated?' to assess regulatory status, resulting in some genome-edited crops being exempt from regulation. Patent issues can hinder small- and medium-sized enterprises from commercializing gene-edited crops, so reasonable licensing fees are recommended. Additionally, advancements in editing-associated technologies can accelerate commercialization. Compared with GM crops, edited crops pose fewer risks, as they typically involve minor changes in DNA. Advanced editing techniques can remove foreign DNA, further reducing regulatory concerns. However, confirming the absence of exogenous DNA remains challenging and requires extensive genotyping. Recent technical developments offer solutions for identifying transgene-free edited plants, including marker-assisted selection, self-elimination systems, and leaf assays. Careful consideration of target traits and host crop suitability is crucial to avoid unintended negative effects on agronomic traits. Haploid induction editing technology simplifies breeding processes by allowing direct editing of elite inbred lines, facilitating the production of commercial varieties.

12.5 Future Recommendations

Genome-editing technologies have rapidly advanced across various systems, including microorganisms, animals, and plants. CRISPR/Cas9, for example, can induce small insertions/deletions (INDELs) or perform precise single-base editing. However, base editors, comprising a nCas9 (sgRNA-guided Cas9 nickase) attached with a deaminase, enable specific nucleotide conversions, expanding the range of edits achievable in crop genomes. Other endonucleases like newly designed Cas9 variants and Cpf1 can recognize diverse protospacer adjacent motif (PAM) sequences, allowing for PAM-flexible site-specific cleavages. Delivery of Cpf1 proteins with target-specific gRNAs has proven effective for DNA-free GE in soybean and tobacco protoplasts. The step of applying these radical technologies to crops will influence their practical use. Different crops exhibit varying responses to GE, highlighting their unique characteristics. To broaden the range of nutritional and functional traits improved via GE, more target genes must be discovered and their associations with traits in metabolic pathways investigated. Identifying desirable targets for GE requires screening recessive traits among traditional genetic mutants in breeding programs. Fragrant, waxy, and 'rc rice,' as well as 'alc tomato' mutants, have been used to develop genome-edited crops through the restoration of recessive alleles. Efforts to enhance crop species' nutritional and functional qualities involve elucidating metabolic pathway functions and regulatory mechanisms. Screening large-scale mutant libraries constructed using CRISPR may aid in identifying novel genes suitable for crop variety development. Continuous improvement in genome-editing technology and better understanding of gene functions will facilitate the development and commercialization of new genome-edited crops with improved traits.

References

Ascherio, A., Katan, M.B., Zock, P.L., Stampfer, M.J. and Willett, W.C. (1999) Trans fatty acids and coronary heart disease. *New England Journal of Medicine* 340, 1994–1998. DOI: 10.1056/NEJM199906243402511.

Calyxt (2019) First Commercial Sale of Calyxt High Oleic Soybean Oil on the U.S. Market. Available at: https://calyxt.com/first-commercial-saleof-calyxt-high-oleic-soybean-oil-on-the-u-s-market/ (accessed 26 February 2019).

Cermák, T., Baltes, N.J., Cegan, R., Zhang, Y. and Voytas, D.F. (2015) High-frequency, precise modification of the tomato genome. *Genome Biology* 16, 1–5. DOI: 10.1186/s13059-015-0796-9.

Chen, K., Wang, Y., Zhang, R., Zhang, H. and Gao, C. (2019) CRISPR/Cas genome editing and precision plant breeding in agriculture. *Annual Review of Plant Biology* 70, 667–697. DOI: 10.1146/annurev-arplant-050718-100049.

Christian, M., Cermak, T., Doyle, E.L., Schmidt, C., Zhang, F. *et al.* (2010) Targeting DNA double-strand breaks with TAL effector nucleases. *Genetics* 186(2), 757–761. DOI: 10.1534/genetics.110.120717.

Clasen, B.M., Stoddard, T.J., Luo, S., Demorest, Z.L., Li, J. *et al.* (2016) Improving cold storage and processing traits in potato through targeted gene knockout. *Plant Biotechnology Journal* 14(1), 169–176. DOI: 10.1111/pbi.12370.

Demorest, Z.L., Coffman, A., Baltes, N.J., Stoddard, T.J., Clasen, B.M. *et al.* (2016) Direct stacking of sequence-specific nuclease-induced mutations to produce high oleic and low linolenic soybean oil. *BMC Plant Biology* 16, 225. DOI: 10.1186/s12870-016-0906-1.

Faure, J. and Tepfer, M. (2016) Camelina, a Swiss knife for plant lipid biotechnology. *Oilseeds and Fats Crops and Lipids* 23(5), D503. DOI: 10.1051/ocl/2016023.

Haun, W., Coffman, A., Clasen, B.M., Demorest, Z.L., Lowy, A. *et al.* (2014) Improved soybean oil quality by targeted mutagenesis of the fatty acid desaturase 2 gene family. *Plant Biotechnology Journal* 12(7), 934–940. DOI: 10.1111/pbi.12201.

Ito, Y., Nishizawa-Yokoi, A., Endo, M., Mikami, M. and Toki, S. (2015) CRISPR/Cas9-mediated mutagenesis of the RIN locus that regulates tomato fruit ripening. *Biochemical and Biophysical Research Communications* 467(1), 76–82. DOI: 10.1016/j.bbrc.2015.09.117.

Jiang, W.Z., Henry, I.M., Lynagh, P.G., Comai, L., Cahoon, E.B. *et al.* (2017) Significant enhancement of fatty acid composition in seeds of the allohexaploid, *Camelina sativa*, using CRISPR/Cas9 gene editing. *Plant Biotechnology Journal* 15(5), 648–657. DOI: 10.1111/pbi.12663.

Jinek, M., Chylinski, K., Fonfara, I., Hauer, M., Doudna, J.A. *et al.* (2012) A programmable dual-RNA-guided DNA endonuclease in adaptive bacterial immunity. *Science* 337(6096), 816–821. DOI: 10.1126/science.1225829.

Jung, Y.J., Lee, G., Bae, S. and Kang, K.K. (2018) Reduced ethylene production in tomato fruits upon CRSPR/Cas9-mediated LeMADS-RIN mutagenesis. *Horticultural Science and Technology* 36(3), 396–405. DOI: 10.12972/kjhst.20180039.

Kim, Y.G., Cha, J. and Chandrasegaran, S. (1996) Hybrid restriction enzymes: zinc finger fusions to Fok I cleavage domain. *Proceedings of the National Academy of Sciences* 93(3), 1156–1160. DOI: 10.1073/pnas.93.3.1156.

Ku, H.K. and Ha, S.H. (2020) Improving nutritional and functional quality by genome editing of crops: status and perspectives. *Frontiers in Plant Science* 11, 577313. DOI: 10.3389/fpls.2020.577313.

Liang, Z., Zhang, K., Chen, K. and Gao, C. (2014) Targeted mutagenesis in *Zea mays* using TALENs and the CRISPR/Cas system. *Journal of Genetics and Genomics* 41(2), 63–68. DOI: 10.1016/j.jgg.2013.12.001.

Li, T., Yang, X., Yu, Y., Si, X., Zhai, X. *et al.* (2018) Domestication of wild tomato is accelerated by genome editing. *Nature Biotechnology* 36(12), 1160–1163. DOI: 10.1038/nbt.4273.

Morineau, C., Bellec, Y., Tellier, F., Gissot, L., Kelemen, Z. *et al.* (2017) Selective gene dosage by CRISPR-Cas9 genome editing in hexaploid *Camelina sativa*. *Plant Biotechnology Journal* 15(6), 729–739. DOI: 10.1111/pbi.12671.

Nakayasu, M., Akiyama, R., Lee, H.J., Osakabe, K., Osakabe, Y. *et al.* (2018) Generation of α-solanine-free hairy roots of potato by CRISPR/Cas9 mediated genome editing of the St16DOX gene. *Plant Physiology and Biochemistry* 131, 70–77. DOI: 10.1016/j.plaphy.2018.04.026.

Nonaka, S., Arai, C., Takayama, M., Matsukura, C. and Ezura, H. (2017) Efficient increase of G-aminobutyric acid (GABA) content in tomato fruits by targeted mutagenesis. *Scientific Reports* 7(1), 7057. DOI: 10.1038/s41598-017-06400-y.

Pacher, M. and Puchta, H. (2017) From classical mutagenesis to nuclease-based breeding - directing natural DNA repair for a natural end-product. *The Plant Journal* 90(4), 819–833. DOI: 10.1111/tpj.13469.

Prado, J.R., Segers, G., Voelker, T., Carson, D., Dobert, R. *et al.* (2014) Genetically engineered crops: from idea to product. *Annual Review of Plant Biology* 65, 769–790. DOI: 10.1146/annurev-arplant-050213-040039.

Qi, W., Zhu, T., Tian, Z., Li, C., Zhang, W. *et al.* (2016) High-efficiency CRISPR/Cas9 multiplex gene editing using the glycine tRNA-processing system-based strategy in maize. *BMC Biotechnology* 16, 1–8. DOI: 10.1186/s12896.

Ren, C., Liu, X., Zhang, Z., Wang, Y., Duan, W. *et al.* (2016) CRISPR/Cas9-mediated efficient targeted mutagenesis in chardonnay (*Vitis vinifera* L.). *Scientific Reports* 6(1), 32289. DOI: 10.1038/srep32289.

Sapone, A., Lammers, K.M., Casolaro, V., Cammarota, M., Giuliano, M.T. *et al.* (2011) Divergence of gut permeability and mucosal immune gene expression in two gluten-associated conditions: celiac disease and gluten sensitivity. *BMC Medicine* 9, 1–11. DOI: 10.1186/1741-7015-9-23.

Shan, Q., Zhang, Y., Chen, K., Zhang, K. and Gao, C. (2015) Creation of fragrant rice by targeted knockout of the OsBADH2 gene using TALEN technology. *Plant Biotechnology Journal* 13(6), 791–800. DOI: 10.1111/pbi.12312.

Shukla, V.K., Doyon, Y., Miller, J.C., DeKelver, R.C., Moehle, E.A. *et al.* (2009) Precise genome modification in the crop species *Zea mays* using zinc-finger nucleases. *Nature* 459(7245), 437–441. DOI: 10.1038/nature07992.

Su, Z., Hao, C., Wang, L., Dong, Y. and Zhang, X. (2011) Identification and development of a functional marker of TaGW2 associated with grain weight in bread wheat (*Triticum aestivum* L.). *Theoretical and Applied Genetics* 122, 211–223. DOI: 10.1007/s00122-010-1437-z.

Surabhi, G.K. and Badajena, B. (2020) Recent advances in plant heat stress transcription factors. In: *Transcription Factors for Abiotic Stress Tolerance in Plants*. pp. 153–200. Available at: https://doi.org/10.1016/B978-0-12-819334-1.00010-1

Tareke, E., Rydberg, P., Karlsson, P., Eriksson, S. and Tornqvist, M. (2002) Analysis of acrylamide, a carcinogen formed in heated foodstuffs. *Journal of Agricultural and Food Chemistry* 50(17), 4998–5006.

USDA-APHIS (2020) Regulated article letters of inquiry “am I regulated?” process. In: Gipson, C.A. (ed.) *United States Department of Agriculture Animal and Plant Health Inspection Service*. United States Department of Agriculture, Washington, DC.

USDA-FAS (2020) Oilseeds: World Markets and Trade. United States Department of Agriculture Foreign Agricultural Service. Available at: https://apps.fas.usda.gov/psdonline/circulars/oilseeds.pdf (accessed 10 September 2023).

Waltz, E. (2016) Gene-edited CRISPR mushroom escapes US regulation. *Nature* 532(7599), 293. DOI: 10.1038/nature.2016.19754.

Yasumoto, S., Umemoto, N., Lee, H.J., Nakayasu, M., Sawai, S. *et al.* (2019) Efficient genome engineering using platinum TALEN in potato. *Plant Biotechnology* 36(3), 167–173. DOI: 10.5511/plantbiotechnology.19.0805a.

Zhang, Y., Li, D., Zhang, D., Zhao, X., Cao, X. *et al.* (2018) Analysis of the functions of TaGW2 homoeologs in wheat grain weight and protein content traits. *The Plant Journal* 94(5), 857–866. DOI: 10.1111/tpj.13903.

Zsögön, A., Čermák, T., Naves, E.R., Notini, M.M., Edel, K.H. *et al.* (2018) De novo domestication of wild tomato using genome editing. *Nature Biotechnology* 36(12), 1211–1216. DOI: 10.1038/nbt.4272.

13 Genome Editing to Accelerate Biotic Stress Resistance in Crop Plants

Ambreen Gul[1]*, Khadija Aaliya[1] and Sehrish Iftikhar[2]

[1]Centre of Excellence in Molecular Biology, University of the Punjab, Lahore, Pakistan; [2]Microfluidics Innovation Centre, Elvesys, Paris

Abstract

Since the advancements in genome editing techniques, a tremendous amount of research has been going on to improve crop plants against various biotic stresses. Plant defense mechanisms to cope with various types of biotic stresses are complex. As sessile organisms, plants exhibit remarkable intelligence in addressing stresses through diverse physical adaptations, via the emission of volatile organic chemicals and by regulating intricate hormonal signaling pathways. Recent advancements in genetic editing of organisms' genomes have made it possible to edit the specific pathway protein, effector, or physical barriers and re-regulate the plant's genome to make it resilient. The CRISPR/Cas system is now leading this field among others (ZFNs, TALENs) to specifically edit the plant genome and enhance resistance against different types of biotic stresses. Moreover, the current focus is on the development of transgene-free gene-edited crops, which has become possible using state-of-the-art base editing systems that can edit and replace the single base pair in the genome without making double-stranded breaks. The contents of this chapter are focused on how genome editing techniques have led to the development of plants resistant to the devastating impact of different types of biotic stresses. This chapter will provide an overview of the techniques and mechanisms behind biotic stress resistance in plants, applications of genome editing in biotic stress resistance, and its potential impact on agriculture and global food security.

Keywords: Biotic stress resistance, Plant genome editing, Crop improvement, Transgene-free editing CRISPR/Cas system, Zinc-finger nucleases (ZFNs), Transcription activator-like effector nucleases (TALENs)

13.1 Introduction

Crops are frequently exposed to different biotic factors/stressors during their lifespan, resulting in significant reduction in crop yield and quality. The main drivers of biotic stresses are living organisms like weeds, insects, fungi, bacteria, and nematodes and nonliving particles like viruses and viroids, which feed on the plants, alter plant defense pathways, transmit diseases, and compete for nutrients and water. Plant pests and pathogens are a serious threat to food security, affecting nearly all its main components, like quality and nutrition, product yield, physical availability, distribution, and economics (Savary *et al.*, 2019). The application of conventional methods in plant breeding has facilitated the development of resistance against biotic stresses, but these methods are time consuming and often ineffective. The process typically involves crossing plant varieties to produce offspring that exhibit desirable traits. Additionally, desirable traits can be selected and

*Corresponding author: ambreen.gul@cemb.edu.pk; ambreengulpu@yahoo.com

DOI: 10.1079/9781800622517.0013

amplified in a process known as selective breeding. While plant breeding has been very effective over the years, it has its limitations, primarily due to the long time required for developing and testing new plant varieties. It can also be challenging to identify the genes responsible for conferring improved resistance traits.

Genome editing (GE), which resembles cut, copy, and paste, has emerged as a promising tool for accelerating biotic stress resistance in crop plants (Makarova *et al.*, 2020). Genome editing refers to the precise and targeted modification of an organism's genetic code. The utilization of advanced GE tools—that is, ZFN (zinc-finger nucleases), TALENs (transcription activator-like effector nucleases), and CRISPR/Cas9 (clustered regularly interspaced short palindromic repeats)—enable scientists to manipulate biotic stress–associated genes in crops for enhanced resistance (Mushtaq *et al.*, 2019). GE has several advantages over traditional breeding methods. For example, it is faster, more efficient, and more precise. It also allows researchers to study the fundamental mechanisms behind biotic stress resistance by identifying the genes and molecular pathways involved (Li *et al.*, 2022a). The modification and trait improvement of crop plants are indispensable for sustainable food production. Moreover, they are also important as sustainable resources for feed, fiber, and fuel production. Not surprisingly, the challenges to sustainable crop production are numerous and are collectively referred to as biotic and abiotic stressors. Biotic stressors can significantly compromise the yield and quality of the produce, fuelling economic slump and food insecurity. Therefore, developing crops with resistance to biotic stress is a critical goal of crop improvement programs for sustainable agriculture.

This chapter will provide an overview of the techniques and mechanisms behind biotic stress resistance in plants, applications of GE in biotic stress resistance, and its potential impact on agriculture and global food security.

13.2 Endogenous Plant Pathways to Handle Biotic Stresses

Plants respond to different stress signals by activating different sets of immune layers. These immune layers involve complex molecular pathways that allow plants to control pests and pathogens effectively and specifically.

13.2.1 Physical barriers

The plant's cuticle and cell wall constitute an important physical barrier to protect it against insects, fungi, bacteria, and other herbivorous organisms as well as from water damage. The cuticle layer is a complex network of waxy and hydrophobic compounds, such as cutin and epicuticular wax, which interact with biological membranes and discourage insect herbivory and microbial infestations (Bernard and Joubès, 2013). Cell walls, which are primarily composed of cellulose, hemicellulose, and pectin, surround plant cells and form a structural defense barrier against both mechanical damage and herbivores (Kaur *et al.*, 2022). Plant cells can also reinforce their cell walls by producing lignin, a polymer that can make the cell wall tougher and hard to digest by insects' salivary enzymes, thereby preventing herbivory (Kumar *et al.*, 2016).

13.2.2 Chemical defenses

Plants produce more than 200,000 bioactive compounds that can defend against biotic stresses, particularly pathogens (Kumar *et al.*, 2023). These compounds, which include secondary metabolites, such as phenolics, glucosinolates, amines, terpenoids, cyanogenic glucosides, alkaloids, and peptides, are stored in specialized cells, such as glandular trichomes, and are released upon herbivore or pathogen attack (Divekar *et al.*, 2022). These compounds can disrupt the metabolism of herbivores or pathogens, inhibit their ability to consume nutrients, or even poison and kill them directly (Pieterse *et al.*, 2014). The biosynthesis of these compounds is regulated by the plant's primary and secondary metabolism, as well as by specific hormone-triggered signaling cascades (salicylic acid and jasmonic acid) (Mithofer and Boland, 2012).

13.2.3 Programmed cell death

In an aggressive immune response like programmed cell death (PCD), the plant kills its own cells (infective, compromised) to prevent the spread of pathogens across the entire plant body (Jones and Dangl, 2006). PCD is initiated by the recognition of pathogen-associated molecular patterns (PAMPs) or effectors, leading to the activation of either PAMP-triggered immunity (PTI) or effector-triggered immunity (ETI). These molecules are recognized by specialized cell surface receptors, including receptor-like kinases (RLKs) and receptor-like proteins (RLPs), for downstream signaling pathway activation. These signaling events trigger various defense responses, such as activation of mitogen-activated protein kinase (MAPK), generation of reactive oxygen species (ROS), calcium influx, and inhibition of pathogen growth by the induction of PCD (Lei *et al.*, 2023; Tian *et al.*, 2024).

13.2.4 Hormonal signaling

Plants have developed various defense mechanisms against biotic stresses. Phytohormones, such as salicylic acid (SA), jasmonic acid (JA), brassinosteroids (BR), and ethylene (ET), play important roles in regulating plant defenses against biotic stresses. Plant hormones interact with diverse molecular responses affecting the upregulation or downregulation of defense-responsive gene expression, thereby regulating between plant growth and development and/or activation of PCD (Robert-Seilaniantz *et al.*, 2007). In this context, the defense responses vary per the hormonal moiety. The SA signaling pathway primarily confers resistance against biotrophic pathogens – that is, fungi – while the JA and ET pathways are primarily responsible for providing resistance against necrotrophic pathogens, including bacteria, fungi, and oomycetes, as well as herbivores (Barna *et al.*, 2012; Tian *et al.*, 2024). These hormone-signaling pathways can cross talk and synergize, leading to a comprehensive defense strategy against biotic stresses. The molecular interventions in the complex cross talks between genes and hormone-signaling pathways could be a more effective strategy to enhance crop resistance to several biotic stresses (Li *et al.*, 2019).

13.3 Traditional Methods of Developing Biotic Stress Resistance in Crops

These methods comprise conventional breeding, genetic engineering, and marker-assisted selection (MAS) (Dixit *et al.*, 2020). Traditional breeding techniques entail the deliberate mating and selection of plants with desirable characteristics to create disease-resistant cultivars. Nevertheless, this approach is frequently plagued by unpredictability and prolonged timelines (Mbinda and Masaki, 2021).

Genetic engineering, which involves the insertion of genes from other organisms into a plant's genome, has been used to develop crops with resistance to pests and diseases (Zhang *et al.*, 2017a). For example, the insertion of the *Bt* toxin gene into crops such as cotton and maize has resulted in resistance to pests such as the cotton bollworm and corn borer (James, 2012). However, this method has faced criticism over concerns about the safety and environmental impact of genetically modified organisms (GMOs) (Zhang *et al.*, 2017a).

Marker-assisted selection involves the identification of genetic markers that are associated with desired traits and the use of these markers to select plants with those traits (Dixit *et al.*, 2020). This method has been successful in developing crops with disease resistance, such as rice resistant to both abiotic and biotic stresses (Das *et al.*, 2017) and wheat resistant to powdery mildew (Li *et al.*, 2022a). However, MAS requires prior knowledge of the genetic basis of the trait, which can be limiting for traits with complex genetic interactions. While traditional methods of developing biotic stress resistance in crops have been successful, they have limitations and challenges. Genome editing techniques such as the CRISPR/Cas systems provide a promising alternative for developing crops with improved resistance to biotic stressors.

13.4 GE Techniques to Develop Biotic Stress Resistance

GE techniques have been developed based on their ability to effectively manipulate the target DNA sequence and ease their construction and delivery into the organism. GE techniques such as ZFNs and TALENs, CRISPR/Cas9 system, prime editing (PE), and base editing (BE) systems have been in use for effective manipulation of host plant DNA (Gaj *et al.*, 2016). The two ZFNs recognize zinc finger binding sites (triplet code) on the two strands of DNA in opposite directions separated by 5–7 nt spacer, and the FokI nuclease domain causes a double-stranded break (DSB). Just like ZFNs, the TALENs have DNA-binding domains, TALE, which recognize the single nucleotide followed by cleavage with FokI (Kim and Kim, 2014). The broad adoption of these nucleases for routine use is, however, constrained by challenges with protein synthesis, design, and confirmation required for TALENs and ZFNs. The CRISPR/Cas9 system works by using a guide RNA to target a specific DNA sequence in the plant genome, and the Cas9 enzyme cuts the DNA at that location. The plant then repairs the cut using either non-homologous end-joining (NHEJ), which results in a deletion or insertion of DNA sequences, or homology-directed repair (HDR), which can be used to replace one gene with another. TALENs and ZFNs have lower off-target effects than CRISPR/Cas but are more complex and expensive to design and construct (Ahn *et al.*, 2020; Negi *et al.*, 2023). Recently, the CRISPR/Cas9, PE, and BE systems have been gaining significant attention to modify plants' genomes due to their simplicity, specificity, and versatility in plant GE. GE offers a faster and more efficient way to develop biotic stress-resistant crops (Fig. 13.1).

13.5 Techniques for Genome Editing

13.5.1 Zinc-finger nucleases for plant breeding

ZFNs for eukaryotic site-specific GE became available in 2002, although they were discovered in the 1990s. ZFNs can target longer DNA sequences than the CRISPR/Cas system, which can be useful for modifying large genomic regions. The chimeric ZFN has two DNA-binding ZFs (specific recognition of three base pairs) coupled with an endonuclease FokI, effectively creating DSBs at the specific recognition site. This targeted genetic editing method has been extensively used for genetic editing in plants ever since Bibikova *et al.* (2002) discovered it in *Drosophila melanogaster*. Over time, this method has become less desirable due to the intricacy

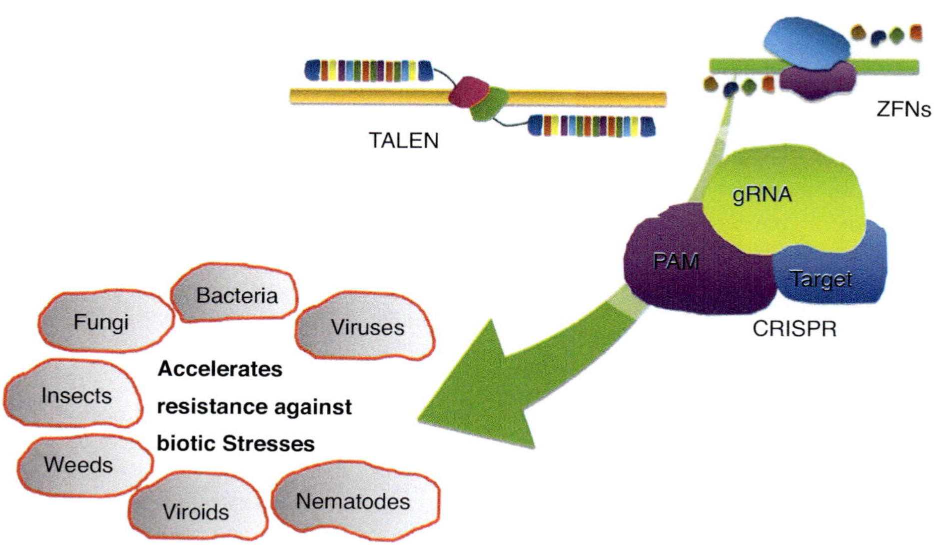

Fig. 13.1. GE approaches to develop biotic stress-resistant plants. Author's own figure.

of ZFN design and the scarce variety of ZFNs as well as limited target sequences (Negi *et al.*, 2023). The introduction of DSBs in DNA initiates the DNA repair mechanism of the cell, and hence, the target of gene editing is fulfilled (Qi *et al.*, 2023).

Targeted integration of transgenes was possible in soybean through the use of ZFNs via NHEJ and production of fertile transgenic plants (Bonawitz *et al.*, 2019). ZFNs are also widely explored for genetic modification in tobacco (Townsend *et al.*, 2009) and *Arabidopsis* (Osakabe *et al.*, 2010). Shukla *et al.* (2009) targeted the inositol phosphate pathway genes *IPK1* and *IPK2*, which are involved in herbicide tolerance and cell growth and development in maize. The study demonstrated that the ZFN-induced genetic modifications were heritable and stable across generations. Other examples include the use of ZFNs to impart viral resistance in tomato (Takenaka *et al.*, 2007), imidazolinone and sulfonylurea resistance in tobacco by editing the acetolactate synthase gene (Townsend *et al.*, 2009), and targeted mutations in the protoporphyrinogen oxidase gene of *Arabidopsis thaliana* (de Pater *et al.*, 2013).

These findings suggest that ZFN technology could be a valuable tool for improving crop traits in different crop species.

13.5.2 TALENs as a useful tool for plant breeding

The TALE nuclease recognizes a single-nucleotide target, making it highly specific for DSBs and genetic editing. TALEN comprises a FokI domain and a DNA-binding domain (Vats *et al.*, 2019; Das *et al.*, 2023). The DNA-binding domain contains a central repeat domain, and the recognition and binding of DNA segments that establish recognition specificity are carried out by 34 highly conserved amino acids (the 12th and 13th amino acids are highly variable). This powerful GE system became available in 2009 after its discovery from phytopathogenic bacteria *Xanthomonas* (Nemudryi *et al.*, 2014).

TALEN has been used to create plants resistant to bacteria such as *Xanthomonas oryzae*. TALEN-mediated deletion of a particular promoter region stopped the expression of the susceptibility gene *Os11N3*, giving resilience to the bacterial leaf blight infection of rice (Li *et al.*, 2012). Similarly, the powdery mildew susceptibility gene *Mlo* knockout through TALEN led to disease resilience in *Triticum aestivum* (Wang *et al.*, 2014). The use of TALEN-based gene editing has also been reported in other crop plants such as potato and soybean (Iqbal *et al.*, 2020). TALENs and ZFNs both work by inducing DSBs in DNA at specific genomic locations, which are then repaired by the cell's endogenous damage repair mechanisms, resulting in the generation of specific mutations or deletions.

13.5.3 CRISPR/Cas system and its applications in plant breeding

CRISPR/Cas9 is a powerful GE tool that has been widely applied in the field of plant breeding since its discovery in 2012 (Doudna and Charpentier, 2014). The CRISPR/Cas system functions using RNA-guided endonucleases (Cas9, cpf1) that can specifically target and cleave DNA sequences. A ribonucleoprotein (RNP) complex is produced when a Cas9 enzyme and a guide RNA (gRNA) interact. It is crucial for the Cas9-gRNA to attach to the target site and recognize the 5′-NGG-3′ protospacer adjacent motif (PAM) site. Protospacer refers to the target DNA sequence, and spacer refers to the portion of gRNA that attaches to the target DNA sequence. Once the target sequence is properly bound, the RuvC and HNH nuclease domains cleave it after the third nucleotide base upstream of the PAM. This targeted DNA cleavage can result in the generation of specific mutations, deletions, or insertions. In plants, the CRISPR/Cas system has been used for various purposes such as the improvement of crop yield, modification of plant architecture, and the development of disease-resistant plants (Das *et al.*, 2023).

13.5.4 Base editors for plant genome editing

Precise GE through the CRISPR/Cas9 system is very complex in plants and results in off-target effects. Nevertheless, the advanced BE technique, based on the CRISPR/Cas9 system, can do precise editing of a single base at the targeted genome locus without inducing DSBs. The BE system holds substantial promise for facilitating gene correction and enhancing genetic diversity in a range of organisms, including yeast, plants, mammalian, and human cells (Xie *et al.*, 2022). In the BE system, the catalytically inactive Cas9, dead Cas9, or Cas9 nickase is fused with either adenosine or cytosine deaminase. This fusion initiates a targeted nick at the genomic locus, facilitating precise GE without inducing unintended deletions or insertions (Mishra *et al.*, 2020). The precise base transitions from C→T, G→A, A→G, and T→C are all possible through the cytosine base editor (CBE) and adenosine base editors (ABE), respectively. Both CBE and ABE have found extensive application in plant systems, serving dual roles in gene functional annotation and precise gene modification. Their widespread adoption is attributed to their inherent simplicity and high efficacy and no off-target effects. Precisely, the CBE is created by a fusion of cytosine deaminase with Cas9 nickase, uracil glycosylase inhibitor, and a nuclear localization signal (Lam *et al.*, 2023). The CBE-mediated C→T transitions have been successfully reported in various plants with few or no off-target effects. The plants that have successfully been edited with CBE include *A. thaliana*, *Solanum lycopersicum*, *Solanum tuberosum*, *Oryza sativa*, *T. aestivum*, and *Brassica napus* (Xie *et al.*, 2022). Similarly, ABEs are constructed by fusing tRNA adenosine deaminase with n/dCas9. ABEs have also been successfully used in plants like cotton, rice, tomato, watermelon, strawberry, and oilseed rape (Zong *et al.*, 2022; Wang *et al.*, 2022a; Li *et al.*, 2023; Li *et al.*, 2024).

13.5.5 Prime editor for crop improvement

Another more advanced GE system for crop improvement is the PE system, developed recently for plant genome editing and successfully for wheat and rice in 2020 (Lin *et al.*, 2020). This precise PE system is capable of transversions, deletions, or insertions, which BEs are unable to perform, without the requirement of donor DNA. The PE consists of a prime editor, reverse transcriptase enzyme fused with nCas9, and prime editing guide RNA (pegRNA). The pegRNA component is of particular importance, as it provides the sgRNA and primer binding site for the reverse transcriptase enzyme. The PE, upon finding the target PAM site, nicks the DNA, and the primer binding site of the PE complex hybridizes at the 3′ end of the nicked DNA. The reverse transcriptase then completes the edit using the template of the pegRNA, and the repaired single-strand DNA is integrated through the DNA repair mechanism (Lin *et al.*, 2020; Xie *et al.*, 2022). There have been many improvements in the PE system since its introduction for improved editing efficiency and multiplexing. According to the investigations by Anzalone *et al.* (2019), the PE system not only performs all types of transition (T→C, C→T, G→A, A→G) and transversion mutations (A→C, A→T, T→G, T→A, G→T, G→C, C→A, and C→G) but also up to 80 bp deletions and 40 bp insertions (Anzalone *et al.*, 2019).

There have been rapid developments to improve the efficiency of prime editing in plants via engineered pegRNA (epegRNA), engineered plant prime editor (ePPE), codon-optimized RT (PEmax), or PE complex fused with a mismatch repair protein (MLH1dn) (in PE4 and PE5 systems). The improved versions of PE systems have thus enabled us to edit the plant genome up to an editing frequency of 78%. (Chen *et al.*, 2021; Zong *et al.*, 2022).

13.6 Applications of Genome Editing in Biotic Stress Resistance

One of the primary applications of GE in biotic stress resistance is developing crops with disease resistance. Plant diseases caused by fungi, bacteria, viruses (Table 13.1), and other pathogens (Table 13.2) are a significant threat to crop production worldwide. Traditional breeding methods for developing disease-resistant crops are often limited by the availability of resistant germplasm and the complexity of plant-pathogen interactions. By modifying genes implicated in plant-pathogen interactions, GE can be used to create crops that are resistant to disease (Boubakri, 2023). Sugar transport genes, pathogen susceptibility genes, and negative regulators of pathogenicity are key gene targets for GE in various crops, such as tomato, potato, and rice, among others, with the aim of introducing resistance against various pathogens.

13.6.1 Fungal resistance

Magnaporthe oryzae is a devastating rice pathogen that causes rice blast disease. Efficient development of rice varieties resistant to *M. oryzae* can be achieved via CRISPR/Cas9 technology targeting various pathogen-interacting pathway genes, such as the polyketide synthase gene (*ALB1*), transcription factors like ethylene responsive factor (*OsERF922* and *OsERF104*), and the heat shock protein HSP70's co-chaperone J-protein (OsDjA2) (Wang *et al.*, 2016; Foster *et al.*, 2018; Távora *et al.*, 2022). The Downy mildew resistance 6 (*DMR6*) gene is implicated in pathogen-induced response, and its knockout results in the acquisition of broad-spectrum resistance against pathogens in *A. thaliana*. A CRISPR/Cas9-based knockdown of the orthologous gene in tomato (*SlDMR6*) has been shown to confer resistance against *Xanthomonas spp.*, *Phytophthora capsici*, and *Pseudomonas syringae* in this crop (Paula de Toledo Thomazella *et al.*, 2016). Similarly, targeted knockouts of the *StCHL1* and *StDMR6* susceptibility genes through CRISPR/Cas9 have been demonstrated to confer resistance against *Phytophthora infestans* in potato (Kieu *et al.*, 2021). TALEN-mediated mildew resistance in wheat was developed by targeting the *TaMlo* gene (Wang *et al.*, 2014).

13.6.2 Bacterial resistance

According to estimates, 32 of the 40 species of *X. oryzae* activate the *Os11N3* gene and divert the plant's sugar transport to the pathogen's benefit (Li *et al.*, 2012). TALEN-based interruption of the susceptibility gene *Os11N3* could be used to manage the bacterial blight disease of rice caused by *X. oryze*. The CRISPR/Cas9 system has been used to target various genes involved in plant defense/immune responses against pathogens. Plant cell surface receptors (pattern recognition receptors, PRRs), on recognition of PAMPs, initiate the first layer of defense against pathogens, termed PTI. The power of gene editing lies in specifically targeting these PAMPs or plant immunoreceptors. For instance, the flagellin receptor *FLS2* (flagellin-sensing-2) of *A. thaliana* recognizes the flg-22 (conserved flagellin epitope) of *Escherichia coli*. *FLS2* homolog in tomato recognizes the flg-15 from *E. coli*, and *FLS2XL* of the wild grape species *Vitis riparia* recognizes a highly divergent flagellin epitope flg-22 (not detected by *FLS2* of other plant species) of *Agrobacterium tumefaciens*, thereby protecting it from the devastating crown gall disease (Fürst *et al.*, 2020; Wu *et al.*, 2022). The CRISPR/Cas9 system has been utilized to knock out the *FLS2* gene in several plant species, such as *Brachypodium distachyon*, *Nicotiana benthamiana*, and *S. lycopersicum*. By mutating *FLS2*, these plants are no longer capable of recognizing flagellin and initiating a defense response against bacterial pathogens, resulting in increased resistance to certain bacterial infections (Zhu *et al.*, 2020; Wang and Han, 2022; Wu *et al.*, 2022). CRISPR/Cas9-mediated mutation in *Fir-1* (*FLS2*/*FLS3*-interacting receptor-like cytoplasmic kinase-1) induced resistance to *P. syringae* in tomato (Sobol *et al.*, 2022). *P. syringae* pv *tomato* (Pto), DC3000, once localized into the tomato apoplast through wounds or stomata, triggers the stomatal reopening by releasing a chemical coronatine (COR). The tomato co-receptor for COR, *SlJAZ2* (Jasmonate ZIM domain), which is localized in guard cells, when mutated by CRISPR/Cas9 to develop *SlJAZ2*Δjas (repressor lacking C-terminal jas domain), no longer initiates the coronatine-triggered response and prevents stomatal opening, thereby preventing the development of bacterial speck disease (Ortigosa *et al.*, 2019).

13.6.3 Viral resistance

Viral resistance in crop plants is a critical factor in agricultural productivity and food security. Plant viruses can cause significant yield losses and reduce the quality of crops. Traditional breeding methods for viral resistance have been slow and labor intensive, but recent advancements in GE technologies have enabled the development of precise and efficient methods to enhance viral resistance in crop plants. ZFN-edited tomato, tobacco, and *Arabidopsis* plants resistant to viruses were developed by targeting the conserved sequence motif of begomoviruses having multiple resistance against tobacco curly shoot virus (TbCSV), tomato yellow leaf curl virus (TYLCV), and tomato yellow leaf curl China virus (TYLCCNV) (Takenaka *et al.*, 2007; Chen *et al.*, 2014).

Viral resistance through CRISPR/Cas9 involves targeting the genes involved in virus entry, replication, and spread within the plant. Research conducted by Aman *et al.* (2018) employed CRISPR/Cas13 to aim at the RNA turnip mosaic virus (TuMV) via the suppression of the helper component proteinase (HC-Pro) (Aman *et al.*, 2018). CRISPR/Cas9 was utilized in a separate research to target the eukaryotic translation initiation factor (eIF4E). In the loss-of-function quadruple mutants (eIF4E1-S, eIF4E1-T, eIF4E2-S, and eIF4E2-T), heritable resistance to potato virus Y was observed in tobacco (Li *et al.*, 2022a). In an effort to combat tomato brown rugose fruit virus (ToBRFV), scientists utilized CRISPR/Cas9 to target four tomato homologs of TOBAMOVIRUS MULTIPLICATION1 (TOM1), conferring resistance against ToBRFV in tomato plants. The quadruple-mutant plants displayed no detectable ToBRFV coat protein (CP) accumulation and demonstrated resistance to other tobamovirus species as well. The study indicates that genome-edited plants could be instrumental in developing ToBRFV-resistant tomato crops (Ishikawa *et al.*, 2022). CRISPR/Cas9-based targeting of large intergenic regions (LIR) and replication regions of CpCDV (chickpea chlorotic dwarf virus) successfully reduced the virus titer (Malik *et al.*, 2022). The CRISPR/Cas system could be efficiently utilized to control gene expression at RNA level instead of DNA. One example is the southern rice black-streaked dwarf virus, which is a devastating pest of rice. CRISPR/Cas13a was successfully implicated to induce resistance in monocotyledonous rice plant by targeting and degrading viral RNA (Zhang *et al.*, 2019). Host plant-pathogen susceptibility genes are also potential target sites for CRISPR/Cas9. CRISPR/Cas9 targeting *SlPelo* (mRNA surveillance factor) and *SlMlo1* (Mildew resistance locus1) susceptibility genes in tomato resulted in suppressed TYLCV disease and powdery mildew fungus accumulation. Results demonstrated the efficiency of CRISPR/Cas9 for rapid development of pathogen-resistant tomato varieties (Pramanik *et al.*, 2021). These studies demonstrate the potential of CRISPR/Cas9 technology to provide resistance against viral infections in crop plants, which could have significant implications for global food security.

The CRISPR/Cas system can effectively target and cleave various viral pathogenic regions, leading to reduced viral infection and increased resistance in plants. For example, in tomato, targeting the viral CP gene of TYLCV using CRISPR/Cas9 resulted in virus-resistant plants (Wang *et al.*, 2022b). In another approach, the CRISPR/Cas9 system produced wheat dwarf virus-resistant rice, whereas the HDR produced 19.4% knock-in efficiency (Wang *et al.*, 2017).

13.6.4 Insects and nematode resistance

Another application of GE in biotic stress resistance is developing crops with pest resistance. Insect pests are a significant threat to crop production, resulting in 20–40% loss to agriculturally important crops, causing both direct damage and transmitting plant diseases. Genome editing is used to develop designer crops by modifying genes involved in insect-plant interactions or by editing insect genomes (Moon *et al.*, 2022). The genomes of both a model insect, *D. melanogaster*, and major phytophagous insect genera—namely, *Spodoptera*, *Helicoverpa*, *Nilaparvata*, *Locusta*, *Tribolium*, *Agrotis*—were successfully edited using CRISPR/Cas tool kits (Moon *et al.*, 2022). In a study, CYP71A1 (an enzyme that converts tryptamine to serotonin in rice) was knocked out using CRISPR/Cas9, resulting in reduced growth of brown planthopper and striped stem borer pests in rice due to their inability to produce sufficient serotonin, essential for larval growth (Lu *et al.*, 2018). Another study targeted the olfactory receptors of insects. By knocking out the *Orco* (odor receptor co-receptor) gene in *Spodoptera littoralis* using CRISPR/Cas9 technology, impaired odor production and sex pheromones in homozygous individuals were observed (Koutrumpa *et al.*, 2016). CRISPR/Cas9 knockout of the *stylin-01* (cuticular protein)

Table 13.1. GE techniques to develop viral, bacterial, and fungal resistance. Author's own table.

Target plant	Targeted gene	Resistance against	Reference
Zinc-finger nucleases			
Soybean	*DCL4a*, *DCL4b*	Targeted mutagenesis	Curtin *et al.*, 2011
Tomato	TYLCV rep protein	Virus	Takenaka *et al.*, 2007
Arabidopsis	Protoporphyrinogen oxidase (*ppo*), ABA-Insensitive4 (*ABI4*), Replication protein of beet severe curly top virus	Herbicide Abscisic acid Virus	de Pater *et al.*, 2013; Osakabe *et al.*, 2010; Sera, 2005
Tobacco	*AC1* (Rep protein of TYLCCNV), Acetolactate synthase (*ALS SuRA* and *SurB*)	Virus Herbicide	Chen *et al.*, 2014; Townsend *et al.*, 2009
Maize	Inositol phosphate (*ipk1*), Phosphinothricin acetyltransferase(*pat*), Aryloxyalkanoate dioxygenase (*aad1*) *Cry34*, *Cry35*, *aad1*	Herbicide Insecticide and herbicide	Shukla *et al.*, 2009; Ainley *et al.*, 2013; Kumar *et al.*, 2015
TALENs			
Rice	SWEET sucrose-efflux transporter family protein (*Os11N3* or *OsSWEET11*), Transcription factor (*TFIIAy5*)	Bacterial	Li *et al.*, 2012; Jin *et al.*, 2020
Wheat	Mildew resistance locus (*TaMlo*)	Fungal	Wang *et al.*, 2014
Tobacco	*AC*1 of tobacco curly shoot virus	Virus	
CRISPR/Cas9			
Tobacco	*FLS2*	Bacterial	Fürst *et al.*, 2020; Wu *et al.*, 2022
Rice	*Os8N3* or *OsSWEET11*, *TFIIAy5* and *Xa23*, Scytalone dehydratase (*SDH*), Polyketide synthase gene (*ALB1*), Ethylene responsive factor (*OsERF922* and *OsERF104*), Co-chaperone J-protein (*OsDjA2*), BSRD1 (Transcription factor), Conserved region of eIF4	Bacteria Bacteria Fungi Fungi Virus	Kim *et al.*, 2019; Gupta *et al.*, 2023; Arazoe *et al.*, 2015; Foster *et al.*, 2018; Wang *et al.*, 2016; Távora *et al.*, 2022; Zhu *et al.*, 2020; Macovei *et al.*, 2018
Arabidopsis	Flagellin-sensing-2(*FLS2*), *AtEDR1*	Bacteria Powdery mildew	Wu *et al.*, 2022; Zhang *et al.*, 2017c
Wheat	Enhanced disease resistance (*TaEDR1*)	Powdery mildew	Zhang *et al.*, 2017c
Potato	*StCHL1* and *StDMR6*, eLF4E	Fungi Virus	Kieu *et al.*, 2021; Li *et al.*, 2022a

Continued

Table 13.1. Continued

Target plant	Targeted gene	Resistance against	Reference
Tomato	*SlDMR6*, *SlMlo1*, *SlPelo*, *Fir-1*, *SlJAZ2*, *TOM1* Large intergenic repeats (LIR) Virus coat protein	Fungi Fungi Virus Virus Virus Virus Virus Virus	Paula de Toledo Thomazella *et al.*, 2016; Pramanik *et al.*, 2021; Sobol *et al.*, 2023; Ortigosa *et al.*, 2019; Ishikawa *et al.*, 2022; Malik *et al.*, 2022; Wang *et al.*, 2022b
Grapefruit	*CsLOB1* (Citrus sinensis lateral organ boundaries)	Bacteria	Jia *et al.*, 2016
Grape vine	*TAS4* (Trans-acting small-interfering locus4) and *MYBA7*	Fungi	Sunitha and Rock, 2020

Table 13.2. GE techniques to develop insect and nematode resistance. Author's own table.

GE technique	Targeted gene	Targeted insect	Reference
CRISPR/Cas9	*Orco* (Odor receptor co-receptor)	*Spodoptera littoralis*	Koutrumpa *et al.*, 2016
	Stylin-01 (cuticular protein)	*Acyrthosiphon pisum*	Le Trionnaire *et al.*, 2019
	DpWnt-1 (cell proliferation)	*Dendrolimus punctatus*	Liu *et al.*, 2017
	HAP (HAIRPLUS)	*Helicoverpa armigera*	Fonseca *et al.*, 2022
	MYB (Transcription factor)	*Spodoptera litura* and *H. armigera*	Misra *et al.*, 2010
	GmCDPK38 (Calcium-dependent protein kinase)	*S. litura*	Li *et al.*, 2022b
	UDP-glycosyltransferase	*H. armigera*	Zhang *et al.*, 2022
	CYP9A186 gene, a-6-nicotinic acetylcholine receptor (nAchR), and P-glycoprotein	Insecticide	Zuo *et al.*, 2018
	SfABCC2	*Spodoptera frugiperda*	Jin *et al.*, 2021
	ABC transporter	*S. frugiperda, H. armigera, Helicoverpa punctigera*	Jin *et al.*, 2020; Wang *et al.*, 2017
	FAR (Fatty acyl-CoA reductase)	*Chilo suppressalis*	Sun *et al.*, 2022
	Tryptamine 5-hydroxylase	*Nilaparvata lugens*	Lu *et al.*, 2018
	W112Rhg1 (Wound-inducing protein)	Soybean cyst nematode	Dong and Hudson, 2022
	ARF8 (Auxin responsive Factor)	Root-knot nematode	Noureddine *et al.*, 2023
	OsHPP04 (Metal-associated plant protein 04)	*Meloidogyne graminicola*	Huang *et al.*, 2023
TALEN	*Orco*	*Ostrinia furnacalis*	Yang *et al.*, 2016
	BmABCB1 (ABC transporter)	*Bombyx mori*	Iwabuchi *et al.*, 2023

affects the virus-transmitting ability of the pea aphid *Acyrthosiphon pisum* (Le Trionnaire *et al.*, 2019). CRISPR/Cas9-based knockout of *DpWnt-1* gene from *Dendrolimus punctatus* led to abnormal embryonic development and embryonic mortality (Liu *et al.*, 2017).

Indirect control from insect herbivory could be achieved through modification of physical barriers of a plant such as trichomes or plant color. The HAIRPLUS (HAP) gene is implicated in the glandular trichomes regulation. CRISPR/Cas9-mediated knockout of HAP increased the trichome density by severalfold, which ultimately developed herbivory resistance in tomato against *Helicoverpa armigera* (Fonseca *et al.*, 2022). Similarly, in another study, tobacco leaf was altered from green to red through MYB transcription factor modification, developing resistance against *Spodoptera litura* and *H. armigera* (Misra *et al.*, 2010). *S. litura* resistance was increased in soybean by knocking out a calcium-dependent protein kinase (GmCDPK38) through CRISPR/Cas9 (Li *et al.*, 2022b). An upstream regulatory gene of flavonoid biosynthetic pathway, UDP-glycosyltransferase (UGT), when mutated through CRISPR/Cas9, also imparts resistance to *H. armigera* in soybean by metabolic alteration of flavonoid and defense pathway-related genes (Zhang *et al.*, 2022). The use of plant-derived protease inhibitors such as lectins (Napoleão *et al.*, 2019), tannins, and flavonoids has been identified as a promising method for insect control (Sultana *et al.*, 2022). Based on this, it is plausible to explore the potential of CRISPR/Cas-based gene editing of these molecules *in planta* to further investigate the efficacy of these protease inhibitors. Elicitors, which are of insect origin, represent a promising class of target molecules for CRISPR/Cas-based GE systems that are aimed at developing crop resistance. These molecules, including fatty acid-amino acid conjugate (FACs), inceptins, B-glucosides, caelifirin, bruchin, and sex pheromones, have been suggested as potential targets for GE, as reported by Jones *et al.* (2021). The gene-drive concept using CRISPR/Cas9 has demonstrated its effectiveness in eradicating invasive pest species through a nearly 20-generation process that involves disturbing the sex chromosome and altering the balance of males and females, thereby completely eradicating the species (McFarlane *et al.*, 2017). Nonetheless, this approach raises significant ethical concerns owing to its spread in unintended populations and non-reversal into the wild-type organisms; therefore, an antagonistic system must be first put into place before releasing this system into the environment (McFarlane *et al.*, 2017). Plant-derived specialized metabolites (phenolics, terpenoids, and alkaloids) are other potential targets to achieve insect resistance in crops (Kortbeek *et al.*, 2019). The CRISPR/Cas9-mediated knockout of *Fatty acyl-CoA reductase* (*FAR*) gene in rice yielded more than 80% resistance to *Chilo suppressalis* (Sun *et al.*, 2022).

Plant parasitic nematodes (PPN) are a class of pathogen that invades plant roots, causing plant rot and resulting in significant crop damage to the host plant. Numerous genes that could potentially impede nematode infection by interfering with the plant-nematode interaction have been suggested (Abd-Elgawad, 2022). To comprehend its resistance mechanism against the soybean cyst nematode (SCN), researchers investigated a wound-inducible protein called WI12Rhg1. This protein is triggered by hormones from the defense signaling pathway: namely, JAs, SA, and gibberellic acid (GA). Using CRISPR/Cas9 technology to edit the soybean's hairy roots, knockout mutants with reduced levels of SA were produced, while GA levels increased (Dong and Hudson, 2022). While CRISPR/Cas9-mediated GE is well established in *Caenorhabditis elegans*, its application for developing nematode-resistant plants is currently limited, despite the abundance of information available on nematode gene targets (Ibrahim *et al.*, 2019; Petitot *et al.*, 2020; Abd-Elgawad, 2022). The SCN-resistant genes *Rhg1* and *Rhg4* were subjected to CRISPR/Cas9 gene editing, and it was found that SCN infectivity was significantly reduced by editing the serine hydroxymethyltransferase (*GmSHMT08*) gene located at the *Rhg4* locus (Kang, 2016). Recently, Huang *et al.* (2023) demonstrated the potential of CRISPR/Cas9 technology for creating PPN-resistant plant varieties. They modified the *OsHPP04S* (susceptibility) gene in rice, creating genetically stable homozygous mutants resistant to the rice root-knot nematode. Mutants lacking transgenic elements showed enhanced immune responses triggered by flg-22—that is, defense pathway genes activation, callose, and ROS burst—and suggesting *OsHPP04* negatively regulates host immunity (Huang *et al*., 2023).

13.6.5 Weed control

Weeds compete with crops for nutritional resources, causing significantly reduced crop quality and yield losses. The traditional weed control method, involving the spraying of herbicides, is often expensive, time consuming, and environmentally damaging. Herbicides target the key enzymes of different metabolic pathways of plants, ultimately leading to plant death. Some of the key enzyme targets for herbicides are acetolactate synthase (ALS) of branched-chain amino acid synthesis, acetyl-CoA carboxylase (*ACCase*) of fatty acid biosynthesis, 5-enolpyru vylshikimate-3-phosphate (*EPSPS*) for aromatic acid biosynthesis, photosystem I in chloroplasts, and so on (Dong *et al*., 2021). CRISPR/Cas9-mediated gene editing of ALS2 gene in *Zea mays* yielded chlorsulfuron-resistant maize plants (Svitashev *et al*., 2015). Chlorsulfuron and bispyribac-sodium resistant rice were generated through CRISPR/Cas9 base gene replacement in ALS gene (Sun *et al*., 2016). CRISPR/Cas-based precise gene editing through base editor technology is now revolutionizing the GE landscape. Wu *et al*. (2020) employed cytosine-mediated base editor (CBE) to create *BnALS* mutants in *B. napus*, which showed resistance against the herbicide tribenuron-methyl (Wu *et al*., 2020). A novel G→T transversion in *OsALS* at position 1882 was achieved through CRISPR/Cas9-based gene editing, showing strong resistance to herbicide and transgene-free rice plant development (Wang *et al*., 2021). CRISPR/Cas9-based editing of *OsALS* in Basmati rice, combined with HDR, also showed strong resistance against broad leaf herbicide bispyribac (Zafar *et al*., 2023). Multiple herbicide-resistant *A. thaliana* plants were generated through CRISPR/Cas9 BE system, and the homozygous plants were resistant against chlorsulfuron, bispyribac-sodium, and flucarbazone-sodium (Wei *et al*., 2022). CRISPR/Cas9-mediated gene editing of polyamine uptake transporter (*PUT*) genes (*OsPUT1/2/3*) successfully produced paraquat-resistant rice plants (Lyu *et al*., 2022). Liang *et al*. (2022) report on the successful development of herbicide-resistant mutant lines of *SiALS* and *SiACCase* genes in a protein-rich foxtail millet crop, using precise CRISPR/Cas9 BE systems, such as cytidine and adenosine base editors (CBE and ABE) (Zhang *et al*., 2017b). The findings of this study demonstrate the effectiveness of these advanced techniques in generating genetic modifications with 100% mutation efficiency and no off-target effects for improved productivity and herbicide resistance (Liang *et al*., 2022). In another study, Wang *et al*. (2022c) produced editing efficiency of 22% in ALS gene of *Brassica oleracea* var *botrytis* through modified CRISPR/Cas9 system BE3 (Wang *et al*., 2022c). Glyphosate-resistant plants like soybean, chili (Ortega *et al*., 2018), flax (Sauer *et al*., 2016), cassava (Hummel *et al*., 2018), and rice were successfully developed through the CRISPR/Cas9 system (Butt *et al*., 2020). The double amino acid substitutions *T102I* and *P106S* (TIPS) in goosegrass (Yu *et al*., 2015) and rice (Zhang *et al*., 2017a) showed strong resistance to glyphosate (Han and Kim, 2019). Strong glyphosate-resistant *O. sativa* plants were recently developed by three amino acid substitutions of EPSPS (5′-enolpyruvylshikimate-3-phosphate synthase) through CRISPR/Cas9 (Sony *et al*., 2023). Genome editing can also be used to develop crops with enhanced competitiveness against weeds. For example, researchers have used CRISPR/Cas9 to modify the gene encoding the *A. thaliana* transcription factor LATERAL ORGAN BOUNDARIES DOMAIN 16 (LBD16), resulting in plants with enhanced root growth and increased competitiveness against weeds (Lee *et al*., 2015).

13.7 Conclusion

GE has the potential to accelerate biotic stress resistance in crop plants by developing crops with enhanced disease resistance, pest resistance, weed control, and other types of biotic stress resistance. The possibility to precisely alter particular genes makes it possible to generate crops with targeted improvements, which decreases the need for chemical inputs like pesticides and fertilizers and shortens the time it takes to introduce new varieties. This increases the sustainability of agriculture. GE has become more and more popular due to the quicker genetic manipulation of crop plants compared to traditional breeding methods that require years to produce the desired features. Many GE tools described in this chapter have their pros and cons depending on various factors like intended target site, availability of resources and skills, and funding opportunities. For instance, TALENs and CRISPR/Cas9 both are designer nucleases; however, the TALENs have been declared as more effective than CRISPR/Cas9 when it comes to clinical applications and mitochondrial gene editing.

There are also challenges and limitations to the widespread adoption of GE in crop plants. One of the major challenges is the introduction of unintended mutation in crop plants (off-target effects), which could lead to harmful mutations. Second, edible plants produced through GE techniques currently fall under the transgenic or GMO category, which are undesirable across many countries due to the non-acceptance of GMOs for feed or food purposes. This regulatory process sometimes delays or hinders the release of GE varieties for commercial production.

Despite these challenges, GE has already shown promising results in developing crops with biotic stress resistance, and ongoing research is likely to address some of these limitations. Researchers are now developing transgene-free GE edible plants, which do not fall under the category of 'GMO' and hence are devoid of the GMO-regulatory approval process. The continued development and application of GE technology has the potential to revolutionize agriculture and contribute to global food security.

References

Abd-Elgawad, M.M. (2022) Understanding molecular plant–nematode interactions to develop alternative approaches for nematode control. *Plants* 11(16), 2141. DOI: 10.3390/plants11162141.

Ahn, C.H., Ramya, M., An, H.R., Park, P.M., Kim, Y.J. *et al.* (2020) Progress and challenges in the improvement of ornamental plants by genome editing. *Plants* 9(6), 687. DOI: 10.3390/plants9060687.

Ainley, W.M., Sastry-Dent, L., Welter, M.E., Murray, M.G., Zeitler, B. *et al.* (2013) Trait stacking via targeted genome editing. *Plant Biotechnology* 11, 1126–1134. DOI: 10.1111/pbi.12107.

Aman, R., Mahas, A., Butt, H., Ali, Z., Aljedaani, F. *et al.* (2018) Engineering RNA virus interference via the CRISPR/cas13 machinery in Arabidopsis. *Viruses* 10(12), 732. DOI: 10.3390/v10120732.

AnjaliKumar, S., Korra, T., Thakur, R. and Arutselvan, R. (2023) Role of plant secondary metabolites in defence and transcriptional regulation in response to biotic stress. *Plant Stress* 8, 100154. DOI: 10.1016/j.stress.2023.100154.

Anzalone, A.V., Randolph, P.B., Davis, J.R., Sousa, A.A., Koblan, L.W. *et al.* (2019) Search-and-replace genome editing without double-strand breaks or donor DNA. *Nature* 576(7785), 149–157. DOI: 10.1038/s41586-019-1711-4.

Arazoe, T., Ogawa, T., Miyoshi, K., Yamato, T., Ohsato, S. *et al.* (2015) Tailor-made TALEN system for highly efficient targeted gene replacement in the rice blast fungus. *Biotechnology and Bioengineering* 112(7), 1335–1342. DOI: 10.1002/bit.25559.

Barna, B., Fodor, J., Harrach, B.D., Pogány, M. and Király, Z. (2012) The Janus face of reactive oxygen species in resistance and susceptibility of plants to necrotrophic and biotrophic pathogens. *Plant Physiology and Biochemistry* 59, 37–43. DOI: 10.1016/j.plaphy.2012.01.014.

Bernard, A. and Joubès, J. (2013) Arabidopsis cuticular waxes: advances in synthesis, transport and regulation of wax biosynthesis. *Progress in Lipid Research* 52, 110–129. DOI: 10.1016/j.plipres.2012.10.002.

Bibikova, M., Golic, M., Golic, K.G. and Carroll, D. (2002) Targeted chromosomal cleavage and mutagenesis in Drosophila using zinc-finger nucleases. *Genetics* 161(3), 1169–1175. DOI: 10.1093/genetics/161.3.1169.

Bonawitz, N.D., Ainley, W.M., Itaya, A., Chennareddy, S.R., Cicak, T. *et al.* (2019) Zinc finger nuclease-mediated targeting of multiple transgenes to an endogenous soybean genomic locus via non-homologous end joining. *Plant Biotechnology Journal* 17(4), 750–761. DOI: 10.1111/pbi.13012.

Boubakri, H. (2023) Recent progress in CRISPR/cas9-based genome editing for enhancing plant disease resistance. *Gene* 866, 147334. DOI: 10.1016/j.gene.2023.147334.

Butt, H., Rao, G.S., Sedeek, K., Aman, R., Kamel, R. *et al.* (2020) Engineering herbicide resistance via prime editing in rice. *Plant Biotechnology Journal* 18, 2370–2372. DOI: 10.1111/pbi.13399.

Chen, P.J., Hussmann, J.A., Yan, J., Knipping, F., Ravisankar, P. *et al.* (2021) Enhanced prime editing systems by manipulating cellular determinants of editing outcomes. *Cell* 184(22), 5635–5652. DOI: 10.1016/j.cell.2021.09.018.

Chen, W., Qian, Y., Wu, X., Sun, Y., Wu, X. *et al.* (2014) Inhibiting replication of begomoviruses using artificial zinc finger nucleases that target viral-conserved nucleotide motif. *Virus Genes* 48, 494–501. DOI: 10.1007/s11262-014-1041-4.

Curtin, S.J., Zhang, F., Sander, J.D., Haun, W.J., Starker, C. *et al.* (2011) Targeted mutagenesis of duplicated genes in soybean with zinc-finger nucleases. *Plant Physiology* 156, 466–473. DOI: 10.1104/pp.111.172981.

Das, G., Patra, J.K. and Baek, K.H. (2017) Insight into MAS: a molecular tool for development of stress resistant and quality of rice through gene stacking. *Frontiers in Plant Science* 8, 985. DOI: 10.3389/fpls.2017.00985.

Das, N., Dhar, G.D. and Dhar, P. (2023) Editing the genome of common cereals (rice and wheat): techniques, applications, and industrial aspects. *Molecular Biology Reports* 50, 739–747. DOI: 10.1007/s11033-022-07664-y.

de Pater, S., Pinas, J.E., Hooykaas, P.J.J. and van der Zaal, B.J. (2013) ZFNmediated gene targeting of the arabidopsis protoporphyrinogen oxidase gene through Agrobacterium-mediated floral dip transformation. *Plant Biotechnology Journal* 11, 510–515.

Divekar, P.A., Narayana, S., Divekar, B.A., Kumar, R., Gadratagi, B.G. *et al.* (2022) Plant secondary metabolites as defense tools against herbivores for sustainable crop protection. *International Journal of Molecular Sciences* 23(5), 2690. DOI: 10.3390/ijms23052690.

Dixit, S., Singh, U.M., Singh, A.K., Alam, S., Venkateshwarlu, C. *et al.* (2020) Marker assisted forward breeding to combine multiple biotic-abiotic stress resistance/tolerance in rice. *Rice* 13, 19. DOI: 10.1186/s12284-020-00391-7.

Dong, H., Huang, Y. and Wang, K. (2021) The development of herbicide resistance crop plants using CRISPR/cas9-mediated gene editing. *Genes* 12(6), 912. DOI: 10.3390/genes12060912.

Dong, J. and Hudson, M.E. (2022) WI12 rhg1 interacts with DELLAs and mediates soybean cyst nematode resistance through hormone pathways. *Plant Biotechnology Journal* 20(2), 283–296. DOI: 10.1111/pbi.13709.

Doudna, J.A. and Charpentier, E. (2014) The new frontier of genome engineering with CRISPR-cas9. *Science* 346(6213), 1258096. DOI: 10.1126/science.1258096.

Fonseca, R., Capel, C., Yuste-Lisbona, F.J., Quispe, J.L., Gómez-Martín, C. *et al.* (2022) Functional characterization of the tomato HAIRPLUS gene reveals the implication of the epigenome in the control of glandular trichome formation. *Horticulture Research* 9, uhab015. DOI: 10.1093/hr/uhab015.

Foster, A.J., Martin-Urdiroz, M., Yan, X., Wright, H.S., Soanes, D.M. *et al.* (2018) CRISPR-Cas9 ribonucleoprotein-mediated co-editing and counterselection in the rice blast fungus. *Scientific Reports* 8, 14355. DOI: 10.1038/s41598-018-32702-w.

Fürst, U., Zeng, Y., Albert, M., Witte, A.K., Fliegmann, J. *et al.* (2020) Perception of *Agrobacterium tumefaciens* flagellin by FLS2XL confers resistance to crown gall disease. *Nature Plants* 6(1), 22–27. DOI: 10.1038/s41477-019-0578-6.

Gaj, T., Sirk, S.J., Shui, S.L. and Liu, J. (2016) Genome-editing technologies: principles and applications. *Cold Spring Harbor Perspectives in Biology* 8(12), 023754. DOI: 10.1101/cshperspect.a023754.

Gupta, A., Liu, B., Chen, Q.J. and Yang, B. (2023) High-efficiency prime editing enables new strategies for broad-spectrum resistance to bacterial blight of rice. *Plant Biotechnology Journal* 21(7), 1454–1464. DOI: 10.1111/pbi.14049.

Han, Y.J. and Kim, J.I. (2019) Application of CRISPR/cas9-mediated gene editing for the development of herbicide-resistant plants. *Plant Biotechnology Reports* 13, 447–457. DOI: 10.1007/s11816-019-00575-8.

Huang, Q., Lin, B., Cao, Y., Zhang, Y., Song, H. *et al.* (2023) CRISPR/Cas9-mediated mutagenesis of the susceptibility gene OsHPP04 in rice confers enhanced resistance to rice root-knot nematode. *Frontiers in Plant Science* 14, 1134653. DOI: 10.3389/fpls.2023.1134653.

Hummel, A.W., Chauhan, R.D., Cermak, T., Mutka, A.M., Vijayaraghavan, A. *et al.* (2018) Allele exchange at the EPSPS locus confers glyphosate tolerance in cassava. *Plant Biotechnology Journal* 16(7), 1275–1282. DOI: 10.1111/pbi.12868.

Ibrahim, H.M., Ahmad, E.M., Martínez-Medina, A. and Aly, M.A. (2019) Effective approaches to study the plant-root knot nematode interaction. *Plant Physiology and Biochemistry* 141, 332–342. DOI: 10.1016/j.plaphy.2019.06.009.

Iqbal, Z., Iqbal, M.S., Ahmad, A., Memon, A.G. and Ansari, M.I. (2020) New prospects on the horizon: genome editing to engineer plants for desirable traits. *Current Plant Biology* 24, 100171. DOI: 10.1016/j.cpb.2020.100171.

Ishikawa, M., Yoshida, T., Matsuyama, M., Kouzai, Y., Kano, A. *et al.* (2022) Tomato brown rugose fruit virus resistance generated by quadruple knockout of homologs of TOBAMOVIRUS MULTIPLICATION1 in tomato. *Plant Physiology* 189(2), 679–686. DOI: 10.1093/plphys/kiac103.

Iwabuchi, K., Miyamoto, K., Jouraku, A., Takasu, Y., Iizuka, T. *et al.* (2023) ABC transporter subfamily B1 as a susceptibility determinant of *Bombyx mori* larvae to Cry1Ba, Cry1Ia and Cry9Da toxins. *Insect Biochemistry and Molecular Biology* 163, 104030. DOI: 10.1016/j.ibmb.2023.104030.

James, C. (2012) ISAAA briefs. In: *Global Status of Commercialized Biotech/GM Crops*, Vol. 44. ISAAA.

Jia, H., Orbovic, V., Jones, J.B. and Wang, N. (2016) Modification of the ptha4 effector binding elements in type I cslob 1 promoter using cas9/sgrna to produce transgenic duncan grapefruit alleviating xccδptha4:dcslob 1.3 infection. *Plant Biotechnology Journal* 14(5), 1291–1301. DOI: 10.1111/pbi.12495.

Jin, M., Yang, Y., Shan, Y., Chakrabarty, S., Cheng, Y. *et al.* (2020) Two ABC transporters are differentially involved in the toxicity of two *Bacillus thuringiensis* Cry1 toxins to the invasive crop-pest *Spodoptera frugiperda* (JE Smith). *Pest Management Science* 77(3), 1492–1501.

Jin, M.H., Tao, J.H., Qi, L.I., Cheng, Y., Sun, X.X. *et al.* (2021) Genome editing of the SfABCC2 gene confers resistance to Cry1F toxin from *Bacillus thuringiensis* in *Spodoptera frugiperda*. *Journal of Integrative Agriculture* 20(3), 815–820. DOI: 10.1016/S2095-3119(19)62772-3.

Jones, A.C., Felton, G.W. and Tumlinson, J.H. (2021) The dual function of elicitors and effectors from insects: Reviewing the ‘arms race’ against plant defenses. *Plant Molecular Biology* 109, 427–445. DOI: 10.1007/s11103-021-01203-2.

Jones, J.D. and Dangl, J.L. (2006) The plant immune system. Nature 444(7117), 323–329. DOI: 10.1038/nature05286.

Kang, J. (2016) Application of CRISPR/Cas9-mediated genome editing for studying soybean resistance to soybean cyst nematode (Doctoral dissertation University of Missouri-Columbia). Available at: https://hdl.handle.net/10355/62395

Kaur, S., Samota, M.K., Choudhary, M., Choudhary, M., Pandey, A.K. *et al.* (2022) How do plants defend themselves against pathogens-biochemical mechanisms and genetic interventions. *Physiology and Molecular Biology of Plants* 28(2), 485–504. DOI: 10.1007/s12298-022-01146-y.

Kieu, N.P., Lenman, M., Wang, E.S., Petersen, B.L. and Andreasson, E. (2021) Mutations introduced in susceptibility genes through CRISPR/Cas9 genome editing confer increased late blight resistance in potatoes. *Scientific Reports* 11(1), 4487. DOI: 10.1038/s41598-021-83972-w.

Kim, H. and Kim, J.S. (2014) A guide to genome engineering with programmable nucleases. *Nature Reviews Genetics* 15(5), 321–334. DOI: 10.1038/nrg3686.

Kim, Y.A., Moon, H. and Park, C.J. (2019) CRISPR/cas9-targeted mutagenesis of Os8N3 in rice to confer resistance to *Xanthomonas oryzae* pv. *oryzae*. *Rice* 12(1), 67. DOI: 10.1186/s12284-019-0325-7.

Kortbeek, R.W., Gragt, M. and Bleeker, P.M. (2019) Endogenous plant metabolites against insects. *European Journal of Plant Pathology* 154, 67–90. DOI: 10.1007/s10658-018-1540-6.

Koutrumpa, F.A., Monsempes, C., François, M.C., de Cian, A., Royer, C. *et al.* (2016) Heritable genome editing with CRISPR/cas9 induces anosmia in a crop pest moth. *Scientific Reports* 6(1), 29620. DOI: 10.1038/srep29620.

Kumar, M., Campbell, L. and Turner, S. (2016) Secondary cell walls: biosynthesis and manipulation. *Journal of Experimental Botany* 67(2), 515–531. DOI: 10.1093/jxb/erv533.

Kumar, S., AlAbed, D., Worden, A., Novak, S., Wu, H. *et al.* (2015) A modular gene targeting system for sequential transgene stacking in plants. *Journal of Biotechnology* 207, 12–20. DOI: 10.1016/j.jbiotec.2015.04.006.

Lam, D.K., Feliciano, P.R., Arif, A., Bohnuud, T., Fernandez, T.P. *et al.* (2023) Improved cytosine base editors generated from TadA variants. *Nature Biotechnology* 41(5), 686–697. DOI: 10.1038/s41587-022-01611-9.

Le Trionnaire, G., Tanguy, S., Hudaverdian, S., Gléonnec, F., Richard, G. *et al.* (2019) An integrated protocol for targeted mutagenesis with CRISPR-Cas9 system in the pea aphid. *Insect Biochemistry and Molecular Biology* 110, 34–44. DOI: 10.1016/j.ibmb.2019.04.016.

Lee, H.W., Cho, C. and Kim, J. (2015) Lateral organ boundaries domain16 and 18 act downstream of the AUXIN1 and LIKE-AUXIN3 auxin influx carriers to control lateral root development in Arabidopsis. *Plant Physiology* 168(4), 1792–1806. DOI: 10.1104/pp.15.00578.

Lei, G., Zhou, KH., Chen, XJ., Huang, Y.Q., Yuan, X.J. *et al.* (2023) Transcriptome and metabolome analyses revealed the response mechanism of pepper roots to *Phytophthora capsici* infection. *BMC Genomics* 24, 626. DOI: 10.1186/s12864-023-09713-7.

Li, N., Han, X., Feng, D., Yuan, D. and Huang, L.J. (2019) Signaling crosstalk between Salicylic acid and ethylene/jasmonate in plant defense: do we understand what they are whispering. *International Journal of Molecular Sciences* 20(3), 671. DOI: 10.3390/ijms20030671.

Li, T., Liu, B., Spalding, M.H., Weeks, D.P. and Yang, B. (2012) High-efficiency TALEN-based gene editing produces disease-resistant rice. *Nature Biotechnology* 30(5), 390–392. DOI: 10.1038/nbt.2199.

Li, X., Hu, D., Cai, L., Wang, H., Liu, X. *et al.* (2022b) Calcium-dependent protein kinase38 regulates flowering time and common cutworm resistance in soybean. *Plant Physiology* 190(1), 480–499. DOI: 10.1093/plphys/kiac260.

Li, X., Xie, J., Dong, C., Zheng, Z., Shen, R. *et al.* (2024) Efficient and heritable A-to-K base editing in rice and tomato. *Horticulture Research* 11(1), uhad250. DOI: 10.1093/hr/uhad250.

Li, Y., Liang, J., Deng, B., Jiang, Y., Zhu, J. *et al.* (2023) Applications and prospects of CRISPR/Cas9-mediated base editing in plant breeding. *Current Issues in Molecular Biology* 45(2), 918–935. DOI: 10.3390/cimb45020059.

Li, Y., Wu, X., Zhang, Y. and Zhang, Q. (2022a) CRISPR/CAS genome editing improves abiotic and biotic stress tolerance of crops. *Frontiers in Genome Editing* 4, 987817. DOI: 10.3389/fgeed.2022.987817.

Liang, Z., Wu, Y., Ma, L., Guo, Y. and Ran, Y. (2022) Efficient genome editing in *Setaria italica* using CRISPR/Cas9 and base editors. *Frontiers in Plant Science* 12, 815946. DOI: 10.3389/fpls.2021.815946.

Lin, Q., Zong, Y., Xue, C., Wang, S., Jin, S. *et al.* (2020) Prime genome editing in rice and wheat. *Nature Biotechnology* 38(5), 582–585. DOI: 10.1038/s41587-020-0455-x.

Liu, H., Liu, Q., Zhou, X., Huang, Y. and Zhang, Z. (2017) Genome editing of Wnt-1, a gene associated with segmentation, via CRISPR/Cas9 in the pine caterpillar moth, *Dendrolimus punctatus*. *Frontiers in Physiology* 7, 666. DOI: 10.3389/fphys.2016.00666.

Lu, H.P., Luo, T., Fu, H.W., Wang, L., Tan, Y.Y. *et al.* (2018) Resistance of rice to insect pests mediated by suppression of serotonin biosynthesis. *Nature Plants* 4(6), 338–344. DOI: 10.1038/s41477-018-0152-7.

Lyu, Y.S., Cao, L.M., Huang, W.Q., Liu, J.X. and Lu, H.P. (2022) Disruption of three polyamine uptake transporter genes in rice by CRISPR/cas9 gene editing confers tolerance to herbicide paraquat. *aBIOTECH* 3(2), 140–145. DOI: 10.1007/s42994-022-00075-4.

Macovei, A., Sevilla, N.R., Cantos, C., Jonson, G.B., Slamet-Loedin, I. *et al.* (2018) Novel alleles of rice eIF4G generated by CRISPR/cas9-targeted mutagenesis confer resistance to rice tungro spherical virus. *Plant Biotechnology Journal* 16(11), 1918–1927. DOI: 10.1111/pbi.12927.

Makarova, K.S., Wolf, Y.I., Iranzo, J., Shmakov, S.A., Alkhnbashi, O.S. *et al.* (2020) Evolutionary classification of CRISPR–Cas systems: A burst of class 2 and derived variants. *Nature Reviews Microbiology* 18(2), 67–83. DOI: 10.1038/s41579-019-0299-x.

Malik, M.A.M., Haider, M.S., Zhai, Y., Khan, M.A.U. and Pappu, H.R. (2022) Towards developing resistance to chickpea chlorotic dwarf virus through CRISPR/Cas9-mediated gene editing using multiplexed grnas. *Journal of Plant Diseases and Protection* 130, 23–33. DOI: 10.1007/s41348-022-00677-6.

Mbinda, W. and Masaki, H. (2021) Breeding strategies and challenges in the improvement of blast disease resistance in finger millet. A current review. *Frontiers in Plant Science* 11, 602882. DOI: 10.3389/fpls.2020.602882.

McFarlane, G.R., Whitelaw, C.B.A. and Lillico, S.G. (2017) CRISPR-based gene drives for pest control. *Trends in Biotechnology* 36(2), 130–133. DOI: 10.1016/j.tibtech.2017.10.001.

Mishra, R., Joshi, R.K. and Zhao, K. (2020) Base editing in crops: current advances, limitations and future implications. *Plant Biotechnology Journal* 18, 20–31. DOI: 10.1111/pbi.13225.

Misra, P., Pandey, A., Tiwari, M., Chandrashekar, K., Sidhu, O.P. *et al.* (2010) Modulation of transcriptome and metabolome of tobacco by Arabidopsis transcription factor, AtMYB12, leads to insect resistance. *Plant Physiology* 152(4), 2258–2268. DOI: 10.1104/pp.109.150979.

Mithofer, A. and Boland, W. (2012) Plant defense against herbivores: chemical aspects. *Annual Reviews of Plant Biology* 63, 431–450. DOI: 10.1146/annurev-arplant-042110-103854.

Moon, T.T., Maliha, I.J., Khan, A.A.M., Chakraborty, M., Uddin, M.S. *et al.* (2022) CRISPR-Cas genome editing for insect pest stress management in crop plants. *Stresses* 2(4), 493–514. DOI: 10.3390/stresses2040034.

Mushtaq, M., Sakina, A., Wani, S.H., Shikari, A.B., Tripathi, P. *et al.* (2019) Harnessing genome editing techniques to engineer disease resistance in plants. *Frontiers in Plant Science* 10, 550. DOI: 10.3389/fpls.2019.00550.

Napoleão, T.H., Albuquerque, L.P., Santos, N.D., Nova, I.C., Lima, T.A. *et al.* (2019) Insect midgut structures and molecules as targets of plant-derived protease inhibitors and lectins. *Pest Management Science* 75(5), 1212–1222. DOI: 10.1002/ps.5233.

Negi, S., Imanishi, M., Hamori, M., Nakagawa, Y.K., Nomura, W. *et al.* (2023) The past, present, and future of artificial zinc finger proteins: design strategies and chemical and biological applications. *Journal of Biological Inorganic Chemistry* 28, 249–261. DOI: 10.1007/s00775-023-01991-6.

Nemudryi, A.A., Valetdinova, K.R., Medvedev, S.P. and Zakian, S.M. (2014) TALEN and CRISPR/CAS genome editing systems: tools of discovery. *Acta Naturae* 6(3), 19–40.

Noureddine, Y., Rocha, M., An, J., Médina, C., Mejias, J. *et al.* (2023) AUXIN RESPONSIVE FACTOR8 regulates development of the feeding site induced by root-knot nematodes in tomato. *Journal of Experimental Botany* 74(18), 5752–5766. DOI: 10.1093/jxb/erad208.

Ortega, J.L., Rajapakse, W., Bagga, S., Apodaca, K., Lucero, Y. *et al.* (2018) An intragenic approach to confer glyphosate resistance in chile (*Capsicum annuum*) by introducing an *in vitro* mutagenized chile EPSPS gene encoding for a glyphosate resistant EPSPS protein. *PLoS One* 13(4), e0194666. DOI: 10.1371/journal.pone.0194666.

Ortigosa, A., Gimenez-Ibanez, S., Leonhardt, N. and Solano, R. (2019) Design of a bacterial speck resistant tomato by CRISPR/Cas9-mediated editing of SlJAZ2. *Plant Biotechnology Journal* 17(3), 665–673. DOI: 10.1111/pbi.13006.

Osakabe, K., Osakabe, Y. and Toki, S. (2010) Site-directed mutagenesis in Arabidopsis using custom-designed zinc finger nucleases. In: *Proceedings of the National Academy of Sciences*, pp. 12034–12039 (Vol. 107). DOI: 10.1073/pnas.1000234107.

Paula de Toledo Thomazella, D., Brail, Q., Dahlbeck, D. and Staskawicz, B. (2016) CRISPR-Cas9 mediated mutagenesis of a DMR6 ortholog in tomato confers broad-spectrum disease resistance. *BioRxiv* 064824. DOI: 10.1101/064824.

Petitot, A.S., Dereeper, A., Da Silva, C., Guy, J. and Fernandez, D. (2020) Analyses of the root-knot nematode (*Meloidogyne graminicola*) transcriptome during host infection highlight specific gene expression profiling in resistant rice plants. *Pathogens* 9(8), 644. DOI: 10.3390/pathogens9080644.

Pieterse, C.M., Zamioudis, C., Berendsen, R.L., Weller, D.M., Van Wees, S.C. *et al.* (2014) Induced systemic resistance by beneficial microbes. *Annual Review of Phytopathology* 52(1), 347–375. DOI: 10.1002/9781118371848.ch4.

Pramanik, D., Shelake, R.M., Park, J., Kim, M.J., Hwang, I. *et al.* (2021) CRISPR/Cas9-mediated generation of pathogen-resistant tomato against tomato yellow leaf curl virus and powdery mildew. *International Journal of Molecular Sciences* 22(4), 1878. DOI: 10.3390/ijms22041878.

Qi, Q., Hu, B., Jiang, W., Wang, Y., Yan, J. *et al.* (2023) Advances in plant epigenome editing research and its application in plants. *International Journal of Molecular Sciences* 24(4), 3442. DOI: 10.3390/ijms24043442.

Robert-Seilaniantz, A., Navarro, L., Bari, R. and Jones, J.D. (2007) Pathological hormone imbalances. *Current Opinion in Plant Biology* 10(4), 372–379. DOI: 10.1016/j.pbi.2007.06.003.

Sauer, N.J., Narváez-Vásquez, J., Mozoruk, J., Miller, R.B., Warburg, Z.J. *et al.* (2016) Oligonucleotide-mediated genome editing provides precision and function to engineered nucleases and antibiotics in plants. *Plant Physiology* 170(4), 1917–1928. DOI: 10.1104/pp.15.01696.

Savary, S., Willocquet, L., Pethybridge, S.J., Esker, P., McRoberts, N. *et al.* (2019) The global burden of pathogens and pests on major food crops. *Nature Ecology & Evolution* 3(3), 430–439. DOI: 10.1038/s41559-018-0793-y.

Sera, T. (2005) Inhibition of virus DNA replication by artificial zinc finger proteins. *Journal of Virology* 79, 2614–2619. DOI: 10.1128/jvi.79.4.2614-2619.2005.

Shukla, V.K., Doyon, Y., Miller, J.C., DeLelver, R.C., Moehle, E.A. *et al.* (2009) Precise genome modification in the crop species *Zea mays* using zinc-finger nucleases. *Nature* 459, 437–441. DOI: 10.1038/nature07992.

Sobol, G., Majhi, B.B., Pasmanik-Chor, M., Zhang, N., Roberts, H.M. *et al.* (2022) Tomato receptor-like cytoplasmic kinase Fir1 is involved in flagellin signaling and preinvasion immunity. *Plant Physiology* 192, 565–581. DOI: 10.1093/plphys/kiac577.

Sobol, G., Majhi, B.B., Pasmanik-Chor, M., Zhang, N., Roberts, H.M. *et al.* (2023) Tomato receptor-like cytoplasmic kinase Fir1 is involved in flagellin signaling and preinvasion immunity. *Plant Physiology* 192(1), 565–581. DOI: 10.1093/plphys/kiac577.

Sony, S.K., Kaul, T., Motelb, K.F.A., Thangaraj, A., Bharti, J. *et al.* (2023) CRISPR/Cas9-mediated homology donor repair base editing confers glyphosate resistance to rice (*Oryza sativa* L.). *Frontiers in Plant Science* 14, 1122926. DOI: 10.3389/fpls.2023.1122926.

Sultana, M.S., Millwood, R.J., Mazarei, M. and Stewart, C.N. (2022) Proteinase inhibitors in legume herbivore defense: from natural to genetically engineered protectants. *Plant Cell Reports* 1–13. DOI: 10.1007/s00299-021-02800-7.

Sun, Y., Gong, Y., He, Q., Kuang, S., Gao, Q. *et al.* (2022) FAR knockout significantly inhibits *Chilo suppressalis* survival and transgene expression of double-stranded FAR in rice exhibits strong pest resistance. *Plant Biotechnology Journal* 20(12), 2272–2283. DOI: 10.1111/pbi.13906.

Sun, Y., Li, J. and Xia, L. (2016) Precise genome modification via sequence-specific nucleases-mediated gene targeting for crop improvement. *Frontiers in Plant Science* 7, 1928. DOI: 10.3389/fpls.2016.01928.

Sunitha, S. and Rock, C.D. (2020) CRISPR/Cas9-mediated targeted mutagenesis of TAS4 and MYBA7 Loci in grapevine rootstock 101-14. *Transgenic Research* 29(3), 355–367. DOI: 10.1007/s11248-020-00196-w.

Svitashev, S., Young, J.K., Schwartz, C., Gao, H., Falco, S.C. *et al.* (2015) Targeted mutagenesis, precise gene editing, and site-specific gene insertion in maize using Cas9 and guide RNA. *Plant Physiology* 169(2), 931–945. DOI: 10.1104/pp.15.00793.

Takenaka, K., Koshino-Kimura, Y., Aoyama, Y. and Sera, T. (2007) Inhibition of tomato yellow leaf curl virus replication by artificial zinc-finger proteins. *Nucleic Acids Symposium Series* 51(1), 429–430.

Távora, F.T., Meunier, A.C., Vernet, A., Portefaix, M., Milazzo, J. *et al.* (2022) CRISPR/Cas9-targeted knockout of rice susceptibility genes OsDjA2 and OsERF104 reveals alternative sources of resistance to *Pyricularia oryzae*. *Rice Science* 29(6), 535–544. DOI: 10.1016/j.rsci.2022.04.001.

Tian, J., Zhang, J. and Francis, F. (2024) The role and pathway of VQ family in plant growth, immunity, and stress response. *Planta* 16, 259. DOI: 10.1007/s00425-023-04292-z.

Townsend, J.A., Wright, D.A., Winfrey, R.J., Fu, F., Maeder, M.L. *et al.* (2009) High-frequency modification of plant genes using engineered zinc-finger nucleases. *Nature* 459(7245), 442–445. DOI: 10.1038/nature07845.

Vats, S., Kumawat, S., Kumar, V., Patil, G.B., Joshi, T. *et al.* (2019) Genome editing in plants: exploration of technological advancements and challenges. *Cells* 8(11), 1386. DOI: 10.3390/cells8111386.

Wang, F., Wang, C., Liu, P., Lei, C., Hao, W. *et al.* (2016) Enhanced rice blast resistance by CRISPR/Cas9-targeted mutagenesis of the ERF transcription factor gene OsERF922. *PloS One* 11(4), e0154027. DOI: 10.1371/journal.pone.0154027.

Wang, F., Xu, Y., Li, W., Chen, Z., Wang, J. *et al.* (2021) Creating a novel herbicide-tolerance *OsALS* allele using CRISPR/Cas9-mediated gene editing. *The Crop Journal* 9(2), 305–312. DOI: 10.1016/j.cj.2020.06.001.

Wang, G., Xu, Z., Wang, F., Huang, Y., Xin, Y. *et al.* (2022a) Development of an efficient and precise adenine base editor (ABE) with expanded target range in allotetraploid cotton (*Gossypium hirsutum*). *BMC Biology* 20(1), 1–15. DOI: 10.1186/s12915-022-01232-3.

Wang, G., Zong, M., Liu, D., Wu, Y., Tian, S. *et al.* (2022c) Efficient generation of targeted point mutations in the *Brassica oleracea* var. *botrytis* genome via a modified CRISPR/Cas9 system. *Horticultural Plant Journal* 8(4), 527–530. DOI: 10.1016/j.hpj.2022.01.005.

Wang, J., Wang, H., Liu, S., Liu, L., Tay, W.T. *et al.* (2017) CRISPR/Cas9 mediated genome editing of Helicoverpa armigera with mutations of an ABC transporter gene HaABCA2 confers resistance to *Bacillus thuringiensis* Cry2A toxins. *Insect Biochemistry and Molecular Biology* 87, 147–153. DOI: 10.1016/j.ibmb.2017.07.002.

Wang, M., Lu, Y., Botella, J.R., Mao, Y., Hua, K. *et al.* (2017) Gene targeting by homology-directed repair in rice using a geminivirus-based CRISPR/Cas9 system. *Molecular Plant* 10(7), 1007–1010. DOI: 10.1016/j.molp.2017.03.002.

Wang, R. and Han, L.B. (2022) Generation of bdfls2-knockout mutant in *Brachypodium distachyon* mediated by CRISPR/Cas9. *Biotechnology Bulletin* 38(1), 70. DOI: 10.13560/j.cnki.biotech.bull.1985.2021-0326.

Wang, Y., Cheng, X., Shan, Q., Zhang, Y., Liu, J. *et al.* (2014) Simultaneous editing of three homoeoalleles in hexaploid bread wheat confers heritable resistance to powdery mildew. *Nature Biotechnology* 32(9), 947–951. DOI: 10.1038/nbt.2969.

Wang, Y., Zafar, N., Ali, Q., Manghwar, H., Wang, G. *et al.* (2022b) CRISPR/Cas genome editing technologies for plant improvement against biotic and abiotic stresses: advances, limitations, and future perspectives. *Cells* 11(23), 3928. DOI: 10.3390/cells11233928.

Wei, T., Wen, X., Niu, C., An, S., Wang, D. *et al.* (2022) Design of acetohydroxyacid synthase herbicide-resistant germplasm through MB-QSAR and CRISPR/Cas9-mediated base-editing approaches. *Journal of Agricultural and Food Chemistry* 70(9), 2817–2824. DOI: 10.1021/acs.jafc.1c07180.

Wu, J., Chen, C., Xian, G., Liu, D., Lin, L. *et al.* (2020) Engineering herbicide-resistant oilseed rape by CRISPR/Cas9-mediated cytosine base-editing. *Plant Biotechnology Journal* 18(9), 1857–1859. DOI: 10.1111/pbi.13368.

Wu, L., Xiao, H., Zhao, L. and Cheng, Q. (2022) CRISPR/Cas9-mediated generation of fls2 mutant in *Nicotiana benthamiana* for investigating the flagellin recognition spectrum of diverse FLS2 receptors. *Plant Biotechnology Journal* 20(10), 1853–1855. DOI: 10.1111/pbi.13898.

Xie, Y., Haq, S.I.U., Jiang, X., Zheng, D., Feng, N. *et al.* (2022) Plant genome editing: CRISPR, base editing, prime editing, and beyond. *Grassland Research* 1(4), 234–243. DOI: 10.1002/glr2.12034.

Yang, B., Fujii, T., Ishikawa, Y. and Matsuo, T. (2016) Targeted mutagenesis of an odorant receptor co-receptor using TALEN in *Ostrinia furnacalis*. *Insect Biochemistry and Molecular Biology* 70, 53–59. DOI: 10.1016/j.ibmb.2015.12.003.

Yu, Q., Jalaludin, A., Han, H., Chen, M., Sammons, R.D. *et al.* (2015) Evolution of a double amino acid substitution in the 5-enolpyruvylshikimate-3-phosphate synthase in *Eleusine indica* conferring high-level glyphosate resistance. *Plant Physiology* 167(4), 1440–1447. DOI: 10.1104/pp.15.00146.

Zafar, K., Khan, M.Z., Amin, I., Mukhtar, Z., Zafar, M. *et al.* (2023) Employing template directed CRISPR-based editing of the OsALS gene to create herbicide tolerance in basmati rice. *AoB PLANTS* 15(2), lac059. DOI: 10.1093/aobpla/plac059.

Zhang, K., Raboanatahiry, N., Zhu, B. and Li, M. (2017a) Progress in genome editing technology and its application in plants. *Frontiers in Plant Science* 8, 177. DOI: 10.3389/fpls.2017.00177.

Zhang, L., Liu, B., Zheng, G., Zhang, A. and Li, R. (2017b) Genome-wide characterization of the SiDof gene family in foxtail millet (*Setaria italica*). *Biosystems* 151, 27–33. DOI: 10.1016/j.biosystems.2016.11.007.

Zhang, Y., Bai, Y., Wu, G., Zou, S., Chen, Y. *et al.* (2017c) Simultaneous modification of three homoeologs of TaEDR1 by genome editing enhances powdery mildew resistance in wheat. *The Plant Journal* 91(4), 714–724. DOI: 10.1111/tpj.13599.

Zhang, T., Zhao, Y., Ye, J., Cao, X., Xu, C. *et al.* (2019) Establishing CRISPR/Cas13a immune system conferring RNA virus resistance in both dicot and monocot plants. *Plant Biotechnology Journal* 17(7), 1185. DOI: 10.1111/pbi.13095.

Zhang, Y., Guo, W., Chen, L., Shen, X., Yang, H. *et al.* (2022) CRISPR/Cas9-mediated targeted mutagenesis of GmUGT enhanced soybean resistance against leaf-chewing insects through flavonoids biosynthesis. *Frontiers in Plant Science* 13, 74. DOI: 10.3389/fpls.2022.802716.

Zhu, Z., Yin, J., Chern, M., Zhu, X., Yang, C. *et al.* (2020) New insights into bsr-d1-mediated broad-spectrum resistance to rice blast. *Molecular Plant Pathology* 21(7), 951–960. DOI: 10.1111/mpp.12941.

Zong, Y., Liu, Y., Xue, C., Li, B., Li, X. *et al.* (2022) An engineered prime editor with enhanced editing efficiency in plants. *Nature Biotechnology* 40, 1394–1402. DOI: 10.1038/s41587-022-01254-w.

Zuo, Y.Y., Huang, J.L., Wang, J., Feng, Y., Han, T.T. *et al.* (2018) Knockout of a P-glycoprotein gene increases susceptibility to abamectin and emamectin benzoate in *Spodoptera exigua*. *Insect Molecular Biology* 27(1), 36–45. DOI: 10.1111/imb.12338.

14 Genome Editing to Improve Abiotic Stress Tolerance in Crop Plants

Sehrish Iftikhar[1]*, Khadija Aaliya[2] and Ambreen Gul[2]
[1]*Microfluidics Innovation Unit, Elvesys, Paris, France;* [2]*Centre of Excellence in Molecular Biology, University of the Punjab, Lahore, Pakistan*

Abstract
The chapter explores the imperative role of genome editing techniques in enhancing tolerance against abiotic stress in crop plants, focusing on major stressors such as salt stress, drought, extreme temperatures, heavy metals, UV-B radiation, and mechanical stress. Abiotic stresses pose substantial challenges to global food security and necessitate sustainable and innovative solutions. Although conventional breeding has contributed significantly to addressing these challenges, the urgency for precise and effective gene editing tools is paramount. Recent breakthroughs in genome editing tools, notably CRISPR/Cas-based techniques, offer unprecedented opportunities to enhance crop traits efficiently. The chapter delves into the applications of genome editing tools to address specific stressors, illustrating the potential to modify single or multiple genes with high precision. Multiplex gene editing techniques enable simultaneous modifications, thereby conferring tolerance to numerous abiotic stresses. The chapter concludes by highlighting the immense potential of genome editing as a transformative tool to develop abiotic stress-tolerant crops, contributing significantly to sustainable agriculture and global food security.

Keywords: Genetic engineering, Crop improvement, Environmental stress adaptation, Plant abiotic tolerance, Stress-responsive genes

14.1 Introduction

With the world population projected to increase up to 10 billion by 2050, it is imperative to increase food crop production by 60% for food security (Jaganathan *et al.*, 2018; Bhatta *et al.*, 2021). However, abiotic stresses, including salt stress, extremely high/low temperatures, drought, and heavy metal stress, present a major threat in the 21st century, causing challenges in supplying sufficient food to an increasing population (Nadeem *et al.*, 2018; 2019a, b; Sharma *et al.*, 2020; Biswas *et al.*, 2021). These stresses are accountable for nearly 50% of crop yield losses, amounting to annual economic losses of $14–$19 million worldwide (Kaur and Pati, 2019). Furthermore, the anticipated impact of these strains is expected to worsen as a result of obscure climatic changes in the coming years (Vaughan *et al.*, 2018; Bhat *et al.*, 2021). Consequently, the management of abiotic stresses in various crops emerges as a key objective for researchers striving to ensure sustainable agriculture.

To confront such challenges, breeding climate-smart crops, which are able to tolerate

*Corresponding author: sehrishiftikhar05@gmail.com

DOI: 10.1079/9781800622517.0014

frequent abiotic stresses, is a more sustainable and eco-friendly approach. Though conventional breeding approaches have contributed significantly in achieving the goal of increased production and improving crop tolerance against abiotic stresses, the effectiveness of these approaches has been constrained by the intricate nature of traits linked to abiotic stress tolerance, leading to low success rates (Azimi *et al.*, 2018). There is an urgent need to employ precise and effective gene editing tools with immediate impacts to cope with the threats posed by abiotic stresses (Driedonks *et al.*, 2016). Over the last decade, the progress made in molecular biology has opened up avenues for significant advancements in plant science. The development in genetic engineering tools has presented promising avenues to improve plants' tolerance against stress, offering substantial opportunities for agricultural improvement (Gao, 2021).

Among the most prominent gene editing tools are zinc finger nucleases (ZFNs), transcription activator-like effector nucleases (TALENs), base editors, and clustered regularly interspaced short palindromic repeats (CRISPR)/CRISPR-associated protein 9 (Cas9). CRISPR/Cas9, in particular, has come forward as a widely used tool due to its simplicity, efficiency, and multiplexing capabilities. It relies on the guidance of a single-guide RNA (sgRNA) to direct the Cas9 nuclease to specific DNA sequences, where it induces double-strand breaks that can be repaired via error-prone non-homologous end-joining or precise homology-directed repair mechanisms. ZFNs represent artificially synthesized endonucleases capable of cleaving double-stranded DNA in a sequence-specific fashion (Iqbal *et al.*, 2020). They allow the introduction of various genetic alterations, such as point mutations, insertions, deletions, and translocations within the specific genetic target. TALENs, on the other hand, utilize customizable DNA-binding domains derived from transcription activator-like effectors fused to the FokI nuclease that targets specific DNA sequences. Base editors, including cytosine and adenine base editors, enable precise single-nucleotide substitutions without inducing double-strand breaks, offering greater control over nucleotide changes. These genome engineering tools have significantly accelerated research in diverse fields, ranging from fundamental genetics to practical biotechnology applications, and hold immense potential for addressing various challenges in agriculture, medicine, and beyond. These highly specific gene editing techniques present an exquisite opportunity to improve various crop traits, especially targeting abiotic stresses in a more efficient way (Dalla Costa *et al.*, 2017; Klap *et al.*, 2017; Hussain *et al.*, 2018). During the last decade, the breakthroughs made in genome editing tools have enabled highly effective gene modifications, to the extent of a single base (Zong *et al.*, 2017; Lu and Zhu, 2017). Multiplex gene editing was applied in rice crop to edit *MPK1*, *2*, *5*, and 6 genes, resulting in modified biotic and abiotic stress signaling (Minkenberg *et al.*, 2017). Li *et al.* (2017) overexpressed *UGT79-B2* and *B3* in Arabidopsis, leading to enhanced antioxidant responses, high levels of anthocyanin, and tolerance against salt, cold, and other abiotic stresses. The rapid development in the field of genome editing permits multiplex gene editing to alter numerous single stress-related genes; nonetheless, it can also be applied to modify diverse genes at the same time, conferring tolerance to different abiotic stresses. This chapter explores the applications of genome editing in addressing specific abiotic stresses, providing insights into the modification of single or multiple genes to confer stress tolerance in diverse crop species (Fig. 14.1). It elucidates the molecular mechanisms that govern plant responses to abiotic stress, emphasizing pivotal genes and pathways implicated in stress detection, signal transmission, and stress tolerance mechanisms. Subsequently, it explores diverse genome editing techniques, including CRISPR/Cas9, TALENs, and base editors and their applications in crop improvement.

14.2 Salt Stress Tolerance

One of the major constraints in the agricultural lands of coastal or semi-arid regions is salt stress (Flowers, 2004). Salt stress results in necrosis and ion disturbance in cells, as well as premature leaf death (Julkowska and Testerink, 2015). The excessive salt concentration has

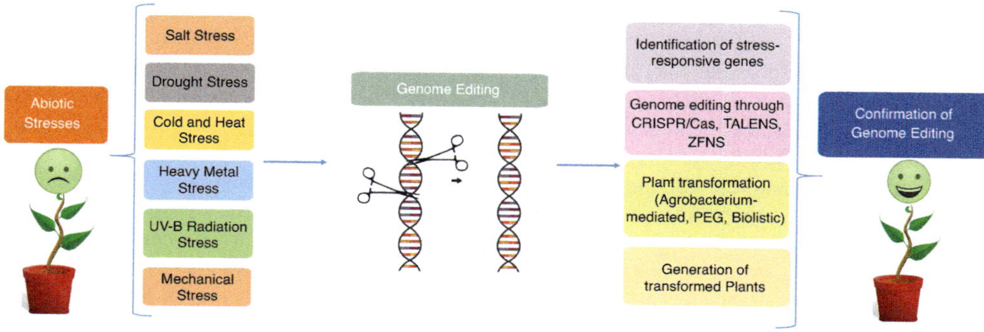

Fig. 14.1. Applications of genome editing in crop improvement against abiotic stresses. (Author's own figure.)

affected 20% of irrigated arable land and 7% of total cultivable land (Li *et al.*, 2014), resulting in reduced crop yields. Plants adapt to various morphological and physiological shifts under salt stress by modifying gene expression and signaling pathways (Julkowska and Testerink, 2015). Numerous genome editing methods are already applied to increase tolerance against salt stress in plants, with the objective of tackling one of the foremost challenges in agriculture. CRISPR/Cas9, TALENs, and base editing technologies stand out as prominent tools in this endeavor. Several genes have already been altered using the CRISPR/Cas system to increase plants' tolerance under salt stress conditions. CRISPR/Cas9 offers unparalleled precision and efficiency in introducing targeted mutations in the plant genome, enabling the alteration of key genes involved in salt stress responses, ion transport, osmotic regulation, and hormonal signaling pathways. In rice, the *OsRR22* gene, which encodes a transcription factor (TF) crucial for governing signaling pathways and cytokinin metabolism in plants, was subjected to editing to confer salt stress tolerance (Takagi *et al.*, 2015; Zhang *et al.*, 2019a). Furthermore, genome editing was applied to investigate the molecular regulation of multiple genes associated with adaptation to abiotic stress. RAV2, a transcription factor, is instrumental in conferring tolerance to abiotic stress (Duan *et al.*, 2015). However, the mechanisms governing its regulation under salt stress conditions remained unclear. Through genome editing of the RAV2 gene's promoter sequence using CRISPR/Cas9, it was discovered that the GT-1 element is essential for its activation under salt stress conditions (Duan *et al.*, 2015). In a separate study, mutants with loss-of-function of the *SAPK2* gene, pivotal for ABA signaling under osmotic stress conditions, displayed high sensitivity to salt stress conditions. This emphasizes the substantial role of *SAPK2* in facilitating adaptation to salt stress (Lou *et al.*, 2017). CRISPR/Cas was employed to mutate the drought and salt tolerance (DST) gene in rice, aiming to boost the plants' tolerance against salt and drought stresses. The resulting mutant plants displayed a decrease in stomatal density, and water retention levels in their leaves were also improved (Kumar *et al.*, 2020). In tomatoes, modifications to the eight cysteine-motif (8CM) and proline-rich domain (PRD) domains of hybrid proline-rich protein 1 (HyPRP1) using CRISPR/Cas-mediated genome editing led to enhanced salinity stress tolerance (Tran *et al.*, 2021). In another study, CRISPR/Cas-mediated mutation of *OsmiR535* resulted in the knockout osmir535 rice mutant plants. These mutant plants performed better under salt stress conditions as compared to control (Yue *et al.*, 2020). These findings underscore the capacity of genome editing to develop tolerance against salt stress in plants. Henceforth, future research initiatives should prioritize the manipulation of pivotal genes involved in regulating the response to salinity stress, especially those involved in transcription factors, through genome editing methodologies.

In some studies, *OsRR22* and *OsNAC041* genes in rice plants have also been targeted to increase tolerance against salt stress (Osakabe and Osakabe, 2017; Zhang *et al.*, 2019a). Additionally, *OsRR22* and *OsNAC041* genes in rice were also stated to confer tolerance to salt (Lan *et al.*, 2019; Sun *et al.*, 2019; Ganie *et al.*, 2021). Wang *et al.* (2017) knocked out the *SlMAP3* gene by CRISPR/Cas-target genome editing, and the resulting tomato mutants showed decreased expression of SlDREB, SlGST, and SlLOX, consequently becoming tolerant to salt stress. Inhibition or knockout of *OsmiR535* in rice by transgenic and CRISPR/Cas9 techniques enhanced tolerance against NaCl, ABA, PEG, and dehydration stresses in mutant plants (Yue *et al.*, 2020). CRISPR/Cas9 has been also used to target the promoter region of rice OsRAV2 for its functional involvement in treating salt stress (Duan *et al.*, 2015). Gao *et al.* (2011) reported that osa-MIR393 regulates the expression of *LOC_Os02g06260*, *LOC_Os05g41010*, and *LOC_Os05g05800* genes, which are stress-responsive genes and negatively regulate the tolerance against salt stress in transgenic Arabidopsis. Similarly, CRISPR has also been applied successfully to improve agronomic traits as well as salt tolerance (Zhang *et al.*, 2019a, 2019b). Genome editing techniques offer remarkable opportunities to improve tolerance against salt stress in plants by precisely modifying key genes involved in stress response pathways. Continuous advancements in these technologies hold promise for developing resilient crop varieties, capable of thriving in saline environments. While each genome editing technology has its own strengths and limitations, the CRISPR/Cas9 technology offers valuable tools to increase tolerance against salt stress. The choice of editing tool depends on factors such as target gene complexity, desired editing precision, and experimental requirements, with each technology providing unique opportunities for advancing crop resilience in salt-affected environments.

14.3 Drought Stress Tolerance

Drought stress poses a significant challenge to global agriculture, impacting crop yield and quality. It poses major threats to plant survival, reducing crop production by interfering with biochemical and physiological processes at various stages during the development of the plant (Golldack *et al.*, 2014; Xu *et al.*, 2014; Siddiqi and Husen, 2017, 2019). To alleviate the adverse impacts of drought, researchers have explored various strategies, including leveraging genome editing tools to boost tolerance against drought in plants. The potential of CRISPR/Cas9-targeted mutagenesis has been established to enhance tolerance against drought stress in plants. By targeting key genes—for example, genes involved in drought response pathways, TFs, stomatal regulation, dehydration-responsive proteins, osmoprotectants, and aquaporins—CRISPR/Cas9 enables precise modifications to enhance plant water use efficiency, osmotic regulation, and stress signaling. For example, editing genes associated with stomatal regulation can help plants maintain adequate water levels under drought conditions, while modifications to genes involved in the biosynthesis of osmolytes like proline and glycine betaine can improve plant osmotic adjustment and stress tolerance. Shi *et al.* (2017) produced *ARGOS8*-v1 and *ARGOS8*-v2 maize plants, resulting in substantial increase in yield of grain in transgenic plants exposed to drought stress. This gene acts as a suppressor within the ethylene signaling pathway, and its high levels of expression are recognized to impart resistance to drought stress in plants.

Genome engineering has proven effective in enhancing various traits in diploid plants. Nonetheless, its application to complex genomes poses a substantial challenge. In recent advancements, the CRISPR/Cas9 technology has been employed to modify stress-responsive TF genes, encoding dehydration-responsive element binding protein 2 (TaDREB2) and ethylene-responsive factor 3 (TaERF3) in wheat protoplasts (Kim *et al.*, 2018). This approach resulted in drought stress-resistant plants, presenting novel avenues for enhancing tolerance against stress in polyploid plants through genome editing techniques. Furthermore, applying CRISPR/Cas to edit the *trehalase* gene, crucial in trehalose catabolism, led to increased tolerance in *Arabidopsis thaliana* when exposed to drought stress. The mutant plants, in which substrate-binding domains of the trehalase enzyme were edited, exhibited phenotypic characteristics of drought tolerance (Nuñez-Muñoz *et al.*, 2021). CRISPR/Cas-based targeted mutation was applied, and slnpr1-drought-tolerant mutant tomato plants were developed (Li *et al.*, 2019). Wang *et al.* (2018) explained the functional involvement of *ZmWRKY40* gene in abiotic stress-related response by the CRISPR/Cas9 system, and mutant lines conferred tolerance to drought in corn. A rice drought and stress tolerance (dst) mutant cultivar was developed by CRISPR/Cas-based targeted editing of *OsDST* gene and led to improved tolerance to drought stress (Kumar *et al.*, 2020; Ganie *et al.*, 2021). *CBP80*, also referred to as the abscisic acid hypersensitive 1 gene, contributes significantly to drought tolerance in *Arabidopsis thaliana*. Pieczynski *et al.* (2013) silenced *CBP80* in potato by CRISPR/Cas technology, and the resulting lines showed enhanced tolerance against drought stress conditions. The PEN STOMATA 2 (*OST2*) gene encodes for an H^+–ATPase and plays a crucial role in generating proton gradients within plant cells. Through precise editing, facilitated by the CRISPR/Cas9 technology, alterations to this gene have been documented to regulate stomatal closure under water-deficient conditions, consequently imparting tolerance to drought (Osakabe *et al.*, 2016). Park *et al.* (2017) applied CRISPR/Cas9-mediated transcriptional activation of vacuolar H^+–pyrophosphatase (AVP1), and mutant Arabidopsis plants displayed an increase of two- to fivefold in APV1 expression and exhibited enhanced drought tolerance. Lou *et al.* (2017) deciphered the potential role of *SAPK2* gene in imparting tolerance against drought stress in rice plants. They applied CRISPR/Cas editing to generate SAPK2-mutant rice, which were more susceptible to drought stress as compared to the wild plants. Wang *et al.* (2018) applied CRISPR/Cas to overexpress *ZmWRKY106* to enhance drought and heat tolerance in transgenic plants. Zhao *et al.* (2016) applied sgRNA/Cas9 mutation to knockout of *MIR169a* in Arabidopsis, and mutant lines conferred tolerance against drought stress. After 12 days of drought exposure, followed by rewatering, approximately 54–57% of mutated plants exhibited survival, whereas no survival was noted among wild-type plants.

Similarly, TALENs offer another avenue for targeted genome editing, leading to increased tolerance against drought stress in plants. TALENs allow researchers to precisely modify specific genomic loci associated with drought-responsive genes, thereby modulating plant physiological and molecular responses to water deficit. Base editing technologies, such as cytosine and adenine base editors, provide additional tools for fine-tuning plant responses to drought stress. These technologies enable precise single-nucleotide substitutions without inducing double-strand breaks, offering greater control over nucleotide changes in target genes associated with drought stress tolerance mechanisms. By introducing specific point mutations in genes encoding drought-responsive proteins or regulatory elements, base editors can optimize plant adaptation to water-limited environments while minimizing unintended genetic alterations. Overall, genome editing tools offer promising strategies for enhancing drought stress tolerance in plants by precisely modulating key genes and regulatory elements involved in drought response pathways. By utilizing the capabilities of CRISPR/Cas9, TALENs, and base editing technologies, researchers can accelerate the development of drought-resistant crop varieties with improved water use efficiency, osmotic regulation, and stress adaptation mechanisms, thereby contributing to global food security and agricultural sustainability in the face of climate change–induced water scarcity.

14.4 Cold and Heat Stress Tolerance

Cold and heat stress represent significant challenges to plant growth and productivity. Uncontrolled emission of greenhouse gases in the environment results in heat stress in various crops (Asseng *et al.*, 2015). It has been reported that as the atmospheric temperature rises by 1°C, it is estimated to reduce the yield of the wheat crop by 6% and by 21–31% for maize, increasing concerns of food security issues (Yang *et al.*, 2017; Wang *et al.*, 2019). As the temperature increases above the normal range, crop plants begin to show different responses to heat stress by diverse mechanisms, involving altered expression of heat shock encoding genes, scavenger proteins, and reactive oxygen species (ROS) (Shah *et al.*, 2013; Fragkostefanakis *et al.*, 2015). On the other hand, cold stress also triggers various responses in plants, including alterations in membrane fluidity and the activation of cold-responsive genes. Chang *et al.* (2017) reported that co-expression of kinase (SAPK6) protein and bZIP transcription factor (OsbZIP46CA1) in rice improved tolerance to heat and cold stresses. Lettuce plants with knockout mutants of the 9-cis-EPOXYCAROTENOID DIOXYGENASE4 (*NCED4*) gene, created through CRISPR/Cas9 technology, have been observed to exhibit improved thermotolerance during the germination stage (Bertier *et al.*, 2018). CRISPR/Cas9-mediated loss-of-function of *CBF1*, *CBF2*, and *CBF3* genes by single, double, and triple mutations in Arabidopsis resulted in extremely sensitive plants to freezing conditions. The CBF proteins play a crucial role in transcribing downstream cold-responsive genes, ultimately increasing the tolerance of plants against freezing (Zhao *et al.*, 2016). In another study, TIFY1a and TIFY1b genes in rice were edited, aiming to elucidate the specific roles in developing tolerance under cold stress (Huang *et al.*, 2017). Biswal *et al.* (2019) and Nguyen *et al.* (2018) reported CRISPR/Cas gene editing technique as the most successful one in the development of thermotolerant crops by modulating genes. Crop plants like tomatoes are susceptible to cold, and its fruits are easily damaged by cold and chilling stress. By using CRISPR/Cas-targeted mutagenesis, Li *et al.* (2018) produced cbf1 mutants and confirmed the function of C-repeat binding factor 1 (*CBF1*) in conferring tolerance to cold injury. Thermosensitive male, sterile maize plants were produced by modifying the *TMS5* gene via CRISPR/Cas-based editing (Li *et al.*, 2017).

Shen *et al.* (2017) produced mutant rice cultivar japonica for a gene encoding Ca^{2+}–dependent phospholipid binding protein (OsANN3) by CRISPR/Cas-mediated mutagenesis, resulting in plants with cold tolerance. CRISPR-bzr1-mutant and BZR1-overexpressing lines of tomato plants showed the role of *BZR1* in thermotolerance in mutant plants (Yin *et al.*, 2018). Another investigation, using CRISPR/Cas9 technology, where three rice genes—namely, *OsPIN5b*, *GS3*, and *OsMYB30*—were modified, led to enhanced cold stress tolerance and improved yield (Zeng *et al.*, 2020). Elimination of OsAnn5 function through CRISPR/Cas9 editing significantly increases survival rates during the seedling stage under cold stress compared with wild-type rice, indicating that OsAnn5 plays a positive role in regulating cold stress tolerance during the seedling stage (Shen *et al.*, 2020). CRISPR/Cas9-mediated deletion mutants of heat-stress-sensitive albino 1 (OsHSA1) rice plants exhibit high susceptibility to heat stress as compared to wild-type rice plants (Qiu *et al.*, 2018). OsHSA1 codes for fructokinase-like protein two and plays essential roles in chloroplast development and safeguarding during heat stress in rice. Yin *et al.* (2018) applied CRISPR/Cas9 to produce knockout plants of SlBZR1, which displayed high sensitivity under heat stress conditions.

Likewise, mutants of tomato calcium-dependent protein kinase28 (Slcpk28), generated through CRISPR/Cas9 technology, display high accumulation of ROS and protein oxidation levels when exposed to heat stress. These mutants also display decreased activity of ascorbate peroxidase (APX), catalase, and glutathione reductase enzymes, resulting in high sensitivity to heat (Hu *et al.*, 2021). The studies underscore the versatility and efficacy of genome editing tools in tailoring plant responses to cold and heat stress, offering promising avenues for developing climate-resilient crops that can thrive in the face of temperature fluctuations.

14.5 Heavy Metal Tolerance

Heavy metal pollution in soil is a global concern, as it not only affects the health of living organisms but also reduces agricultural productivity by reducing crop yield and quality. Heavy metal pollution in soil is a growing problem, with industrialization and urbanization being the main culprits. Heavy metals such as copper (Cu), nickel (Ni), lead (Pb), arsenic (As), zinc (Zn), and cadmium (Cd) are some of the common soil pollutants that can accumulate in plant tissues (Fasani *et al.*, 2018). Heavy metal toxicity may cause numerous plant disorders, like reduced seed germination, stunted growth, chlorosis, and even death. The accumulation of these heavy metals in plants can interfere with key cellular functions, leading to decreased growth and ultimately reduced agricultural productivity. In response to heavy metal toxicity, plants possess various defense mechanisms, including heavy metal chelation by metalloids and their sequestration in vacuoles so that they do not interact with other cellular components. However, these mechanisms have limitations and may be affected under adverse conditions.

Several promising studies have shown that genome editing in plants can improve their tolerance to heavy metals. For instance, Ye *et al.* (2017) reported enhanced tolerance against heavy metal stress in rice by CRISPR/Cas9 system to knockout the *OsPT4* gene. The *OsPT4* gene encodes a heavy metal transport protein that can transport arsenic into rice roots. The knockout of this gene resulted in reduced accumulation of arsenic and increased rice tolerance to heavy metal stress. Songmei *et al.* (2019) applied CRISPR/Cas9-targeted mutagenesis to alter the expression of two heavy metal transporters, OsLCT1 and OsNramp5, in rice under heavy cadmium stress conditions. The study showed that both transporters positively regulate the accumulation of Cd in the plant, while their knockout significantly improved plant stress tolerance. Similarly, Nieves-Cordones *et al.* (2017) inactivated the Cs^+–permeable K^+ transporter OsHAK1 in plants through CRISPR/Cas. Consequently, cesium uptake and accumulation were dramatically reduced in genetically engineered plants.

Although these studies represent a substantial achievement in improving plant performance under heavy metal stress, they also emphasize the need to develop a precise and cost-effective genome editing technique that is applicable to various plant species. Genome editing offers promising opportunities to improve plant tolerance under heavy metal stress conditions. With the advent of advanced genome editing technologies like CRISPR/Cas9, TALENs, and ZFNs, it has become feasible to induce specific alterations in genes and improve their performance under adverse conditions like heavy metal stress.

14.6 UV-B Radiation Stress Tolerance

Another major environmental stress that plants face is ultraviolet B (UV-B) radiation, which is harmful to plant growth, development, and productivity. UV-B radiation is part of the electromagnetic spectrum that lies between 280 and 320 nm wavelengths, and it is known to disrupt various biochemical and physiological processes in plants, subsequently resulting in several negative outcomes. Exposure to UV-B radiation may cause damage to DNA and proteins, resulting in oxidative stress, growth inhibition, and even plant death (Fatima *et al.*, 2021). Moreover, with the increasing problem of global warming, along with other anthropogenic activities, UV-B radiation has become a challenging issue and requires an efficient solution to mitigate its deleterious effects (Galvão and Fankhauser, 2015). One approach that has been proposed to improve plant response to UV-B radiation is genome editing. Harnessing the power of genome editing, researchers can possibly create plants that are more resilient to UV-B radiation by manipulating genes that are involved in the plant's defensive mechanisms.

For example, the *UVR8* gene in Arabidopsis is reported to reduce responsiveness against UV-B radiation (Favory *et al.*, 2009). *UVR8* is a photoreceptor, which is responsible for acclimatization to UV-B rays, hence improving plant tolerance. Genetic engineering of photoreceptors may provide promising results for improved tolerance in plants (Rai *et al.*, 2021). Liu *et al.* (2020) knocked out the photoreceptor SlUVR8 in tomato. Knocked-out mutants exhibited enhanced tolerance to stress under UV-B radiation exposure. Apart from UVR8 and TFs, other genes have also been identified as playing important roles in plants' response to UV-B radiation. For example, overexpression of moss flavone synthase I gene (*PnFNSI*) is reported to enhance tolerance against the stress by UV-B in *Arabidopsis* (Wang *et al.*, 2020). Genome editing provides a great prospect to increase plant UV-B stress tolerance by precisely manipulating genes that are involved in defensive mechanisms. The use of a targeted approach to induce mutations in specific regions of the genome has been demonstrated to be successful, with various studies showing that it is a viable method to develop plants that are more resilient to UV-B radiation.

14.7 Mechanical Stress Tolerance

Throughout their lifespan, plants encounter a multitude of environmental stresses, among which mechanical stress is included. Mechanical stress can result from strong winds, heavy rain, hail, snow, or animals feeding on crops. Mechanical stress can lead to breakage or lodging, reduce yield, and cause significant losses in agriculture. Improving the tolerance against mechanical stress in crops is therefore crucial for sustainable farming (Lamers *et al.*, 2020). Mechanical stress in plants can result in tissue stretching, deformation, or rupture. Mechanical stress tolerance in plants is determined by the strength and elasticity of plant tissues. Plants can respond to mechanical stress through several mechanisms, including cell wall modification, hormonal signaling, and gene expression regulation (Brenya *et al.*, 2022). Cell wall modification is critical for mechanical stress tolerance, as it determines the strength and elasticity of plant tissues, including structural and enzymatic changes to cell wall components, including cellulose, hemicellulose, and lignin (Wu *et al.*, 2001).

One approach to increase tolerance against this stress is to modify genes involved in cell wall biosynthesis of plants. The plant cell wall is composed of several polysaccharides, including cellulose, hemicellulose, and pectin. The key structural component of the cell wall, cellulose, is responsible for providing strength and rigidity to plant tissues. Hemicellulose and pectin are involved in the cross-linking and elasticity of the cell wall. Therefore, modifying the genes involved in cellulose biosynthesis or cross-linking of hemicellulose and pectin can improve the mechanical strength and elasticity of plant tissues (Kouhen *et al.*, 2023). Plants can respond to mechanical stress by activating stress response pathways ethylene (ET) pathways, which regulate plant growth and development, resulting in better stress responses. Modifying the genes involved in stress response pathways can improve mechanical stress tolerance by enhancing stress signaling and responses (Zeng *et al.*, 2021). Genome editing holds significant promise in helping to improve mechanical stress tolerance in plants. By targeting specific genes and implementing precise modifications, researchers can create novel plant varieties with improved mechanical strength and tolerance against a range of stressors impacting plants. While there are concerns about the long-term risks and ethical implications of this technology, proponents argue that the benefits of genome editing outweigh the potential risks, especially in achieving sustainable agricultural practices.

14.8 Conclusion

Genome engineering tools have transformed the crop improvement domain by offering precise and targeted modifications to plant genomes. As climate change magnifies abiotic stresses, such as cold, heat, heavy metals, drought, salinity, and UV radiation, the need to improve tolerance in crop plants becomes increasingly urgent. Genome editing emerges as a versatile technique for crop enhancement initiatives owing to its adaptability, ease of use, and precision. It has introduced novel avenues in functional genomics and plant breeding. Initial research indicates that genome editing has the potential to develop abiotic stress-resistant crops, offering possible solutions to address food insecurity. However, several technological and regulatory challenges must be addressed to effectively revolutionize global agriculture through genome editing. Nonetheless, continued research efforts, interdisciplinary collaborations, and regulatory frameworks are essential to harness the full benefits of genome editing while addressing societal concerns and ethical considerations.

References

Asseng, S., Ewert, F., Martre, P., Rötter, R.P., Lobell, D.B. *et al.* (2015) Rising temperatures reduce global wheat production. *Nature Climate Change* 5, 143–147. DOI: 10.1038/nclimate2470.

Azimi, M.H., Alvijeh, M.K. and Zarei, A. (2018) Intervarietal hybridization and observation of heterosis in the new hybrids of *Iris germanica*. *International Journal of Horticultural Science and Technology* 5(1), 65–79. DOI: 10.22059/ijhst.2018.254015.225.

Bertier, L.D., Ron, M., Huo, H., Bradford, K.J., Britt, A.B. *et al.* (2018) High-resolution analysis of the efficiency, heritability, and editing outcomes of CRISPR/Cas9-induced modifications of NCED4 in lettuce (*Lactuca sativa*). *G3: Genes, Genomes Genetics* 8(5), 1513–1521. DOI: 10.1534/g3.117.300396.

Bhat, M.A., Mir, R.A., Kumar, V., Shah, A.A., Zargar, S.M. *et al.* (2021) Mechanistic insights of CRISPR/CAS-mediated genome editing towards enhancing abiotic stress tolerance in plants. *Physiologia Plantarum* 172, 1255–1268. DOI: 10.1111/ppl.13359.

Bhatta, M., Sandro, P., Smith, M.R., Delaney, O., Voss-fels, K.P. *et al.* (2021) Need for speed: manipulating plant growth to accelerate breeding cycles. *Current Opinion in Plant Biology* 60, 101986. DOI: 10.1016/j.pbi.2020.101986.

Biswal, A.K., Mangrauthia, S.K., Reddy, M.R. and Yugandhar, P. (2019) CRISPR mediated genome engineering to develop climate smart rice: challenges and opportunities. *Seminars in Cell & Developmental Biology* 96, 100–106. DOI: 10.1016/j.semcdb.2019.04.005.

Biswas, D., Saha, S.C. and Dey, A. (2021) CRISPR-CAS genome-editing tool in plant abiotic stress-tolerance. *Plant Gene* 26, 100286. DOI: 10.1016/j.plgene.2021.100286.

Brenya, E., Pervin, M., Chen, Z.H., Tissue, D.T., Johnson, S. *et al.* (2022) Mechanical stress acclimation in plants: linking hormones and somatic memory to thigmomorphogenesis. *Plant Cell & Environment* 45(4), 989–1010. DOI: 10.1111/pce.14252.

Chang, Y., Nguyen, B.H., Xie, Y., Xiao, B., Tang, N. *et al.* (2017) Co-overexpression of the constitutively active form of OsbZIP46 and ABA-activated protein kinase SAPK6 improves drought and temperature stress resistance in rice. *Frontiers in Plant Science* 8, 1102. DOI: 10.3389/fpls.2017.01102.

Dalla Costa, L., Malnoy, M. and Gribaudo, I. (2017) Breeding next generation tree fruits: technical and legal challenges. *Horticulture Research* 4, 17067. DOI: 10.1038/hortres.2017.67.

Driedonks, N., Rieu, I. and Vriezen, W.H. (2016) Breeding for plant heat tolerance at vegetative and reproductive stages. *Plant Reproduction* 29, 67–79. DOI: 10.1007/s00497-016-0275-9.

Duan, Y.B., Li, J., Qin, R.Y., Xu, R.F., Li, H. *et al.* (2015) Identification of a regulatory element responsible for salt induction of rice OsRAV2 through *ex* situ and *in situ* promoter analysis. *Plant Molecular Biology* 90, 49–62. DOI: 10.1007/s11103-015-0393-z.

Fasani, E., Manara, A., Martini, F., Furini, A. and DalCorso, G. (2018) The potential of genetic engineering of plants for the remediation of soils contaminated with heavy metals. *Plant, Cell & Environment* 41(5), 1201–1232. DOI: 10.1111/pce.12963.

Fatima, A., Singh, M. and Prasad, S.M. (2021) Regulatory role of UV-B in modulating antioxidant defence system of plants. In: *Organic Solutes, Oxidative Stress, and Antioxidant Enzymes under Abiotic Stressor*. CRC Press, Florida, pp. 339–352.

Favory, J.J., Stec, A., Gruber, H., Rizzini, L., Oravecz, A. *et al.* (2009) Interaction of COP1 and UVR8 regulates UV-B-induced photomorphogenesis and stress acclimation in Arabidopsis. *The EMBO Journal* 28, 591–601. DOI: 10.1038/emboj.2009.4.

Flowers, T.J. (2004) Improving crop salt tolerance. *Journal of Experimental Botany* 55(396), 307–319. DOI: 10.1093/jxb/erh003.

Fragkostefanakis, S., Simm, S., Paul, P., Bublak, D., Scharf, K.D. *et al.* (2015) Chaperone network composition in *Solanum lycopersicum* explored by transcriptome profiling and microarray meta-analysis. *Plant, Cell and Environment* 38(4), 693–709. DOI: 10.1111/pce.12426.

Galvão, V.C. and Fankhauser, C. (2015) Sensing the light environment in plants: photoreceptors and early signalling steps. *Current Opinion in Neurobiology* 34, 46–53. DOI: 10.1016/j.conb.2015.01.013.

Ganie, S.A., Wani, S.H., Henry, R. and Hensel, G. (2021) Improving rice salt tolerance by precision breeding in a new era. *Current Opinion in Plant Biology* 60, 101996. DOI: 10.1016/j.pbi.2020.101996.

Gao, C. (2021) Genome engineering for crop improvement and future agriculture. *Cell* 184(6), 1621–1635. DOI: 10.1016/j.cell.2021.01.005.

Gao, P., Bai, X., Yang, L., Lv, D., Pan, X. *et al.* (2011) Osa-MIR393: a salinity- and alkaline stress-related microRNA gene. *Molecular Biology Reports* 38, 237–242. DOI: 10.1007/s11033-010-0100-8.

Golldack, D., Li, C., Mohan, H. and Probst, N. (2014) Tolerance to drought and salt stress in plants: unravelling the signalling networks. *Frontiers in Plant Science* 5, 151. DOI: 10.3389/fpls.2014.00151.

Hu, Z., Li, J., Ding, S., Cheng, F., Li, X. *et al.* (2021) The protein kinase CPK28 phosphorylates ascorbate peroxidase and enhances thermotolerance in tomato. *Plant Physiology* 186, 1302–1317. DOI: 10.1093/plphys/kiab120.

Huang, X.Z., Zeng, X.F., Li, J.R. and Zhao, D.G. (2017) Construction and analysis of tify1a and tify1b mutants in rice (*Oryza sativa*) based on CRISPR/Cas9 technology. *Journal of Agricultural Biotechnology* 25(6), 1003–1012.

Hussain, S., Cao, X., Zhong, C., Zhu, L., Khaskheli, M.A. *et al.* (2018) Sodium chloride stress during early growth stages altered physiological and growth characteristics of rice. *Chilean Journal of Agricultural Research* 78, 183–197. DOI: 10.4067/S0718-58392018000200183.

Iqbal, Z., Iqbal, M.S., Ahmad, A., Memon, A.G. and Ansari, M.I. (2020) New prospects on the horizon: genome editing to engineer plants for desirable traits. *Current Plant Biology* 24, 100171. DOI: 10.1016/j.cpb.2020.100171.

Jaganathan, D., Ramasamy, K., Sellamuthu, G., Jayabalan, S. and Venkataraman, G. (2018) CRISPR for crop improvement: an update review. *Frontiers in Plant Science* 9, 364675. DOI: 10.3389/fpls.2018.00985.

Julkowska, M.M. and Testerink, C. (2015) Tuning plant signalling and growth to survive salt. *Trends in Plant Science* 20(9), 586–594. DOI: 10.1016/j.tplants.2015.06.008.

Kaur, N. and Pati, P.K. (2019) Harnessing the potential of Brassinosteroids in abiotic stress tolerance in plants. *Brassinosteroids: Plant Growth and Development* 407–423. DOI: 10.1007/978-981-13-6058-9_15.

Kim, D., Alptekin, B. and Budak, H. (2018) CRISPR/Cas9 genome editing in wheat. *Functional and Integrative Genomics* 18(1), 31–41. DOI: 10.1007/s10142-017-0572-x.

Klap, C., Yeshayahou, E., Bolger, A.M., Arazi, T., Gupta, S.K. *et al.* (2017) Tomato facultative parthenocarpy results from SlAGAMOUS-LIKE 6 loss of function. *Plant Biotechnology Journal* 15, 634–647. DOI: 10.1111/pbi.12662.

Kouhen, M., Dimitrova, A., Scippa, G.S. and Trupiano, D. (2023) The course of mechanical stress: types, perception, and plant response. *Biology* 12(2), 217. DOI: 10.3390/biology12020217.

Kumar, S.V.V., Verma, R.K., Yadav, S.K., Yadav, P., Watts, A. *et al.* (2020) CRISPR-Cas9 mediated genome editing of drought and salt tolerance (OsDST) gene in *Indica* mega rice cultivar MTU_{1010}. *Physiology and Molecular Biology of Plants* 26, 1099–1110. DOI: 10.1007/s12298-020-00819-w.

Lamers, J., Meer, T.V.D. and Testerink, C. (2020) How plants sense and respond to stressful environments. *Plant Physiology* 182(4), 1624–1635. DOI: 10.1104/pp.19.01464.

Lan, T., Zheng, Y., Su, Z., Yu, S., Song, H. *et al.* (2019) OsSPL10, a SBP-box gene, plays a dual role in salt tolerance and trichome formation in rice (*Oryza sativa* L.). *Genes, Genomes, Genetics* 9, 4107–4114. DOI: 10.1534/g3.119.400700.

Li, J., Pu, L., Han, M., Zhu, M., Zhang, R. *et al.* (2014) Soil salinization research in china: advances and prospects. *Journal of Geographical Sciences* 24, 943–960. DOI: 10.1007/s11442-014-1130-2.

Li, J., Zhang, H., Si, X., Tian, Y., Chen, K. *et al.* (2017) Generation of thermosensitive male-sterile maize by targeted knockout of the ZmTMS5 gene. *Journal of Genetics and Genomics* 44(9), 465–468. DOI: 10.1016/j.jgg.2017.02.002.

Li, R., Liu, C., Zhao, R., Wang, L., Chen, L. *et al.* (2019) CRISPR/Cas9-mediated SlNPR1 mutagenesis reduces tomato plant drought tolerance. *BMC Plant Biology* 19(1), 38. DOI: 10.1186/s12870-018-1627-4.

Li, T., Yang, X., Yu, Y., Si, X., Zhai, X. *et al.* (2018) Domestication of wild tomato is accelerated by genome editing. *Nature Biotechnology* 36, 1160–1163. DOI: 10.1038/nbt.4273.

Liu, X., Zhang, Q., Yang, G., Zhang, C., Dong, H. *et al.* (2020) Pivotal roles of tomato photoreceptor SlUVR8 in seedling development and UV-B stress tolerance. *Biochemical and Biophysical Research Communications* 522(1), 177–183. DOI: 10.1016/j.bbrc.2019.11.073.

Lou, D., Wang, H., Liang, G. and Yu, D. (2017) OsSAPK2 confers abscisic acid sensitivity and tolerance to drought stress in rice. *Frontiers in Plant Science* 8, 993. DOI: 10.3389/fpls.2017.00993.

Lu, Y. and Zhu, J.K. (2017) Precise editing of a target base in the rice genome using a modified CRISPR/Cas9 system. *Molecular Plant* 10, 523–525.

Minkenberg, B., Xie, K. and Yang, Y. (2017) Discovery of rice essential genes by characterizing a CRISPR-edited mutation of closely related rice MAP kinase genes. *The Plant Journal* 89(3), 636–648. DOI: 10.1111/tpj.13399.

Nadeem, M., Li, J., Wang, M., Shah, L., Lu, S. *et al.* (2018) Unravelling field crops sensitivity to heat stress: mechanisms, approaches and future prospects. *Agronomy* 8(7), 128. DOI: 10.3390/agronomy8070128.

Nadeem, M., Li, J., Yahya, M., Sher, A., Ma, C. *et al.* (2019a) Research progress and perspective on drought stress in legumes: a review. *International Journal of Molecular Sciences* 20(10), 2541. DOI: 10.3390/ijms20102541.

Nadeem, M., Li, J., Yahya, M., Wang, M., Ali, A. *et al.* (2019b) Grain legumes and fear of salt stress: focus on mechanisms and management strategies. *International Journal of Molecular Sciences* 20(4), 799. DOI: 10.3390/ijms20040799.

Nguyen, H.C., Lin, K.H., Ho, S.L., Chiang, C.M. and Yang, C.M. (2018) Enhancing the abiotic stress tolerance of plants: from chemical treatment to biotechnological approaches. *Physiologia Plantarum* 164, 452–466. DOI: 10.1111/ppl.12812.

Nieves-Cordones, M., Mohamed, S., Tanoi, K., Kobayashi, N.I., Takagi, K. *et al.* (2017) Production of low-cs+ rice plants by inactivation of the K+ transporter oshak1 with the CRISPR-cas system. *The Plant Journal* 92(1), 43–56. DOI: 10.1111/tpj.13632.

Nuñez-Muñoz, L., Vargas-Hernández, B., Hinojosa-Moya, J., Ruiz-Medrano, R. and Xoconostle-Cázares, B. (2021) Plant drought tolerance provided through genome editing of the trehalase gene. *Plant Signaling & Behavior* 16(4), 1877005. DOI: 10.1080/15592324.2021.1877005.

Osakabe, Y. and Osakabe, K. (2017) Genome editing to improve abiotic stress responses in plants. *Progress in Molecular Biology and Translational Science* 149, 99–109. DOI: 10.1016/bs.pmbts.2017.03.007.

Osakabe, Y., Watanabe, T., Sugano, S.S., Ueta, R., Ishihara, R. *et al.* (2016) Optimization of CRISPR/Cas9 genome editing to modify abiotic stress responses in plants. *Scientific Reports* 6, 26685. DOI: 10.1038/srep26685.

Park, J.J., Dempewolf, E., Zhang, W. and Wang, Z.Y. (2017) RNA-guided transcriptional activation via CRISPR/Dcas9 mimics overexpression phenotypes in Arabidopsis. *PLoS One* 12(6), e0179410. DOI: 10.1371/journal.pone.0179410.

Pieczynski, M., Waldemar, M., Hennig, J., Dolata, J., Bielewicz, D. *et al.* (2013) Down-regulation of CBP 80 gene expression as a strategy to engineer a drought-tolerant potato. *Plant Biotechnology Journal* 11(4), 459–469. DOI: 10.1111/pbi.12032.

Qiu, Z., Kang, S., He, L., Zhao, J., Zhang, S. *et al.* (2018) The newly identified heat-stress sensitive albino 1 gene affects chloroplast development in rice. *Plant Science* 267, 168–179. DOI: 10.1016/j.plantsci.2017.11.015.

Rai, N., Morales, L.O. and Aphalo, P.J. (2021) Perception of solar UV radiation by plants: photoreceptors and mechanisms. *Plant Physiology* 186, 1382–1396. DOI: 10.1093/plphys/kiab162.

Shah, K., Singh, M. and Rai, A.C. (2013) Effect of heat-shock induced oxidative stress is suppressed in BcZAT12 expressing drought tolerant tomato. *Phytochemistry* 95, 109–117. DOI: 10.1016/j.phytochem.2013.07.026.

Sharma, P., Sharma, M.M.M., Patra, A., Vashisth, M., Mehta, S. *et al.* (2020) The role of key transcription factors for cold tolerance in plants. In: *Transcription Factors for Abiotic Stress Tolerance in Plants*. Academic Press, London, pp. 123–152. DOI: 10.1016/B978-0-12-819334-1.00009-5.

Shen, C., Que, Z., Xia, Y., Tang, N., Li, D. *et al.* (2017) Knock out of the annexin gene OsAnn3 via CRISPR/Cas9-mediated genome editing decreased cold tolerance in rice. *Journal of Plant Biology* 60, 539–547. DOI: 10.1007/s12374-016-0400-1.

Shen, C., Que, Z., Lu, Q., Liu, T., Li, S. *et al.* (2020) The rice annexin gene OsAnn5 is a positive regulator of cold stress tolerance at the seedling stage. *Research Square, Durham*. DOI: 10.21203/rs.3.rs-21726/v1.

Shi, J., Gao, H., Wang, H., Lafitte, H.R., Archibald, R.L. *et al.* (2017) ARGOS8 variants generated by CRISPR-cas9 improve maize grain yield under field drought stress conditions. *Plant Biotechnology Journal* 15, 207–216. DOI: 10.1111/pbi.12603.

Siddiqi, K.S. and Husen, A. (2017) Plant response to strigolactones: current developments and emerging trends. *Applied Soil Ecology* 120, 247–253. DOI: 10.1016/j.apsoil.2017.08.020.

Siddiqi, K.S. and Husen, A. (2019) Plant response to jasmonates: current developments and their role in changing environment. *Bulletin of the National Research Centre* 43, 153. DOI: 10.1186/s42269-019-0195-6.

Songmei, L., Jie, J., Yang, L., Jun, M., Shouling, X. *et al.* (2019) Characterization and evaluation of OsLCT1 and OsNramp5 mutants generated through CRISPR/Cas9-mediated mutagenesis for breeding low Cd rice. *Rice Science* 26(2), 88–97. DOI: 10.1016/j.rsci.2019.01.002.

Sun, B.R., Fu, C.Y., Fan, Z.L., Chen, Y., Chen, W.F. *et al.* (2019) Genomic and transcriptomic analysis reveal molecular basis of salinity tolerance in a novel strong salt-tolerant rice landrace Changmaogu. *Rice* 12(1), 99. DOI: 10.1186/s12284-019-0360-4.

Takagi, H., Tamiru, M., Abe, A., Yoshida, K., Uemura, A. *et al.* (2015) MutMap accelerates breeding of a salt-tolerant rice cultivar. *Nature Biotechnology* 33(5), 445–449. DOI: 10.1038/nbt.3188.

Tran, M.T., Doan, D.T.H., Kim, J., Song, Y.J., Sung, Y.W. *et al.* (2021) CRISPR/Cas9-based precise excision of SlHyPRP1 domain (s) to obtain salt stress-tolerant tomato. *Plant Cell Reports* 40(6), 999–1011. DOI: 10.1007/s00299-020-02622-z.

Vaughan, M.M., Block, A., Christensen, S.A., Allen, L.H. and Schmelz, E.A. (2018) The effects of climate change associated abiotic stresses on maize phytochemical defences. *Phytochemistry Reviews* 17(1), 37–49. DOI: 10.1007/s11101-017-9508-2.

Wang, C.T., Ru, J.N., Liu, Y.W., Yang, J.F., Li, M. *et al.* (2018) The maize WRKY transcription factor ZmWRKY40 confers drought resistance in transgenic Arabidopsis. *International Journal of Molecular Sciences* 19(9), 2580. DOI: 10.3390/ijms19092580.

Wang, H., Liu, S., Wang, T., Liu, H., Xu, X. *et al.* (2020) The moss flavone synthase I positively regulates the tolerance of plants to drought stress and UV-B radiation. *Plant Science* 298, 110591. DOI: 10.1016/j.plantsci.2020.110591.

Wang, L., Chen, L., Li, R., Zhao, R., Yang, M. *et al.* (2017) Reduced drought tolerance by CRISPR/Cas9-mediated SlMAPK3 mutagenesis in tomato plants. *Journal of Agricultural and Food Chemistry* 65(39), 8674–8682. DOI: 10.1021/acs.jafc.7b02745.

Wang, Y., Wang, L., Zhou, J., Hu, S., Chen, H. *et al.* (2019) Research progress on heat stress of rice at flowering stage. *Rice Science* 26, 1–10. DOI: 10.1016/j.rsci.2018.06.009.

Wu, L., Joshi, C.P. and Chiang, V.L. (2001) A xylem-specific cellulose synthase gene from aspen (*Populus tremuloides*) is responsive to mechanical stress. *The Plant Journal* 22(6), 495–502. DOI: 10.1046/j.1365-313x.2000.00758.x.

Xu, J., Yuan, Y., Xu, Y., Zhang, G., Guo, X. *et al.* (2014) Identification of candidate genes for drought tolerance by whole-genome resequencing in maize. *BMC Plant Biology* 14, 83. DOI: 10.1186/1471-2229-14-83.

Yang, H., Huang, T., Ding, M., Lu, D. and Lu, W. (2017) High temperature during grain filling impacts on leaf senescence in waxy maize. *Agronomy Journal* 109, 906–916. DOI: 10.2134/agronj2016.08.0452.

Ye, Y., Li, P., Xu, T., Zeng, L., Cheng, D. *et al.* (2017) OsPT4 contributes to arsenate uptake and transport in rice. *Frontiers in Plant Science* 8, 2197. DOI: 10.3389/fpls.2017.02197.

Yin, Y., Qin, K., Song, X., Zhang, Q., Zhou, Y. *et al.* (2018) BZR1 transcription factor regulates heat stress tolerance through FERONIA receptor-like kinase-mediated reactive oxygen species signalling in tomato. *Plant and Cell Physiology* 59(11), 2239–2254. DOI: 10.1093/pcp/pcy146.

Yue, E., Cao, H. and Liu, B. (2020) OsmiR535, a potential genetic editing target for drought and salinity stress tolerance in *Oryza sativa*. *Plants* 9(10), 1337. DOI: 10.3390/plants9101337.

Zeng, J., Li, X., Ge, Q., Dong, Z., Luo, L. *et al.* (2021) Endogenous stress-related signal directs shoot stem cell fate in *Arabidopsis thaliana*. *Nature Plants* 7(9), 1276–1287. DOI: 10.1038/s41477-021-00985-z.

Zeng, Y., Wen, J., Zhao, W., Wang, Q. and Huang, W. (2020) Rational improvement of rice yield and cold tolerance by editing the three genes OsPIN5b, GS3, and OsMYB30 with the CRISPR–Cas9 system. *Frontiers in Plant Science* 10, 1663. DOI: 10.3389/fpls.2019.01663.

Zhang, A., Liu, Y., Wang, F., Li, T., Chen, Z. *et al.* (2019a) Enhanced rice salinity tolerance via CRISPR/Cas9-targeted mutagenesis of the OsRR22 gene. *Molecular Breeding* 39(47), 1–10. DOI: 10.1007/s11032-019-0954-y.

Zhang, R., Liu, J., Chai, Z., Chen, S., Bai, Y. *et al.* (2019b) Generation of herbicide tolerance traits and a new selectable marker in wheat using base editing. *Nature Plants* 5, 480–485. DOI: 10.1038/s41477-019-0405-0.

Zhao, C., Zhang, Z., Xie, S., Si, T., Li, Y. *et al.* (2016) Mutational evidence for the critical role of CBF genes in cold acclimation in Arabidopsis. *Plant Physiology* 171, 2744–2759. DOI: 10.1104/pp.16.00533.

Zong, Y., Wang, Y., Li, C., Zhang, R., Chen, K. *et al.* (2017) Precise base editing in rice, wheat and maize with a Cas9–cytidine deaminase fusion. *Nature Biotechnology* 35, 438–440. DOI: 10.1038/nbt.

15 Genome Editing for Herbicide Tolerance in Crop Plants

Asadullah[1]* and S.R.U. Shah[2]

[1]Department of Plant Breeding and Genetics, Directorate of Agriculture Research Wayaro Farm, Lasbela, Pakistan; [2]Department of Soil and Environment, Swedish University of Agricultural Sciences, Uppsala, Sweden

Abstract

Herbicides are widely used to eradicate or control undesirable weeds. However, the major problems associated with this are the capability of weeds to develop resistance against herbicides, toxic effects on non-target organisms, and residue perseverance within the soil and its seepage into groundwater. Cultivating herbicide-tolerant (HT) crops provides farmers with valuable options for effective weed control, ultimately leading to increased crop yields. Therefore, introducing HT crops is one of the most resourceful strategies for controlling the proliferation of weeds via genetic engineering. In recent decades, traditional breeding, combined with transgenic methods and mutagenesis, has been pivotal in driving the progress of herbicide-resistant (HR) crops. The current, most suitable alternative to conventional genetic engineering approaches for developing HT plants is the creation of transgenic varieties through targeted genome editing. Presently, scientists are finding numerous mutations associated with HT worldwide. This chapter provides an in-depth exploration of various gene editing tools used to make HR crops, with a focus on target-based editing (ABEs and CBEs), CRISPR ribonucleoprotein complexes/Cas9, PLSM, PE, SNPs, NHEJ, BEMGE, and HDR, along with 2 DNA targeting systems (protein based): namely, TALENs, ZFNs—that are utilized for site-directed genome mutagenesis. With these valuable evidential details and relevant information, the creation of new, essential crop varieties, such as rice, soybean, and maize, exhibiting broad-spectrum resistance to several herbicides, can be achieved through successive rounds of genetic engineering.

Keywords: Herbicides, Tolerance, Genome editing, CRISPR/Cas9

15.1 Introduction

Since humans first started to cultivate plants, they have been dealing with weed invasions in crop-growing areas. These unwanted weeds are considered the key problem that the agriculture sector faces today due to their negative impact on crop yield and resource allocation, as they compete with the crop plants for water, space, fertilizer, sunlight, soil nutrients, and so on. Undeniably, great losses in crop productivity can occur if suitable approaches for weed control are not applied (Chauhan, 2020). Nowadays, the old method to control weed proliferation, either through mechanical disruption or weeding by hand, is still practiced in modern agriculture. Herbicides emerged as a significant tool used by farmers to manage weeds in the latter half of the 20th century, particularly in the industrialized world. Herbicides contain chemical elements that are applied to agricultural land during or

*Corresponding author: asadkhanagri24@gmail.com

 Genome Editing for Crop Improvement: Theory and Methodology (eds J.M. Al-Khayri *et al.*)
DOI: 10.1079/9781800622517.0015

sometimes before plantation to get maximum yield, by reducing the competition against naturally growing vegetation.

The most recent gene editing technique, clustered regularly interspaced short palindromic repeats (CRISPR/Cas9), is presently the most prominent means of genetic engineering in plants, by providing options for crop enhancement through specific modification of endogenous genes within the genome. The CRISPR/Cas system is the foremost, resilient, and presently predominant gene editing mechanism, as it is highly efficient, accurate, and cost effective (Manghwar *et al.*, 2020). These include prime editors (PEs), base editors, and nuclease systems that are simple and highly efficient in developing novel germplasms with HT crops. Similarly, CRISPR/Cas9-mediated gene editing tools are broadly utilized to create new germplasm having vast variations in their gene pool for agronomic parameters, as the primary basis of such technologies lies in Cas9 nickase, enabling the precise introduction of changes to a targeted genome at a single-base resolution (Molla *et al.*, 2021).

CRISPR/Cas9 is an endonuclease that operates under the direction of guide RNA (gRNA). It attaches to the designated location next to the protospacer adjacent motif or PAM and initiates the formation of a DSB (short for double-stranded break). After which, the resultant break is repaired with the help of two techniques: non-homologous end-joining (NHEJ), leading to nucleotide substitution or insertions/deletions (InDels) resulting in mutation, and HDR (short for homology-directed repair) that repairs DSBs, by introducing a template DNA, leading to the rectification of existing mutations (Zhang *et al.*, 2020).

15.1.1 Significance

The cultivation of HR plants along with efficient herbicides is one of the most resourceful strategies in controlling the proliferation of weeds. Emerging HT crops and breeding methods to keep abreast of new herbicides are promising strategies to combat unwanted weeds (Wang *et al.*, 2018). Such tolerant crops provide more flexible and improved weed control with reduced tillage application. Therefore, there is a dire need to develop HT varieties that would empower farmers with cost-effective weed management tools.

Genome editing in plants has recently been used to achieve site-directed genome mutagenesis, which provides an effective way to modify nucleotides in target genes through diverse DNA targeting tools: namely, zinc finger nuclease (short for ZFN) and transcription activator-like effector nuclease (abbreviated as TALEN), which are precisely based on proteins. Moreover, in plant breeding, targeted base editing with ABEs (stands for adenine base editors), cytosine base editors or CBEs, and an approach of HDR has been utilized, which includes HT crops as well (Li *et al.*, 2020b). CRISPR/Cas9 systems stand out as the most efficient and extensively utilized among all gene editing methods for enhancing traits in crop plants, including HR (Butt *et al.*, 2020). Furthermore, the potential of CRISPR/Cas9—specifically, base editing—lies in the creation of HT crops without genetic modification, which are free from genetically modified organism (GMO) regulations in numerous countries (Zhang *et al.*, 2020). Hence, such tolerant crops emerge as the most viable alternative to transgenic and traditional methods, and this approach presents a cost-effective solution for growers to effectively control weeds.

15.2 Genome Editing with Engineered Endonucleases

Engineered endonucleases, including CRISPR, TALENs, and ZFNs, have been documented for introducing HR single-nucleotide polymorphisms or SNPs within the designated genome, predominantly relying on the HR-mediated pathway to repair DNA (Table 15.1). Generally, ZFNs are targetable cleaving proteins employed to precisely cut DNA sequences at specific locations, whereas TALENs instigate DSBs at predetermined target sites, thus initiating DNA damage response pathways, ultimately resulting in modifications to the genome. Once DSB has been incorporated, a fragment of DNA (donor), which contains the homologous arms and required nucleotide mutations, is transported along with the engineered endonuclease into the plant cells. Then the donor is injected within the designated site during DNA repair through

Table 15.1. Herbicide-tolerant crops with engineered endonucleases. Author's own table.

Genome editing system	Plant	Target gene	Delivery method	Type of mutation	Target herbicide	Bibliography
CRISPR/Cas9	Rice	*OsALS1*	*Agrobacterium*-mediated transformation	G628W	Imazethapyr, imazapic	Wang *et al*. (2021a)
		OsSF3b1		ΔK1049, H1048Q / Δ1049/ H1048Q, ΔD223 ~ G232	Herboxidiene	Butt *et al*. (2019)
		OsPPO1, OsHPPD	Particle bombardment	Inversion, duplication	FCD, bipyrazone	Lu *et al*. (2021)
	Cassava	*MeEPSPS*	*Agrobacterium*-mediated transformation	T102I + P06A, 35S replaces MeEPSPS promotor	Glyphosate	Hummel *et al*. (2018)
	Flax	*LuEPSPS*	Protoplast transfection	T178I/P182A		Sauer *et al*. (2016)
	Rapeseed	*BnEPSPS*	Particle bombardment	L169F/ G173A/ A175G/ M176C/ R177L		Wang *et al*. (2021b)
	Soybean	*GmALS1*		P178S	Chlorsulfuron	Li *et al*. (2015)
	Sugarcane	*SoALS*		Amino acid substitutions W574L or S653I	Sulfonylureas, imidazolinones	Oz *et al*. (2021)
	Maize	*ZmALS2*		P165S	Chlorsulfuron	Svitashev *et al*. (2015)
TALEN	Rice	*OsALS1*	Ballistic delivery	W548L/S627I	Bispyribac sodium	Li *et al*. (2016b)
TALEN, CRISPR/Cas9	Potato	*SlALS1*	*Agrobacterium*-mediated transformation	W563L/S642T	Imidazolinone	Butler *et al*. (2016)
ZFN	Wheat	*TaALS*	Particle bombardment	S653N		Ran *et al*. (2018)

homologous recombination, which results in an exchange within the plant genome for a precise DNA fragment (Townsend *et al.*, 2009; Li *et al.*, 2016a; Sun *et al.*, 2016). For instance, genetic mutations within the acetolactate-synthase (*ALS*) gene such as S647T, W568L, and P191A are highly recognized for tolerance to herbicides—that is, sulfonamide and imidazolinones, which have been introduced independently to the tobacco acetohydroxyacid synthase genes (*ALS SuRA* and *SuRB*) through ZFN, having various donors (Townsend *et al.*, 2009).

Furthermore, homologous recombination through TALENs has been utilized to inject two-point mutations, by replacing two amino acids (S627I and W548L) in rice gene (*OsALS*), which resulted in bispyribac sodium-resistant genotypes with the inherent capability to T_1 generation along with standard morphological features (Li *et al.*, 2016b). Moreover, a novel HR allele was created by Wang *et al.* (2021a) in rice using a pointing gene (*OsALS*), which reveals the probability of such genetic engineering tools in forming mutations when breeding for plants. A similar technique of target base editing (from C to T at *NRT1.1B*) that altered its 627th amino acid (asparagine replaced by serine) generated a novel mutant for HR rice, targeting imazethapyr, which permitted the precise replacement of genes to get a desirable allele (Wang *et al.*, 2019).

15.3 Genome Editing with Cytosine Base Editors

The recently devised CBEs present an efficient substitute for CRISPR/Cas9 to manipulate genes associated with HT in plants (Komor *et al.*, 2016). This involves utilizing either Cas9 nickase (nCas9) or deactivated Cas9 (dCas9) fused with an enzyme possessing base conversion capabilities (Gaudelli *et al.*, 2017). First of all, a gene is identified in the plant genome that, when mutated, confers HT. A guide RNA (gRNA) is then created to mark a specific sequence within the identified gene, which guides the CBE, consisting of a modified Cas9 protein (rAPOBEC1) fused to a cytidine deaminase enzyme, toward the desired locus to induce precise base editing (Fig. 15.1). The cytidine deaminase enzyme catalyzes the substitution of uridine (U) by cytidine (C) nucleotides, therefore causing a G→A (or C→T) replacement within the target gene, leading to specific point mutations that simultaneously result in changes to the protein encoded by the target gene, rendering it less susceptible or resistant to the herbicide (Komor *et al.*, 2016). Thus far, diverse HT crops, including oilseed rape, wheat, Arabidopsis, rice, watermelon, maize, and Solanaceae plants, have been effectively generated through the utilization of this technique (Table 15.2).

15.4 Genome Editing with Adenine Base Editors

Besides CBEs, a similar approach of ABE has recently been developed, which comprises adenine deaminases that are proficient in inducing adenine to guanine replacement and have successfully been used to generate HR in crop plants as well. A gene is identified in the plant genome that confers HT when mutated. Then a gRNA marks the specific DNA sequence within an identified gene, which guides the ABE, consisting of a modified Cas9 protein (ecTadA) fused with an enzyme, toward the desired locus to induce precise base editing (Gaudelli *et al.*, 2017). Moreover, this enzyme catalyzes a substitution to inosine (I) from adenine (A) nucleotides within the target gene, which are later considered as guanine nucleotides during replication of DNA (Fig. 15.2), thus leading to specific point mutations that simultaneously result in changes to the protein encoded by the target gene, rendering it less susceptible or resistant to the herbicide (Richter *et al.*, 2020). Consequently, plants harboring these mutations gain tolerance to the herbicide, enabling them to withstand herbicide application. Thus far, diverse HT crops, including wheat, rice, maize, and tomato plants, have been efficiently optimized through the ABE approach (Table 15.3).

15.5 Genome Editing with Prime Editors

Prime editors, representing additional advancement in genetic engineering, have been

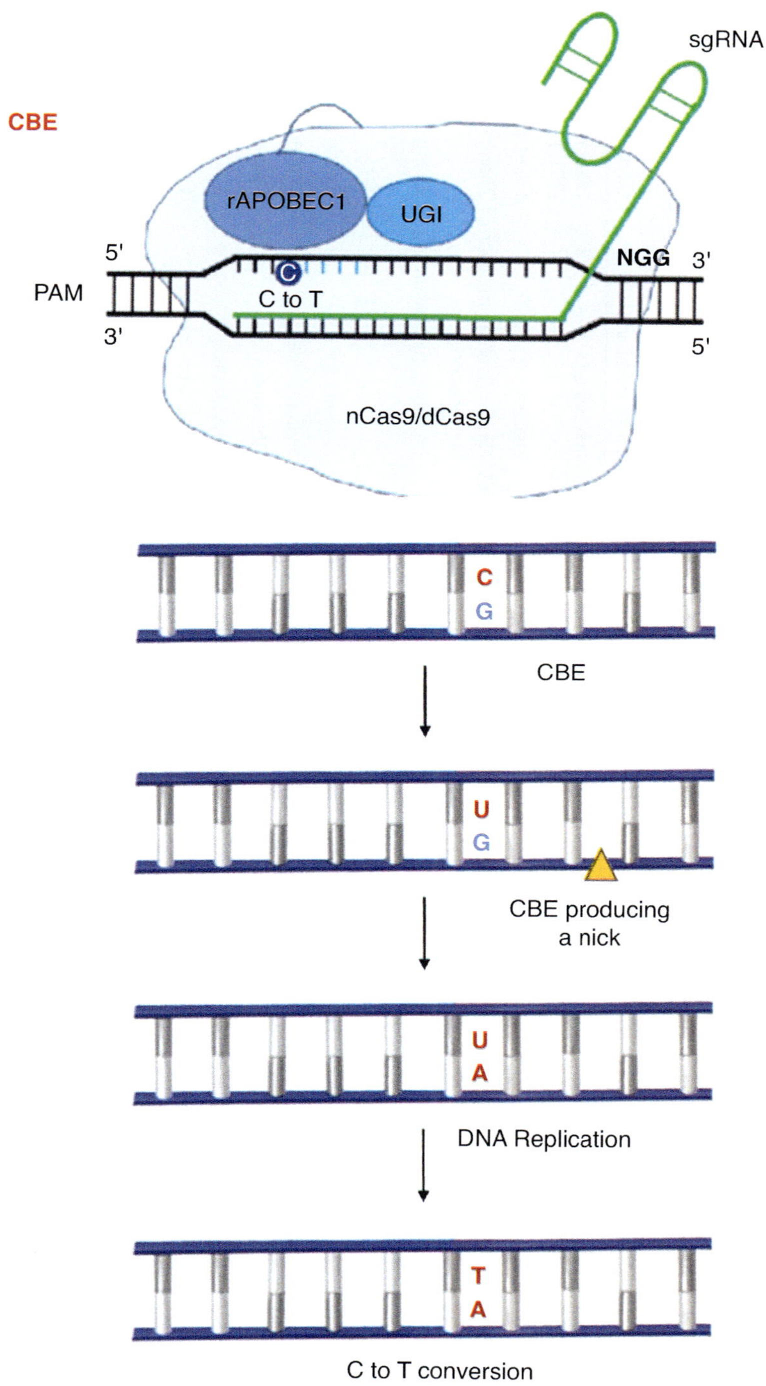

Fig. 15.1. Workflow of cytosine base editing. (Figure constructed by Asadullah.)

Table 15.2. Herbicide-tolerant crops with cytosine base editors (CBEs). Author's own table.

Genome editing system	Plant	Target gene	Delivery method	Type of mutation	Target herbicide	Citations
CBE (Target-AID)	Rice	*OsALS1*	Agrobacterium-mediated transformation	A96V	Imazamox	Shimatani *et al.* (2017)
CBE		*OsACCase*		I1789V	Gallant	Liu *et al.* (2020)
		OsALS1		P171F/G628E/G629S	Nicosulfuron, pyroxsulam, flucarbazone	Zhang *et al.* (2021)
	Wheat	*TaALS*		G632N / G631D, G632S / G631D, P174 F/S/C	Imazapic, nicosulfuron	Zhang *et al.* (2019)
		TaACCase		A1992V	Quizalofop	Zhang *et al.* (2019)
	Maize	*ZmALS1*		P165A/S/L/W	Sulfonylurea	Li *et al.* (2020b)
	Tomato, potato	*SlALS1*, *SlALS2*		P186, P184	Chlorsulfuron	Veillet *et al.* (2019)
	Watermelon	*ClALS*		P190S	Tribenuron	Tian *et al.* (2018)
	Arabidopsis	*AtALS1*		P197L/S/F	Tribenuron-methyl	Chen *et al.* (2017)
	Tobacco	*ALS*		Pro194	Imazapic	Veillet *et al.* (2019)
	Oilseed rape	*BnALSs*		P197S	Tribenuron-methyl	Wu *et al.* (2020)

Table 15.3. Herbicide-tolerant crops with adenine base editors (ABEs) and prime editors (PEs). Author's own table.

Genome editing system	Crops	Targeted gene	Delivery method	Type of mutation	Targeted herbicide	Bibliography
ABE	Rice	*OsGS1, OsACCase, OsALS1*	Agrobacterium-mediated transformation	H249Y, C2186R, S627G	Glufosinate, haloxyfop-R-methyl, imidazolinone, pendimethalin	Packer and Liu (2015), Yan *et al*. (2021)
		OsTubA2		M268T	Dinitroaniline, trifluralin, pendimethalin	Liu *et al*. (2021)
		OsACCase		C2186R	Haloxyfop-R-methyl	Li *et al*. (2018)
	Wheat	*TaTubs*		M268T	Dinitroaniline	Han *et al*. (2022)
Prime editing	Maize	*ZmALS*	Agrobacterium-mediated transformation	W542L and S621I	\	Jiang *et al*. (2020)
	Rice	*OsALS*		A170V	Imidazolinone	Hua *et al*. (2020)
		OsACCase		W2125C	Haloxyfop-R-methyl	Xu *et al*. (2020)
		OsALS1		W548L	Bispyribac sodium	Butt *et al*. (2020)
		OsEPSPS		T169I/P173S (TIAVPS)	Glyphosate	Li *et al*. (2020c)
	Potato	*StALS1*		P186S	Chlorsulfuron	Veillet *et al*. (2020)

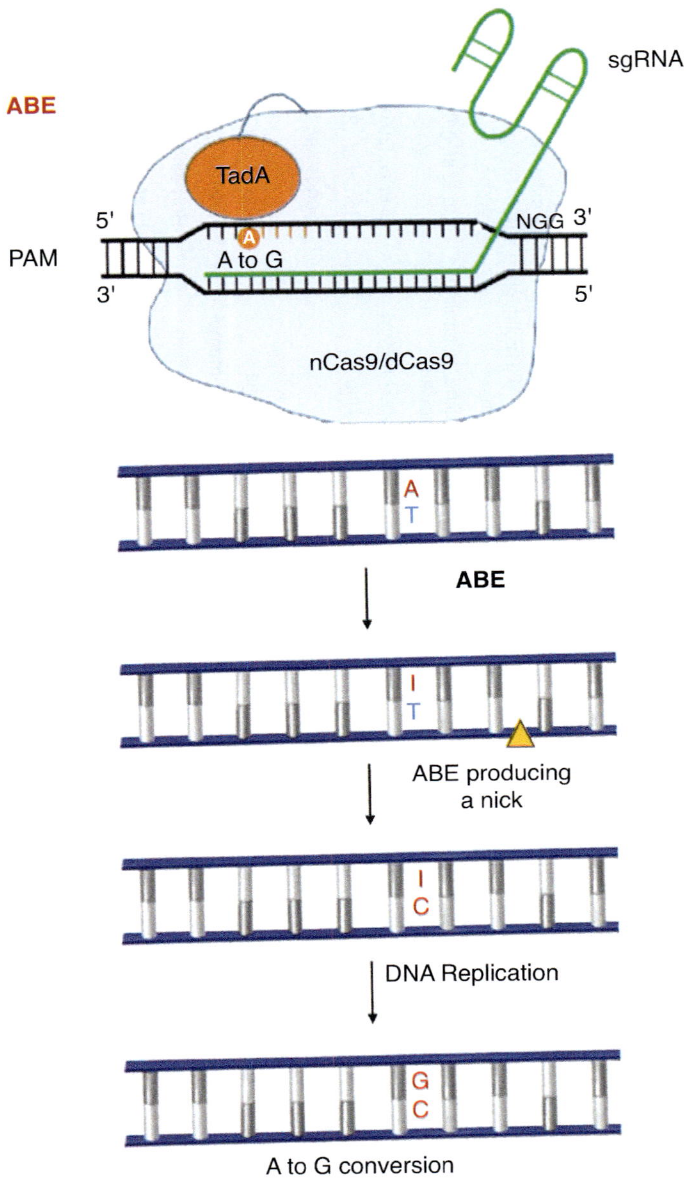

Fig. 15.2. Mechanism of adenine base editing. (Figure constructed by Asadullah.)

successfully developed for generating HT crops. This exploits an RT (abbreviated as reverse-transcriptase) associated with a prime editing guide RNA or pegRNA and CRISPR/Cas9 nickase (H840A), that facilitated all 12 possible base-to-base replacements, as well as InDels, without requiring the presence of donor DNA. Once inside the plant cells, the PE complex, composed of the PE protein, locates the target gene, guided by the pegRNA, and then creates a nick in one strand of the DNA at the target site (Anzalone *et al.*, 2019). Afterward, the protein

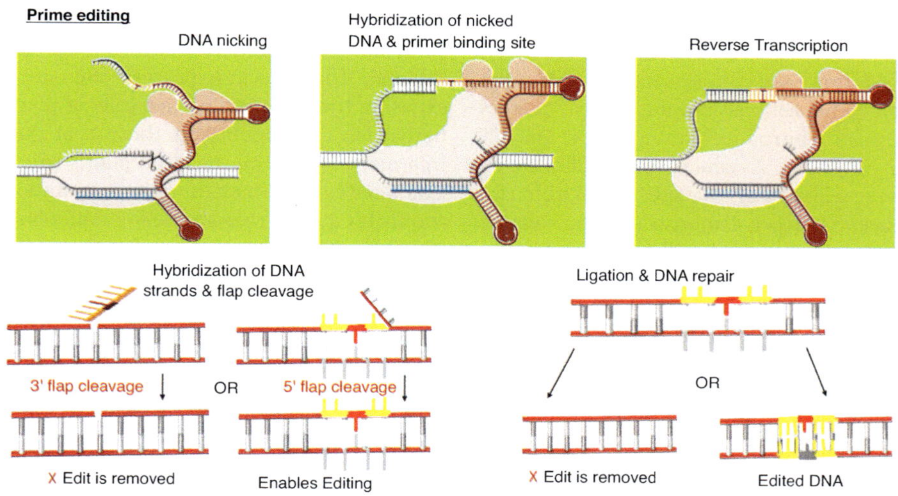

Fig. 15.3. Work mechanism of prime editing. (Figure constructed by Asadullah.)

uses its integrated RT activity to copy the information from the pegRNA template and insert it into the nicked DNA strand (Fig. 15.3). This results in precise nucleotide changes that can modify the protein encoded by the target gene, making it HT. Prime editing demonstrates either higher or comparable efficiency to HDR, with fewer by-products, and it possesses strengths and weaknesses that complement those of base editing (Anzalone *et al.*, 2019). Additionally, at recognized Cas9 sites, PE induces significantly lower off-target mutation rates than Cas9 nucleases.

It is important to note that these PEs are currently in their early stages of development, as indicated by the significantly low editing reported by many research groups. However, there have been recent reports of notable progress, with 92% efficiency in these PEs of the *StALS* gene in tetraploid potatoes (Veillet *et al.*, 2020).

15.6 Base-editing-mediated Gene Evolution

The spontaneous and natural mutations have been fundamental to the evolutionary processes shaping the diversity of life over time. However, with advancements in gene editing technologies, researchers are now able to simulate and accelerate the natural evolution process in plants. Indeed, a method known as base-editing-mediated gene evolution (BEMGE) has recently been designed and applied to modify the *OsALS1* gene within the cells of rice plants, employing particle bombardment along with Agrobacterium-mediated transformation. Consequently, the analysis revealed the replacement of four amino acids at two specific loci (R190 and P171), imparting HT to bispyribac sodium, whereas these loci were unrecognized for their role in HT previously and had never undergone human and natural selection during rice domestication. Moreover, amino acid substitution at P171F, found within the robust allele of gene *OsALS1* was promptly integrated in rice variety (Nangeng-46) using specific base-editing facilitated by relevant single guide RNA or sgRNA (Kuang *et al.*, 2020).

15.7 Saturated Targeted Endogenous Mutagenesis Editors

Targeted saturation mutagenesis of crop genes holds the potential for generating genetic variants with enhanced agronomic traits. Nevertheless, the required equipment for the directed evolution of plant genes, including shuffling of DNA or error-prone PCR, are currently constrained (Engqvist and Rabe, 2019). Li *et al.* (2020a) employed dual base editors—namely, saturated targeted endogenous mutagenesis editors (STEMEs)—to introduce mutations at the CT domain of *OsACC2*, which was facilitated by employing a library of 200 sgRNAs. Within rice protoplast, STEME-1 demonstrated the capability to edit cytosine and adenine concurrently at identical target sites, achieving a C > T efficiency of up to 62%, along with simultaneous C > T and A > G efficiency reaching up to 15%. Similarly, STEME-NG, employing the nickase Cas9-NG PAM variant, was applied alongside 20 distinct sgRNAs in rice protoplast to achieve nearly saturated mutagenesis (73%) across a 56-amino-acid segment of the rice acetyl-coenzyme A carboxylase (*OsACC*). Through this approach, new variants of the *OsACC2* gene were generated, which imparted HT specifically to haloxyfop.

15.8 Prime-editing-library-mediated Saturation Mutagenesis

A genetic strategy called prime-editing-library-mediated saturation mutagenesis (PLSM) has recently been created to introduce diverse substitutions of amino acids at specifically designated positions within the rice genome. This process typically begins with the creation of a library containing a diverse range of potential edits in the *OsACC1* gene (D2176G, W2097G/S/L, P1927Y/F, I2139N/S/V, G2194S/A, and 11,879L/M/S/T/V), through prime editing technology. These residues were independently targeted using pegRNA, encompassing all potential mutations in the amino acid at precise positions. Once the library is generated, it is introduced into rice cells or plants, and those carrying potentially resistant mutations are selected. This selection process often involves exposing the transformed rice cells or plants to herbicides and identifying those that survive treatment. Surviving cells or plants are then examined to determine the specific mutations in the *OsACC1* gene responsible for conferring HR (Xu *et al.*, 2021).

15.9 Conclusions and Prospects

Cultivating HT crops provides farmers with a valuable option for effective weed control, ultimately leading to increased crop yields. In recent decades, traditional breeding, combined with transgenic methods and mutagenesis, has been instrumental in advancing the generation of HR crops. The optimal replacement for conventional gene editing methods for developing HT plants is the creation of transgene-free varieties through targeted genome editing. CRISPR technologies, specifically prime and base editing, hold significant potential for creating HR plants without the incorporation of transgenes. It is certainly believable that concerned genes and their monitoring sites are crucial for the metabolism or interaction of particular chemicals (exclusively, non-selective herbicides and environmentally friendly) that can be artificially evolved within plants. This could be achieved through the utilization of gRNA in conjunction with prime and base-editing systems. Presently, scientists are finding numerous mutations associated with HT worldwide. With this valuable evidence and relevant information, the creation of new, essential crop varieties, such as rice, soybean, and maize, exhibiting broad-spectrum resistance to several herbicides, can be achieved through successive rounds of genetic engineering.

References

Anzalone, A.V., Randolph, P.B., Davis, J.R., Sousa, A.A., Koblan, L.W. *et al.* (2019) Search-and-replace genome editing without double-strand breaks or donor DNA. *Nature* 576(7785), 149–157.

Butler, N.M., Baltes, N.J., Voytas, D.F. and Douches, D.S. (2016) Geminivirus-mediated genome editing in potato (*Solanum tuberosum* L.) using sequence-specific nucleases. *Frontiers in Plant Science* 7, 1045.

Butt, H., Eid, A., Momin, A.A., Bazin, J., Crespi, M. *et al.* (2019) CRISPR directed evolution of the spliceosome for resistance to splicing inhibitors. *Genome Biology* 20(1), 1–9.

Butt, H., Rao, G.S., Sedeek, K., Aman, R., Kamel, R. *et al.* (2020) Engineering herbicide resistance via prime editing in rice. *Plant Biotechnology Journal* 18(12), 2370–2372.

Chauhan, B.S. (2020) Grand challenges in weed management. *Frontiers in Agronomy* 1, 3.

Chen, Y.Y., Wang, Z.P., Ni, H.W., Xu, Y., Chen, Q.J. *et al.* (2017) CRISPR/Cas9- mediated base-editing system efficiently generates gain-of-function mutations in arabidopsis. *Science China Life Sciences* 60(5), 520–523.

Engqvist, M.K. and Rabe, K.S. (2019) Applications of protein engineering and directed evolution in plant research. *Plant Physiology* 179(3), 907–917.

Gaudelli, N.M., Komor, A.C., Rees, H.A., Packer, M.S., Badran, A.H. *et al.* (2017) Programmable base editing of A•T to G•C in genomic DNA without DNA cleavage. *Nature* 551(7681), 464–471.

Han, H., Wu, Z., Zheng, L., Han, J., Zhang, Y. *et al.* (2022) Generation of a high-efficiency adenine base editor with TadA8e for developing wheat dinitroaniline-resistant germplasm. *The Crop Journal* 10(2), 368–374.

Hua, K., Jiang, Y., Tao, X. and Zhu, J.K. (2020) Precision genome engineering in rice using prime editing system. *Plant Biotechnology Journal* 18(11), 2167.

Hummel, A.W., Chauhan, R.D., Cermak, T., Mutka, A.M., Vijayaraghavan, A. *et al.* (2018) Allele exchange at the EPSPS locus confers glyphosate tolerance in cassava. *Plant Biotechnology Journal* 16(7), 1275–1282.

Jiang, Y.Y., Chai, Y.P., Lu, M.H., Han, X.L., Lin, Q.P. *et al.* (2020) Prime editing efficiently generates W542L and S621I double mutations in two ALS genes in maize. *Genome Biology* 21, 1–10.

Komor, A.C., Kim, Y.B., Packer, M.S., Zuris, J.A. and Liu, D.R. (2016) Programmable editing of a target base in genomic DNA without double-stranded DNA cleavage. *Nature* 533(7603), 420–424.

Kuang, Y., Li, S., Ren, B., Yan, F., Spetz, C. *et al.* (2020) Base-editing-mediated artificial evolution of *OsALS1* in planta to develop novel herbicide-tolerant rice germplasms. *Molecular Plant* 13(4), 565–572.

Li, C., Zong, Y., Wang, Y.P., Jin, S., Zhang, D.B. *et al.* (2018) Expanded base editing in rice and wheat using a Cas9-adenosine deaminase fusion. *Genome Biology* 19, 1–9.

Li, C., Zhang, R., Meng, X., Chen, S., Zong, Y. *et al.* (2020a) Targeted, random mutagenesis of plant genes with dual cytosine and adenine base editors. *Nature Biotechnology* 38(7), 875–882.

Li, H., Li, J., Chen, J., Yan, L. and Xia, L. (2020c) Precise modifications of both exogenous and endogenous genes in rice by prime editing. *Molecular Plant* 13(5), 671–674.

Li, J., Meng, X.B., Zong, Y., Chen, K.L., Zhang, H.W. *et al.* (2016a) Gene replacements and insertions in rice by intron targeting using CRISPR-Cas9. *Nature Plants* 2(10), 1–6.

Li, T., Liu, B., Chen, C.Y. and Yang, B. (2016b) TALEN-mediated homologous recombination produces site-directed DNA base change and herbicide-resistant rice. *Journal of Genetics and Genomics* 43(5), 297–305.

Li, Y.M., Zhu, J.J., Wu, H., Liu, C.L., Huang, C.L. *et al.* (2020b) Precise base editing of non-allelic acetolactate synthase genes confers sulfonylurea herbicide resistance in maize. *The Crop Journal* 8(3), 449–456.

Li, Z.S., Liu, Z.B., Xing, A.Q., Moon, B.P., Koellhoffer, J.P. *et al.* (2015) Cas9-guide RNA directed genome editing in soybean. *Plant Physiology* 169(2), 960–970.

Liu, L., Kuang, Y., Yan, F., Li, S., Ren, B. *et al.* (2021) Developing a novel artificial rice germplasm for dinitroaniline herbicide resistance by base editing of OsTubA2. *Plant Biotechnology Journal* 19(1), 5–7.

Liu, X.S., Qin, R.Y., Li, J., Liao, S.X., Shan, T.F. *et al.* (2020) A CRISPR-Cas9-mediated domain-specific base-editing screen enables functional assessment of accase variants in rice. *Plant Biotechnology Journal* 18(9), 1845–1847.

Lu, Y., Wang, J.Y., Chen, B., Mo, S.D., Lian, L. *et al.* (2021) A donor-DNA-free CRISPR/CAS-based approach to gene knock-up in rice. *Nature Plants* 7(11), 1445–1452.

Manghwar, H., Li, B., Ding, X., Hussain, A., Lindsey, K. *et al.* (2020) CRISPR/CAS systems in genome editing: methodologies and tools for sgRNA design, off-target evaluation, and strategies to mitigate off-target effects. *Advanced Science* 7(6), 1902312.

Molla, K.A., Sretenovic, S., Bansal, K.C. and Qi, Y.P. (2021) Precise plant genome editing using base editors and prime editors. *Nature Plants* 7(9), 1166–1187.

Oz, M.T., Altpeter, A., Karan, R., Merotto, A. and Altpeter, F. (2021) CRISPR/Cas9-mediated multi-allelic gene targeting in sugarcane confers herbicide tolerance. *Frontiers in Genome Editing* 3, 673566.

Packer, M.S. and Liu, D.R. (2015) Methods for the directed evolution of proteins. *Nature Reviews Genetics* 16, 379–394.

Ran, Y.D., Patron, N., Kay, P., Wong, D., Buchanan, M. *et al.* (2018) Zinc finger nuclease-mediated precision genome editing of an endogenous gene in hexaploid bread wheat (*Triticum aestivum*) using a DNA repair template. *Plant Biotechnology Journal* 16(12), 2088–2101.

Richter, M.F., Zhao, K.V.T., Eton, E., Lapinaite, A., Newby, G.A. *et al.* (2020) Phage-assisted evolution of an adenine base editor with improved Cas domain compatibility and activity. *Nature Biotechnology* 38(7), 883–884.

Sauer, N.J., Narvaez-Vasquez, J., Mozoruk, J., Miller, R.B., Warburg, Z.J. *et al.* (2016) Oligo-nucleotide mediated genome editing provides precision and function to engineered nucleases and antibiotics in plants. *Plant Physiology* 170(4), 1917–1928.

Shimatani, Z., Kashojiya, S., Takayama, M., Terada, R., Arazoe, T. *et al.* (2017) Targeted base editing in rice and tomato using a CRISPR-Cas9 cytidine deaminase fusion. *Nature Biotechnology* 35(5), 441–443.

Sun, Y.W., Zhang, X., Wu, C.Y., He, Y.B., Ma, Y.Z. *et al.* (2016) Engineering herbicide-resistant rice plants through CRISPR/Cas9-mediated homologous recombination of acetolactate synthase. *Molecular Plants* 9(4), 628–631.

Svitashev, S., Young, J.K., Schwartz, C., Gao, H.R., Falco, S.C. *et al.* (2015) Targeted mutagenesis, precise gene editing, and site-specific gene insertion in maize using Cas9 and guide RNA. *Plant Physiology* 169(2), 931–945.

Tian, S.W., Jiang, L.J., Cui, X.X., Zhang, J., Guo, S.G. *et al.* (2018) Engineering herbicide-resistant watermelon variety through CRISPR/Cas9-mediated base-editing. *Plant Cell Reports* 37(9), 1353–1356.

Townsend, J.A., Wright, D.A., Winfrey, R.J., Fu, F., Maeder, M.L. *et al.* (2009) High-frequency modification of plant genes using engineered zinc-finger nucleases. *Nature* 459(7245), 442–445.

Veillet, F., Perrot, L., Chauvin, L., Kermarrec, M.P., Guyon-Debast, A. *et al.* (2019) Transgene-free genome editing in tomato and potato plants using agrobacterium-mediated delivery of a CRISPR/Cas9 cytidine base editor. *International Journal of Molecular Sciences* 20(2), 402.

Veillet, F., Kermarrec, M.P., Chauvin, L., Guyon-Debast, A., Chauvin, J.E. *et al.* (2020) Prime editing is achievable in the tetraploid potato, but needs improvement. *BioRxiv*. DOI: 10.1101/2020.06.18.159111.

Wang, F., Xu, Y., Li, W., Chen, Z., Wang, J. *et al.* (2021a) Creating a novel herbicide-tolerance OsALS allele using CRISPR/Cas9-mediated gene editing. *The Crop Journal* 9(2), 305–312.

Wang, F.Q., Yang, J., Fan, F.J., Li, W.Q., Wang, J. *et al.* (2018) Development and application of the functional marker for imidazolinone herbicides resistant ALS gene in rice. *Acta Agronomica Sinica* 44(3), 324–331.

Wang, M., Wang, Z., Mao, Y., Lu, Y., Yang, R. *et al.* (2019) Optimizing base editors for improved efficiency and expanded editing scope in rice. *Plant Biotechnology Journal* 17(9), 1697–1699.

Wang, Z.A.R., Wan, L.L., Xin, Q., Zhang, X.H., Song, Y.X. *et al.* (2021b) Optimizing glyphosate tolerance in rapeseed by CRISPR/Cas9-based geminiviral donor DNA replicon system with Csy4-based single-guide RNA processing. *Journal of Experimental Botany* 72(13), 4796–4808.

Wu, J., Chen, C., Xian, G.Y., Liu, D.X., Lin, L. *et al.* (2020) Engineering herbicide-resistant oilseed rape by CRISPR/Cas9-mediated cytosine base-editing. *Plant Biotechnology Journal* 18(9), 1857–1859.

Xu, R., Li, J., Liu, X., Shan, T., Qin, R. *et al.* (2020) Development of plant prime-editing systems for precise genome editing. *Plant Communications* 1(3), 100043.

Xu, R., Liu, X., Li, J., Qin, R. and Wei, P. (2021) Identification of herbicide resistance OsACC1 mutations via in planta prime-editing-library screening in rice. *Nature Plants* 7(7), 888–892.

Yan, D., Ren, B., Liu, L., Yan, F., Li, S. *et al.* (2021) High-efficiency and multiplex adenine base editing in plants using new tada variants. *Molecular Plants* 14(5), 722–731.

Zhang, D., Hussain, A., Manghwar, H., Xie, K., Xie, S. *et al.* (2020) Genome editing with the CRISPR-Cas system: an art, ethics, and global regulatory perspective. *Plant Biotechnology Journal* 18(8), 1651–1669.

Zhang, R., Liu, J.X., Chai, Z.Z., Chen, S., Bai, Y. *et al.* (2019) Generation of herbicide tolerance traits and a new selectable marker in wheat using base editing. *Nature Plants* 5(5), 480–485.

Zhang, R., Chen, S., Meng, X.B., Chai, Z.Z., Wang, D.L. *et al.* (2021) Generating broad-spectrum tolerance to ALS-inhibiting herbicides in rice by base editing. *Science China Life Sciences* 64(10), 1624–1633.

16 Genome Editing in Improving Photosynthesis in Crop Plants

Sherien Bukhat[1], Sumaira Rasul[1], Habib-ur-Rehman Athar[2] and Hamid Manzoor[1*]

[1]*Plant Cell Signaling Laboratory, Institute of Molecular Biology and Biotechnology, Bahauddin Zakariya University, Multan, Pakistan;* [2]*Institute of Botany, Bahauddin Zakariya University, Multan, Pakistan*

Abstract

To meet the needs of the growing global population, there is an urgent need to increase agricultural production. However, this task has become more challenging due to changing climatic conditions resulting from human activities. The traditional breeding has led to steady enhancements in crop yield; a "yield ceiling" may be reached in the near future. Photosynthesis is the fundamental process that underpins primary productivity on Earth. Increasing evidence suggests that enhancing photosynthetic efficiency can pave the way for increased crop yields. These improvements in yield have been achieved through strategies that target specific photosynthetic processes, including manipulation of photosystems, electron transport, ribulose-1,5-biphosphate carboxylase/oxygenase (RuBisCO) for fixation of CO_2, the photorespiratory pathway, and by altering the distribution of photosynthates within the plant. Improving photosynthetic machinery represents a promising strategy for boosting crop production, as simulations have shown that it has the potential to significantly enhance crop productivity. Genome editing, a rapidly advancing technology, has the potential to facilitate such improvements by allowing precise modification of plant genes for enhancing the efficiency of photosynthesis in plants. This technology enables precise alterations of plant genes involved in photosystem I (PSI) and photosystem II (PSII), potentially improving their light-harvesting capacity and non-photochemical quenching. Moreover, transgenic plants incorporating such modified genes have shown significant improvements in photosynthetic efficiency. Genome editing techniques have not only made advancements in photosynthesis research but will also offer exciting possibilities for crop improvement and increasing global food security. Thus, this chapter discusses various genome editing tools employed for targeted manipulation of photosynthetic pathways, including zinc-finger nucleases (ZFNs), clustered regularly interspaced palindromic repeat (CRISPR)-CRISPR-associated (Cas) (CRISPR/Cas9), and transcription activator-like effector nucleases (TALENs). The potential of genome editing to improve photosynthesis by identifying the targets that can be engineered to improve crop productivity is explored alongside the unintended consequences of these manipulations and the inherent limitations of these techniques.

Keywords: Genome editing, Photosynthesis, RuBisCO, PSI and PSII, ZFNs, TALENs, CRISPR/Cas9

*Corresponding author: dr.hamid@bzu.edu.pk

DOI: 10.1079/9781800622517.0016

16.1 Introduction

Crop productivity is a critical component of food security and economic development worldwide. Photosynthesis is a fundamental physiological process governing plant growth and yield. Despite its critical importance for plant growth, photosynthesis is not an efficient process, and crop yields are often limited by the rate of photosynthesis (Archer and Barber, 2004). However, crop photosynthetic capacity and its translation into yield are limited by various (a)biotic stresses (Spiertz and Ewert, 2009). Abiotic stresses can cause photooxidative damage or photoinhibition to photosystems, reduction in carboxylation efficiency of RuBisCO, activities of Calvin cycle key enzymes, export of soluble sugars, and so on, leading to reduced photosynthetic efficiency in plants. Several adaptive mechanisms have evolved in plants for mitigating the detrimental effects of abiotic stress on photosynthesis, such as developing photoprotection of photosystems (non-photochemical quenching, cyclic electron transport, activities of antioxidant enzymes, etc.), stomatal regulation, photorespiration, and CO_2-concentrating mechanisms (CCMs). With the advancement of molecular biology-based techniques and system biology approaches, it is possible to understand the link between crop yield and photosynthesis under abiotic stress conditions. Such approaches will help in developing stress-tolerant crop cultivars, stress management practices, and the application of growth regulators to enhance plant tolerance against abiotic stresses.

Genome editing technologies are emerging as powerful techniques for improving crop productivity by introducing desirable traits and enhancing the resilience of crops to environmental stresses (Gao, 2021). Genome editing technique could be used to optimize plant photosynthetic efficiency in crops (Karavolias *et al.*, 2021a). This includes optimization of antenna size by manipulating photosynthetic pigments, activity of RuBisCO and RuBisCO activase for enhancement of CO_2 fixation, regulating linear and cyclic electron transport from PSII to PSI by manipulating components of electron transport such as *cyt*b_6f complex, PGR5, and PTOX. These interventions aim to prevent the generation of reactive oxygen species. Additionally, the introduction of metabolic pathways like the CCM or re-routing photorespiratory pathways further contribute to the optimization of plant photosynthesis in crops (Simkin *et al.*, 2022). For example, by introducing the C_4 photosynthetic pathway in rice, researchers were able to increase its photosynthetic efficiency and yield (Hibberd *et al.*, 2008). In addition, researchers are using CRISPR/Cas9 to target genes involved in the C_4 photosynthesis pathway, which is more effective than the more common C_3 pathway (Zheng *et al.*, 2021b). By modifying genes related to this pathway, scientists hope to develop crops that can grow more efficiently in environments with limited resources, including water and nitrogen. Similarly, by introducing a mutation in the *RuBisCO* gene, which is the rate-limiting enzyme of CO_2 fixation, researchers were able to enhance yield and photosynthetic efficiency in soybean (Yang *et al.*, 2022). Another area of research involves improving the light-harvesting capacity of plant cells through modifications to genes involved in the antenna complex of chloroplasts (Friedland *et al.*, 2019). This could lead to crop plants with improved photosynthetic efficiency and higher yields.

In this chapter, a comprehensive overview of the current state of genome editing in improving photosynthesis, the various genome editing techniques being used to improve photosynthesis in plants, including the modification of photosynthetic pigments, the optimization of the photosynthetic apparatus, and the manipulation of CO_2 fixation pathway, will be discussed.

16.2 Genome Editing Technologies

Genome editing tools allow for precise, efficient, and targeted gene modification in plants, thus revolutionizing the development of transgenic plants with ease. Sequence-specific nucleases (SSNs) recently have gained popularity in genome editing of plants (Wang and Doudna, 2023). There are currently four types of SSNs-based technologies: TALENs, ZFNs, CRISPR/Cas, and mega-nucleases (Abdallah *et al.*, 2015; Saeed *et al.*, 2022). These technologies work by recognizing the targeted-DNA sequence and creating double-strand breaks (DSBs) in DNA at the specific target site. These DSBs are

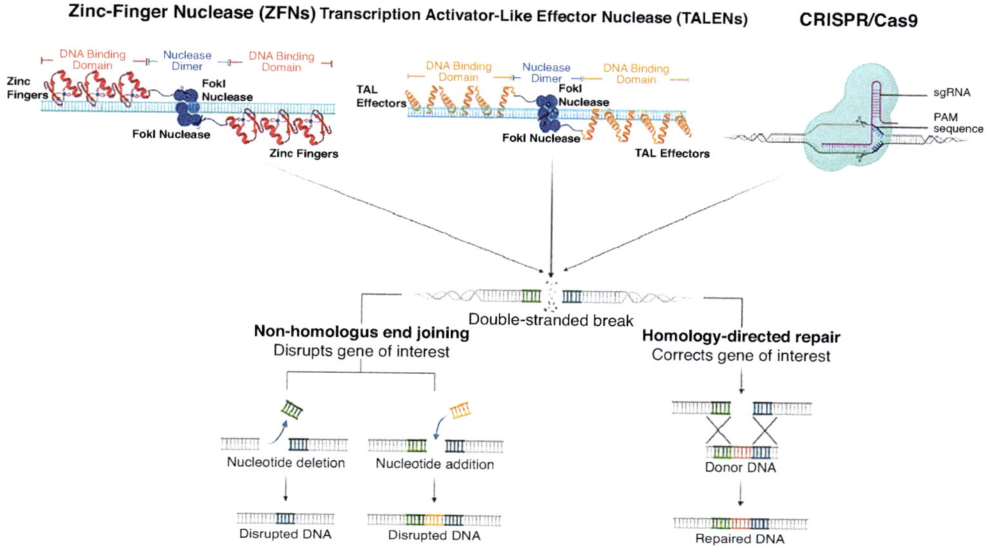

Fig. 16.1. Genome editing process in plants. ZFNs have two distinct parts, a DNA-binding domain and a DNA-cleavage domain, joined by a short linker sequence. The cleavage site can be regulated by modifying this recognition domain. Each zinc-finger domain in the DNA-binding domain is specialized to identify three specific base pairs. In contrast, TALENs feature the DNA-binding domain and the *Flavobacterium okeanokoites* (*Fok*I) non-specific cleavage domain that could be programmed. Two transcription activator-like effectors (TALE) DNA-binding sites with repeat sequences of 34 amino acids are found in the TALENs' recognition domain. The identification of a single nucleotide is facilitated by the amino acids at positions 12 and 13 of each repeat sequence. A gRNA sequence upstream of a protospacer neighbouring motif is fused to a gRNA scaffold sequence in the CRISPR/Cas9 system. To create small deletions or insertions, nucleotide changes, or knock-ins, gRNAs lead Cas9 RNA-guided endonucleases to the target location, where they induce a double-stranded DNA break that is repaired via NHEJ or homology-directed repair (HDR). Author's own figure.

repaired by cells' own repair machinery using homologous recombination or with the help of exogenous DNA using non-homologous end-joining (NHEJ) (Fig. 16.1). These NHEJs with the help of exogenous DNA produce the desired modification in the sequence (Gaj *et al.*, 2013; Saeed *et al.*, 2023). Editing the chloroplast genome to improve photosynthesis remains a significant challenge. However, researchers are actively pursuing advancements in this field, utilizing genome editing techniques as shown in Table 16.1.

TALENs and ZFNs both possess tandem repeats within their DNA-binding domains, which can be modified to identify specific sequences of DNA. This enables the creation of chimeric nucleases that can be directed to target sequences within the genome, inducing DSBs. However, the major challenge in the widespread application of both of these gene editing methods lies in the need to engineer a new TALEN or ZFN chimeric protein for each target site, making the process complex, costly, and labor intensive. Moreover, the major drawback of both these techniques has been the cytotoxicity resulting from increased off-target effects. The complexities associated with ZFNs and TALENs have limited their utilization in plant genome editing (Shamshirgaran *et al.*, 2022). The introduction of the CRISPR/Cas system in 2012 marked a breakthrough in overcoming the limitations of the TALENs and ZFNs systems in gene editing. Ongoing efforts to improve genetic engineering efficiency have significantly

Table 16.1. Chloroplast genome editing for photosynthesis improvement. Author's own table.

GE technique	Chloroplast gene	Plant	References
Plastid-targeted TALE deaminase (ptTALED)	D1 photosynthetic protein (*psbA*)	Arabidopsis Lettuce Rapeseed	Kang *et al.*, 2021; Nakazato *et al.*, 2021; Mok *et al.*, 2022
ptTALED	CP-47 photosynthetic protein (ps*bB*)	Rapeseed Lettuce	Kang *et al.*, 2021
ptTALED	rbcL (RuBisCO large subunit)	Arabidopsis Lettuce	Ahmad *et al.*, 2021
ptTALED	*rpoC1* (RNA polymerase β subunit)	Arabidopsis	Nakazato *et al.*, 2021
CRISPR/Cas9 and ptTALED	PSI chlorophyll *a* (*psaA*)	Arabidopsis Rice Lettuce Chlamydomonas	Yoo *et al.*, 2020; Li *et al.*, 2021

Table 16.2. Comparison of different genome editing techniques. Author's own table.

Feature	ZFNs	TALENs	CRISPR/Cas
Off-target effects	Low-to-moderate	Low	Low-to-moderate
Targeted editing	Complex and time consuming	Moderately complex and laborious	Easy and efficient
Multiplexing capability	Limited	Moderate	High throughput
Scalability for large libraries	Complex and laborious for protein engineering	Extensive cloning required	Simple and easy oligo synthesis

enhanced the applications of the CRISPR/Cas system across various domains of plant science. As a result, it is emerging as a highly promising tool for genome engineering (Shakeel *et al.*, 2021). Comparison of all three techniques is given in Table 16.2.

The gene editing mechanism in CRISPR/Cas shares similarities with ZFNs and TALENs. However, the CRISPR/Cas system uses a sequence-specific binding mechanism based on the complementary interaction between guide RNA (gRNA) and the targeted sequence of DNA (Fig. 16.1). This complex possesses nuclease activity and can precisely cleave double-stranded DNA at the targeted site, making the system site-specific and programmable (Fonfara *et al.*, 2016; Karvelis *et al.*, 2020). On the basis of these advantages, the CRISPR/Cas system stands out as more robust, easy to design, cost-effective, and efficient compared to TALEN- and ZFN-based genome editing techniques (Periwal, 2017). Recent progress in CRISPR/Cas technologies, such as prime editing and base editing, has further enhanced its application in genome engineering. These advancements have increased the precision of CRISPR/Cas technology, reducing or eliminating off-target effects (Miladinovic *et al.*, 2021).

This system finds applications in the genetic modification of cells in living organisms or cultures (Manghwar *et al.*, 2019). To modify cells in culture, viral or plasmid vectors are used for delivering the components of this system for consistent and high expression. For modification of plants, protoplasts are cultured and a plasmid containing CRISPR/Cas components is introduced (Kuluev *et al.*, 2019). Another method, employed for plants, involves *Agrobacterium*, a natural 'genetic engineer' containing a specific plasmid that harbors the system of CRISPR/Cas9 (Zhang *et al.*, 2020). CRISPR system has also been used to improve photosynthesis in plants by modifying genes that regulate the photosynthetic process, as discussed later.

16.3 CRISPR-mediated Improvement in Photosynthetic Efficiency

CRISPR/Cas technology has successfully enhanced agronomic characteristics in various crops. Improving productivity through the increased photosynthetic efficiency serves as a key example. Numerous approaches, often combined, have been proposed to improve photosynthetic efficiency, which include improving light capture capacity, increasing flux through the Calvin cycle, and reducing photorespiration (Zhu *et al.*, 2007; Ort *et al.*, 2015; South *et al.*, 2018). Recent studies conducted in both field and laboratory environment using genetically modified plants, demonstrate that improving photosynthetic efficiency stands as a transformative strategy with the potential to substantially enhance crop yields (López-Calcagno *et al.*, 2019).

The rapid screening of multiplexing approaches made possible by CRISPR/Cas is more efficiently conducted in plants, accelerating the advancement of more complex approaches, particularly in processes that involve the regulation of numerous genes. For example, in C_3 plants, the process of photosynthetic carbon assimilation is supported by 38 different enzymes. Moreover, multiple regulatory elements and chaperones, encoded by diverse gene families, play essential roles in fine tuning and orchestrating this fundamental pathway (Zhu *et al.*, 2007).

Studies have shown that modifying the activity of certain enzymes can improve photosynthesis. These enzymes include RuBisCO in the photorespiratory pathway and Calvin–Benson (CB) enzymes (Zhu *et al.*, 2007). One approach to increase photosynthesis involves overexpressing a single CB enzyme (sedoheptulose-1, 5-bisphosphatase; SBPase). This has been shown to be effective in various plants, including tomato and wheat (Lefebvre *et al.*, 2005; Driever *et al.*, 2017). Further research has explored overexpressing SBPase in combination with another CB enzyme such as fructose-1,6-bisphosphate aldolase, in tobacco and Arabidopsis (Simkin *et al.*, 2015; Simkin *et al.*, 2017). These successful examples of manipulating multiple enzymes for photosynthesis improvement pave the way for similar strategies in crops using CRISPR/Cas. This technology offers exciting possibilities, like directly modifying existing enzyme activity or influencing their expression through promoter manipulation. However, to fully harness CRISPR/Cas for commercial applications, a deeper understanding of other photosynthesis-related pathways is crucial. This includes C_4 and CAM photosynthesis, photorespiration, and CCM. CRISPR/Cas can be a powerful tool in model species to rapidly explore the activity and regulation of these pathways, ultimately leading to the identification of new targets for editing or manipulation in crops.

Genetic engineering faces a significant challenge when dealing with enzymes encoded by gene families. Their complex nature, involving multiple isoforms and complex regulation, makes them difficult to target and manipulate (Abdallah *et al.*, 2015). One such example is RuBisCO; this key enzyme fixes CO_2 and dictates the overall photosynthetic rate. However, its activity is limited by its low catalytic efficiency (k_{cat}), affinity for CO_2 (K_c), and preference for oxygen over CO_2 ($S_{c/o}$) under various environmental conditions. RuBisCO is composed of large and small subunits encoded by separate genes: rbcL in the chloroplast and a multigene family, rbcS, in the nucleus. The small subunit of RuBisCO (rbcS) exhibits a high degree of diversity, with several isoforms ranging from four in Arabidopsis to as many as 22 in other species (Crafts-Brandner and Salvucci, 2000). Additionally, its assembly with the large subunit (rbcL), encoded by the plastid, requires multiple species-specific chaperones in plants (Aigner *et al.*, 2017). CRISPR/Cas technology offers promising solutions for manipulating enzymes encoded by gene families.

For example, CRISPR/Cas presents an alternative approach in modifying the catalytic site directly on the plastid-encoded larger subunit, which is currently limited to specific plants like tobacco. This approach focuses on the nuclear-encoded rbcS subunit, which also plays a role in RuBisCO catalysis (Atkinson *et al.*, 2017). Thus, by targeting rbcS with CRISPR/Cas, we can potentially overcome limitations associated with plastid transformation and engineer RuBisCO activity even in species where it is challenging.

Researchers recently discovered a new group of RuBisCO small subunits called rbcS-T.

These subunits are found uniquely in plant organs with specialized metabolic functions, such as trichomes (Pottier *et al.*, 2018). Studies suggest that naturally occurring rbcS-T isoforms likely interact with their existing chaperones but may influence the RuBisCO catalytic activity (Laterre *et al.*, 2017). This opens up the possibility that replacing mesophyll cells located in rbcS with the rbcS-T isoform could potentially enhance leaf CO_2 uptake. However, not all plant species possess rbcS-T isoforms (Pottier *et al.*, 2018). In such cases, expressing isoforms significantly different from the native family could be another strategy. While studies suggest that chaperones involved in the assembly of rbcS are highly species-specific, the exact mechanisms underlying this specificity in plants remain largely unknown. CRISPR/Cas techniques like base editing or homology-directed repair (HDR) could be powerful tools to replace or modify existing chaperones within plant cells (Aigner *et al.*, 2017). This approach has the potential to significantly advance our understanding of how these chaperones influence the assembly of RuBisCO and its large subunit (rbcL).

CRISPR/Cas technology has the potential to revolutionize existing methods for improving photosynthesis. It could not only enhance current strategies like enzyme overexpression (as HDR techniques become more efficient) but also address critical challenges like manipulating RuBisCO and its subunits. Recent advancements in plastid transformation protocols across various plant species suggest the possibility of reliably engineering both nuclear and plastid-encoded photosynthetic enzymes and their regulatory proteins (Martin Avila *et al.*, 2016; Yu *et al.*, 2017). This opens doors to fine-tuning the entire photosynthetic pathway for improved efficiency. Studies utilizing CRISPR/Cas editing have shown promising results in enhancing plant photosynthesis. For instance, researchers targeted the negative regulator of photosynthesis 1 (*NRP1*) gene, which acts like a transcription factor regulating genes involved in photosynthesis. By editing *NRP1* with CRISPR/Cas, scientists were able to improve rice grain yield, biomass production, and photosynthetic efficiency (Chen *et al.*, 2021).

A recent study employed CRISPR/Cas technology in rice (a C_3 plant) to disrupt the rbcS multigene family and replace its RuBisCO machinery with that of sorghum, a C_4 plant. This resulted in rice plants with C_4-like photosynthetic rates and increased productivity. The genetically modified plants displayed improved k_{cat}, K_c, and reduced $S_{c/o}$, reflecting enhanced overall efficiency. This successful manipulation highlights the potential of CRISPR/Cas for engineering C_4-like traits in C_3 crops, potentially leading to significant improvements in photosynthetic activity. CRISPR/Cas technology also has the potential for enhancing photosynthetic efficiency in both polyploid and diploid crop plants. Studies have shown successful applications in diverse species, including tobacco, a tetraploid model crop (Matsumura *et al.*, 2020 #2318). For instance, researchers simultaneously targeted and knocked out multiple RuBisCO small subunit homologs (*rbcS_S1b*, *rbcS_T1*, and *rbcS_S1a*) in tobacco using CRISPR/Cas. The resulting mutant plants exhibited significantly higher photosynthetic rates compared to their wild-type counterparts (Batista-Silva *et al.*, 2020; Matsumura *et al.*, 2020).

Researchers have used CRISPR/Cas9 to knock out genes that inhibit photosynthesis in plants. This system has also been used for the modification of genes coding for a protein called PSII subunit S (PsbS) in tobacco plants. PsbS is a protein that regulates the amount of light energy absorbed by plants during photosynthesis. By knocking out the PsbS gene, the researchers were able to improve the efficiency of photosynthesis in tobacco plants by up to 15% (Głowacka *et al.*, 2018). These successful manipulations exemplify the ability of CRISPR/Cas to engineer diverse plant genomes, opening doors for the development of highly efficient photosynthetic systems across both polyploid and diploid crops. Several other potential targets for photosynthetic improvement using CRISPR/Cas are discussed further.

16.4 Components of Photosynthesis: Targets of CRISPR/Cas for Improving Photosynthetic Efficiency

Initially, genetic engineering aimed at improving the photosynthetic rate and efficiency focused on upregulating the expression of just one enzyme (SBPase) within the Calvin–Benson cycle (CB

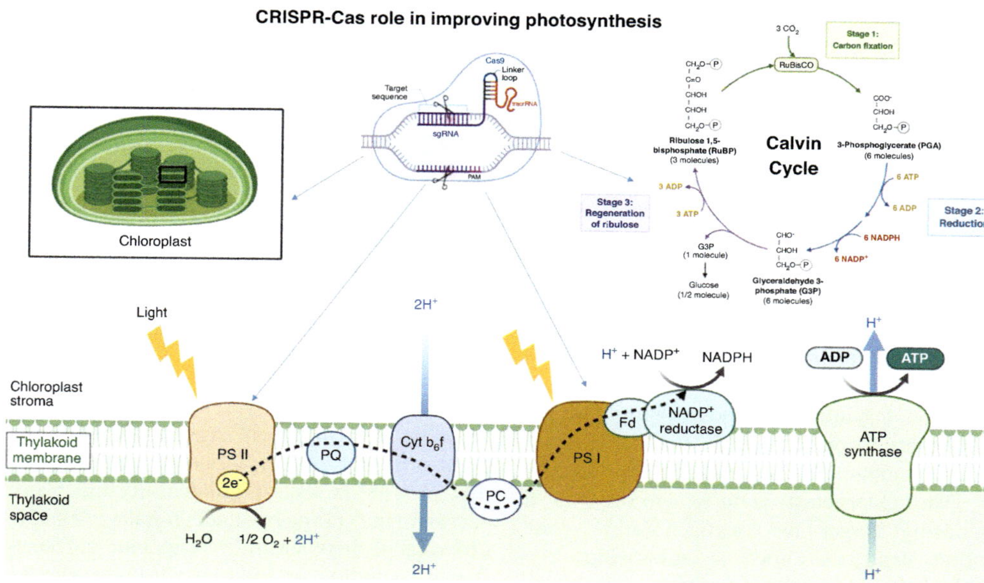

Fig. 16.2. Photosynthetic components: Targets of CRISPR-Cas gene editing. Photosynthesis ranks among the most intricate biological processes, encompassing a multitude of interconnected biochemical reactions mediated by sophisticated multi-enzyme mechanisms. In higher plants, the thylakoid membranes exhibit a complex architecture. These membranes consist of folded layers, with appressed stacks of grana interconnected by non-appressed thylakoids. Embedded within these membranes lie photosystems (PSI and PSII) along with their respective light-harvesting complexes. Thylakoid membrane encloses the lumen, an internal aqueous compartment that facilitates crucial electron and proton transport during photosynthesis. Deciphering the intricate interplay between these mechanisms and components governing photosynthesis is crucial to unlocking its full potential through genome engineering technologies, such as CRISPR/Cas. Author's own figure.

cycle). SBPase enzyme overexpression has been shown to increase biomass yield and photosynthetic carbon assimilation in Arabidopsis, tobacco, and tomato (Simkin *et al.*, 2015; Ding *et al.*, 2016; Simkin *et al.*, 2017). However, with time, different targets were identified and engineered, providing potential targets for CRISPR/Cas-mediated enhancement of photosynthetic efficiency. CRISPR/Cas gene editing technology facilitates the introduction of targeted modifications to genes encoding critical elements of photosynthesis, including the PSI and PSII, RuBisCO, and light-harvesting complex (LHC) as shown in Fig. 16.2 (Oh *et al.*, 2023). These CRISPR-mediated modifications may include gene replacements, knockouts, or strategic mutations designed to enhance the efficacy and performance of the photosynthetic machinery (Sattari Vayghan *et al.*, 2022). CRISPR/Cas9 technology holds promise for large-scale gene screening to identify rice mutants exhibiting reduced respiration rates and photorespiration under low CO_2 conditions, ultimately leading to enhanced photosynthetic performance (Mubarak *et al.*, 2023). In instances where generating knockout mutants are challenging, CRISPR-interference (CRISPRi) offers a valuable alternative. This approach utilizes a DNase-dead Cas nuclease, (specifically dCpf1), to achieve targeted gene suppression. Knoot *et al.* (2019) successfully employed this strategy in the fast-growing *Synechococcus* sp., demonstrating its efficacy in modulating photosynthetic processes.

16.4.1 Thylakoid membrane and chloroplast

Within photosynthetic cells of eukaryotes, chloroplasts serve as the primary site for light energy utilization and carbon fixation. These active organelles possess a well-organized internal membrane system known as thylakoids. These are flattened membrane sacs, stacked together to form stroma lamellae and grana (Bussi *et al.*, 2019). Grana membranes contain PSII, while PSI is localized within stroma lamellae. The sophisticated membrane organization within chloroplasts separates the internal thylakoid lumen, an aqueous compartment, from external stroma. The genetic machinery of chloroplast and CB cycle enzymes reside within the stroma. This complex membrane organization facilitates efficient electron transport and light capturing during photosynthesis, providing plants with the adaptability to respond to fluctuating environmental conditions (Johnson and Wientjes, 2020).

The chloroplast genome emerges as a promising target for next-generation gene editing strategies aimed at enhancing crop productivity under changing climate conditions. Yang *et al.* (2023) demonstrated this potential by employing CRISPR/Cas9 for chloroplast editing. Their work involved disruption of the phosphatidylinositol transfer protein7 (*AtPITP7*) locus in Arabidopsis and its corresponding *OsPITP6* gene in rice. This manipulation led to a decrease in inorganic phosphate uptake and hindered plant growth. However, the same study employed CRISPR/Cas9 to achieve overexpression of these genes, leading to improved plant growth, increased yield, and enhanced inorganic phosphorus uptake in rice. Wang *et al.* (2022a, b) utilized CRISPR/Cas to functionally characterize *GmFtsH25* gene in soybean. This gene encodes a protein belonging to the temperature-sensitive protein H (FtsH) protease family. Their findings revealed *GmFtsH25* as a critical regulator of soybean photosynthesis. Its overexpression in soybean plants resulted in an increased abundance of thylakoid stacks of grana within chloroplasts (Wang *et al.*, 2022a). Another study employed the CRISPR/Cas9 approach to generate cucumber mutant plants, exhibiting an albino phenotype, characterized by a lack of chlorophyll. This approach involved screening an ethyl-methane-sulfonate (EMS) mutagenesis library. Subsequent analyses validated a direct link between the albino phenotype and mutant gene (*CsTIC21*). Furthermore, they demonstrated that disruption of this gene leads to abnormal development of chloroplasts, ultimately resulting in albinism and the death of cucumber plants (Ke *et al.*, 2023).

16.4.2 Manipulation of chlorophyll and anthocyanin

Chlorophyll, a vital constituent of the photosynthetic machinery, in the antennae of both photosystems captures the solar energy and transfers it to the reaction center (Blankenship, 2002). Chlorophyll metabolism (biosynthesis and degradation) is modulated mainly by prevailing environmental conditions. Proper turnover of chlorophyll is essential for functioning of the photosynthetic machinery, as this process is also regulated during plant development (Tanaka and Tanaka, 2011). Chlorophyll metabolism involves four different stages, including protoporphyrin-IX synthesis, Mg^{2+} insertion in the protoporphyrin-IX for the synthesis of chlorophyll *a* (known as the ‘Mg branch’), the incorporation of apoproteins, and the cycle of chlorophyll. Several studies have focused on manipulating different steps of the metabolic pathway of chlorophyll to enhance antennae size or number of chlorophyll molecules per reaction center to improve PSII efficiency, photosynthetic efficiency, and plant yield (Simkin *et al.*, 2019; Miglani *et al.*, 2021). Such strategy will benefit the plants growing in low light intensity or in shade by the other plants. The CRISPR/Cas9 system presents a powerful tool for manipulating genes associated with chlorophyll and pigment synthesis, the essential components of photosynthesis. The mutants of leaf color developed from altered chlorophyll biosynthesis pathways serve as valuable sources for investigating the regulatory genes and their subsequent effect on photosynthesis (Liu *et al.*, 2022). This study employed CRISPR/Cas9 to investigate the regulatory mechanisms governing the balance between chlorophyll levels and heme synthesis within green tissues. Specifically, they focused on ferrochelatase 1 and 2 (FC1 and FC2), enzymes with distinct tissue expression patterns, to study how mutations in these genes affect heme production and consequently influence chlorophyll levels.

Pavese *et al.* (2021) demonstrated the effectiveness of CRISPR/Cas9 by editing the phytoene desaturase (*PDS*) gene, a crucial gene in chlorophyll synthesis. Mutations in *PDS* disrupt chlorophyll production, allowing for visual confirmation of knockout efficacy. This study represents a pioneering application of CRISPR/Cas9 for targeted *PDS* gene modification, providing valuable insights into the evaluation of knockout efficiency and chlorophyll biosynthesis. Furthermore, Eid *et al.* (2021) successfully employed CRISPR/Cas9 to edit Mg^{2+} chelatase subunit I, a crucial enzyme involved in chlorophyll biogenesis in sugarcane (polyploid crop). Sanger sequencing confirmed editing of all 49 copies of the targeted gene. Edited plants exhibited significantly reduced chlorophyll content, evident by yellow leaves during regeneration. Moreover, heat treatment enhanced coediting of various alleles and editing frequency. Despite these modifications, the modified sugarcane plants grew properly in soil without displaying growth retardation. This highlights the robustness of CRISPR technology, particularly for challenging targets like polyploid crops. Its multiplexing capability, design simplicity, and overall versatility contribute to its success in such applications (Hillary and Ceasar, 2023).

Anthocyanins, a class of soluble pigments, contribute significantly as antioxidants in plants. They are well-recognized for their protective role against photooxidative and photoinhibition damage induced by high light conditions (Zheng *et al.*, 2021a). Studies have identified a grapevine gene, *VvbZIP36* (bZIP transcription factor), acting as a negative regulator for anthocyanin synthesis. Utilizing the CRISPR/Cas9 system, they successfully knocked out one allele from this gene (*VvbZIP36*), leading to enhanced accumulation of anthocyanin in grapevine plants (Tu *et al.*, 2022). Another study explored the potential of interspecies genetic modification to enhance the anthocyanin content in rice. Their research involved the transfer of *GzMYB-7D1* gene from wheat plants into the rice varieties. This approach not only yielded rice enriched with anthocyanin but also led to a significant increase in grain yield in the overexpressed lines. This yield improvement was attributed to a combination of factors, including an increase in the number of tillers, increase in secondary panicle branching, and a significant rise in the total amino acid content (Luo *et al.*, 2022).

Such findings highlight the potential for manipulating chlorophyll metabolism through genome editing techniques, particularly CRISPR/Cas, as a promising approach for enhancing agricultural productivity and promoting sustainable crop production.

16.4.3 Manipulation of antenna

To maintain efficient photosynthesis, an optimal level of sunlight quantum flux is required, as excess photons can lead to decreased photosynthetic efficiency. One approach to mitigate the damage caused by excessive light is through the utilization of extra quantum flux absorbed by the LHC for fluorescence and thermal dissipation (Miglani *et al.*, 2021). Molecular level adjustments, like reducing the chlorophyll antenna size, have been shown to enhance efficiency of solar energy conversion, resulting in increased productivity (Polle *et al.*, 2003; Melis, 2009). Ort *et al.* (2015) proposed modifying the size of antennae and substituting the photosystem with a more efficient cyanobacterial reaction center. It has been reported that approximately 730 nm light can be utilized by *Acaryochloris marina* when the Chl *a* PSII system was replaced with the Chl *d* PSII of cyanobacteria.

A study highlighted the potential application of CRISPR/Cas-generated mutants in enhancing crop productivity through improved light use efficiency (LUE). Their focus lies on trimeric LHCII, a crucial component for the photosynthetic apparatus plasticity. This plasticity enables rapid acclimation to fluctuating light conditions, a key factor influencing plant yield and growth (Sattari Vayghan *et al.*, 2022). They utilized CRISPR/Cas9 technology to generate knockout mutants in Arabidopsis by targeting all five genes encoding LHCB1. They observed that knockout of LHCB1 resulted in a minimal increase of several other isoforms of LHCII. Furthermore, the absence of LHCB1 led to alterations in thylakoid structure, causing a decrease in both the number of membrane layers and the overall grana width. These stable knockout lines of LHCB1 present a valuable tool for researchers to further elucidate the role of this protein in plant acclimation to light conditions and light harvesting. Additionally, they open doors for further *in vivo* mutational analyses aimed at a more comprehensive understanding of LHCII function.

To dissect the specific roles of both Lhcb1 and Lhcb2 within LHCII, another study generated *amiLhcb1* and *amiLhcb2* Arabidopsis mutants using artificial microRNA (amiRNA). Their work suggests that Lhcb1 plays a critical role in maintaining the flexibility of thylakoid membranes, while Lhcb2 is essential for mediating the association of LHC with PSI (Pietrzykowska *et al.*, 2014).

Rudenko *et al.* (2022) employed the CRISPR/Cas9 system in Arabidopsis to elucidate the function of carbonic anhydrase during photosynthetic activity. Their research involved generating the knockout of the thylakoid gene of carbonic anhydrase αCA4 (AT4g20990). This targeted mutation resulted in altered expression patterns, including all light-harvesting antennae proteins of PSII, both thylakoid and cytoplasmic carbonic anhydrases, and even plant immune defense signaling molecules.

16.4.4 Manipulation of electron transport chain

Current limitations to photosynthesis include CO_2 assimilation at high CO_2 concentrations and photodamage at high light intensities (Cen and Sage, 2005; Liu *et al.*, 2019). To address these issues and improve photosynthetic activity and crop yields, researchers propose plastocyanin, a key protein in the ETC (Schottler *et al.*, 2004), as a target for genome editing. Pesaresi *et al.* (2009) demonstrated that overexpression of plastocyanin isoforms such as *plastocyanin 1* (PETE 1 and PETE 2) in transgenic lines of *Arabidopsis thaliana* led to increased biomass production. Another approach involved NAD kinase (NADK) overexpression of Arabidopsis in rice, which resulted in improved CO_2 assimilation and ET rates (Takahara *et al.*, 2010). The alternative flow of electrons, which involves the cyclic electron flow and water-water cycle near PSI (CEF-PSI), is known to contribute significantly to the alleviation of photoinhibition and photosynthesis regulation (Miyake, 2010). However, the precise role of CEF-PSI in photoinhibition remains uncertain, although some studies have suggested its involvement (Yamori and Shikanai, 2016; Huang *et al.*, 2018). While several studies have examined modifications to the ETC, promising opportunities for improving its efficiency lie with plastocyanin. Nevertheless, further research using genome editing tools is still needed to enhance the solar energy conversion rate into biomass and develop photosystems that are less prone to photodamage and reactive oxygen species production (Schmermund *et al.*, 2019).

16.4.5 Manipulation of RuBisCO

Manipulation of RuBisCO is a challenging task, as it is a multimeric enzyme with its larger subunit encoded by the plastid genome while the smaller subunit is encoded by the nuclear genome. Red algae RuBisCO has been reported to have high catalytic rates and specificity but is unable to assemble correctly in higher plants (Parry *et al.*, 2013). Despite challenges due to the multimeric nature of RuBisCO, it is still a promising target for genome engineering to improve photosynthesis. The working strategy includes the incorporation of genes for more efficient RuBisCO subunits from other plants. For example, when RuBisCO smaller subunit of sorghum was expressed in rice, transgenic lines showed improved catalytic turnover rate (Ishikawa *et al.*, 2011) or to engineer RuBisCO variants with higher carboxylation turnover and specificity rates. Prins *et al.* (2016) demonstrated the potential of such variants in wheat for CO_2-enriched crops having high turnover and specificity. Alternatively, engineering the L-subunit (larger subunit) of RuBisCO (Miglani *et al.*, 2021) or introducing thermostable forms of RCA (RuBisCO activase) enzyme provide promising avenues to improve photosynthetic activity (Kumar *et al.*, 2009; Carmo-Silva and Salvucci, 2012; Yamori *et al.*, 2012; Yamori *et al.*, 2016; Scafaro *et al.*, 2019). However, the thermostable RCA overexpression alone has been reported to reduce the RuBisCO concentration in rice (Fukayama *et al.*, 2018; Suganami *et al.*, 2020). To address this issue, Qu *et al.* (2021) conducted an experiment in which both the RCA and RuBisCO were overexpressed in rice to protect photosynthesis and increase biomass during heat stress.

Moreover, the CRISPR/Cas9 approach successfully knocked out multiple RuBisCO small subunit genes in tobacco. Two guide RNAs targeted three key rbcS homologs, achieving a significant reduction (around 93%) in RuBisCO content. This study demonstrates the effectiveness of CRISPR for *rbcS* mutagenesis in polyploid plants, paving the way for future efforts to introduce and express superior RuBisCO enzymes for improved photosynthesis (Donovan *et al.*, 2020).

16.4.6 Manipulation of CCM or increase in CO_2 influx around RuBisCO

The process of carbon assimilation in photosynthetic organisms is a crucial biological activity. However, it is limited by the dual activity of RuBisCO (having both carboxylase and oxygenase activities), leading to photorespiration and reduced efficiency. In addition, this enzyme has a limitation in differentiating between CO_2 and O_2, resulting in both carboxylation and RuBP oxygenation (Flamholz *et al.*, 2019; Iniguez *et al.*, 2020). Manipulating RuBisCO and C_3 pathway manipulation offers promising avenues for enhancing carbon assimilation to improve photosynthesis (Hügler and Sievert, 2011).

C_4 plants concentrate CO_2 around RuBisCO, offering an alternative strategy to increase photosynthesis (Kajala *et al.*, 2011). Researchers have attempted to enhance photosynthetic efficiency by overexpressing PEP carboxylase in C_3 plants such as Arabidopsis and rice, resulting in increased CO_2 assimilation and decreased oxygenation (Fukayama *et al.*, 2003; Kandoi *et al.*, 2016). However, challenges remain in optimizing the expression of enzymes involved in photorespiration (Covshoff and Hibberd, 2012). Altering stomatal conductance and cell properties improves CO_2 diffusion, while aquaporins facilitate membrane permeability for CO_2 transport (Gaastra, 1959; Hanba *et al.*, 2004; Kawase *et al.*, 2013; Yamori *et al.*, 2016). Another strategy is to utilize bacterial glycolate metabolic pathway, which involves glycolate oxidation through the glycolate oxidase enzyme, producing H_2O_2, and converting to tartronate-semialdehyde, releasing the photorespiratory CO_2 directly into the Arabidopsis chloroplast (Maier *et al.*, 2012). However, this method has showed limited success. Algae and cyanobacteria utilize the single-cell carbon-concentrating process, where RuBisCO is enclosed in pyrenoids and carboxysomes, exploring innovative approaches to improve the process of photosynthesis (Moroney and Ynalvez, 2007; Kerfeld and Melnicki, 2016).

In addition, CRISPR/Cas9 editing successfully knocked out thylakoid carbonic anhydrase in Arabidopsis. This resulted in lower non-photochemical quenching of chlorophyll fluorescence compared to wild-type plants, suggesting improved protection against photoinhibition (Rudenko *et al.*, 2022). However, in another study, CRISPR-edited tobacco plants with inactivated activity of stromal carbonic anhydrase showed normal photosynthesis but displayed abnormal development, elevated stromal pH, and increased reactive oxygen species production. This suggests that carbonic anhydrase might not directly supply CO_2, but it plays a role in providing bicarbonate for other biosynthetic processes within the chloroplast, similar to its function in microorganisms (Hines *et al.*, 2021). According to McGrath and Long (2014), the implementation of a carboxysome-type CCM in chloroplasts of plants resulted in a 36% to 60% increase in the growth and yield. In tobacco, when the endogenous gene of RuBisCO was replaced with the RuBisCO comprising carboxysome, the resulting transgenic plants exhibited stunted growth even at 2% CO_2 concentration (Long *et al.*, 2018).

16.4.7 Manipulation of C_3 cycle (Calvin–Benson cycle)

The C_3 cycle utilizes the ADP phosphorylated and $NADP^+$ reduced products of the electron transport chain (ETC) to fix CO_2, essential for starch and sucrose production (Bhagwat, 2005; Sheoran and Singh, 2012). This process involves 11 different enzymes, with RuBisCO initiating carbon fixation. Enzymes such as fructose-1,6-biphosphatase (FBPase), sedoheptulose-1,7-biphosphatase (SBPase), transketolase, and aldolase have been identified as potential enhancers of productivity, providing a thermodynamic push and improving flux control (Uematsu *et al.*, 2012; Kubis and Bar-Even, 2019; Kumar *et al.*, 2023). SBPase, particularly, exerts significant control over RuBisCO activity even at higher CO_2 concentrations compared to all other enzymes (Raines, 2011). Several researchers have reported that SBPase overexpression in plants such as Arabidopsis and tomato increases biomass production and photosynthesis, with potential benefits in abiotic stress tolerance (Rosenthal *et al.*, 2011; Ding *et al.*, 2018).

Researchers at the University of Essex employed CRISPR technology to address a bottleneck in the C_3 cycle, the pathway where plants fix CO_2 into sugars. They achieved this by increasing the amount of a key enzyme, SBPase, through the incorporation of genetic material from cyanobacteria and another plant. This combined approach resulted in a significant boost to photosynthetic activity and crop productivity, with greenhouse studies demonstrating a 52% increase and field conditions showing a 27% improvement (Ding *et al.*, 2018; Manning, 2019). These studies provide clear evidence that SBPase plays a leading role in regulating the C_3 cycle, and could serve as a prime target for genetic manipulation aimed at improving biomass, photosynthetic rates, and overall crop productivity.

Other enzymes involved in the CB cycle, including photorespiratory GDH-H (glycine-decarboxylase-H), FBPase, and FBPA have also been targeted to improve biomass yield and photosynthesis in several studies (Simkin *et al.*, 2015; Simkin *et al.*, 2017). Co-overexpression studies involving enzymes like FBPA, GDC-H, and SBPase in tobacco and Arabidopsis have been reported to enhance photosynthesis and biomass (Simkin *et al.*, 2015; Simkin *et al.*, 2017).

Conversely, overexpressing or silencing enzymes catalyzing reversible reactions, such as transketolase and FBPase, has been found to enhance and decrease photosynthesis, respectively. Several studies have reported that integrating metabolic pathways from microorganisms such as cyanobacteria and algae in to rice significantly improved biomass and crop yield (Maier *et al.*, 2012; Andralojc *et al.*, 2018).

16.4.8 Manipulation of photorespiration pathway

The process of photorespiration, crucial for C_3 plants, involves the conversion of 2-phosphoglycolate (2 PG) to phosphoglycerate (PGA) to regenerate the latter; however, 2 PG acts as an inhibitor for several enzymes involved in the starch degradation (phospho-fructokinase) and C_3 cycle (triose-phosphate isomerase, sedo-heptulose-1,7-bisphosphate-phosphatase, and RuBisCO), making effective 2 PG recycling crucial for C_3 plants (Timm and Hagemann, 2020; Zhou *et al.*, 2021). Nevertheless, photorespiration causes carbon loss and can lead to yield losses of approximately 25% to 50% (Bauwe *et al.*, 2010). To enhance crop productivity, genome engineering approaches aim to bypass photorespiration by optimizing metabolic flux and reducing RuBP oxygenation, thus minimizing 2 PG production and its inhibitory effects (South *et al.*, 2018). Overexpressing glycine decarboxylase has shown promise in increasing photosynthesis (Timm *et al.*, 2015; Simkin *et al.*, 2017; López-Calcagno *et al.*, 2019).

One potential approach to improve plant productivity is to genetically engineer the photorespiration pathway to occur solely in the chloroplast organelle (South *et al.*, 2019) as demonstrated in *Camelina sativa* and Arabidopsis; this resulted in enhanced productivity and photosynthesis (Maier *et al.*, 2012; Dalal *et al.*, 2015).

However, silencing the plastidal-glycolate-glycerate translocator 1 (*PLGG1*) gene led to reduced glycolate flux and affected CO_2 fixation (South *et al.*, 2019; Shim *et al.*, 2020). Hence, *PLGG1* plays a critical role in maintaining levels of glycerate/glycolate within the chloroplast and presents a promising target for enhancing photosynthesis and productivity. Roell *et al.* (2021) presented a novel strategy in *Arabidopsis thaliana* by introducing a synthetic bypass mechanism inspired by the β-hydroxyaspartate cycle found in marine proteobacteria. This innovative approach has the potential to transform photorespiration into a carbon-concentrating pathway by creating a synergistic link between C_4 metabolism and photorespiration. Although genome engineering has offered several ways to improve the photosynthesis rate, there have been limited achievements in improving crop yield (Betti *et al.*, 2016; Flexas, 2016). Therefore, other opportunities to enhance crop productivity and photosynthetic rate, such as modification of regulatory sequences, gene expression fine-tuning, substrate affinity modification, and utilization of the inducible system might be considered to reduce photorespiratory loss (Miglani *et al.*, 2021).

16.4.9 Manipulation of non-photochemical quenching

During the process of plant biomass production, the photochemical reactions and absorption of light result in significant energy loss (Stitt, 2013). Enhancing the rate of photoprotection relaxation or NPQ can lead to a remarkable increase in growth of tobacco in field experiments (Kromdijk *et al.*, 2016; Manzoor *et al.*, 2022). NPQ dissipates excess light as heat when plants receive more light than needed, and modeling studies have suggested that improving NPQ could be an effective approach to enhance plant productivity (Zhu *et al.*, 2004; Murchie and Niyogi, 2011). Tobacco plants were genetically modified with different genes related to photoprotection, resulting in the acceleration of NPQ relaxation, which increased the biomass production rate by approximately 20% in the glasshouse experiments and 15% in replicated field trials (Kromdijk *et al.*, 2016). Introducing genes related to photoprotection, including those encoding elements of the PSII PsbS subunit and xanthophyll cycle or abiotic stress (*AVP1*), facilitated rapid adaptation to changes in light levels, directing more light towards photosynthesis (Kromdijk *et al.*, 2016; Manzoor *et al.*, 2022). However, strategies like NPQ, although conserving energy, may decrease energy storage efficiency, necessitating exploration of methods to enhance the photosynthetic apparatus's capacity to process energy influx and reduce losses. One potential strategy involves introducing bacteriochlorophylls into plant photosystems to broaden their light absorption spectrum up to the far-red region, approximately up to 1100 nm (Blankenship *et al.*, 2011).

One possible strategy to reduce NPQ and improve efficiency of solar conversion is by changing the PS antenna size, which has been demonstrated to be successful in cyanobacteria and microalgae (Polle *et al.*, 2003; Kirst *et al.*, 2014). Recent research indicates that truncated light-harvesting antenna (TLA) can increase both plant canopy photosynthesis and biomass accumulation under high-density conditions in rice and tobacco (Gu *et al.*, 2017; Kirst *et al.*, 2017). This strategy represents intrinsic practical challenges and limitations (Ort *et al.*, 2011); however, current research indicates that continued efforts may eventually optimize this promising biotechnological process. A study utilized CRISPR/Cas9 to engineer rice plants with TLA by knocking out the genes involved in antenna assembly. The results showed improved photosynthetic rate with decreased NPQ (Caddell *et al.*, 2023).

16.4.10 Manipulation of balance of sink and source

Environmental factors and developmental changes often limit the full potential of source and sink in the manipulation of the sink:source balance. The limitations on photosynthesis and productivity are well-known when there is either an excess or insufficient amount of light, nutrients, or water. The source activity can be altered by an increase in CO_2 concentration, defoliation, or shading, while the activity of sinks could be changed by sink removal or chilling (White *et al.*, 2016). Stimulation of photosynthesis due to higher levels of CO_2 can enhance efficiency of plants in utilizing carbon sources, thereby leading to accelerated growth as evidenced by previous research (Leakey *et al.*, 2009). Additionally, larger sink sizes, achieved through growing plants in larger containers, promote carbon-sink activity and improve plant growth (Poorter *et al.*, 2012; Aranjuelo *et al.*, 2013). Defoliation triggers compensatory responses in the remaining leaves to maintain source:sink balance, suggesting that plant growth may be limited by sink capacity (Eyles *et al.*, 2013). This phenomenon may suggest that the growth of the plant is limited by the sink's capacity, as leaves in undisturbed conditions may not achieve their maximum photosynthetic rates.

The combination of experimental treatments targeting both the source and sink has demonstrated the interdependent relationship between these factors, as they work in conjunction to provide feedback to one another (White *et al.*, 2016). The phenomenon of photosynthetic acclimation, observed in response to an increase in CO_2 concentration, entails a reduction in photosynthetic potential that counteracts the stimulatory effect of elevated CO_2 on the photosynthetic rate. The process of acclimation serves to decrease the ratio of source-to-sink activity, thereby promoting an equilibrium in the source-to-sink balance (Dong and Beckles, 2019). As a result, genome editing targeting both the sink and source at an appropriate developmental stage of the plant is necessary to manipulate these processes and potentially boost crop yield.

16.5 CRISPR-mediated Indirect Improvement in Photosynthesis

Genetic engineering and traditional breeding have limitations in addressing the challenges posed by climate change and the ever-increasing food demand worldwide. These established methods often lack the precision and efficiency required for rapid crop improvement. CRISPR/Cas technology emerged as a promising tool to overcome these limitations. Its ability to precisely target single-nucleotide mutations is associated with key agronomic traits (Azameti and Dauda, 2021). The high accuracy, ease of use, and multiplexing capabilities make it a powerful tool for crop improvement applications.

16.5.1 Crop yield

Crop yield is a complex trait determined by the interplay of numerous genetic and environmental factors (Xu *et al.*, 2022). Quantitative trait loci (QTLs) and various genes contribute to the overall yield of a crop. These genetic factors influence a multitude of plant characteristics, such as pod number, seed/grain size, branching patterns, leaf angle, seeds per pod, and overall plant architecture. Ultimately, the combined effects of these factors determine the final crop yield. CRISPR/Cas9 technology offers a powerful approach for manipulating genes involved in biomass accumulation. By targeting specific transcription factors, researchers can not only elucidate their functional roles but also edit them to improve crop production. An *et al.* (2020) demonstrated this concept through the application of CRISPR/Cas9 to mutate PdGATA19/PdGNC transcription factors in plants. Their findings revealed that mutant lines displayed underdeveloped primary growth but enhanced secondary growth. Conversely, overexpression of these genes led to increased photosynthetic rate, chlorophyll content, and ultimately, biomass accumulation.

16.5.2 Disease resistance and stress tolerance

CRISPR/Cas9 technology holds promise for generating disease-resistant crops through targeted genetic modifications (Table 16.3). This approach allows scientists to introduce specific mutations or perform gene knockouts, thereby enhancing the plant's intrinsic immune response and mitigating the risk of yield losses due to viral, bacterial, or fungal pathogens (Kuroiwa *et al.*, 2022).

Moreover, CRISPR/Cas technology extends beyond disease resistance, offering the potential to develop crops with enhanced tolerance against various abiotic stresses including drought, cold, heavy metals, salinity, and heat. By strategically modifying genes involved in defense mechanism pathways, researchers have developed crops better adapted to cope with a changing climate as shown in Table 16.3 (Wahab *et al.*, 2022). Targeted mutagenesis using CRISPR/Cas, particularly within structural or regulatory genes alongside cis-regulatory elements, has been proposed as an efficient, practical, and robust approach for generating climate-resilient crop varieties (Zafar *et al.*, 2020; Du *et al.*, 2023). Table 16.3 provides a comprehensive overview of how CRISPR/Cas technology has been utilized for targeted improvement of different traits in various crop species.

Gene knockout methods using CRISPR/Cas technology are demonstrating promise in enhancing plant tolerance to abiotic stresses. Studies in millet, sorghum, maize, and rice have shown that knocking out Alkaline Tolerance 1 (AT1) gene significantly reduces sensitivity to alkaline conditions (Zhang *et al.*, 2023a, b).

Table 16.3. CRISPR-mediated improvement in different crops. Author's own table.

Crop plant	Target gene(s)	Trait improved	Reference
Arabidopsis	*eIF4E1*	Potyvirus resistance	Bastet *et al.*, 2019
Maize	*ZmALS* *ZmCENH3*	Herbicide tolerance Improved chromosome segregation	Zong *et al.*, 2017; Li *et al.*, 2020
Potato	*StGBSS* *StALS*	Herbicide tolerance Starch biosynthesis	Zong *et al.*, 2018
Rice	*OsACC-T1* *OsTubA2* *OsWaxy* *OsSPL14* *OsNAC050*	Herbicide tolerance Decreased amylose content Improved grain yield and plant height Improved low-temperature tolerance	Monsur *et al.*, 2021 Hua *et al.*, 2018
Rapeseed	*RGA* *IAA7* *ALS*	Herbicide tolerance	Cheng *et al.*, 2021
Tomato	*SLALS1*	Herbicide tolerance	Veillet *et al.*, 2019
Watermelon	*ALS*	Herbicide tolerance	Tian *et al.*, 2018
Wheat	*TaLOX2* *TaDEP1* *TaEPSPS* *TaGW2*	Altered lipid metabolism Improved panicle length Improved grain weight Herbicide tolerance	Zong *et al.*, 2017; Li *et al.*, 2020

Furthermore, Badhan *et al.* (2021) pioneered the application of CRISPR/Cas9 technology for DNA-free genome engineering in protoplasts of chickpea plants. They targeted two genes associated with drought tolerance: *Reveille-7* (*RVE7*) and 4-coumarate ligase (*4-CL*). *4-CL* encodes a key enzyme in the metabolic pathway of phenylpropanoid, specifically involved in lignin synthesis, whereas *RVE7* functions as a transcription factor involved in regulating the plants' circadian rhythm.

Rathnasamy *et al.* (2023) investigated the potential of manipulating stomatal function utilizing CRISPR/Cas9 technology. They engineered *OsEPF1* gene and its alleles, a key gene that negatively regulates stomatal frequency, in the widely cultivated rice variety ASD 16. Their findings revealed that targeted modifications within *OsEPF1* led to alterations in stomatal conductance and density, ultimately influencing photosynthetic rate in rice plants. To further elucidate the functional role of the *VvEPFL9-1* in grapevine, Karavolias *et al.* (2021b) employed CRISPR/Cas9 to generate a knockout mutation in the Epidermal Patterning Factor Like 9 (*EPFL9*) gene, called STOMAGEN. This targeted mutagenesis resulted in decreased stomatal density in both rice and grapevine (Clemens *et al.*, 2022).

16.5.3 Improved nitrogen-use efficiency (NUE)

Studies have reported CRISPR/Cas-mediated improvement of nitrogen uptake and assimilation (Table 16.4), which is crucial because nitrogen deficiency directly restricts chlorophyll production in plants, consequently limiting their photosynthetic efficiency and sugar yield (Lebedev *et al.*, 2021).

16.6 CRISPR-mediated Improvement of Photosynthesis in Photosynthetic Microbes

Following the successful implementation of the CRISPR/Cas system in both plant and animal cells, researchers are actively evaluating its potential for photosynthetic microbe's genome

Table 16.4. CRISPR/Cas-mediated improvement in nitrogen-use efficiency. Author's own table.

Crop	Gene	Function	Effect on NUE	Reference
Rice	*OsNPF6.1^{HapB}*	Nitrate transporter	Increased NO_3^- uptake and yield	Tang *et al.*, 2019
	OsNPF4.5	Mycorrhizal nitrate transporter	Increased symbiotic N uptake	Wang *et al.*, 2020a
	OsFd-GOGAT (ES7)	Ferredoxin-dependent-glutamate synthase	Essential for N utilization and normal growth	Bi *et al.*, 2017
	myb61	Transcription factor	Affects NUE and yield through interaction with growth-regulating factor 4 (GRF4)	Gao *et al.*, 2020
	OsAAP3 *OsAAP4* *OsAAP5*	Amino acid permeases	Regulation of bud growth and amino acid accumulation	Lu *et al.*, 2018; Ji *et al.*, 2020
	OsNLP1	NIN-like protein 1 (NLP)	Positive regulator of N assimilation and uptake genes	Alfatih *et al.*, 2020
	OsASN1	Asparagine synthetase	Improved grain yield, protein content, and amino acid content	Lee *et al.*, 2020
Soybean	*GS1*	Glutamate synthase	Crucial for N remobilization and assimilation	Michno *et al.*, 2015
Wheat	*TaGS1.1*			Wang *et al.*, 2020b
Wheat Rice	*GRF4*	Transcription factor	Increased yield and nitrogen uptake	Li *et al.*, 2018
Poplar	*PdGNC* (GATA transcription factor)	Involved in carbon metabolism and root NO_3^- transport	Increased biomass	An *et al.*, 2020
	ICDH	Cytosolic $NADP^+$-isocitrate dehydrogenase	Increased plant height and amino acid content	Pascual *et al.*, 2018

editing (Naduthodi *et al.*, 2018). Also, preliminary studies using Cas9 in photosynthetic bacteria and model green algae have demonstrated some initial success (Patel *et al.*, 2019). However, significant advancements are needed to improve editing efficiency and DNA delivery methods. CRISPR/Cas technology enables targeted genetic manipulations in microorganisms, leading to enhanced efficiency in CO_2 absorption and conversion. This breakthrough holds significant promise for improving photosynthetic efficiency of microbes (Tarafdar *et al.*, 2023). For example, scientists at Harvard University employed CRISPR/Cas system to genetically modify cyanobacteria, aiming to enhance their CO_2 absorption capabilities and photosynthetic efficiency. Their strategy involved targeted knockout of the *isiA* gene, encoding the protein for light harvesting, previously identified as inhibiting photosynthetic efficacy under specific conditions in cyanobacteria. Analysis of the genetically modified cyanobacteria revealed a significant 37% enhancement in biomass compared with their wild-type control. This finding suggests a successful enhancement in photosynthetic efficiency due to the targeted knockout of the *isiA* gene (Xie *et al.*, 2021). Moreover, the mutant strains exhibited enhanced resilience to low CO_2 levels and high light intensity, two common stresses encountered by cyanobacteria in their natural habitats (Cassier-Chauvat *et al.*, 2021).

Lee *et al.* (2023) further explored the potential of CRISPR/Cas technique in cyanobacteria by engineering a novel CO_2 fixation pathway. Their research focused on the *rbcL* gene, essential for photosynthetic CO_2 fixation. Using CRISPR/Cas, they successfully introduced a new pathway derived from various bacterial strains. This innovative approach resulted in significant enhancements, as evidenced by increased biomass production and CO_2 uptake in the engineered cyanobacteria. These findings highlight the CRISPR/Cas potential for manipulation of metabolic mechanisms within cyanobacteria, paving the way for advancements in CO_2 fixation and photosynthetic efficiency.

Moreover, CRISPR/Cas technology is also being explored for engineering algae for carbon storage and capture, a potential strategy to improve photosynthetic efficiency. Previous studies in *Chlamydomonas reinhardtii*, a model organism for plant research, employed DNA insertion mutagenesis to demonstrate the potential of a reduced chlorophyll antenna for enhanced photosynthetic productivity (Polle *et al.*, 2000, 2003). These experiments resulted in mutant strains *tla*, harboring mutations in genes encoding CpSRP (chloroplast-signal recognition particle). The mutated strain exhibited increased efficiency in converting solar energy and displayed higher photosynthetic output, particularly under high light conditions (Kirst *et al.*, 2012; Jeong *et al.*, 2017). These findings suggest the CpSRP complex as a promising target for manipulating the chlorophyll antenna size because it plays a critical role in the proper folding and thylakoid membrane targeting of LHCs without disrupting essential electron transport processes (Kirst and Melis, 2014). Furthermore, recent research demonstrates the effectiveness of CRISPR/Cas9 technology in generating similar antenna mutants (*tla* mutants) with targeted precision (Baek *et al.*, 2016; Jeong *et al.*, 2018).

Yamano *et al.* (2022) employed CRISPR/Cas to manipulate genes in *C. reinhardtii*, aiming to enhance its CO_2 conversion into biomass. They targeted *RBCS2* gene, encoding a protein involved in algal and plant CO_2 fixation. Modifying this gene using CRISPR/Cas resulted in a significant 50% increase in the algae's CO_2 fixation efficiency. Additionally, they observed increased biomass, suggesting the gene's potential role in enhancing the photosynthetic rate.

Bestrophins (BSTs), a conserved anion transporter family, are crucial for CCMs in green algae. The study conducted by Nigishi *et al.* (2024) investigated the expression and subcellular localization of putative BSTs in diatoms by generating mutants using CRISPR/Cas9 nickase to reveal their potential role in low-carbon environments. Their results suggested that *PtBST1* from the marine diatom *Phaeodactylum tricornutum* localized outside the pyrenoid and increased under low-carbon conditions, suggesting its role in photoprotection and CCM function.

16.7 Challenges of Genome Editing Technologies

The introduction of revolutionary concepts for the redesigning of photosynthetic machinery and its regulation (as discussed in the aforementioned targets) will need the incorporation of multiple transgenes, maybe on synthetic chromosomes, and will involve genetic engineering on a massive scale. Furthermore, the development of such organisms will entail public discourse on the benefits and costs of those transgenics (McClung, 2014). The genetic engineering involved on such a large-scale attempt presents numerous technical challenges, as detailed in Table 16.5 (Ort *et al.*, 2015).

Over the past 30 years, significant progress has been made in transgenic research, resulting in a vast array of tools for genetic engineering of eukaryotic algae, land plants, and cyanobacteria. Emerging synthetic biology and biotechnology tools include artificial chromosomes, synthetic promoters, and large-scale assembly of DNA fragments, including complete eukaryotic chromosomes and bacterial genomes (Gibson *et al.*, 2010; Dymond *et al.*, 2011). Enhanced transformation techniques offer promising opportunities for redesigning various components of photosynthesis. However, despite these advancements, there remain unresolved challenges and critical bottlenecks (as outlined in Table 16.5). The plastid genome of photosynthetic eukaryotes encodes several crucial elements of RuBisCO larger subunit and ETC. Therefore, various approaches aimed at radical redesign of photosynthesis would greatly benefit from the ability to engineer the plastid genome in different species. Moreover, transformation of the plastid genome offers several additional advantages, including (i) numerous transgenes can be conveniently stacked in artificial operons, (ii) the ability to express transgenes at high levels, (iii) due to an effective homologous recombination mechanism, plastid engineering can achieve a high degree of precision, and (iv) the maternal method of plastid inheritance, which prevents plastid (trans)genes from being transmitted via pollen in many crops, thus allowing for much greater transgene containment (Ort *et al.*, 2015).

However, the technology for effective plastid transformation is currently accessible only for a limited number of species. Despite intensive research, no successful plastid transformation techniques have been established for cereals, the most widely consumed crop worldwide (Maliga, 2004; Bock, 2007). Significant technological obstacles include the establishment of appropriate molecular markers and effective selection systems and tissue culture. Furthermore, since plastid transgenes often exhibit very high levels of expression, it is crucial to manage such levels for optimal results. As a result, more research is required to improve plastid transformation methods. This includes enhancing the effective expression of multiple transgenes and ensuring correct posttranslational modification and protein folding within the plastid (Ort *et al.*, 2015). To overcome these challenges, substantial investments in research programs is necessary. Unfortunately, at the present time, such investments are not being provided, especially by the government.

During this period, it would be suitable to perform experiments with innovative redesign ideas in developed model systems like green algae and cyanobacteria. Additionally, it is crucial to create an intermediate model of a land plant to serve as a foundation for implementing methods to enhance photosynthesis and fine-tune the redesigned systems for transformation to essential crops once the genetic engineering technologies have been established. One feasible model might be a diploid *Nicotiana sylvestris* species with transformable nuclear and chloroplast genomes (Thyssen, 2012). By using an intermediate model plant, it would be possible to evaluate the effectiveness of various photosynthetic modifications in an enclosed canopy environment similar to that existing in important crops.

Recent advancements in technology hold significant potential to enhance the feasibility and efficiency of nuclear-based transformation (Gibson *et al.*, 2010). Nobel and new techniques like TALENs, CRISPR/Cas9 system, and ZFNs as discussed above now allow for precise genome engineering, including gene editing and replacement of sequences directly in their regulatory regions (Durai *et al.*, 2005; Bedell *et al.*, 2012; Cong *et al.*, 2013). To fully utilize the potential of the latest tools available, it is crucial that our understanding of the regulatory mechanisms involved in the functioning and assembly of the photosynthetic machinery advances accordingly.

16.8 Conclusion

In this chapter, we explored various genome editing-based strategies used to enhance plant photosynthesis in order to improve crop

Table 16.5. Tools for engineering photosynthesis: pros and cons. Author's own table.

Tools	Potential application	Limitations
Nuclear transformation	• Manipulation of nucleus-encoded elements of the photosynthetic machinery • Expression of new pathways and genes • Artificial chromosomes development	• Currently, there is a lack of efficient techniques to prevent transgene silencing. • This method cannot achieve high expression levels of transgenes. • This method cannot identify well-characterized chloroplast-transporting signals, promoters, and terminators. • There is a lack of appropriate techniques for site-directed manipulation. • There are limited mature technologies for the stable transformation of large DNA segments into plants. • There is insufficient knowledge about the sequences of telomeres and centromeres.
Mitochondrial transformation	• Engineer the components of the respiratory chain encoded by mitochondria in a way that reduces respiratory losses • To express new carbon metabolism pathways	• This procedure cannot be done with any plant that produces seeds yet. • *Chlamydomonas reinhardtii* (algae) is the only known exception.
Plastid transformation	• Modifications to the photosynthetic machinery by tailoring plastid-encoded elements • Expression of metabolic pathways and novel genes of carbon metabolism	• There is a lack of flexible and robust regulatory techniques. • There are a limited number of transformable crop plants, none of which are economically important crops or cereals.
Bacterial transformation	• Genetic manipulation of photosynthetic machinery in cyanobacteria	• Most significant model species of cyanobacteria do not exhibit any significant technical limitations.
Multigene engineering	• Manipulation of the electron transport chain and carbon fixation processes by engineering protein complexes	• There are no major constraints, as this process can be carried out using synthetic operons in plastids and cyanobacteria, or through combinational nuclear transformation. • However, advanced techniques are required to introduce larger fragments, containing more than ten genes, into plants.

Continued

Table 16.5. Continued

Tools	Potential application	Limitations
Synthetic genomics	• Photosynthetic apparatus radical redesigning using mini-artificial chromosomes within the nucleus • Synthetic genomes of plastid	• List of components for the photosynthetic apparatus, including its regulators and assembly factors, is incomplete. • Understanding of the dynamic quantitative variations that occur due to environmental factors is also incomplete. • There are limitations in the efficiency and synthesis of assembling larger DNA molecules. • Mini-chromosomes are not very stable.
Protein design	• Redesigning enzymes involved in carbon fixation • Modifications in electron transport chain and RuBisCO engineering	• Development of computer-based modelling is still insufficient, with limited knowledge of protein dynamics. • There is limited success in rational optimization and design of proteins using this approach.
Field-based phenotyping	• Design concepts are evaluated in the field. • Further optimization through the process of mutagenesis	• Current imaging approaches lack sufficient sensitivity. • There is a shortage of sensitive spectrometric techniques for continuous and large-scale measurement of photosynthetic parameters, radiation quantity and quality within canopies, as well as the growth of multiple transformants or genotypes in field conditions.
Logic circuit designing accompanied with sensor development for CO_2 concentration, temperature, light quality, and intensity	• Idea of smart canopy	• List of components for photosynthesis is incomplete due to an insufficient understanding of the structure of the metabolic and genetic networks that regulate it. • There is limited knowledge about the sizes of metabolite pools in various subcellular and cellular compartments. • There is incomplete information regarding transporters.

yield. Different genome editing technologies like CRISPR, TALEN, and ZFN hold immense potential for improving photosynthesis by precisely modifying specific targets, such as optimizing light utilization and absorption by manipulating the chlorophyll metabolism, ETC, light-harvesting antenna/complex, and non-photochemical quenching (NPQ). Additionally, carbon assimilation can be improved by engineering the CB cycle, RuBisCO activity, and minimizing photorespiratory loss through manipulation of the bacterial carboligase pathway. These technologies offer a faster, more precise, and efficient way of introducing genetic modifications, reducing the need for lengthy and laborious conventional breeding programs. However, there are potential unexpected outcomes of genetic manipulation, such as the off-target effects and the difficulty of achieving simultaneous modification of multiple targets, which must be acknowledged. To overcome these limitations, it is essential to develop well-organized and more precise methods such as un-biased and biased off-target detection approaches, improved variants of Cas system, effective delivery systems, anti-CRISPR proteins, guided-RNA engineering, effective prime-editing and base-editing systems, and well-established nuclear, plastid, and mitochondrial transformation protocols of plants (Naeem *et al.*, 2020). Currently, our knowledge about assembling, maintaining, and breaking down of thylakoid protein complex is not comprehensive, and several essential components are yet to be identified. Moreover, limited information is available regarding the proteins and chemical compounds that regulate carbon metabolism pathways and electron transfer mechanisms. However, cutting-edge techniques in systems biology, quantitative biology, and novel modeling approaches are expected to enhance our knowledge about biochemical and genetic networks of photosynthesis. Such knowledge will facilitate in developing innovative regulatory circuits for optimizing photosynthesis (Nogales *et al.*, 2012). Despite several limitations, genome editing technologies have the potential to revolutionize agriculture by enhancing crop yields and increasing food security, especially in the face of global climate change. Continued research and development in this area will be crucial in overcoming current limitations and optimizing the use of these tools to improve photosynthesis and address the challenges of feeding a growing global population in a changing environment.

References

Abdallah, N.A., Prakash, C.S. and McHughen, A.G. (2015) Genome editing for crop improvement: Challenges and opportunities. *GM Crops and Food* 6(4), 183–205.

Ahmad, N., Nielsen, B. and Mansoor, S. (2021) Editing the plastid genome of recalcitrant plant species. *Trends in Genetics* 37(11), 955–957.

Aigner, H., Wilson, R., Bracher, A., Calisse, L., Bhat, J. *et al.* (2017) Plant RuBisCo assembly in *E. coli* with five chloroplast chaperones including BSD2. *Science* 358(6368), 1272–1278.

Alfatih, A., Wu, J., Zhang, Z.S., Xia, J.Q., Jan, S.U. *et al.* (2020) Rice NIN-LIKE PROTEIN 1 rapidly responds to nitrogen deficiency and improves yield and nitrogen use efficiency. *Journal of Experimental Botany* 71(19), 6032–6042.

An, Y., Zhou, Y., Han, X., Shen, C., Wang, S. *et al.* (2020) The GATA transcription factor GNC plays an important role in photosynthesis and growth in poplar. *Journal of Experimental Botany* 71(6), 1969–1984.

Andralojc, P.J., Carmo-Silva, E., Degen, G.E. and Parry, M.A. (2018) Increasing metabolic potential: C-fixation. *Essays in Biochemistry* 62(1), 109–118.

Aranjuelo, I., Sanz-Sáez, Á., Jauregui, I., Irigoyen, J.J., Araus, J.L. *et al.* (2013) Harvest index, a parameter conditioning responsiveness of wheat plants to elevated CO_2. *Journal of Experimental Botany* 64(7), 1879–1892.

Archer, M.D. and Barber, J. (2004) Photosynthesis and photoconversion. *Molecular to Global Photosynthesis* 2, 1–42.

Atkinson, N., Leitão, N., Orr, D.J., Meyer, M.T., Carmo-Silva, E. *et al.* (2017) Rubisco small subunits from the unicellular green alga *Chlamydomonas* complement Rubisco-deficient mutants of Arabidopsis. *New Phytologist* 214(2), 655–667.

Azameti, M.K. and Dauda, W.P. (2021) Base editing in plants: Applications, challenges, and future prospects. *Frontiers in Plant Science* 12, 664997.

Badhan, S., Ball, A.S. and Mantri, N. (2021) First report of CRISPR/Cas9 mediated DNA-free editing of *4CL* and *RVE7* genes in chickpea protoplasts. *International Journal of Molecular Sciences* 22(1), 396.

Baek, K., Kim, D.H., Jeong, J., Sim, S.J., Melis, A. *et al.* (2016) DNA-free two-gene knockout in *Chlamydomonas reinhardtii* via CRISPR-Cas9 ribonucleoproteins. *Scientific Reports* 6(1), 30620.

Bastet, A., Zafirov, D., Giovinazzo, N., Guyon-Debast, A., Nogué, F. *et al.* (2019) Mimicking natural polymorphism in *eIF4E* by CRISPR-Cas9 base editing is associated with resistance to potyviruses. *Plant Biotechnology Journal* 17(9), 1736–1750.

Batista-Silva, W., Fonseca-Pereira, P., Martins, A.O., Zsögön, A., Nunes-Nesi, A. *et al.* (2020) Engineering improved photosynthesis in the era of synthetic biology. *Plant Communications* 1(2), 100032.

Bauwe, H., Hagemann, M. and Fernie, A.R. (2010) Photorespiration: Players, partners and origin. *Trends in Plant Science* 15(6), 330–336.

Bedell, V.M., Wang, Y., Campbell, J.M., Poshusta, T.L., Starker, C.G. *et al.* (2012) *In vivo* genome editing using a high-efficiency TALEN system. *Nature* 491(7422), 114–118.

Betti, M., Bauwe, H., Busch, F.A., Fernie, A.R., Keech, O. *et al.* (2016) Manipulating photorespiration to increase plant productivity: Recent advances and perspectives for crop improvement. *Journal of Experimental Botany* 67(10), 2977–2988.

Bhagwat, A.S. (2005) Photosynthetic carbon assimilation of C3, C4, and CAM pathways. *Handbook of Photosynthesis* 2, 367–389.

Bi, Z., Zhang, Y., Wu, W., Zhan, X., Yu, N. *et al.* (2017) *ES7*, encoding a ferredoxin-dependent glutamate synthase, functions in nitrogen metabolism and impacts leaf senescence in rice. *Plant Science* 259, 24–34.

Blankenship, R. (2002) *Molecular mechanisms of photosynthesis*. Blackwell Science, Wiley, Maiden, MA.

Blankenship, R.E., Tiede, D.M., Barber, J., Brudvig, G.W., Fleming, G. *et al.* (2011) Comparing photosynthetic and photovoltaic efficiencies and recognizing the potential for improvement. *Science* 332(6031), 805–809.

Bock, R. (2007) Plastid biotechnology: Prospects for herbicide and insect resistance, metabolic engineering and molecular farming. *Current Opinion in Biotechnology* 18(2), 100–106.

Bussi, Y., Shimoni, E., Weiner, A., Kapon, R., Charuvi, D. *et al.* (2019) Fundamental helical geometry consolidates the plant photosynthetic membrane. *Proceedings of the National Academy of Sciences* 116(44), 22366–22375.

Caddell, D., Langenfeld, N.J., Eckels, M.J., Zhen, S., Klaras, R. *et al.* (2023) Photosynthesis in rice is increased by CRISPR/Cas9-mediated transformation of two truncated light-harvesting antenna. *Frontiers in Plant Science* 14, 1050483.

Carmo-Silva, A.E. and Salvucci, M.E. (2012) The temperature response of CO_2 assimilation, photochemical activities and Rubisco activation in *Camelina sativa*, a potential bioenergy crop with limited capacity for acclimation to heat stress. *Planta* 236, 1433–1445.

Cassier-Chauvat, C., Blanc-Garin, V. and Chauvat, F. (2021) Genetic, genomics, and responses to stresses in cyanobacteria: Biotechnological implications. *Genes* 12(4), 500.

Cen, Y.P. and Sage, R.F. (2005) The regulation of Rubisco activity in response to variation in temperature and atmospheric CO_2 partial pressure in sweet potato. *Plant Physiology* 139(2), 979–990.

Chen, F., Zheng, G., Qu, M., Wang, Y., Lyu, M.J.A. *et al.* (2021) Knocking out NEGATIVE REGULATOR OF PHOTOSYNTHESIS 1 increases rice leaf PHOTOSYNTHESIS and biomass production in the field. *Journal of Experimental Botany* 72(5), 1836–1849.

Cheng, H., Hao, M., Ding, B., Mei, D., Wang, W. *et al.* (2021) Base editing with high efficiency in allotetraploid oilseed rape by A3A-PBE system. *Plant Biotechnology Journal* 19(1), 87–97.

Clemens, M., Faralli, M., Lagreze, J., Bontempo, L., Piazza, S. *et al.* (2022) *VvEPFL9-1* knock-out via CRISPR/Cas9 reduces stomatal density in grapevine. *Frontiers in Plant Science* 13, 878001.

Cong, L., Ran, F.A., Cox, D., Lin, S., Barretto, R. *et al.* (2013) Multiplex genome engineering using CRISPR/Cas systems. *Science* 339(6121), 819–823.

Covshoff, S. and Hibberd, J.M. (2012) Integrating C4 photosynthesis into C3 crops to increase yield potential. *Current Opinion in Biotechnology* 23(2), 209–214.

Crafts-Brandner, S.J. and Salvucci, M.E. (2000) Rubisco activase constrains the photosynthetic potential of leaves at high temperature and CO_2. *Proceedings of the National Academy of Sciences* 97(24), 13430–13435.

Dalal, J., Lopez, H., Vasani, N.B., Hu, Z., Swift, J.E. *et al.* (2015) A photorespiratory bypass increases plant growth and seed yield in biofuel crop *Camelina sativa*. *Biotechnology for Biofuels* 8, 1–22.

Ding, F., Wang, M., Zhang, S. and Ai, X. (2016) Changes in SBPase activity influence photosynthetic capacity, growth, and tolerance to chilling stress in transgenic tomato plants. *Scientific Reports* 6(1), 32741.

Ding, F., Hu, Q., Wang, M. and Zhang, S. (2018) Knockout of SlSBPASE suppresses carbon assimilation and alters nitrogen metabolism in tomato plants. *International Journal of Molecular Sciences* 19(12), 4046.

Dong, S. and Beckles, D.M. (2019) Dynamic changes in the starch-sugar interconversion within plant source and sink tissues promote a better abiotic stress response. *Journal of Plant Physiology* 234, 80–93.

Donovan, S., Mao, Y., Orr, D.J., Carmo-Silva, E. and McCormick, A.J. (2020) CRISPR-Cas9-mediated mutagenesis of the Rubisco small subunit family in *Nicotiana tabacum*. *Frontiers in Genome Editing* 2, 605614.

Driever, S.M., Simkin, A.J., Alotaibi, S., Fisk, S.J., Madgwick, P.J. *et al.* (2017) Increased SBPase activity improves photosynthesis and grain yield in wheat grown in greenhouse conditions. *Philosophical Transactions of the Royal Society B: Biological Sciences* 372(1730), 20160384.

Du, N., Xue, L., Xue, D., Dong, X., Yang, Q. *et al.* (2023) The transcription factor SlNAP1 increases salt tolerance by modulating ion homeostasis and ROS metabolism in *Solanum lycopersicum*. *Gene* 849, 146906.

Durai, S., Mani, M., Kandavelou, K., Wu, J., Porteus, M.H. *et al.* (2005) Zinc finger nucleases: Custom-designed molecular scissors for genome engineering of plant and mammalian cells. *Nucleic Acids Research* 33(18), 5978–5990.

Dymond, J.S., Richardson, S.M., Coombes, C.E., Babatz, T., Muller, H. *et al.* (2011) Synthetic chromosome arms function in yeast and generate phenotypic diversity by design. *Nature* 477(7365), 471–476.

Eid, A., Mohan, C., Sanchez, S., Wang, D. and Altpeter, F. (2021) Multiallelic, targeted mutagenesis of magnesium chelatase with CRISPR/Cas9 provides a rapidly scorable phenotype in highly polyploid sugarcane. *Frontiers in Genome Editing* 3, 654996.

Eyles, A., Pinkard, E.A., Davies, N.W., Corkrey, R., Churchill, K. *et al.* (2013) Whole-plant versus leaf-level regulation of photosynthetic responses after partial defoliation in *Eucalyptus globulus* saplings. *Journal of Experimental Botany* 64(6), 1625–1636.

Flamholz, A.I., Prywes, N., Moran, U., Davidi, D., Bar-On, Y.M. *et al.* (2019) Revisiting trade-offs between Rubisco kinetic parameters. *Biochemistry* 58(31), 3365–3376.

Flexas, J. (2016) Genetic improvement of leaf photosynthesis and intrinsic water use efficiency in C3 plants: Why so much little success. *Plant Science* 251, 155–161.

Fonfara, I., Richter, H., Bratovič, M., Le Rhun, A. and Charpentier, E. (2016) The CRISPR-associated DNA-cleaving enzyme Cpf1 also processes precursor CRISPR RNA. *Nature* 532(7600), 517–521.

Friedland, N., Negi, S., Vinogradova-Shah, T., Wu, G., Ma, L. *et al.* (2019) Fine-tuning the photosynthetic light harvesting apparatus for improved photosynthetic efficiency and biomass yield. *Scientific Reports* 9(1), 13028.

Fukayama, H., Hatch, M.D., Tamai, T., Tsuchida, H., Sudoh, S. *et al.* (2003) Activity regulation and physiological impacts of maize C4-specific phosphoenol pyruvate carboxylase overproduced in transgenic rice plants. *Photosynthesis Research* 77, 227–239.

Fukayama, H., Mizumoto, A., Ueguchi, C., Katsunuma, J., Morita, R. *et al.* (2018) Expression level of Rubisco activase negatively correlates with Rubisco content in transgenic rice. *Photosynthesis Research* 137, 465–474.

Gaastra, P. (1959) *Photosynthesis of Crop Plants as Influenced by Light, Carbon Dioxide, Temperature, and Stomatal Diffusion*. Veenman.

Gaj, T., Gersbach, C.A. and Barbas III, C.F. (2013) ZFN, TALEN, and CRISPR/Cas-based methods for genome engineering. *Trends in Biotechnology* 31(7), 397–405.

Gao, C. (2021) Genome engineering for crop improvement and future agriculture. *Cell* 184(6), 1621–1635.

Gao, Y., Xu, Z., Zhang, L., Li, S., Wang, S. *et al.* (2020) MYB61 is regulated by GRF4 and promotes nitrogen utilization and biomass production in rice. *Nature Communications* 11(1), 5219.

Gibson, D.G., Glass, J.I., Lartigue, C., Noskov, V.N., Chuang, R.Y. *et al.* (2010) Creation of a bacterial cell controlled by a chemically synthesized genome. *Science* 329(5987), 52–56.

Głowacka, K., Kromdijk, J., Kucera, K., Xie, J., Cavanagh, A.P. *et al.* (2018) Photosystem II subunit S overexpression increases the efficiency of water use in a field-grown crop. *Nature Communications* 9(1), 868.

Gu, J., Zhou, Z., Li, Z., Chen, Y., Wang, Z. *et al*. (2017) Rice (*Oryza sativa* L.) with reduced chlorophyll content exhibit higher photosynthetic rate and efficiency, improved canopy light distribution, and greater yields than normally pigmented plants. *Field Crops Research* 200, 58–70.

Hanba, Y.T., Shibasaka, M., Hayashi, Y., Hayakawa, T., Kasamo, K. *et al*. (2004) Overexpression of the barley aquaporin HvPIP2; 1 increases internal CO2 conductance and CO2 assimilation in the leaves of transgenic rice plants. *Plant and Cell Physiology* 45(5), 521–529.

Hibberd, J.M., Sheehy, J.E. and Langdale, J.A. (2008) Using C4 photosynthesis to increase the yield of rice – rationale and feasibility. *Current Opinion in Plant Biology* 11(2), 228–231.

Hillary, V.E. and Ceasar, S.A. (2023) A review on the mechanism and applications of CRISPR/Cas9/Cas12/Cas13/Cas14 proteins utilized for genome engineering. *Molecular Biotechnology* 65(3), 311–325.

Hines, K.M., Chaudhari, V., Edgeworth, K.N., Owens, T.G. and Hanson, M.R. (2021) Absence of carbonic anhydrase in chloroplasts affects C3 plant development but not photosynthesis. *Proceedings of the National Academy of Sciences* 118(33), e2107425118.

Hua, K., Tao, X., Yuan, F., Wang, D. and Zhu, J.K. (2018) Precise A·T to G·C base editing in the rice genome. *Molecular Plant* 11(4), 627–630.

Huang, W., Yang, Y.J., Zhang, S.B. and Liu, T. (2018) Cyclic electron flow around photosystem I promotes ATP synthesis possibly helping the rapid repair of photodamaged photosystem II at low light. *Frontiers in Plant Science* 9, 239.

Hügler, M. and Sievert, S.M. (2011) Beyond the calvin cycle: Autotrophic carbon fixation in the ocean. *Annual Review of Marine Science* 3, 261–289.

Iniguez, C., Capó-Bauçà, S., Niinemets, Ü., Stoll, H., Aguiló-Nicolau, P. *et al*. (2020) Evolutionary trends in Rubisco kinetics and their CO-evolution with CO_2 concentrating mechanisms. *The Plant Journal* 101(4), 897–918.

Ishikawa, C., Hatanaka, T., Misoo, S., Miyake, C. and Fukayama, H. (2011) Functional incorporation of sorghum small subunit increases the catalytic turnover rate of Rubisco in transgenic rice. *Plant Physiology* 156(3), 1603–1611.

Jeong, J., Baek, K., Kirst, H., Melis, A. and Jin, E. (2017) Loss of CpSRP54 function leads to a truncated light-harvesting antenna size in *Chlamydomonas reinhardtii*. *Biochimica et Biophysica Acta (BBA)-Bioenergetics* 1858(1), 45–55.

Jeong, J., Baek, K., Yu, J., Kirst, H., Betterle, N. *et al*. (2018) Deletion of the chloroplast LTD protein impedes LHCI import and PSI–LHCI assembly in *Chlamydomonas reinhardtii*. *Journal of Experimental Botany* 69(5), 1147–1158.

Ji, Y., Huang, W., Wu, B., Fang, Z. and Wang, X. (2020) The amino acid transporter AAP1 mediates growth and grain yield by regulating neutral amino acid uptake and reallocation in *Oryza sativa*. *Journal of Experimental Botany* 71(16), 4763–4777.

Johnson, M.P. and Wientjes, E. (2020) The relevance of dynamic thylakoid organisation to photosynthetic regulation. *Biochimica et Biophysica Acta (BBA)-Bioenergetics* 1861(4), 148039.

Kajala, K., Covshoff, S., Karki, S., Woodfield, H., Tolley, B.J. *et al*. (2011) Strategies for engineering a two-celled C4 photosynthetic pathway into rice. *Journal of Experimental Botany* 62(9), 3001–3010.

Kandoi, D., Mohanty, S. and Tripathy, B.C. (2016) Towards efficient photosynthesis: Overexpression of *Zea mays* phosphoenolpyruvate carboxylase in *Arabidopsis thaliana*. *Photosynthesis Research* 130, 47–72.

Kang, B.C., Bae, S.J., Lee, S., Lee, J.S., Kim, A. *et al*. (2021) Chloroplast and mitochondrial DNA editing in plants. *Nature Plants* 7(7), 899–905.

Karavolias, N.G., Horner, W., Abugu, M.N. and Evanega, S.N. (2021a) Application of gene editing for climate change in agriculture. *Frontiers in Sustainable Food Systems* 5, 685801.

Karavolias, N.G., Patel, D., Seong, K., Tjahjadi, M., Gueorguieva, G.A. *et al*. (2021b) CRISPR/Cas9 knockout of EPFL10 reduces stomatal density while maintaining photosynthesis and enhancing water conservation in rice. *bioRxiv* 9, 14. DOI: 10.1101/2021.12.21.473329.

Karvelis, T., Bigelyte, G., Young, J.K., Hou, Z., Zedaveinyte, R. *et al*. (2020) PAM recognition by miniature CRISPR–Cas12f nucleases triggers programmable double-stranded DNA target cleavage. *Nucleic Acids Research* 48(9), 5016–5023.

Kawase, M., Hanba, Y.T. and Katsuhara, M. (2013) The photosynthetic response of tobacco plants overexpressing ice plant aquaporin McMIPB to a soil water deficit and high vapor pressure deficit. *Journal of Plant Research* 126, 517–527.

Kerfeld, C.A. and Melnicki, M.R. (2016) Assembly, function and evolution of cyanobacterial carboxysomes. *Current Opinion in Plant Biology* 31, 66–75.

Kirst, H. and Melis, A. (2014) The chloroplast signal recognition particle (CpSRP) pathway as a tool to minimize chlorophyll antenna size and maximize photosynthetic productivity. *Biotechnology Advances* 32(1), 66–72.

Ke, X., Shen, J., Niu, Y., Zhao, H., Guo, Y. *et al.* (2023) Cucumber NUCLEAR FACTOR-YC2/-YC9 target translocon component CsTIC21 in chloroplast photomorphogenesis. *Plant Physiology* 192(4), 2822–2837.

Kirst, H., Garcia-Cerdan, J.G., Zurbriggen, A., Ruehle, T. and Melis, A. (2012) Truncated photosystem chlorophyll antenna size in the green microalga *Chlamydomonas reinhardtii* upon deletion of the TLA3-CpSRP43 gene. *Plant Physiology* 160(4), 2251–2260.

Kirst, H., Formighieri, C. and Melis, A. (2014) Maximizing photosynthetic efficiency and culture productivity in cyanobacteria upon minimizing the phycobilisome light-harvesting antenna size. *Biochimica et Biophysica Acta (BBA)-Bioenergetics* 1837(10), 1653–1664.

Kirst, H., Gabilly, S.T., Niyogi, K.K., Lemaux, P.G. and Melis, A. (2017) Photosynthetic antenna engineering to improve crop yields. *Planta* 245, 1009–1020.

Knoot, C.J., Biswas, S. and Pakrasi, H.B. (2019) Tunable repression of key photosynthetic processes using Cas12a CRISPR interference in the fast-growing cyanobacterium *Synechococcus* sp. UTEX 2973. *ACS Synthetic Biology* 9(1), 132–143.

Kromdijk, J., Głowacka, K., Leonelli, L., Gabilly, S.T., Iwai, M. *et al.* (2016) Improving photosynthesis and crop productivity by accelerating recovery from photoprotection. *Science* 354(6314), 857–861.

Kubis, A. and Bar-Even, A. (2019) Synthetic biology approaches for improving photosynthesis. *Journal of Experimental Botany* 70(5), 1425–1433.

Kuluev, B., Gumerova, G., Mikhaylova, E., Gerashchenkov, G., Rozhnova, N. *et al.* (2019) Delivery of CRISPR/Cas components into higher plant cells for genome editing. *Russian Journal of Plant Physiology* 66, 694–706.

Kumar, A., Li, C. and Portis, A.R. (2009) *Arabidopsis thaliana* expressing a thermostable chimeric Rubisco activase exhibits enhanced growth and higher rates of photosynthesis at moderately high temperatures. *Photosynthesis Research* 100, 143–153.

Kumar, A., Pandey, S.S., Kumar, D. and Tripathi, B.N. (2023) Genetic manipulation of photosynthesis to enhance crop productivity under changing environmental conditions. *Photosynthesis Research* 155(1), 1–21.

Kuroiwa, K., Thenault, C., Nogué, F., Perrot, L., Mazier, M. *et al.* (2022) CRISPR-based knock-out of *elf4E2* in a cherry tomato background successfully recapitulates resistance to pepper veinal mottle virus. *Plant Science* 316, 111160.

Laterre, R., Pottier, M., Remacle, C. and Boutry, M. (2017) Photosynthetic trichomes contain a specific Rubisco with a modified pH-dependent activity. *Plant Physiology* 173(4), 2110–2120.

Leakey, A.D., Ainsworth, E.A., Bernacchi, C.J., Rogers, A., Long, S.P. *et al.* (2009) Elevated CO_2 effects on plant carbon, nitrogen, and water relations: Six important lessons from FACE. *Journal of Experimental Botany* 60(10), 2859–2876.

Lebedev, V.G., Popova, A.A. and Shestibratov, K.A. (2021) Genetic engineering and genome editing for improving nitrogen use efficiency in plants. *Cells* 10(12), 3303.

Lee, M., Heo, Y.B. and Woo, H.M. (2023) Cytosine base editing in cyanobacteria by repressing archaic type IV uracil-DNA glycosylase. *The Plant Journal* 113(3), 610–625.

Lee, S., Park, J., Lee, J., Shin, D., Marmagne, A. *et al.* (2020) OsASN1 overexpression in rice increases grain protein content and yield under nitrogen-limiting conditions. *Plant and Cell Physiology* 61(7), 1309–1320.

Lefebvre, S., Lawson, T., Fryer, M., Zakhleniuk, O.V., Lloyd, J.C. *et al.* (2005) Increased sedoheptulose-1, 7-bisphosphatase activity in transgenic tobacco plants stimulates photosynthesis and growth from an early stage in development. *Plant Physiology* 138(1), 451–460.

Li, R., Char, S.N., Liu, B., Liu, H., Li, X. *et al.* (2021) High-efficiency plastome base editing in rice with TAL cytosine deaminase. *Molecular Plant* 14(9), 1412–1414.

Li, S., Tian, Y., Wu, K., Ye, Y., Yu, J. *et al.* (2018) Modulating plant growth–metabolism coordination for sustainable agriculture. *Nature* 560(7720), 595–600.

Li, Y., Zhu, J., Wu, H., Liu, C., Huang, C. *et al.* (2020) Precise base editing of non-allelic acetolactate synthase genes confers sulfonylurea herbicide resistance in maize. *The Crop Journal* 8(3), 449–456.

Liu, J., Lu, Y., Hua, W. and Last, R.L. (2019) A new light on photosystem II maintenance in oxygenic photosynthesis. *Frontiers in Plant Science* 10, 975.

Liu, M., Ma, W., Su, X., Zhang, X., Lu, Y. *et al.* (2022) Mutation in a chlorophyll-binding motif of brassica ferrochelatase enhances both heme and chlorophyll biosynthesis. *Cell Reports* 41, 10.

Long, B.M., Hee, W.Y., Sharwood, R.E., Rae, B.D., Kaines, S. *et al.* (2018) Carboxysome encapsulation of the CO_2-fixing enzyme Rubisco in tobacco chloroplasts. *Nature Communications* 9(1), 1–14.

López-Calcagno, P.E., Fisk, S., Brown, K.L., Bull, S.E., South, P.F. *et al.* (2019) Overexpressing the H-protein of the glycine cleavage system increases biomass yield in glasshouse and field-grown transgenic tobacco plants. *Plant Biotechnology Journal* 17(1), 141–151.

Lu, K., Wu, B., Wang, J., Zhu, W., Nie, H. *et al.* (2018) Blocking amino acid transporter Os AAP 3 improves grain yield by promoting outgrowth buds and increasing tiller number in rice. *Plant Biotechnology Journal* 16(10), 1710–1722.

Luo, J., Huang, W., Yan, J., Fang, Z. and Ren, M. (2022) The *GzMYB-7D1* gene of Guizimai no.1 wheat is essential for seed anthocyanins accumulation and yield regulation. *Plant Science* 320, 111293.

Maier, A., Fahnenstich, H., Caemmerer, S., Engqvist, M.K., Weber, A.P. *et al.* (2012) Transgenic introduction of A glycolate oxidative cycle into *A. thaliana* chloroplasts leads to growth improvement. *Frontiers in Plant Science* 3, 38.

Maliga, P. (2004) Plastid transformation in higher plants. *Annual Review of Plant Biology* 55, 289–313.

Manghwar, H., Lindsey, K., Zhang, X. and Jin, S. (2019) CRISPR/CAS system: Recent advances and future prospects for genome editing. *Trends in Plant Science* 24(12), 1102–1125.

Manning, T. (2019) *Photosynthesis Improvement in Crop Plants Through Gene Replacement*. RMIT University.

Manzoor, H., Mehwish, S., Bukhat, S., Rasul, S. and Rehmani, M.I.A. (2022) Methyl jasmonate alleviated the adverse effects of cadmium stress in pea (*Pisum sativum* L.): A nexus of photosystem II activity and dynamics of redox balance. *Frontiers in Plant Science* 13, 860664.

Manzoor, H., Anjam, M.S., Saeed, F., Rasul, S., Yousaf, S. *et al.* (2022) Photosynthetic efficiency and antioxidant defense potential are key players in inducing drought tolerance in transgenic tobacco plants over-expressing *AVP1*. *Journal of Plant Growth Regulation* 41(7), 2653–2668.

Martin Avila, E., Gisby, M.F. and Day, A. (2016) Seamless editing of the chloroplast genome in plants. *BMC Plant Biology* 16(1), 1–13.

Matsumura, H., Shiomi, K., Yamamoto, A., Taketani, Y., Kobayashi, N. *et al.* (2020) Hybrid Rubisco with complete replacement of rice Rubisco small subunits by sorghum counterparts confers C4 plant-like high catalytic activity. *Molecular Plant* 13(11), 1570–1581.

McClung, C.R. (2014) Making hunger yield. *Science* 344(6185), 699–700.

McGrath, J.M. and Long, S.P. (2014) Can the cyanobacterial carbon-concentrating mechanism increase photosynthesis in crop species? A theoretical analysis. *Plant Physiology* 164(4), 2247–2261.

Melis, A. (2009) Solar energy conversion efficiencies in photosynthesis: Minimizing the chlorophyll antennae to maximize efficiency. *Plant Science* 177(4), 272–280.

Michno, J.M., Wang, X., Liu, J., Curtin, S.J., Kono, T.J. *et al.* (2015) CRISPR/CAS mutagenesis of soybean and *Medicago truncatula* using a new web-tool and a modified Cas9 enzyme. *GM Crops and Food* 6(4), 243–252.

Miglani, G.S., Kaur, R., Sharma, P. and Gupta, N. (2021) Leveraging photosynthetic efficiency toward improving crop yields. *Journal of Crop Improvement* 35(3), 361–402.

Miladinovic, D., Antunes, D., Yildirim, K., Bakhsh, A., Cvejić, S. *et al.* (2021) Targeted plant improvement through genome editing: From laboratory to field. *Plant Cell Reports* 40, 935–951.

Miyake, C. (2010) Alternative electron flows (water–water cycle and cyclic electron flow around PSI) in photosynthesis: Molecular mechanisms and physiological functions. *Plant and Cell Physiology* 51(12), 1951–1963.

Mok, Y.G., Hong, S., Bae, S.J., Cho, S.I. and Kim, J.S. (2022) Targeted A-to-G base editing of chloroplast DNA in plants. *Nature Plants* 8(12), 1378–1384.

Monsur, M.B., Ni, C., Xiangjin, W., Lihong, X., Guiai, J. *et al.* (2021) Improved eating and cooking quality of indica rice cultivar YK17 via adenine base editing of wx allele of granule-bound starch synthase I (GBSS I). *Rice Science* 28(5), 427–430.

Moroney, J.V. and Ynalvez, R.A. (2007) Proposed carbon dioxide concentrating mechanism in *Chlamydomonas reinhardtii*. *Eukaryotic Cell* 6(8), 1251–1259.

Mubarak, A., Burgess, A., Pyke, K., Quick, W. and Murchie, E. (2023) Mass screening of rice mutant populations at low CO_2 for identification of lowered photorespiration and respiration rates. *Frontiers in Plant Science* 14, 1125770.

Murchie, E.H. and Niyogi, K.K. (2011) Manipulation of photoprotection to improve plant photosynthesis. *Plant Physiology* 155(1), 86–92.

Naduthodi, M.I.S., Barbosa, M.J. and Oost, J. (2018) Progress of CRISPR-Cas based genome editing in photosynthetic microbes. *Biotechnology Journal* 13(9), 1700591.

Naeem, M., Majeed, S., Hoque, M.Z. and Ahmad, I. (2020) Latest developed strategies to minimize the off-target effects in CRISPR-Cas-mediated genome editing. *Cells* 9(7), 1608.

Nakazato, I., Okuno, M., Yamamoto, H., Tamura, Y., Itoh, T. *et al.* (2021) Targeted base editing in the plastid genome of *Arabidopsis thaliana*. *Nature Plants* 7(7), 906–913.

Nigishi, M., Shimakawa, G., Yamagishi, K., Amano, R., Ito, S. *et al.* (2024) Low-Co2-inducible bestrophins outside the pyrenoid sustain high photosynthetic efficacy in diatoms. *Plant Physiology* 195(2), 1432–1445.

Nogales, J., Gudmundsson, S., Knight, E.M., Palsson, B.O. and Thiele, I. (2012) Detailing the optimality of photosynthesis in cyanobacteria through systems biology analysis. *Proceedings of the National Academy of Sciences* 109(7), 2678–2683.

Oh, Z.G., Askey, B. and Gunn, L.H. (2023) Red Rubiscos and opportunities for engineering green plants. *Journal of Experimental Botany* 74(2), 520–542.

Ort, D.R., Zhu, X. and Melis, A. (2011) Optimizing antenna size to maximize photosynthetic efficiency. *Plant Physiology* 155(1), 79–85.

Ort, D.R., Merchant, S.S., Alric, J., Barkan, A., Blankenship, R.E. *et al.* (2015) Redesigning photosynthesis to sustainably meet global food and bioenergy demand. *Proceedings of the National Academy of Sciences* 112(28), 8529–8536.

Parry, M.A., Andralojc, P.J., Scales, J.C., Salvucci, M.E., Carmo-Silva, A.E. *et al.* (2013) Rubisco activity and regulation as targets for crop improvement. *Journal of Experimental Botany* 64(3), 717–730.

Pascual, M.B., Molina-Rueda, J.J., Cánovas, F.M. and Gallardo, F. (2018) Overexpression of a cytosolic NADP+-isocitrate dehydrogenase causes alterations in the vascular development of hybrid poplars. *Tree Physiology* 38(7), 992–1005.

Patel, V.K., Soni, N., Prasad, V., Sapre, A., Dasgupta, S. *et al.* (2019) CRISPR–Cas9 system for genome engineering of photosynthetic microalgae. *Molecular Biotechnology* 61, 541–561.

Pavese, V., Moglia, A., Corredoira, E., Martínez, M.T., Torello Marinoni, D. *et al.* (2021) First report of CRISPR/Cas9 gene editing in *Castanea sativa* mill. *Frontiers in Plant Science* 12, 728516.

Periwal, V. (2017) A comprehensive overview of computational resources to aid in precision genome editing with engineered nucleases. *Briefings in Bioinformatics* 18(4), 698–711.

Pesaresi, P., Scharfenberg, M., Weigel, M., Granlund, I., Schröder, W.P. *et al.* (2009) Mutants, overexpressors, and interactors of *Arabidopsis* plastocyanin isoforms: Revised roles of plastocyanin in photosynthetic electron flow and thylakoid redox state. *Molecular Plant* 2(2), 236–248.

Pietrzykowska, M., Suorsa, M., Semchonok, D.A., Tikkanen, M., Boekema, E.J. *et al.* (2014) The light-harvesting chlorophyll a/b binding proteins LHcb1 and LHcb2 play complementary roles during state transitions in Arabidopsis. *The Plant Cell* 26(9), 3646–3660.

Polle, J.E., Kanakagiri, S.D. and Melis, A. (2003) Tla1, a DNA insertional transformant of the green alga *Chlamydomonas reinhardtii* with a truncated light-harvesting chlorophyll antenna size. *Planta* 217, 49–59.

Polle, J.E., Benemann, J.R., Tanaka, A. and Melis, A. (2000) Photosynthetic apparatus organization and function in the wild type and a chlorophyll b-less mutant of *Chlamydomonas reinhardtii*. Dependence on carbon source. *Planta* 211, 335–344.

Poorter, H., Bühler, J., Dusschoten, D., Climent, J. and Postma, J.A. (2012) Pot size matters: A meta-analysis of the effects of rooting volume on plant growth. *Functional Plant Biology* 39(11), 839–850.

Pottier, M., Gilis, D. and Boutry, M. (2018) The hidden face of Rubisco. *Trends in Plant Science* 23(5), 382–392.

Prins, A., Orr, D.J., Andralojc, P.J., Reynolds, M.P., Carmo-Silva, E. and Parry, M.A.J. (2016) Rubisco catalytic properties of wild and domesticated relatives provide scope for improving wheat photosynthesis. *Journal of Experimental Botany* 67(6), 1827–1838.

Qu, Y., Sakoda, K., Fukayama, H., Kondo, E., Suzuki, Y. *et al.* (2021) Overexpression of both Rubisco and Rubisco activase rescues rice photosynthesis and biomass under heat stress. *Plant, Cell and Environment* 44(7), 2308–2320.

Raines, C.A. (2011) Increasing photosynthetic carbon assimilation in C3 plants to improve crop yield: Current and future strategies. *Plant Physiology* 155(1), 36–42.

Rathnasamy, S.A., Kambale, R., Elangovan, A., Mohanavel, W., Shanmugavel, P. *et al.* (2023) Altering stomatal density for manipulating transpiration and photosynthetic traits in rice through CRISPR/Cas9 mutagenesis. *Current Issues in Molecular Biology* 45(5), 3801–3814.

Roell, M.S., Schada von Borzyskowskl, L., Westhoff, P., Plett, A., Paczia, N. *et al.* (2021) A synthetic C4 shuttle via the β-hydroxyaspartate cycle in C3 plants. *Proceedings of the National Academy of Sciences* 118(21), e2022307118.

Rosenthal, D.M., Locke, A.M., Khozaei, M., Raines, C.A., Long, S.P. *et al.* (2011) Over-expressing the C3 photosynthesis cycle enzyme sedoheptulose-1-7 bisphosphatase improves photosynthetic carbon gain and yield under fully open air CO_2 fumigation (FACE). *BMC Plant Biology* 11(1), 1–12.

Rudenko, N.N., Permyakova, N.V., Ignatova, L.K., Nadeeva, E.M., Zagorskaya, A.A. *et al.* (2022) The role of carbonic anhydrase αCA4 in photosynthetic reactions in *Arabidopsis thaliana* studied, using the Cas9 and T-DNA induced mutations in its gene. *Plants* 11(23), 3303.

Saeed, F., Shah, T., Bukhat, S., Munsif, F., Ahmad, I. *et al.* (2022) Genome engineering as a tool for enhancing crop traits: lessons from CRISPR/cas9. In: Wani, S. and Hensel, G. (eds) *Genome Editing: Current Technology Advances and Applications for Crop Improvement*. Springer, Cham, Switzerland, pp. 3–25.

Saeed, F., Bukhat, S., Shah, T., Rasul, S. and Manzoor, H. (2023) Multiplexed genome editing in cereals: Trait improvement using CRISPR/Cas9. In: Shah, T., Filho, M., Nie, L. and Amir, R. (eds) *Cereal Crops*. CRC Press, Boca Raton, Florida, pp. 313–326.

Sattari Vayghan, H., Nawrocki, W.J., Schiphorst, C., Tolleter, D., Hu, C. *et al.* (2022) Photosynthetic light harvesting and thylakoid organization in a CRISPR/Cas9 arabidopsis thaliana LHCB1 knockout mutant. *Frontiers in Plant Science* 13, 833032.

Scafaro, A.P., Bautsoens, N., Boer, B., Rie, J. and Gallé, A. (2019) A conserved sequence from heat-adapted species improves Rubisco activase thermostability in wheat. *Plant Physiology* 181(1), 43–54.

Schmermund, L., Jurka, V., Özgen, F.F., Barone, G.D. and Büchsenschütz, H.C. (2019) Photo-biocatalysis: Biotransformations in the presence of light. *Acs Catalysis* 9(5), 4115–4144.

Schottler, M.A., Kirchhoff, H. and Weis, E. (2004) The role of plastocyanin in the adjustment of the photosynthetic electron transport to the carbon metabolism in tobacco. *Plant Physiology* 136(4), 4265–4274.

Shakeel, A., Tang, L., Shahzad, R., Mawia, A.M., Rao, G.S. *et al.* (2021) CRISPR-based crop improvements: A way forward to achieve zero hunger. *Journal of Agricultural and Food Chemistry* 69(30), 8307–8323.

Shamshirgaran, Y., Liu, J., Sumer, H., Verma, P.J. and Taheri-Ghahfarokhi, A. (2022) Tools for efficient genome editing; ZFN, TALEN, and CRISPR. *Applications of Genome Modulation and Editing* 2495, 29–46.

Sheoran, I. and Singh, R. (2012) Carbon dioxide metabolism in photosynthesis. In: Singhal, G.S., Renger, G., Sopory, S.K., Irrgang, K. and Govindjee, (eds) *Concepts in Photobiology: Photosynthesis and Photomorphogenesis*. Springer Dordrecht, Delhi, India, pp. 430–437.

Shim, S.H., Lee, S.K., Lee, D.W., Brilhaus, D., Wu, G. *et al.* (2020) Loss of function of rice plastidic glycolate/glycerate translocator 1 impairs photorespiration and plant growth. *Frontiers in Plant Science* 10, 1726.

Simkin, A.J., McAusland, L., Headland, L.R., Lawson, T. and Raines, C.A. (2015) Multigene manipulation of photosynthetic carbon assimilation increases CO_2 fixation and biomass yield in tobacco. *Journal of Experimental Botany* 66(13), 4075–4090.

Simkin, A.J., Lopez-Calcagno, P.E., Davey, P.A., Headland, L.R., Lawson, T. *et al.* (2017) Simultaneous stimulation of sedoheptulose 1, 7-bisphosphatase, fructose 1, 6-bisphophate aldolase and the photorespiratory glycine decarboxylase-H protein increases CO_2 assimilation, vegetative biomass and seed yield in Arabidopsis. *Plant Biotechnology Journal* 15(7), 805–816.

Simkin, A.J., López-Calcagno, P.E. and Raines, C.A. (2019) Feeding the world: Improving photosynthetic efficiency for sustainable crop production. *Journal of Experimental Botany* 70(4), 1119–1140.

Simkin, A.J., Kapoor, L., Doss, C.G.P., Hofmann, T.A., Lawson, T. *et al.* (2022) The role of photosynthesis related pigments in light harvesting, photoprotection and enhancement of photosynthetic yield in planta. *Photosynthesis Research* 152(1), 23–42.

South, P., Cavanagh, A., Liu, H. and Ort, D. (2019) Synthetic glycolate metabolism pathways stimulate crop growth and productivity in the field. *Science* 365, 6452.

South, P.F., Cavanagh, A.P., Lopez-Calcagno, P.E., Raines, C.A. and Ort, D.R. (2018) Optimizing photorespiration for improved crop productivity. *Journal of Integrative Plant Biology* 60(12), 1217–1230.

Spiertz, J. and Ewert, F. (2009) Crop production and resource use to meet the growing demand for food, feed and fuel: Opportunities and constraints. *NJAS: Wageningen Journal of Life Sciences* 56(4), 281–300.

Stitt, M. (2013) Progress in understanding and engineering primary plant metabolism. *Current Opinion in Biotechnology* 24(2), 229–238.

Suganami, M., Suzuki, Y., Kondo, E., Nishida, S., Konno, S. *et al.* (2020) Effects of overproduction of Rubisco activase on Rubisco content in transgenic rice grown at different N levels. *International Journal of Molecular Sciences* 21(5), 1626.

Takahara, K., Kasajima, I., Takahashi, H., Hashida, S., Itami, T. *et al.* (2010) Metabolome and photochemical analysis of rice plants overexpressing Arabidopsis NAD kinase gene. *Plant Physiology* 152(4), 1863–1873.

Tanaka, R. and Tanaka, A. (2011) Chlorophyll cycle regulates the construction and destruction of the light-harvesting complexes. *Biochimica et Biophysica Acta (BBA)-Bioenergetics* 1807(8), 968–976.

Tang, W., Ye, J., Yao, X., Zhao, P., Xuan, W. *et al.* (2019) Genome-wide associated study identifies NAC42-activated nitrate transporter conferring high nitrogen use efficiency in rice. *Nature Communications* 10(1), 5279.

Tarafdar, A., Sowmya, G., Yogeshwari, K., Rattu, G., Negi, T. *et al.* (2023) Environmental pollution mitigation through utilization of carbon dioxide by microalgae. *Environmental Pollution* 328, 121623.

Thyssen, G.N. (2012) *Nicotiana Sylvestris, a Model Plant for Cell Biology: Organelle Movement and Retrotransposon Mutagenesis*. Rutgers The State University of New Jersey, School of Graduate Studies.

Tian, S., Jiang, L., Cui, X., Zhang, J., Guo, S. *et al.* (2018) Engineering herbicide-resistant watermelon variety through CRISPR/Cas9-mediated base-editing. *Plant Cell Reports* 37, 1353–1356.

Timm, S. and Hagemann, M. (2020) Photorespiration – how is it regulated and how does it regulate overall plant metabolism. *Journal of Experimental Botany* 71(14), 3955–3965.

Timm, S., Wittmiß, M., Gamlien, S., Ewald, R., Florian, A. *et al.* (2015) Mitochondrial dihydrolipoyl dehydrogenase activity shapes photosynthesis and photorespiration of *Arabidopsis thaliana*. *The Plant Cell* 27(7), 1968–1984.

Tu, M., Fang, J., Zhao, R., Liu, X., Yin, W. *et al.* (2022) CRISPR/Cas9-mediated mutagenesis of *VvbZIP36* promotes anthocyanin accumulation in grapevine (*Vitis vinifera*). *Horticulture Research* 9.

Uematsu, K., Suzuki, N., Iwamae, T., Inui, M. and Yukawa, H. (2012) Increased fructose 1, 6-bisphosphate aldolase in plastids enhances growth and photosynthesis of tobacco plants. *Journal of Experimental Botany* 63(8), 3001–3009.

Veillet, F., Perrot, L., Chauvin, L., Kermarrec, M.P., Guyon-Debast, A. *et al.* (2019) Transgene-free genome editing in tomato and potato plants using agrobacterium-mediated delivery of a CRISPR/Cas9 cytidine base editor. *International Journal of Molecular Sciences* 20(2), 402.

Wahab, A., Abdi, G., Saleem, M.H., Ali, B., Ullah, S. *et al.* (2022) Plants' physio-biochemical and phyto-hormonal responses to alleviate the adverse effects of drought stress: A comprehensive review. *Plants* 11(13), 1620.

Wang, J.Y. and Doudna, J.A. (2023) CRISPR technology: A decade of genome editing is only the beginning. *Science* 379(6629), eadd8643.

Wang, S., Chen, A., Xie, K., Yang, X., Luo, Z. *et al.* (2020a) Functional analysis of the OsNPF4. 5 nitrate transporter reveals a conserved mycorrhizal pathway of nitrogen acquisition in plants. *Proceedings of the National Academy of Sciences* 117(28), 16649–16659.

Wang, Y., Teng, W., Wang, Y., Ouyang, X., Xue, H. *et al.* (2020b) The wheat cytosolic glutamine synthetase Gs1. 1 modulates N assimilation and spike development by characterizing CRISPR-edited mutants. *bioRxiv* 10–14. DOI: 10.1101/2020.09.03.281014.

Wang, C., Gu, Q., Zhao, L., Li, C., Ren, J. *et al.* (2022a) Photochemical efficiency of photosystem II in inverted leaves of soybean [*Glycine max* (L.) merr.] affected by elevated temperature and high light. *Frontiers in Plant Science* 12, 772644.

Wang, Y., Cao, Y., Liang, X., Zhuang, J., Wang, X. *et al.* (2022b) A dirigent family protein confers variation of Casparian strip thickness and salt tolerance in maize. *Nature Communications* 13(1), 2222.

White, A.C., Rogers, A., Rees, M. and Osborne, C.P. (2016) How can we make plants grow faster? A source–sink perspective on growth rate. *Journal of Experimental Botany* 67(1), 31–45.

Xie, Y., Chen, L., Sun, T. and Zhang, W. (2021) Deciphering and engineering high-light tolerant cyanobacteria for efficient photosynthetic cell factories. *Chinese Journal of Chemical Engineering* 30, 82–91.

Xu, Q., Liu, X., Zhang, C., Du, W., Guan, Y. *et al.* (2022) Insights into soybean with high photosynthetic efficiency. In: Lam, H.M. and Li, M.W. (eds) *Advances in Botanical Research*, Vol. 102. Elsevier, London, pp. 121–151.

Yamano, T., Toyokawa, C., Shimamura, D., Matsuoka, T. and Fukuzawa, H. (2022) CO_2-dependent migration and relocation of LCIB, a pyrenoid-peripheral protein in *Chlamydomonas reinhardtii*. *Plant Physiology* 188(2), 1081–1094.

Yamori, W. and Shikanai, T. (2016) Physiological functions of cyclic electron transport around photosystem I in sustaining photosynthesis and plant growth. *Annual Review of Plant Biology* 67, 81–106.

Yamori, W., Masumoto, C., Fukayama, H. and Makino, A. (2012) Rubisco activase is a key regulator of non-steady-state photosynthesis at any leaf temperature and, to a lesser extent, of steady-state photosynthesis at high temperature. *The Plant Journal* 71(6), 871–880.

Yamori, W., Makino, A. and Shikanai, T. (2016) A physiological role of cyclic electron transport around photosystem I in sustaining photosynthesis under fluctuating light in rice. *Scientific Reports* 6(1), 20147.

Yang, M., Sakruaba, Y., Ishikawa, T., Ohtsuki, N., Kawai-Yamada, M. *et al.* (2023) Chloroplastic Sec14-like proteins modulate growth and phosphate deficiency responses in Arabidopsis and rice. *Plant Physiology* 192(4), 3030–3048.

Yang, Y., Wang, L., Che, Z., Wang, R., Cui, R. *et al.* (2022) Novel target sites for soybean yield enhancement by photosynthesis. *Journal of Plant Physiology* 268, 153580.

Yoo, B.C., Yadav, N.S., Orozco, E.M., Jr and Sakai, H. (2020) Cas9/gRNA-mediated genome editing of yeast mitochondria and *Chlamydomonas* chloroplasts. *PeerJ* e8362.

Yu, Q., Lutz, K.A. and Maliga, P. (2017) Efficient plastid transformation in Arabidopsis. *Plant Physiology* 175(1), 186–193.

Zafar, S.A., Zaidi, S.S.E.A., Gaba, Y., Singla-Pareek, S.L., Dhankher, O.P. *et al.* (2020) Engineering abiotic stress tolerance via CRISPR/Cas-mediated genome editing. *Journal of Experimental Botany* 71(2), 470–479.

Zhang, C., Akhlaq, M., Yan, H., Ni, Y., Liang, S. *et al.* (2023a) Chlorophyll fluorescence parameter as a predictor of tomato growth and yield under CO_2 enrichment in protective cultivation. *Agricultural Water Management* 284, 108333.

Zhang, H., Yu, F., Xie, P., Sun, S., Qiao, X. *et al.* (2023b) A Gγ protein regulates alkaline sensitivity in crops. *Science* 379(6638), eade8416.

Zhang, Y., Zhang, Q. and Chen, Q.J. (2020) Agrobacterium-mediated delivery of CRISPR/CAS reagents for genome editing in plants enters an era of ternary vector systems. *Science China Life Sciences* 63, 1491–1498.

Zheng, S., Ye, C., Lu, J., Liufu, J., Lin, L. *et al.* (2021b) Improving the rice photosynthetic efficiency and yield by editing *OsHXK1* via CRISPR/Cas9 system. *International Journal of Molecular Sciences* 22(17), 9554.

Zheng, X.T., Yu, Z.C., Tang, J.W., Cai, M.L., Chen, Y.L. *et al.* (2021a) The major photoprotective role of anthocyanins in leaves of *Arabidopsis thaliana* under long-term high light treatment: Antioxidant or light attenuator. *Photosynthesis Research* 149, 25–40.

Zhou, J., Yang, F., Zhang, F., Meng, H., Zhang, Y. *et al.* (2021) Impairing photorespiration increases photosynthetic conversion of CO_2 to isoprene in engineered cyanobacteria. *Bioresources and Bioprocessing* 8(1), 42.

Zhu, X.G., Ort, D.R., Whitmarsh, J. and Long, S.P. (2004) The slow reversibility of photosystem II thermal energy dissipation on transfer from high to low light may cause large losses in carbon gain by crop canopies: A theoretical analysis. *Journal of Experimental Botany* 55(400), 1167–1175.

Zhu, X.G., De Sturler, E. and Long, S.P. (2007) Optimizing the distribution of resources between enzymes of carbon metabolism can dramatically increase photosynthetic rate: A numerical simulation using an evolutionary algorithm. *Plant Physiology* 145(2), 513–526.

Zong, Y., Wang, Y., Li, C., Zhang, R., Chen, K. *et al.* (2017) Precise base editing in rice, wheat and maize with a Cas9-cytidine deaminase fusion. *Nature Biotechnology* 35(5), 438–440.

Zong, Y., Song, Q., Li, C., Jin, S., Zhang, D. *et al.* (2018) Efficient C-to-T base editing in plants using a fusion of nCcas9 and human APOBEC3A. *Nature Biotechnology* 36(10), 950–953.

17 Plant Genome Editing to Explain Metabolic Pathways in Plants

Vinay Kumar Dhiman[1], Nidhi Singh[2], Devendra Singh[3*], Vivek Kumar Dhiman[4*], Himanshu Pandey[5] and Avinash Sharma[6]

[1]Department of Basic Sciences, Dr. Yashwant Singh Parmar University of Horticulture and Forestry, Solan, India; [2]Medi-caps University, Indore, India; [3]Faculty of Biotechnology, Institute of Biosciences and Technology, Shri Ramswaroop Memorial University, Barabanki, India; [4]Centre for Herbal Pharmacology and Environmental Sustainability, Chettinad Academy of Research and Education, Tamil Nadu, India; [5]Department of Agriculture, Khalsa College, Amritsar, India; [6]Faculty of Agricultural Sciences Arunachal University of Studies, Namsai, India

Abstract

Plants are an abundant source of high-valued metabolites. These metabolites are produced in response to adverse climatic conditions (i.e., cold, salt, heat, drought, heavy metals, nutrient deficiencies, UV radiation, and ozone) and are rare and possess commercial importance. Knowledge of the metabolic pathways and genes involved is essential for metabolic engineering. Metabolite engineering utilizing robust genome-editing tools, bioinformatics, and newer-sequencing technologies has encouraged scientists and researchers to elucidate the complex mechanism of metabolic pathways and the genes encoding them. Metabolomics combined with gene editing, data mining, system biology, and predictive modeling can be used to harness this wealth of metabolites to generate metabolically refined plants. Precise genome-editing tools such as CRISPR (clustered regularly interspaced short palindromic repeats)/Cas9 allow the modification of specific genes to modify intermediate biosynthetic steps and to study the key biocatalysts or intermediates involved in metabolite synthesis. This book chapter discusses the use of plant genome-editing tools to elucidate complex metabolic pathways and facilitate metabolic engineering.

Keywords: Genome editing, Metabolomics, Metabolic engineering, Metabolites

17.1 Introduction

Regulation of metabolic pathways in plants results in the formation of important metabolites with their immense importance in the fields of pharmaceutical, medical, and agro-based industries (Kalia *et al.*, 2020; Zhang *et al.*, 2022; Singh *et al.*, 2022a; Singh *et al.*, 2024b). Plant metabolite biosynthetic pathways are complex and involve many genes that are not yet fully understood. Further exploration and discovery are needed to elucidate the mechanisms of these pathways and harness their potential for human benefit. Continuous development and application of advanced molecular tools and techniques have resulted in success in the exploration of

*Corresponding authors: devendrasingh.ibst@srmu.ac.in; vivekkrdhiman@gmail.com

 Genome Editing for Crop Improvement: Theory and Methodology (eds J.M. Al-Khayri *et al.*)
DOI: 10.1079/9781800622517.0017

plant metabolic pathways. In the present time, genome editing (GE) has come into existence as a part of genetic engineering, leading to the formation of transgenic organisms through free transformation (Kuluev *et al.*, 2019). Genetic modification at the precise site and sequences can be achieved through the technique and, thus, wider prospects and potential to explore hidden genes in plant metabolism (Das *et al.*, 2022; Singh *et al.*, 2022a; Dhiman *et al.*, 2024a). The GE tool has the efficiency to induce multiple gene mutations in plants and thus has opened up prospects to decipher the function of genes involved in metabolic engineering. The combined use of genetic engineering methods and metabolomics has attracted the scientific community toward intensive studies to detect and produce commercial novel metabolites in plants. Metabolic engineering through GE can enhance the amount and, thereby, the utility of the desirable compounds (Singh *et al.*, 2024a). Future agriculture is more challenging and requires genome engineering to develop climate-smart crops more efficiently.

17.2 Plant GE: A Preview of Modern Tools and Techniques

Genome editing serves as an advanced molecular biology technique, providing a wider platform to conduct precise, efficient, and specific alterations at genomic levels (Zhang *et al.*, 2018). Recently, this targeted GE platform has been of keen interest for scientists due to its robustness, specificity, and readability in targeted gene editing. Modern genetic engineering platforms include engineered nucleases, which potentially serve to bring about a gene revolution in plant metabolomics and, most importantly, agricultural crops (Fig. 17.1). These technologies are generally based on sequence-specific nuclease enzymes having the potential to target particular sequences and the creation of breaks in the genome (Molla *et al.*, 2021). Zinc-finger nucleases (ZFNs) are chimerically engineered nucleases containing zinc protein and nuclease motifs. The ZFN unit specifically spots and gets attached to three base pairs or triplet DNA sequences. Upon dimerization of ZFN protein motifs at a specific locus, cleavage is initiated at the targeted site. These zinc-finger-based nucleases have been applied and reported in several plants, including maize, rice, soybean, apple, rapeseed, tobacco, and *Arabidopsis* (Rana *et al.*, 2022). Transcription activator-like effector nucleases (TALENs) are also nucleases with a combination of nucleotide-specific effector proteins and the nuclease domain (Bhardwaj and Nain, 2021). Several TALEN-edited crops have been studied, including potato, sugarcane, rice, maize, and wheat. Calyno™ soybean has been developed through the metabolic engineering of the Delta 12 desaturase gene using TALENs obtained from Calyxr Inc., which contained 80% more oleic acid and 20% reduced FAs (Saini

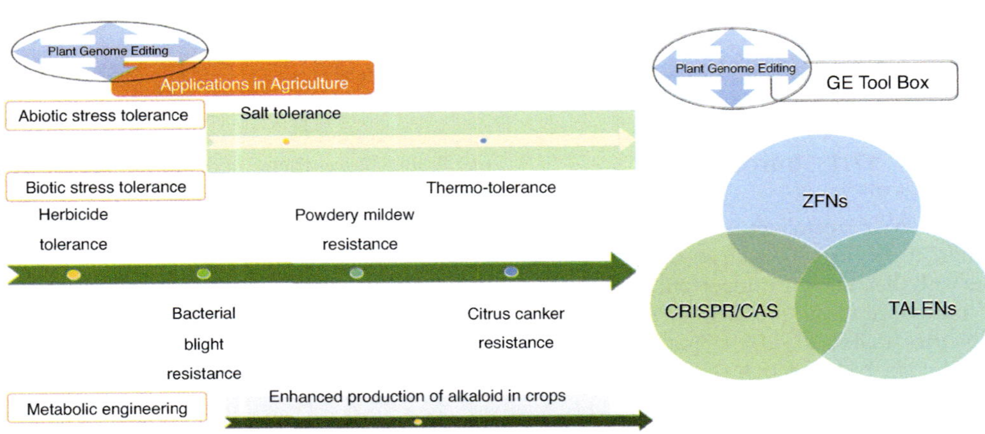

Fig. 17.1. Plant GE preview of modern tools and techniques. Author's own figure.

et al., 2020). Among these GE tools, the CRISPR/Cas9 system is the most versatile editing system with a variety of potential applications. In comparison to others, the CRISPR/Cas system is the better choice because of its simplicity, adaptability, ease of multiplexing, and cost-effectiveness (Dilawari *et al.*, 2021). These advantages of CRISPR/Cas9 have resulted in rapid exploitation of this system in plant genomes, giving effective solutions to plant breeding problems (Montecillo *et al.*, 2020; Prasad *et al.*, 2024). The genomes of plants, including rice, tomato, and cotton, have been studied for quality improvement and enhancement of beneficial traits (Ahmar *et al.*, 2020).

Modern GE is based on an engineered nuclease consisting of a DNA-recognizing protein and a functional nuclease domain attached to it. This engineered nuclease precisely cleaves targeted genes, and resultant cuts get repaired through homology-directed repair (HDR) or nonhomologous end joining (NHEJ) systems (Mussolino and Cathomen, 2012). Targeted GE generates double-stranded nicks (dsNicks) at desired sites, prompting the activation of DNA repair machinery (Ferreira da Silva *et al.*, 2021). The HDR pathway of the DNA repair system is based on the homology repair of the dsNicks, whereas in the NHEJ pathway, the ends of dsNicks get ligated (Xue and Greene, 2021). Even though the NHEJ pathway is more common, it is associated with flaws, due to which HDR pathways are more desirable. The HDR repair mechanism is precise and specific and requires homologous DNA (Tang *et al.*, 2019).

Since the introduction of modern GE tools, splendid work in the field of genome engineering has ensued (Razzaq *et al.*, 2022). Gene inactivation through ZFNs was first done in 2003 (Bibikova *et al.*, 2003), and it was studied successfully in plants in 2005 (Lloyd *et al.*, 2005). Several applications of plant GE have been developed through ZFNs. After ZFNs, the use of TALENs for GE in plants was reported by several researchers (Joung and Sander, 2013). These engineered nuclease-based genome manipulations have fascinated several researchers and were crowned 'Method of the Year' by the end of 2011 (Razzaq *et al.*, 2019).

CRISPR/Cas9-based GE has become the center of attraction and has gained importance in genetic engineering. The first CRISPR/Cas9-based GE was applied in eukaryotes in 2013 (Cong *et al.*, 2013). Several researchers have reported the success achieved through the CRISPR/Cas9 platform. Advancements in the CRISPR/Cas9 platform are growing with time with the introduction of a more accurate system of base editing. All these tools and techniques of GE have potential prospects to bring a revolution in genome engineering.

This trending toolbox of GE has generally been exploited for genome engineering by generating mutations in plant genomes. The first generation of genome manipulation tools included TALENs, ZFNs, and MNs (meganucleases), which were time-consuming and involved tedious procedures. Because of their complexity, the application of these first-generation tools is limited to plants. Upon comparison, the second-generation genome-editing tool, that is, CRISPR/Cas9, has been found to be more powerful and cost-effective. Furthermore, the CRISPR/Cas9 system is simpler because it is based on only sg-RNA and Cas9 nuclease in comparison to previous GE tools (Peng *et al.*, 2016). The CRISPR/Cas modification is a more stable and efficient platform that does not require tedious re-engineering of protein and can be extensively used in plant GE.

17.3 Role of GE in Metabolic Manipulation in Plants

Genome-editing tools have many applications in crop improvement programs in agriculture, especially in plant abiotic and biotic stress tolerance and breeding improvement (Singh and Verma, 2023; Tyagi *et al.*, 2024; Table 17.1). Genome editing has been successfully applied and reported in major cereal crops. The global scenario of climate change has caused increased stress on crop plants, thus leading to decreased agricultural productivity (Sudarshna and Sharma, 2024). Currently, scientific platforms have started using GE techniques to develop resilience in important crops against stress conditions (Yadav *et al.*, 2023).

Metabolic pathways are a sequence of chemical reactions that take place inside a cell to generate energy and create molecules. Metabolic engineering manipulates cellular

Table 17.1. Plant GE in crop improvement. Author's own table.

Crop	Genome-editing tool	Gene affected	Trait improved	Reference
Rice	CRISPR/Cas9	*OsRR22*	Stress tolerance	Zhang *et al.*, 2019
	TALENs	*OsSWEET14*	DR	Li *et al.*, 2012
	CRISPR/Cas9	*OsGS3*	Seed quality	Lacchini *et al.*, 2020
	TALENs	*OsBADH2*	Improved fragrance	Shan *et al.*, 2015
	CRISPR/Cas9	*Gn1a, GS3, DEP1*	Grain quality	Li *et al.*, 2016
	CRISPR/Cas9	*LAZY1*	Increased tillering	Miao *et al.*, 2013
Tomato	CRISPR/Cas9	*SlARF4*	Increased root growth and density	Bouzroud *et al.*, 2020
	CRISPR/Cas9	*SlAGL6 NHEJ*	Parthenocarpy	Klap *et al.*, 2017
	CRISPR/Cas9	*SlMLO1*	DR	Nekrasov *et al.*, 2017
	CRISPR/Cas9	*SlJAZ2*	DR	Ortigosa *et al.*, 2019
Sweet potato	CRISPR/Cas9		Starch quality was improved	Wang *et al.*, 2019
Maize	ZFNs	*ZmIPK1*	HR and low antimetabolite (phytate)	Shukla *et al.*, 2009
	ZFNs	*ZmTLP*	HR	Ainley *et al.*, 2013
	TALENs	*ZmGL2*	Low epicuticular wax	Char *et al.*, 2015
	CRISPR/Cas9	*ALS*	HR	Svitashev *et al.*, 2015
	CRISPR/Cas9	*Wx1*	Increased amylopectin	Aglawe *et al.*, 2022
Wheat	TALENs	*TaMLO*	DR	Wang *et al.*, 2014
	CRISPR/Cas9	*GW2*	Increased grain weight and protein content	Zhang *et al.*, 2018
	CRISPR/Cas9	*EDR1*	DR	Zhang *et al.*, 2017
Sugarcane	TALENs	*COMT*	Higher lignocellulose biomass	Jung and Altpeter, 2016
Potato	TALENs	*VInv*	Reduced sweetening of potatoes	Clasen *et al.*, 2016
	CRISPR/Cas9	*ALS*	HR	Butler *et al.*, 2016
Orange	CRISPR/Cas9	*CsLOB1 promoter*	DR	Peng *et al.*, 2017
Mushroom	CRISPR/Cas9	*PPO*	Browning phenotype	Waltz, 2016
Soybean	CRISPR/Cas9	*ALS*	HT	Li *et al.*, 2016
Flax	CRISPR/Cas9	*EPSPS*	HT	Sauer *et al.*, 2016

HR = Herbicide resistance, DR = Disease resistance, HT = Herbicide tolerance

Table 17.2. Use of GE in plant metabolite production. Author's own table.

Plant species	Target gene	Metabolite	Reference
Oryza sativa	*Osor*	β-Carotene	Endo *et al*., 2019
Camelina sativa	*FAE1*	Oleic acid	Ozseyhan *et al*., 2018
Glycine max	*GmUGT*	Flavonoid	Zhang *et al*., 2022
Taraxacum kok-saghyz	*1-FFT*	Natural rubber	Iaffaldano *et al*., 2016
Salvia miltiorrhiza	*bZIP*	Phenolic compounds	Shi *et al*., 2021
Zea mays	*ZmBADH2a* and *Zm BADH2b*	2-acetyl-1-pyrroline	Wang *et al*., 2021

systems to enhance bioprocess efficiency by redirecting enzymatic reactions within endogenous pathways to generate new compounds or enhance existing ones. A metabolic pathway is a chemical chain reaction, where enzymes convert substrates into target products and by-products (Chauhan *et al.*, 2022; Sharma *et al.*, 2022). Strategies to increase target compound synthesis include increasing upstream-gene expression, inhibiting key enzyme genes, overexpressing transcription factors, and combining methods. The CRISPR/dCas9-based activation/repression systems can also be used in metabolic engineering manipulation to maximize the synthesis of target compounds. Metabolic engineering in plants is an efficient way to provide means for increasing the synthesis of certain beneficial metabolites. The pathways involved in the production of primary and secondary metabolites yield plenty of compounds beneficial for humans. Metabolic engineering redirects enzymatic reactions to produce novel compounds within organisms. In plants, it entails modifying endogenous pathways to amplify the flow toward required molecules or form novel compounds. Even minute alterations can have prevalent effects on the entire pathway. While genomic and proteomic technologies have propelled advancements, future endeavors hinge on acquiring deeper insights into regulation across various levels, including genes, enzymes, compartmentation, transport, and accumulation. At times, it is possible to stimulate the activity of several genes in a pathway by regulating the expression of key regulators such as transcription factors. However, this is dependent on the presence and recognition of efficient regulators. A more effective strategy includes the simultaneous activation of many pathway genes, ideally encompassing every essential enzymatic process. This method eliminates all obstacles, guaranteeing unimpeded metabolic circulation. To do this, it is necessary to transfer many genes that encode enzymes into the recipient plant, preferably all at once, to maximize efficiency.

Synthesis of metabolic engineering offers a solution to counteract the adverse impacts of constitutive metabolite synthesis on plant growth by forming transgenes that express in specific tissues. Molecular techniques such as high-efficiency multi-gene expression vector systems and CRISPR-gene-editing tools can be used to produce such metabolic pathways. These tools facilitate the flexible expression and regulation of both upstream and downstream genes within metabolic pathways (Fig. 17.2) (Sabzehzari *et al.*, 2020).

Within plants, these pathways are categorized into primary and secondary metabolism (Dhiman *et al.*, 2021). Primary metabolism is accountable for the synthesis of essential macromolecules like carbohydrates, proteins, lipids, nucleic acids, and hormones, which are crucial for plant growth and development. Primary metabolites, such as sugars, amino acids, and lipids, are essential for plant growth and development (Singh *et al.*, 2022b; Dhiman *et al.*, 2024b; Verma *et al.*, 2024). Genome editing can optimize pathways involved in their synthesis, leading to increased biomass and improved nutritional content. Secondary metabolites, including alkaloids, flavonoids, and terpenoids, and other bioactive compounds have diverse applications in pharmaceuticals, agriculture, and industry (Dhiman *et al.*, 2022a). Genome editing enables the manipulation of complex biosynthetic pathways to boost

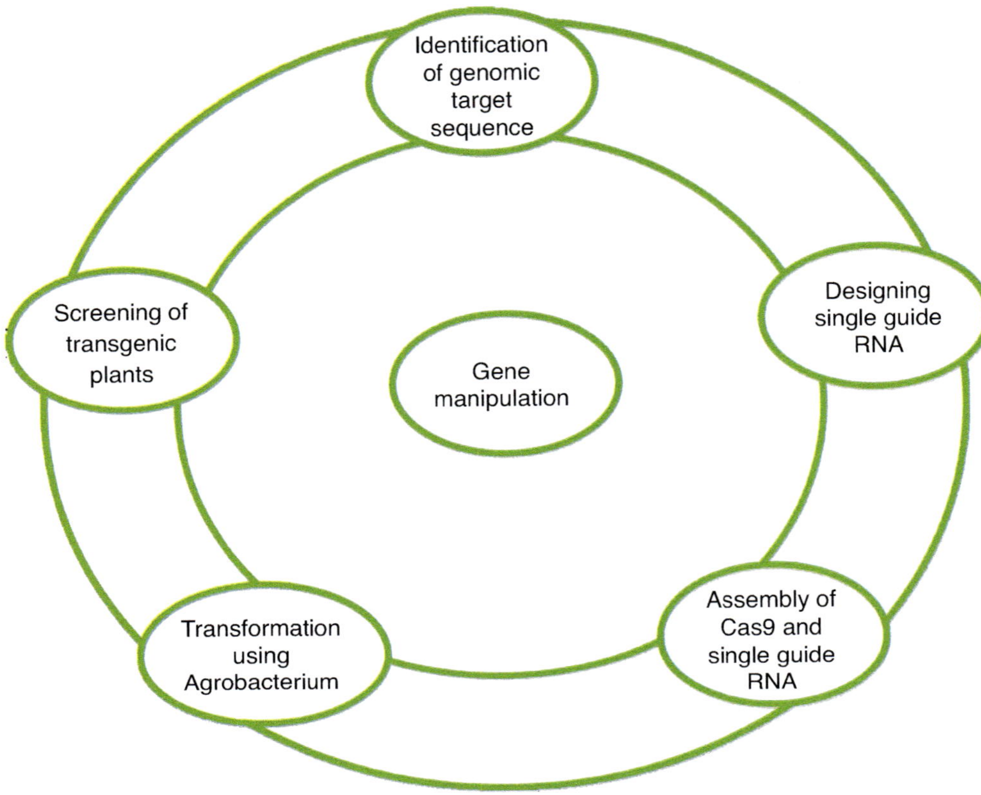

Fig. 17.2. The CRISPR/Cas system is employed for genetic material modification through gene editing. Author's own figure.

the production of these valuable compounds (Dey, 2021; Chauhan *et al.*, 2023). On the other hand, secondary metabolism leads to the production of more advanced and distinctive products found in phytochemical compounds. To regulate the activity of specialized metabolic pathways and generate valuable compounds, metabolic engineering employs GE tools such as phylogenetic profiling and transcription factors (Hyun, 2020; Chauhan *et al.*, 2023). Expression of the edited-*DST* gene showed enhancement in leaf water retention capability and reduction in stomatal density in rice (Mamrutha *et al.*, 2022). These research works have resulted in improvements in abiotic stresses. Improvement in heat stress has been observed by targeting *SlAGL6* in tomatoes (Abdallah *et al.*, 2021). Cold-tolerant rice lines have also been obtained by GE studies (Srikanth *et al.*, 2021).

17.4 Enhancing the Medicinal Value of GE Plants

Plant secondary metabolites, such as flavonoids and stilbenes, are essential for our daily diets due to their antiallergenic, anti-inflammatory, antiviral, and antioxidant properties. Resveratrol, derived from *STS* genes, is a potential phytopharmaceutical found in red wine and has been linked to a lower incidence of heart attacks. It also has biological functions like protection against arteriosclerosis, inhibition of platelet aggregation, regulation of hepatic apolipoprotein and lipid synthesis, and modulation of proatherogenic eicosanoid production. Lv *et al.* (2016) utilized the CRISPR system to inhibit genes encoding enzymes β-farnesene synthase, caryophyllene synthase, and squalene synthase at branching points, thereby increasing the artemisinin content in *Artemisia annua*, a plant with antimalaria properties. The *Rehmannia glutinosa* (Gaertn.) herb was genetically modified using the CRISPR/Cas9 system. The *Agrobacterium*-mediated method successfully regenerated albino and chimeric albino plants. The CRISPR/Cas9 system led to effective site-directed editing, contributing to molecular breeding for desirable *R. glutinosa* DC phenotypes (Mitra *et al.*, 2023).

17.5 Engineering Nutrients From Crops

Phytonutrients and micronutrients play essential roles in human nutrition and well-being, yet they are frequently lacking in staple crops (Dhiman *et al.*, 2022b). Through synthetic metabolic engineering, biofortification can augment the nutritional and health-enhancing properties of food crops. An array of biofortified crops, including golden rice, purple tomatoes, and purple endosperm rice, has been successfully cultivated (Bhambhani *et al.*, 2021; Negi *et al.*, 2021; Chauhan *et al.*, 2024).

Transgenic approaches target elevating iron and zinc levels in crops by amplifying gene expression and diminishing antinutrient factors. Coexpression of ferritin and NAS enhances iron content, while ferritin and NAS combined concurrently bolster zinc levels (Singh *et al.*, 2017). For instance, transgenic rice has been fortified with iron, zinc, and β-carotene, yielding a multinutrient-enriched biofortified crop. Such crops hold promise in addressing global human mineral deficiencies (Sangma *et al.*, 2022).

Research teams have effectively produced eicosapentaenoic acid (EPA) and docosahexaenoic acid (DHA) in *Arabidopsis*, *Brassica juncea*, *Camelina sativa*, and canola employing two methodologies: one involves incorporating desaturase and elongase genes sourced from algae or fungi, while the other entails transferring genes encoding unsaturated-fatty acid (FA) synthase-subunits from microalgae (Qin *et al.*, 2023).

Amino acid synthesis pathways are often targeted for metabolic engineering to increase the abundance of essential amino acids in food and feed crops. Modulating these amino acids can help engineer stress tolerance in transgenic plants. Recent examples show the impact of RNA interference on metabolic engineering and the value of multiple-point intervention. Polyamines, derived from ornithine and arginine, play a crucial role in plant physiology and development. Studies on the polyamine biosynthesis pathway in tobacco and rice have shown increased levels of higher polyamines in seeds and fruits. Inhibition of the arginine decarboxylase gene in rice reduces putrescine and spermine levels in transgenic plants (Chen *et al.*, 2019).

Vitamins are crucial micronutrients vital for animal growth and development, yet humans lack the ability to produce them internally. While primarily sourced from plant foods, many staple crops and edible plants offer insufficient amounts (Avnee *et al.*, 2023). Genetic engineering offers a solution through biofortification, enhancing crop vitamin levels to address micronutrient deficiencies. By employing metabolic engineering techniques to enhance biosynthetic efficiency and bolster storage stability, crops like rice, maize, potato, and tomato can be enriched with essential vitamins. The utilization of the multigene stacking strategy proves highly effective for folate biofortification in crops, elevating folate levels and ensuring stability throughout postharvest storage.

17.6 Understanding a Key Metabolic Pathway in Plants

Deciphering the complexity of plant metabolism is a requisite in its application for genetic engineering. Understanding plant metabolism and its regulation is required to make use of metabolism products (Schwender *et al.*, 2004; Hur *et al.*, 2013). Plant metabolism is highly segmented because of the several organelles present in the plant cell, each carrying out particular physiological and metabolic functions (Araújo *et al.*, 2012). Due to this compartmentation in the plant cell, *in situ*, metabolic pool sizes of plant cells vary within a cell and also depend on the developmental stage. Knowledge about the amount of metabolites in the cell compartment is required to be determined to ascertain its role in the metabolic process of that organelle. Genome editing has numerous applications in metabolic pathway engineering, focusing on enhancing the production of primary and secondary metabolites (Table 17.2).

Yuan *et al.* (2019) employed CRISPR/Cas9 GE to induce *de novo* mutations in *ahFAD2* genes of *Arachis hypogaea* (peanut) and enhanced concentration of oleic acid. Zeng *et al.* (2021) developed *Atropa belladonna* plants with increased yield of hyoscyamine with a targeted approach of CRISPR/Cas9 mutagenesis for the disruption of hyoscyamine 6β-hydroxylase (*AbH6H*). A study conducted by Ma *et al.* (2021) investigated the regulation mechanism of caffeine biosynthesis in *Camellia sinensis* (tea plant) with the aim of regulating the amount of caffeine present by targeting the *CsHB1* gene using CRISPR/Cas9-mediated gene editing.

Furthermore, scientists have employed gene-editing methods to alter the *PDS* gene in *Cannabis sativa* by utilizing the AtU6 promoter, with the aim of enhancing the synthesis of phytocannabinoids. The alteration was accomplished using *Agrobacterium*-mediated transformation, following the protocol outlined by Zhang *et al.* (2021). These examples demonstrate the versatility and effectiveness of CRISPR/Cas9 in enhancing the production of advantageous secondary metabolites in several plant species.

CRISPR/Cas9 GE has been employed to augment secondary metabolite synthesis in a diverse array of plants. The *FAD2* gene in the plant species *Camelina sativa* has been altered through the use of targeted mutagenesis techniques. This was accomplished by employing U3 and U6 promoters and utilizing the *Agrobacterium*-mediated floral dip method. These alterations have led to an increase in the levels of oleic acid and polyunsaturated FAs (PUFAs) in the plant (Morineau *et al.*, 2017). In addition, the *FAE1* gene in *Camelina sativa* has been disrupted by floral vacuum infiltration, leading to alterations in the FA compositions (Ozseyhan *et al.*, 2018). Deleting the *CiGAS* gene in *Cichorium intybus* (chicory) using PEG transfection with the AtU6 promoter has led to increased production of sesquiterpene lactones (Cankar *et al.*, 2021).

The CRISPR/Cas9 system has been employed in numerous studies to selectively target certain genes, such as *GmF3H1*, *GmF3H2*, and *GmFNS-1*, in leguminous plants like *Glycine max* (soybean). Zhang *et al.* (2019) found an augmentation in isoflavone content through *Agrobacterium*-mediated cot-node transformation. Furthermore, the application of PEG-mediated transformation for targeted mutagenesis of the *GmFAD2-1A* and *GmFAD2-1B* genes has led to elevated levels of oleic acid (Kim *et al.*, 2017). Rice, scientifically known as *Oryza sativa*, has undergone genetic modification through the use of CRISPR/Cas9 technology to alter the *OsBADH2* gene. This modification has led to the synthesis of aromatic compounds and 2-acetyl-1-pyrroline. The genetic modification has enhanced the aroma and taste of rice (Ashokkumar *et al.*, 2020).

Targeting genes, such as *SmRAS*, *SmLAC*, and *SmCPS1*, with CRISPR/Cas9 have significantly improved the production of phenolic acids, salvianolic acids, and tanshinones in *Salvia miltiorrhiza*, a medicinal plant. The enhancements were accomplished via *Agrobacterium rhizogenes*-mediated transformation, as evidenced by the investigations carried out by Zhao *et al.* (2024) and Ying *et al.* (2023). Moreover, the modification of the *bZIP2* gene has demonstrated the potential to enhance the phenolic acid content in this traditional Chinese medicinal plant (Shi *et al.*, 2022).

The comprehensive efforts highlight the transformative potential of CRISPR/Cas9 technology in the realm of agricultural biotechnology. By precisely manipulating genes that are involved in the production pathways of secondary metabolites, scientists can significantly boost the nutritional, medicinal, and commercial value of many crops.

The application of CRISPR/Cas9 for GE is revolutionizing the synthesis of secondary metabolites in different plant species, leading to enhancements in both nutritional and medicinal attributes. The CRISPR/Cas9 system has been utilized in *Dendrobium officinale* to selectively deactivate multiple genes, namely *C3H*, *C4H*, *4CL*, *CCR*, and *IRX* that are crucial for the synthesis of alkaloids, phenanthrenes, and polysaccharides. *Agrobacterium tumefaciens* promotes this genetic alteration, resulting in increased production of these crucial compounds (Kui *et al.*, 2017). Another example involves the plant species *Dioscorea zingiberensis*, in which the *Dzfps* gene was altered utilizing the OsU3 promoter to enhance the production of diosgenin and squalene. These molecules hold great importance in the pharmaceutical industry (Feng *et al.*, 2018).

Nitarska *et al.* (2021) documented the genetic modification of the *F3'H* gene in *Euphorbia pulcherrima* (poinsettia) through the use of targeted mutagenesis. The introduction of the AtU6-26 promoter was accomplished using *Agrobacterium tumefaciens* transformation. Consequently, the production of pelargonidin, a sought-after pigment, was greatly increased. The *FtMYB45* gene in *Fagopyrum tataricum* (tartary buckwheat) was disrupted utilizing *Agrobacterium rhizogenes*-mediated transformation, which involved the application of CRISPR/Cas9 technology. According to Wen *et al.* (2022), this modification resulted in an elevation of flavonoid concentrations.

The OsU3 promoter was employed to disrupt the *HvHGGT* and *HvHPT* genes in *Hordeum vulgare* (barley) by the utilization of *Agrobacterium tumefaciens*. Zeng *et al.* (2020) utilized this approach to enhance the levels of tocotrienol and tocopherol, both of which possess significant antioxidant properties. Similarly, researchers have shown that by intentionally modifying the *PDS* gene in the plant species *Humulus lupulus* (hops) utilizing U6-626p and U6-29p promoters by *Agrobacterium*-mediated transformation, the number of carotenoids present can be increased (Awasthi *et al.*, 2021).

The CRISPR/Cas9 technique was used in the solanaceous plant *Nicotiana tabacum* (tobacco) to intentionally damage the *BBL* gene by utilizing the AtU6-26 promoter. This genetic change successfully reduces the concentration of nicotine, offering a potential approach to growing tobacco varieties with lower amounts of nicotine (Schachtsiek and Stehle, 2019). Furthermore, the application of the U6 promoter to specifically target the *XylT* and *FucT* genes in tobacco has enabled the accurate alteration of glycans and other secondary metabolites. This has led to the production of alkaloids and flavonoids that have been improved in terms of their desirability (Mercx *et al.*, 2017).

The CRISPR/Cas9-mediated gene-editing technique has been extensively utilized in *Oryza sativa* (rice) to enhance the synthesis of aromatic chemicals. This is accomplished by selectively focusing on genes such as *OsBADH2* and *Osor*. The enhancement of traits such as fragrance and carotenoid levels has been accomplished through the utilization of both *Agrobacterium* and particle bombardment techniques for genetic modification (Endo *et al.*, 2019).

The CRISPR/Cas9 system was used in *Papaver somniferum* (opium poppy) to deactivate the *4'OMT2* gene using the AtU6 promoter. According to a study conducted by Alagoz *et al.* (2016), the use of *Agrobacterium*-mediated leaf infiltration has caused a reduction in the levels of benzylisoquinoline alkaloids. This modification has the capacity to generate poppy varieties with reduced levels of narcotic alkaloids.

17.7 Future Prospects and Conclusion

Genome editing aims to achieve more precise and efficient genome engineering and has the potential to bring revolution in the field of crop improvement by evading unwanted alterations in genes. Genome editing proved to be an extraordinary platform to meet the enhancing food demands of a growing population and also to meet the challenge of malnutrition by contributing to the development of higher-yielding, biofortified, and climate-smart crops. Genomics studies have created immense knowledge that will help in investigating genes associated with crop improvement, and the hub of information can be better utilized for GE in the targeted genes in a precise manner to generate desirable crops. Not only is this platform providing a breakthrough way to address several challenges in the agriculture community, but it also has more public acceptance. The development of climate-smart crops to face global food security can be achieved through a genome-editing toolbox. A genome-editing toolbox with the potential capability to carry out epi-GE and precise GE has more public acceptance.

The production of secure, cheap, and nutritive food by employing sustainable farming practices is a big challenge for the farming community. Advanced techniques to get improved and better farming practices are some of the basic facets that can be utilized significantly (Pandey *et al.*, 2021). Genome editing is a robust tool that could mitigate increasing food demands. Integration of GE and breeding techniques can be used to increase crop yield with more nutrition and for the development of smart crops. The breakthrough tools in plant GE could alter crops for agronomic traits and have been applied in various breeding works to get desired traits. Integration of CRISPR/Cas9 technology with advanced molecular breeding techniques, including speed breeding and genomic selection, can be done to bring about a gene revolution in agriculture and to achieve the food demands of a growing human population.

Reference

Abdallah, N.A., Hamwieh, A., Radwan, K., Fouad, N. and Prakash, C. (2021) Genome editing techniques in plants: a comprehensive review and future prospects toward zero hunger. *GM Crops and Food* 12, 601–615. DOI: 10.1080/21645698.2021.2021724.

Aglawe, S.B., Magar, N.D., Dhawane, Y., Bhamare, D., Shah, P. *et al.* (2022) Genome editing crops in food and futuristic crops. In: *Recent Advances in Food Biotechnology*. Springer, Berlin, Germany, pp. 401–445.

Ahmar, S., Saeed, S., Khan, M.H.U., Ullah Khan, S., Mora-Poblete, F. *et al.* (2020) A revolution toward gene-editing technology and its application to crop improvement. *International Journal of Molecular Sciences* 21, 5665. DOI: 10.3390/ijms21165665.

Ainley, W.M., Sastry-Dent, L., Welter, M.E., Murray, M.G., Zeitler, B. *et al.* (2013) Trait stacking via targeted genome editing. *Plant Biotechnology Journal* 11, 1126–1134. DOI: 10.1111/pbi.12107.

Alagoz, Y., Gurkok, T., Zhang, B. and Unver, T. (2016) Manipulating the biosynthesis of bioactive compound alkaloids for next-generation metabolic engineering in opium poppy using CRISPR-Cas 9 genome editing technology. *Scientific Reports* 6, 30910. DOI: 10.1038/srep30910.

Araújo, W.L., Nunes-Nesi, A., Nikoloski, Z., Sweetlove, L.J. and Fernie, A.R. (2012) Metabolic control and regulation of the tricarboxylic acid cycle in photosynthetic and heterotrophic plant tissues plant. *Cell and Environment* 35, 1–21. DOI: 10.1111/j.1365-3040.2011.02332.x.

Ashokkumar, S., Jaganathan, D., Ramanathan, V., Rahman, H., Palaniswamy, R. *et al.* (2020) Creation of novel alleles of fragrance gene OsBADH2 in rice through CRISPR/Cas9 mediated gene editing. *PLoS ONE* 15(8), e0237018.

Avnee, Sood, S., Chaudhary, D.R., Jhorar, P. and Rana, R.S. (2023) Biofortification: an approach to eradicate micronutrient deficiency. *Frontiers in Nutrition* 10, 1233070.

Awasthi, P., Kocábek, T., Mishra, A.K., Nath, V.S., Shrestha, A. *et al.* (2021) Establishment of CRISPR/Cas9-mediated targeted mutagenesis in hop (*Humulus lupulus*). *Plant Physiology and Biochemistry* 160, 1–7.

Bhambhani, S., Kondhare, K.R. and Giri, A.P. (2021) Advanced genome editing strategies for manipulation of plant specialized metabolites pertaining to biofortification. *Phytochemistry Reviews* 1–19.

Bhardwaj, A. and Nain, V. (2021) TALENs-an indispensable tool in the era of CRISPR: a mini review. *Genetic Engineering and Biotechnology* 19, 125. DOI: 10.1186/s43141-021-00225-z.

Bibikova, M., Beumer, K., Trautman, J.K. and Carroll, D. (2003) Enhancing gene targeting with designed zinc finger nucleases. *Science* 300, 764. DOI: 10.1126/science.1079512.

Bouzroud, S., Gasparini, K., Hu, G., Barbosa, M.A.M., Rosa, B.L. *et al.* (2020) Down regulation and loss of auxin response factor 4 function using CRISPR/Cas9 alters plant growth, stomatal function and improves tomato tolerance to salinity and osmotic stress. *Genes* 11(3), 272.

Butler, N.M., Baltes, N.J., Voytas, D.F. and Douches, D.S. (2016) Geminivirus-mediated genome editing in potato (*Solanum tuberosum* L.) using sequence-specific nucleases. *Frontiers in Plant Science* 7, 1045.

Cankar, K., Bundock, P., Sevenier, R., Häkkinen, S.T., Hakkert, J.C. *et al.* (2021) Inactivation of the germacrene A synthase genes by CRISPR/Cas9 eliminates the biosynthesis of sesquiterpene lactones in *Cichorium intybus* L. *Plant Biotechnology Journal* 19(12), 2442–2453.

Char, S.N., Unger-Wallace, E., Frame, B., Briggs, S.A., Main, M. *et al.* (2015) Heritable site-specific mutagenesis using talens in maize. *Plant Biotechnology Journal* 13, 1002–1010. DOI: 10.1111/pbi.12.

Chauhan, D., Singh, D., Pandey, H., Khan, Z., Srivastava, R. *et al*. (2023) Impact of transcription factors in plant abiotic stress: a recent advancement for crop improvement. In: *Plant Transcription Factors*. Academic Press, Cambridge, Massachusetts, pp. 271–286.

Chauhan, D., Prasad, G., Singh, D. and Pandey, H. (2024) Role of next-generation sequencing technology in crop improvement. In: *Smart Breeding*. Apple Academic Press, pp. 371–390.

Chauhan, V., Kaushal, D., Dhiman, V.K., Kanwar, S.S., Singh, D. *et al*. (2022) An insight in developing carrier-free immobilized enzymes. *Frontiers in Bioengineering and Biotechnology* 10, 794411.

Chen, C.C., Lin, C., Chen, M.H. and Chiang, P.Y. (2019) Stability and quality of anthocyanin in purple sweet potato extracts. *Foods* 8(9), 393.

Clasen, B.M., Stoddard, T.J., Luo, S., Demorest, Z.L., Li, J. *et al*. (2016) Improving cold storage and processing traits in potato through targeted gene knockout. *Plant Biotechnology Journal* 14(1), 169–176. DOI: 10.1111/pbi.12370.

Cong, L., Ran, F.A., Cox, D., Lin, S., Barretto, R. *et al*. (2013) Multiplex genome engineering using CRISPR/Cas systems. *Science* 339(6121), 819–823. DOI: 10.1126/science.1231143.

Das, T., Ghorai, M., Pandey, D.K., Thakur, M., Rathour, S. *et al*. (2022) CRISPR/Cas genome editing in engineering plant secondary metabolites of therapeutic benefits. In: *Metabolic Engineering in Plants*. Springer, Berlin, Germany, pp. 187–208.

Dey, A. (2021) CRISPR/Cas genome editing to optimize pharmacologically active plant natural products. *Pharmacological Research* 164, 105359.

Dhiman, V.K., Chauhan, V., Kanwar, S.S., Singh, D. and Pandey, H. (2021) Purification and characterization of actinidin from *Actinidia deliciosa* and its utilization in inactivation of α-amylase. *Bulletin of the National Research Centre* 45, 1–9.

Dhiman, V.K., Rana, N., Dhiman, V.K., Pandey, H., Verma, P. *et al*. (2022a) Effect of rhizobial isolates and nitrogen fertilizers on nursery performance, nodulation behavior and nitrogenase activity of *Dalbergia sissoo* Roxb. seedlings. *Plant Stress* 4, 100080.

Dhiman, V.K., Singh, D., Pandey, H., Chauhan, D., Dhiman, V.K. *et al*. (2022b) Mycoremediation of agricultural waste for the cultivation of edible mushroom. In: *Sustainable Management of Environmental Contaminants: Eco-Friendly Remediation Approaches*. Springer International Publishing, Cham, Switzerland, pp. 471–482.

Dhiman, V.K., Singh, D., Dhiman, V.K. and Pandey, H. (2024a) Multi-omics: an advanced bioinformatics approach for crop improvement in agriculture. In: *Bioinformatics for Plant Research and Crop Breeding*. pp. 75–98. DOI: 10.1002/9781394209965.ch3.

Dhiman, V.K., Singh, D., Dhiman, V.K., Pandey, M., Sharma, A. *et al*. (2024b) Insects as biotic stress in agriculture. In: *Plant-Microbe Interaction and Stress Management*. Springer Nature, Singapore, pp. 193–210.

Dilawari, R., Kaur, N., Priyadarshi, N., Kumar, B., Abdelmotelb, K.F. *et al*. (2021) Genome editing: a tool from the vault of science for engineering climate-resilient cereals. In: *Harsh Environment and Plant Resilience*. Springer, Berlin, Germany, pp. 45–72.

Endo, A., Saika, H., Takemura, M., Misawa, N. and Toki, S. (2019) A novel approach to carotenoid accumulation in rice callus by mimicking the cauliflower orange mutation via genome editing. *Rice* 12, 81. DOI: 10.1186/s12284-019-0345-3.

Feng, S., Song, W., Fu, R., Zhang, H., Xu, A. *et al*. (2018) Application of the CRISPR/Cas9 system in *Dioscorea zingiberensis*. *Plant Cell, Tissue and Organ Culture (PCTOC)* 135, 133–141.

Ferreira da Silva, J.F., Meyenberg, M. and Loizou, J.I. (2021) Tissue specificity of DNA repair: the CRISPR compass. *Trends in Genetics* 37, 958–962. DOI: 10.1016/j.tig.2021.07.010.

Hur, M., Campbell, A.A., Almeida-de-Macedo, M., Li, L., Ransom, N. *et al*. (2013) A global approach to analysis and interpretation of metabolic data for plant natural product discovery. *Natural Product Reports* 30, 565–583. DOI: 10.1039/c3np20111b.

Hyun, T.K. (2020) CRISPR/CAS-based genome editing to improve abiotic stress tolerance in plants. *Botanica Serbica* 44(2), 121–127. DOI: 10.2298/BOTSERB2002121H.

Iaffaldano, B., Yingxiao, Z. and Cornish, K. (2016) CRISPR/Cas9 genome editing of rubber producing dandelion *Taraxacum kok-saghyz* using *Agrobacterium rhizogenes* without selection. *Industrial Crops and Products* 89, 356–362. DOI: 10.1016/j.indcrop.2016.05.029.

Joung, J.K. and Sander, J.D. (2013) TALENs: A widely applicable technology for targeted genome editing. *Nature Reviews. Molecular Cell Biology* 14, 49–55. DOI: 10.1038/nrm3486.

Jung, J.H. and Altpeter, F. (2016) TALEN-mediated targeted mutagenesis of the caffeic acid O-methyltransferase in highly polyploid sugarcane improves cell wall composition for production of bioethanol. *Plant Molecular Biology* 92, 131–142. DOI: 10.1007/s11103-016-0499-y.

Kalia, M., Singh, D., Sharma, D., Narvi, S.S. and Agarwal, V. (2020) *Senna alexandriana* Mill as a potential inhibitor for quorum sensing-controlled virulence factors and biofilm formation in *Pseudomonas aeruginosa* PAO1. *Pharmacognosy Magazine* 16, 797–802.

Kim, D., Lim, K., Kim, S.T., Yoon, S.H., Kim, K. *et al.* (2017) Genome-wide target specificities of CRISPR RNA-guided programmable deaminases. *Nature Biotechnology* 35, 475–480. DOI: 10.1038/nbt.3852.

Klap, C., Yeshayahou, E., Bolger, A.M., Arazi, T., Gupta, S.K. *et al.* (2017) Tomato facultative parthenocarpy results from sl AGAMOUS-LIKE 6 loss of function. *Plant Biotechnology Journal* 15(5), 634–647. DOI: 10.1111/pbi.12662.

Kui, L., Chen, H., Zhang, W., He, S., Xiong, Z. *et al.* (2017) Building a genetic manipulation tool box for orchid biology: identification of constitutive promoters and application of CRISPR/Cas9 in the orchid, *Dendrobium officinale*. *Frontiers in Plant Science* 7, 2036.

Kuluev, B.R., Gumerova, G.R., Mikhaylova, E.V., Gerashchenkov, G.A., Rozhnova, N.A. *et al.* (2019) Delivery of CRISPR/Cas components into higher plant cells for genome editing. *Russian Journal of Plant Physiology* 66, 694–706. DOI: 10.1134/S102144371905011X.

Lacchini, E., Kiegle, E., Castellani, M., Adam, H., Jouannic, S. *et al.* (2020) CRISPR-mediated accelerated domestication of African rice landraces. *PLOS ONE* 15, e0229782. DOI: 10.1371/journal.pone.0229782.

Li, M., Li, X., Zhou, Z., Wu, P., Fang, M. *et al.* (2016) Reassessment of the four yield-related genes Gn1A, DEP1, GS3, and IPA1 in rice using a CRISPR/Cas9 system. *Frontiers in Plant Science* 7, 377. DOI: 10.3389/fpls.2016.00377.

Li, T., Liu, B., Spalding, M.H., Weeks, D.P. and Yang, B. (2012) High-efficiency TALEN-based gene editing produces disease-resistant rice. *Nature Biotechnology* 3, 390–392. DOI: 10.1038/nbt.2199.

Lloyd, A., Plaisier, C.L., Carroll, D. and Drews, G.N. (2005) Targeted mutagenesis using zinc-finger nucleases in Arabidopsis. *Proceedings of the National Academy of Sciences of the United States of America* 102(6), 2232–2237. DOI: 10.1073/pnas.0409339102.

Lv, Z., Zhang, F., Pan, Q., Fu, X., Jiang, W. *et al.* (2016) Branch pathway blocking in *Artemisia annua* is a useful method for obtaining high yield artemisinin. *Plant and Cell Physiology* 57(3), 588–602.

Ma, W., Kang, X., Liu, P., Zhang, Y., Lin, X. *et al.* (2021) The analysis of transcription factor CsHB1 effects on caffeine accumulation in tea callus through CRISPR/Cas9-mediated gene editing. *Process Biochemistry* 101, 304–311.

Mamrutha, H.M., Deswal, K., Wadhwa, Z., Singh, R., Kumar, R. *et al.* (2022) Genome editing for stress tolerance in cereals: methods, opportunities, and applications. In: Mallikarjuna, M.G., Nayaka, S.C. and Kaul, T. (eds) *Next-Generation Plant Breeding Approaches for Stress Resilience in Cereal Crops*. Springer, Berlin, Germany, pp. 345–367.

Mercx, S., Smargiasso, N., Chaumont, F., Pauw, E., Boutry, M. *et al.* (2017) Inactivation of the β(1,2)-xylosyltransferase and the α(1,3)-fucosyltransferase genes in *Nicotiana tabacum* BY-2 cells BY a multiplex CRISPR/Cas9 strategy results in glycoproteins without plant-specific glycans. *Frontier in Plant Science* 8. DOI: 10.3389/fpls.2017.00403.

Miao, J., Guo, D., Zhang, J., Huang, Q., Qin, G. *et al.* (2013) Targeted mutagenesis in rice using CRISPR-Cas system. *Cell Research* 23, 1233–1236. DOI: 10.1038/cr.2013.123.

Mitra, S., Anand, U., Ghorai, M., Kant, N. and Kumar, M. (2023) Genome editing technologies, mechanisms and improved production of therapeutic phytochemicals: opportunities and prospects. *Biotechnology and Bioengineering* 120(1), 82–94.

Molla, K.A., Sretenovic, S., Bansal, K.C. and Qi, Y. (2021) Precise plant genome editing using base editors and prime editors. *Nature Plants* 7, 1166–1187. DOI: 10.1038/s41477-021-00991-1.

Montecillo, J.A.V., Chu, L.L. and Bae, H. (2020) CRISPR-Cas9 system for plant genome editing: current approaches and emerging developments. *Agronomy* 10, 1033. DOI: 10.3390/agronomy10071033.

Morineau, C., Bellec, Y., Tellier, F., Gissot, L., Kelemen, Z. *et al.* (2017) Selective gene dosage by CRISPR-Cas9 genome editing in hexaploid *Camelina sativa*. *Plant Biotechnology Journal* 15, 729–739. DOI: 10.1111/pbi.12671.

Mussolino, C. and Cathomen, T. (2012) TALE nucleases: tailored genome engineering made easy. *Current Opinion in Biotechnology* 23, 644–650. DOI: 10.1016/j.copbio.2012.01.013.

Negi, S., Singh, D., Pandey, H. and Singhal, R.K. (2021) Biofortification for high Fe and Zn in various Poaceae crops by using different molecular breeding and biotechnological approaches. *Plant Physiology Reports* 26(4), 636–646.

Nekrasov, V., Wang, C., Win, J., Lanz, C., Weigel, D. *et al.* (2017) Rapid generation of a transgene-free powdery mildew resistant tomato by genome deletion. *Scientific Reports* 7, 482. DOI: 10.1038/s41598-017-00578-x.

Nitarska, D., Boehm, R., Debener, T., Lucaciu, R.C. and Halbwirth, H. (2021) First genome edited poinsettias: Targeted mutagenesis of flavonoid 3'-hydroxylase using CRISPR/Cas9 results in a colour shift. *Plant Cell Tissue Organ Culture* 147(1), 49–60. DOI: 10.1007/s11240-021-02103-5.

Ortigosa, A., Gimenez-Ibanez, S., Leonhardt, N. and Solano, R. (2019) Design of a bacterial speck resistant tomato by CRISPR/Cas9-mediated editing of SlJAZ2. *Plant Biotechnology Journal* 17, 665–673. DOI: 10.1111/pbi.13006.

Ozseyhan, M.E., Kang, J., Mu, X. and Lu, C. (2018) Mutagenesis of the FAE1 genes significantly changes fatty acid composition in seeds of *Camelina sativa*. *Plant Physiology and Biochemistry* 123, 1–7. DOI: 10.1016/j.plaphy.2017.11.021.

Pandey, H., Singh, D., Das, R. and Pandey, D. (2021) Precision farming and its application. In: Choudhury, A., Biswas, A., Singh, T.P. and Ghosh, S.K. (eds) *Smart Agriculture Automation Using Advanced Technologies: Data Analytics and Machine Learning, Cloud Architecture, Automation and IoT*. Springer, Singapore. DOI: 10.1007/978-981-16-6124-2_2.

Peng, A., Chen, S., Lei, T., Xu, L., He, Y. *et al.* (2017) Engineering canker resistant plants through CRISPR/Cas9-targeted editing of the susceptibility gene CsLOB1 promoter in citrus. *Plant Biotechnology Journal* 10, 1011–1013.

Peng, R., Lin, G. and Li, J. (2016) Potential pitfalls of CRISPR/Cas9-mediated genome editing. *FEBS Journal* 283, 1218–1231. DOI: 10.1111/febs.13586.

Prasad, G., Chauhan, D., Pandey, H., Singh, D., Dhiman, V.K. *et al.* (2024) Recent advances in CRISPR-Cas for climate-resilient horticulture in fruits crops. *Climate-Resilient Agriculture* 415–431.

Razzaq, A., Saleem, F., Kanwal, M., Mustafa, G., Yousaf, S. *et al.* (2019) Modern trends in plant genome editing: an inclusive review of the CRISPR/Cas9 toolbox. *International Journal of Molecular Sciences* 20, 4045. DOI: 10.3390/ijms20164045.

Qin, J., Kurt, E., Bassi, T., Sa, L. and Xie, D. (2023) Biotechnological production of omega-3 fatty acids: current status and future perspectives. *Frontiers in Microbiology* 14, 1280296.

Rana, S., Aggarwal, P.R., Shukla, V., Giri, U., Verma, S. *et al.* (2022) Genome editing and designer crops for the future. In: *Plant Gene Silencing*. Humana Press, Totowa, New Jersey, pp. 37–69. DOI: 10.1007/978-1-0716-1875-2_3.

Razzaq, A., Wishart, D.S., Wani, S.H., Hameed, M.K., Mubin, M. *et al.* (2022) Advances in metabolomics-driven diagnostic breeding and crop improvement. *Metabolites* 12, 511. DOI: 10.3390/metabo12060511.

Sabzehzari, M., Zeinali, M. and Naghavi, M.R. (2020) CRISPR-based metabolic editing: next-generation metabolic engineering in plants. *Gene* 759, 144993.

Saini, P., Saini, P., Kaur, J.J., Francies, R.M., Gani, M. *et al.* (2020) Molecular approaches for harvesting natural diversity for crop improvement. In: *Rediscovery of Genetic and Genomic Resources for Future Food Security*. Springer, Berlin, Germany, pp. 67–169.

Sangma, R.R., Manpoong, C., Sharma, A., Devadas, V.S., Singh, D. *et al.* (2022) Performance of black rice (Oryza sativa) varieties grown in Namsai district of Arunachal Pradesh, India. *Research on Crops* 23(1), 11–14.

Sauer, N.J., Narváez-Vásquez, J., Mozoruk, J., Miller, R.B., Warburg, Z.J. *et al.* (2016) Oligonucleotide-mediated genome editing provides precision and function to engineered nucleases and antibiotics in plants. *Plant Physiology* 170, 1917–1928. DOI: 10.1104/pp.15.01696.

Schachtsiek, J. and Stehle, F. (2019) Nicotine-free, nontransgenic tobacco (*Nicotiana tabacum* l.) edited by CRISPR-Cas9. *Plant Biotechnology Journal* 17(12), 2228.

Schwender, J., Ohlrogge, J. and Shachar-Hill, Y. (2004) Understanding flux in plant metabolic networks. *Current Opinion in Plant Biology* 7, 309–317. DOI: 10.1016/j.pbi.2004.03.016.

Shan, Q., Zhang, Y., Chen, K., Zhang, K. and Gao, C. (2015) Creation of fragrant rice by targeted knockout of the *OsBADH2* gene using TALEN technology. *Plant Biotechnology Journal* 13, 791–800. DOI: 10.1111/pbi.12312.

Sharma, S.K., Singh, D., Pandey, H., Jatav, R.B., Singh, V. *et al.* (2022) An overview of roles of enzymatic and nonenzymatic antioxidants in plant. *Antioxidant Defense in Plants: Molecular Basis of Regulation* 1–13.

Shi, M., Du, Z., Hua, Q. and Kai, G. (2021) CRISPR/Cas9-mediated targeted mutagenesis of bZIP2 in *Salvia miltiorrhiza* leads to promoted phenolic acid biosynthesis. *Industrial Crops and Products* 167, 113560. DOI: 10.1016/j.indcrop.2021.113560.

Shi, M., Zhu, R., Zhang, Y., Zhang, S., Liu, T. *et al.* (2022) A novel WRKY34-bZIP3 module regulates phenolic acid and tanshinone biosynthesis in *Salvia miltiorrhiza*. *Metabolic Engineering* 73, 182–191.

Shukla, V.K., Doyon, Y., Miller, J.C., DeKelver, R.C., Moehle, E.A. *et al.* (2009) Precise genome modification in the crop species *Zea mays* using zinc-finger nucleases. *Nature* 459, 437–441. DOI: 10.1038/nature07992.

Singh, D. and Verma, S.K. (2023) Impacts of particulate matter pollution on plants. In: *New Frontiers in Plant-Environment Interactions: Innovative Technologies and Developments*. Springer Nature, Cham, Switzerland, pp. 483–500.

Singh, D., Dhiman, V.K., Pandey, H., Dhiman, V.K. and Pandey, D. (2022a) Crosstalk between salicylic acid and auxins, cytokinins and gibberellins under biotic stress. In: *Auxins, Cytokinins and Gibberellins Signaling in Plants*. Springer International Publishing, Cham, Switzerland, pp. 249–262. DOI: 10.1007/978-3-031-05427-3_11.

Singh, D., Pandey, H., Dhiman, V.K., Bang, N.T.H., Dhiman, V.K. *et al.* (2022b) Secondary metabolite engineering for plant immunity against various pathogens. In: *Metabolic Engineering in Plants*. Springer Nature, Singapore, pp. 123–143.

Singh, D., Verma, S.K. and Kumar, S. (2024a) Trends in secondary metabolites production from plant sources. In: *Secondary Metabolites and Biotherapeutics*. Academic Press, pp. 103–126.

Singh, D., Verma, S.K. and Shyam, P. (2024b) Identification and purification of plant secondary metabolite as medicinal raw materials. In: *Secondary Metabolites and Biotherapeutics*. Academic Press, pp. 9–38.

Singh, S.P., Keller, B., Gruissem, W. and Bhullar, N.K. (2017) Rice NICOTIANAMINE SYNTHASE 2 expression improves dietary iron and zinc levels in wheat. *Theoretical and Applied Genetics* 130, 283–292.

Srikanth, M., Tata, S.S. and Godi, S. (2021) Abiotic stress tolerance enhancement in rice through CRISPR/Cas9 approach. *Research Journal of Agricultural Sciences* 12(5), 1543–1546.

Sudarshna and Sharma, N. (2024) Endophytic bacteria associated with critically endangered medicinal plant *Trillium govanianum* (Wall ex. Royle) and their potential in soil nutrition alleviation. *Plant Stress* 11, 100349.

Svitashev, S., Young, J.K., Schwartz, C., Gao, H., Falco, S.C. *et al.* (2015) Targeted mutagenesis, precise gene editing, and site-specific gene insertion in maize using Cas9 and guide RNA. *Plant Physiology* 169, 931–945. DOI: 10.1104/pp.15.00793.

Tang, X.D., Gao, F., Liu, M.J., Fan, Q.L., Chen, D.K. *et al.* (2019) Methods for enhancing clustered regularly interspaced short palindromic repeats/Cas9-mediated homology-directed repair efficiency. *Frontiers in Genetics* 10, 551. DOI: 10.3389/fgene.2019.00551.

Tyagi, S., Dhiman, V.K., Dhiman, V.K., Pandey, H., Singh, D. *et al.* (2024) Plant defense strategies and biomarkers against heavy metal-induced stress: A comprehensive review. *ACS Agricultural Science & Technology* 4(2), 129–143. DOI: 10.1021/acsagscitech.3c00271.

Verma, S.K., Singh, D. and Shyam, P. (2024) Elicitation of secondary metabolites from plants. In: *Secondary Metabolites and Biotherapeutics*. Academic Press, pp. 127–159.

Waltz, E. (2016) Gene-edited CRISPR mushroom escapes US regulation. *Nature* 532(7599), 293. DOI: 10.1038/nature.2016.19754.

Wang, H., Wu, Y., Zhang, Y., Yang, J., Fan, W. *et al.* (2019) CRISPR/Cas9-based mutagenesis of starch biosynthetic genes in sweet potato (*Ipomoea batatas*) for the improvement of starch quality. *International Journal of Molecular Sciences* 20, 4702. DOI: 10.3390/ijms20194702.

Wang, Y., Cheng, X., Shan, Q., Zhang, Y., Liu, J. *et al.* (2014) Simultaneous editing of three homoeoalleles in hexaploid bread wheat confers heritable resistance to powdery mildew. *Nature Biotechnology* 32, 947–951. DOI: 10.1038/nbt.2969.

Wang, Y., Liu, X., Zheng, X., Wang, W., Yin, X. *et al.* (2021) Creation of aromatic maize by CRISPR/Cas. *Journal of Integrative Plant Biology* 63(9), 1664–1670. DOI: 10.1111/jipb.13105.

Wen, D., Wu, L., Wang, M., Yang, W., Wang, X. *et al.* (2022) CRISPR/Cas9-mediated targeted mutagenesis of FtMYB45 promotes flavonoid biosynthesis in tartary buckwheat (*Fagopyrum tataricum*). *Frontiers in Plant Science* 13, 879390.

Xue, C. and Greene, E.C. (2021) DNA repair pathway choices in CRISPR-Cas9-mediated genome editing. *Trends in Genetics* 37, 639–656. DOI: 10.1016/j.tig.2021.02.008.

Yadav, G., Prajapati, P., Singh, D., Hora, S., Singh, S. *et al.* (2023) Emerging trends in plant metabolomics and hormonomics to study abiotic stress tolerance associated with rhizospheric probiotics. *Plant Hormones in Crop Improvement* 1, 283–306. DOI: 10.1016/B978-0-323-918 86-2.00012-4.

Ying, W., Wen, G., Xu, W., Liu, H., Ding, W. *et al.* (2023) *Agrobacterium rhizogenes*: paving the road to research and breeding for woody plants. *Frontiers in Plant Science* 14, 1196561.

Yuan, M., Zhu, J., Gong, L., He, L., Lee, C. *et al.* (2019) Mutagenesis of FAD2 genes in peanut with CRISPR/Cas9 based gene editing. *BMC Biotechnology* 19, 24. DOI: 10.1186/s12896-019-0516-8.

Zeng, L., Zhang, Q., Jiang, C., Zheng, Y., Zuo, Y. *et al.* (2021) Development of *Atropa belladonna* L. plants with high-yield hyoscyamine and without its derivatives using the CRISPR/Cas9 system. *International Journal of Molecular Sciences* 22, 1731. DOI: 10.3390.

Zeng, Z., Han, N., Liu, C., Buerte, B., Zhou, C. *et al.* (2020) Functional dissection of HGGT and HPT in barley vitamin E biosynthesis via CRISPR/Cas9-enabled genome editing. *Annals of Botany* 126(5), 929–942. DOI: 10.1093/aob/mcaa115.

Zhang, A., Liu, Y., Wang, F., Li, T., Chen, Z. *et al.* (2019) Enhanced rice salinity tolerance via CRISPR/Cas9-targeted mutagenesis of the OsRR22 gene. *Molecular Breeding* 39. DOI: 10.1007/s11032-019-0954-y.

Zhang, C., Xue, J., Yang, X., Ke, Y., Ou, R. *et al.* (2022) From plant phenols to novel bio-based polymers. *Progress in Polymer Science* 125, 101473. DOI: 10.1016/j.progpolymsci.2021.101473.

Zhang, H., Qin, C., An, C., Zheng, X., Wen, S. *et al.* (2021) Application of the CRISPR/Cas9-based gene editing technique in basic research, diagnosis, and therapy of cancer. *Molecular Cancer* 20, 1–22.

Zhang, S., Zhang, R., Song, G., Gao, J., Li, W. *et al.* (2018) Targeted mutagenesis using the *Agrobacterium tumefaciens*-mediated CRISPR-Cas9 system in common wheat. *BMC Plant Biology* 18(1), 302. DOI: 10.1186/s12870-018-1496-x.

Zhang, Y., Bai, Y., Wu, G., Zou, S., Chen, Y. *et al.* (2017) Simultaneous modification of three homoeologs of TaEDR1 by genome editing enhances powdery mildew resistance in wheat. *Plant Journal: For Cell and Molecular Biology* 91, 714–724. DOI: 10.1111/tpj.13599.

Zhao, Y., Zhu, H., Lu, X., Anees, M., He, N. *et al.* (2024) Streamlined *Agrobacterium rhizogenes*-mediated hairy root transformation for efficient CRISPR/Cas9-based gene editing evaluation in diverse citrullus varieties. *Horticultural Plant Journal*. DOI: 10.1016/j.hpj.2023.10.005.

18 Regulatory, Biosafety, and Ethical Perspectives of Plant Genome Editing

Muhammad Nouman Tahir* **and Sana Zahra**
Department of Plant Protection, Ghazi University, Dera Ghazi Khan, Pakistan

Abstract
In agriculture, innovation is vital for increasing crop yield to meet growing demands. Genetic modification has greatly advanced research in agriculture, bioindustry, and biomedicine. In this review, we have analyzed the landscape of the use of plant genome-editing technology through the lens of regulatory frameworks, biosafety considerations, and ethical perspectives. The evolution of genetic modification in agriculture is examined, particularly focusing on the recent advancements in genome-editing tools and breeding techniques. We examine global regulatory oversight of genetically modified organisms (GMOs), highlighting challenges in harmonization and the necessity for transparent, science-driven regulatory practices. Biosafety concerns, including unintended mutations and off-target effects, are thoroughly explored, along with ethical considerations such as the implications of synthetic biology and targeted mutants in animal husbandry. The study aims to offer knowledge-based guidance for policy makers and researchers to facilitate responsible decision making and navigate the challenges and opportunities presented by these transformative technologies in agricultural biotechnology.

Keywords: Plant genome editing, New breeding techniques, CRISPR-Cas9, Biosafety, Ethics, Regulatory frameworks

18.1 Introduction

The rise of genome-editing technologies has revolutionized agriculture, allowing precise modification of plant traits to boost crop productivity and resilience. As the global food demand rises, the demand for sustainable food production drives the exploration of advanced biotechnological solutions. However, alongside these scientific breakthroughs come intricate regulatory challenges and ethical considerations that shape the adoption and governance of these powerful tools.

For a long time, genetic modification has been employed to address issues like pest resistance and climate resilience. From the early days of transgenesis to the revolutionary clustered regularly interspaced short palindromic repeats (CRISPR)-Cas9 system, each advancement has pushed the boundaries of what is possible in crop improvement (El-Mounadi *et al.*, 2020). With the advancement in genome-editing techniques, there must be advanced regulatory frameworks. Existing regulations, which were designed for earlier technologies, need to be upgraded according to the current genome-

*Corresponding author: mntahir@gudgk.edu.pk

 Genome Editing for Crop Improvement: Theory and Methodology (eds J.M. Al-Khayri *et al.*)
DOI: 10.1079/9781800622517.0018

editing techniques. This adaptation requires international cooperation and harmonization of policies to ensure safety standards, striking a balance between fostering innovation and ensuring safety and ethical responsibility (Idris *et al.*, 2023).

Moreover, public discourse and engagement are crucial in shaping the future of genome-editing technologies. Transparent communication and inclusive dialogue among stakeholders are essential to address concerns about biosafety, environmental impact, and ethical implications. Engaging with the local communities, farmers, and consumers helps build trust and acceptance, which is critical for successfully implementing these technologies (Pixley *et al.*, 2022).

This comprehensive exploration covers the intricate interplay of innovation, regulation, and ethics in plant genome editing. From the global distribution of genetically modified (GM) crops to the nuances of regulatory frameworks in different countries, we aim to understand the challenges and opportunities presented by these transformative technologies. Through informed discussion and collaboration, it is possible to navigate this complex landscape effectively.

18.2 Plant Genome Editing and Regulatory Measures

In agriculture, innovation plays a crucial role in enhancing crop productivity to meet the demands of a growing population. Genetic modification of organisms has significantly contributed to research and development in agriculture, industry, and biomedicine. Over the past three decades, since the advent of the first GM products through transgenesis, global regulatory frameworks have struggled to keep pace with advancing technologies and safety concerns. The emergence of new genome-editing techniques is posing challenges to existing frameworks, particularly in the context of crops. Various other methodologies have been employed to exploit the genetic potential of plants, although some are perceived as posing higher risks. Public concerns, often fueled by inadequate communication, have led to mistrust and the establishment of regulatory systems, prompting a reevaluation of the logic and motivations behind existing regulations for more efficient and effective delivery of the benefits offered by these technologies.

On the other hand, global food security also confronts challenges from environmental, social, socioeconomic, and health factors. In response, significant investments in agricultural research have aimed at increasing yields and developing new traits using innovative techniques such as genome editing. However, the diverse spectrum of cutting-edge genome-editing tools has created an international divide in regulatory, ethical, and societal perspectives. This dynamic landscape calls for ongoing discussions to navigate the complexities and ensure the responsible and effective use of these powerful technologies (Khan, 2019).

18.3 Genetic Modifications

Genetic modification of organisms for various purposes has been a key area of research since the advent of modern agriculture. The rapid implementation of genetic engineering, starting with DNA recombination in the 1970s, has profoundly impacted industries such as medicine, food, feed, and scientific research. Throughout history, humanity has employed various plant improvement techniques, including artificial selection, selective crosses, mutagenesis, and genetic engineering, to enhance crop yields, optimize resources, and improve nutritional content. Transgenesis, the transfer of genetic material between different species, is a widely used genetic engineering technique, often applied to confer traits such as resistance to pathogens, tolerance to stress, and herbicide resistance.

Despite the potential of transgenesis, defining genetically modified organisms (GMOs) has been a contentious issue, evolving since the commercial availability of transgenic applications in the 1990s. The Cartagena Protocol on Biosafety broadens the definition to 'living modified organisms' (LMOs), encompassing any organism with a novel genetic combination obtained through modern biotechnology. The protocol outlines 'modern biotechnology' as the application of *in vitro*-nucleic acid techniques and cell fusion beyond taxonomic families. This definition significantly influences global regulatory strategies (Rozas *et al.*, 2022).

18.4 Genome-editing Tools and Breeding Techniques

To address the time-consuming and expensive nature of conventional crop breeding and the urgency posed by climate change, new breeding techniques (NBTs) have emerged, focusing on DNA sequence-specific modifications. Among these, genome editing stands out, utilizing site-directed nucleases (SDNs) and oligo-directed mutagenesis (ODM) to insert, delete, or substitute nucleotides. The SDN technology introduces small, precise modifications (SDNs 1 and 2) or larger DNA pieces and complete foreign genes (SDN3) at predetermined locations, yielding both transgenic and nontransgenic outcomes. The ODM enables point mutations at specific locations. These genome-editing tools offer accuracy, cost-effectiveness, and simplicity, facilitating the introduction of valuable traits into plants and promoting innovation in crop research.

18.5 GMOs at Global Landscape

The majority (85%) of global GMO agriculture is concentrated in four countries in North and South America. These are the USA (40% of global GMO hectares), Brazil (26%), Argentina (12%), and Canada (7%). Worldwide GMO agriculture covers 189.8 million hectares across 24 countries. The number of countries planting GMOs peaked at 29 in 2010 and has been steadily decreasing since then. Most countries (88%) have no GMO hectares, including Australia accounting for 0.4% of global GMO hectares. The dominance of GMO agriculture is evident in North and South America, with Asia and Africa having modest to diminutive presences (Paull and Hennig, 2019). Asia has a modest presence, led by India (cotton), Pakistan (cotton), and China (cotton, papaya). Africa has a diminutive presence led by South Africa (maize, soy, cotton) (Paull and Hennig, 2019).

Table 18.1. Top 10 countries by GM area (in million hectares). Data retrieved from AgBio Investor.

Rank	Country	GM Area (mHa)
1	USA	74.4
2	Brazil	66.9
3	Argentina	23.1
4	India	12.1
5	Canada	11.5
6	Paraguay	4.3
7	South Africa	3.3
8	China	2.8
9	Pakistan	2.3
10	Bolivia	1.5

18.6 Recent Breakthrough: Approval of GM Cavendish Banana by Regulators

Australia and New Zealand have approved the commercial growth and consumption of a GM Cavendish banana. Developed by a team at Queensland University of Technology, Australia, this strain, named QCAV-4, is resistant to a devastating fungal disease caused by *Fusarium*.

Despite widespread global cultivation of GM crops, approval remains limited in regions like the UK and EU. Table 18.1 provides an overview of the top 10 countries ranked by the area of land dedicated to GM crops, showcasing the global distribution of biotech agriculture. These statistics highlight the role of specific nations in adopting and cultivating GMOs for various agricultural purposes (AgBio Investor GM Monitor, 2024).

18.7 Regulatory Measures

Regulation, when appropriately managed, serves as a crucial safeguard for society, the environment, and the economy, particularly in the context of GM crops and foods. At the national level, it aims to ensure human safety, environmental protection, prevent fraud, and address public concerns by instilling confidence in the products and processes used. While acknowledging the need for effective regulation, it is essential to strike a balance to avoid over-regulation and maintain the freedom to operate. This balance ensures that the benefits of new technologies, such as agricultural biotechnology, can be realized without dismissing potential risks.

The concerns with the regulation of GM crops arise from the fact that existing legislation, established in the late 1990s or early 2000s, may no longer align with advancements in molecular biology. Many countries are in the process of reevaluating their regulations, and some have developed positions on managing products from new breeding technologies. Inconsistencies exist among countries, with variations in food safety standards and regulations permitting or restricting the cultivation and production of GM crops. For instance, some countries allow the import and use of GM food with labeling requirements, while others prohibit their farmers from growing GM crops. These regulatory inconsistencies, as observed in New Zealand, Ecuador, and parts of Europe, highlight the need for a comprehensive review and updating of regulatory frameworks to ensure relevance and effectiveness.

The regulatory framework serves the crucial purpose of determining the following aspects:

- The safe exploration of GM technologies in both contained and open environments for crop research.
- The commercial cultivation of GM crops for domestic and international markets, ensuring compatibility with other activities and opportunities.
- Import regulations for GM-crop seeds intended for animal feed, considering that a significant portion of GM crops serves as feedstock for food-producing animals.
- Establishing tolerance thresholds for imported seeds that may contain GM content.
- Assessing the safety and permissibility of importing and consuming food derived from GM crops, with a focus on preventing harmful effects.

Regulation of GM crops varies globally, with rules often embedded in different legislations managed by distinct authorities. For instance, New Zealand employs the HSNO Act for GM crops and the Food Standards Australia New Zealand (FSANZ) Act 1991 for GM food. Australia has the Gene Technology Act 2000 for GM crops, and the FSANZ Act for GM food (Caradus, 2023). In the EU, national laws govern the release of GM crops, while EU-level regulations cover commercial cultivation. Canada splits oversight between the Canadian Food Inspection Agency and Health Canada, which deal with the regulation of novel plants and novel foods, respectively (Ellens *et al.*, 2019). The USA lacks specific GM crop legislation but uses existing laws under the Food and Drug Administration (FDA), Environmental Protection Agency (EPA), and the United States Department of Agriculture (USDA).

18.8 Regulation of Non-GM Seeds and Crops

Seed certification, a widespread practice globally, aims to prevent consumer deception and ensure the quality of seed products. The Organization for Economic Co-operation and Development (OECD) Scheme for Certification, established in 1958 with 61 participating countries, promotes consistently high-quality certified agricultural seeds across 204 plant species (Caradus, 2023). In North America, principles of certification were developed in the early 20th century, with the Association of Official Seed Certifying Agencies (AOSCA) formed in 1969/1970. In Europe, the European Seed Certification Agencies Association (ESCAA) harmonizes seed certification bodies in the EEA and EFTA. Sub-Saharan Africa has adopted harmonized-seed regulations through bodies like COMESA, ECOWAS, and SADC. New Zealand introduced voluntary seed certification in 1929 to maintain cultivar identity across generations for agricultural plant species.

In various countries, including Europe, the UK, Norway, Canada, Ukraine, Kazakhstan, Georgia, Egypt, sub-Saharan African countries, Kenya, Zambia, Brazil, Costa Rica, Cyprus, and Malaysia, marketing new cultivars requires approval from National List Trials, ensuring improvements over the existing ones. Registration on a country's National List grants commercial licensing, ensuring productivity, providing more opportunities to farmers by trade facilitation, and also giving guarantees to the users about the purity of seeds. However, many countries, such as the USA, New Zealand, and Australia, do not mandate cultivar registration before market release, rather depending on demand for variety acceptance.

18.9 Regulatory Focus on Human Health Impacts

Food derived from new technologies must undergo rigorous safety testing, scientifically identifying hazards, and assessing exposure. Regulations are mandatory to test the food from GM crops for safety purposes (Hull *et al.*, 2010; Gizaw, 2019). In this regard, the focus should be on ensuring safety, health, and nutrition without politicization, opinion-based complications, or mistrust (Burke, 2004). While the majority of data regarding GM crops show no negative impacts, there are many doubts among people, which emphasizes the importance of transparent testing. Allergy screening tests against transgenic proteins is a vital factor because of the termination of the GM-crop development program due to cross-reactivity with Brazil nut allergens in soybeans (Nordlee *et al.*, 1996).

In 1994, the World Trade Organization (WTO) attempted to establish global standards for managing intellectual property rights in biotechnology inventions through the Agreement on Trade-Related Aspects of Intellectual Property Rights (TRIPS). However, developing countries criticized TRIPS for neglecting cultural, political–economic, and ecological dimensions.

In 2018, under the WTO Committee on Sanitary and Phytosanitary Measures, delegations from several countries signed an international statement on the agricultural applications of precision biotechnology (Menz *et al.*, 2020). It emphasizes the positive contributions of precision biotechnology, the need for science-based regulatory approaches, and the importance of collaborative efforts for open and fair trade. Despite these efforts, many countries are not signatories to the 2018 Agreement, resulting in uncoordinated regulatory regimes and trade rules for GM crops. Such uncoordinated regulatory rules lead to disorganized international trade processes by zero-tolerance standards for GM material (Hobbs *et al.*, 2014).

The European stance on GM crops significantly impacts acceptance in some developing countries, specifically in Africa as some African governments adopted the European regulatory approaches for GM crops. These African countries also adopt a similar centralized risk-assessment approach following the European Food Safety Authority (EFSA) model to deal with the challenges in establishing biosafety regulations (Adenle *et al.*, 2013). Analysis of EU GM legislation reveals shortcomings in legal certainty, nondiscrimination, and scientific adaptability, highlighting inefficiencies and a lack of alignment (Adenle *et al.*, 2013). The potential benefits of NBT revolutionized crop production. However, the acceptance of such techniques in food will face a similar situation as GM crops. To avoid such circumstances, effective, science-based government regulations are required, which will be achieved by engaging society by giving them access to new technology and resources (Lassoued *et al.*, 2020; Gordon *et al.*, 2021).

The regulatory landscape for GMOs varies globally, with each country implementing its framework to oversee their cultivation and use (Turnbull *et al.*, 2021). A summary of the regulatory frameworks and laws governing GMOs in select countries is given in Table 18.2.

18.10 Navigating Regulatory Challenges in GM Crop Adoption

In the realm of GM crops, regulatory frameworks worldwide are crucial for governing the acceptance, adoption, and trade in different countries. Here is a concise breakdown of key aspects:

18.10.1 Risk evaluation and precautionary approach

- Risk assessment involves hazard identification, characterization, exposure assessment, and risk characterization.

Table 18.2. Regulatory frameworks and laws governing GMOs by country. Author's own table.

Country	Regulatory framework	Regulatory law
USA	Regulated by FDA, EPA, and USDA. No specific GM crop legislation. Existing laws applied for approval and oversight.	Coordinated framework under Coordinated Framework for the Regulation of Biotechnology. Laws include Federal Food, Drug, and Cosmetic Act, Federal Plant Pest Act, and Federal Insecticide, Fungicide, and Rodenticide Act.
European Union (EU)	EU-level regulations for commercial cultivation; national laws govern release. Strict approval process through EFSA. Mandatory labeling.	Regulations include Directive 2001/18/EC and Regulation (EC) No 1829/2003. Mandatory labeling under Regulation (EC) No 1830/2003.
Canada	Oversight by Canadian Food Inspection Agency and Health Canada. Regulations cover novel plants and foods.	Regulated under the Seeds Act and Plant Breeders' Rights Act. Assessment conducted under the Food and Drugs Act and the Canadian Environmental Protection Act, 1999.
China	Regulations under the Ministry of Agriculture and Rural Affairs (MARA) and the Biosafety Committee of GMOs. Mandatory safety assessments for GMOs.	Regulations include the Safety Management Regulations on Agricultural GMOs and the Regulations on the Administration of Agricultural GMOs.
India	Regulated by the Genetic Engineering Approval Committee (GEAC) under the Ministry of Environment, Forest, and Climate Change.	Oversight under the Environment Protection Act, 1986, and Rules thereunder. The Food Safety and Standards Act, 2006 regulates GM food.
Pakistan	Oversight by the National Biosafety Committee (NBC) under the Ministry of Climate Change. Regulations include the Pakistan Biosafety Rules, 2005.	Governed by the Pakistan Environmental Protection Act, 1997. NBC responsible for risk assessment and approval of GMOs.
South Africa	Oversight by the Executive Council for GMOs (ECGMOs) under the Department of Agriculture, Forestry, and Fisheries.	Regulated under the GMO Act, 1997 (Act No. 15 of 1997), and the National Environmental Management: Biodiversity Act, 2004 (Act No. 10 of 2004).

- The Precautionary Principle, while criticized, underscores the need for careful risk management, giving rise to the concept of 'optimal precaution.'
- Decision making should weigh relative knowledge against ignorance, aiming to minimize overall negative consequences.

18.10.2 Process standards versus product standards

- Regulatory frameworks should shift from process based to product based for GM crops, focusing on the unique traits and risks of each cultivar rather than the method of production (Jenkins *et al.*, 2023).
- This approach aims to reduce costs, promote innovation, and accommodate various stakeholders, emphasizing the need to examine target and off-target effects.

18.10.3 Coexistence regulations

- Coexistence refers to farmers' ability to choose between conventional, organic, or approved GM crops, necessitating practical measures at the farm level.
- Isolation distances vary for different crops, requiring careful consideration to manage

cross-fertilization between GM and non-GM variants (Giraldo *et al.*, 2019).

18.10.4 GM tolerance threshold in trade

- The adventitious presence of GM material in non-GM crops can disrupt international trade.
- Zero-tolerance policies, as seen in the EU, may lead to trade barriers and challenges for supply chains, impacting global food security.

18.10.5 Global trade and regulatory harmonization

- Harmonizing international regulations for GM crops and foods is crucial to facilitate open and fair trade, encourage research and innovation, and build trust in regulatory frameworks.
- Negotiating rules for trade in GM crops requires separate, complex negotiations, emphasizing the need for compromise.

18.11 Navigating Challenges and Regulatory Effectiveness in GM Crop Governance

18.11.1 Labeling and consumer right to know

- The call for mandatory GM labeling faces challenges related to costs, stigmatization, feasibility, and potential impacts on food security and innovation (Oh and Ezezika, 2014).
- Over 64 countries mandate some form of labeling, each with varying thresholds, such as the USA and Canada's 5% tolerance level.

18.11.2 Components of effective regulation

- Effective regulatory systems demand a science-based transparent approach with public participation.
- Key components include premarket approval, safety standards, transparency, public involvement, expert advice, postapproval activities, and robust enforcement (Jaffe, 2004).

18.11.3 Challenges and considerations

- Cumbersome regulatory systems jeopardize international competitiveness and pose risks to food security and biodiversity.
- Harmonization within regions, like the EU's model, is a pragmatic goal for streamlined GM regulations.

18.12 Advancements and Challenges in Plant Genome Editing: A Biosafety Perspective

Genome-editing tools offer unprecedented precision in modifying plant traits, yet there are concerns regarding biosafety and social implications. The key focus revolves around target gene site selection, guide RNA design, and off-target effects. The risk of unintended genetic alterations, particularly due to off-target mutants, is a primary apprehension. Strategies like *in vitro*-preassembled CRISPR/Cas9 ribonucleoproteins aim to mitigate these concerns but face challenges such as stability and cost-effectiveness.

Efforts to minimize Cas9-related issues involve enhancing RNA guide-design strategies, employing spatiotemporally controlled Cas9, and utilizing synthetic genetic circuits (Liang *et al.*, 2018). Base editing is improved to enhance specificity, restricting deaminase activity outside Cas9 binding (Shimatani *et al.*, 2017). Concerns about Cas9-induced immune responses and limited target sites due to the PAM requirement prompt ongoing protein engineering and exploration of alternative Cas proteins like Cpf1 nucleases. The quest for precise plant genome editing encounters biosafety challenges, necessitating continual improvements in techniques and vigilance in addressing societal concerns (Eckerstorfer *et al.*, 2019).

18.13 Unintended Effects in Genome Editing: Risks and Mitigation

Genome-editing techniques, including CRISPR/Cas9, may lead to unintended genetic and epigenetic changes. Off-target modifications, where molecular tools act on sequences similar to the target loci, pose a significant risk. Factors influencing off-target activity include the frequency of homologous sequences, nuclease-type characteristics, expression levels, and the accessibility of chromatin. Strategies like high-fidelity Cas9 variants and transient nuclease expression aim to minimize off-target effects. Bioinformatic tools aid in predicting off-target sites, but challenges exist in reliable *in vivo* identification. Various methods, including GUIDE-seq and BLESS, help screen for off-target activity (Naeem and Alkhnbashi, 2023). Testing genome-edited plants through whole-genome sequencing and biochemical methods is essential. Concerns arise about the detectability of subtle changes and the potential long-term impact on agronomic traits. The removal of unintended changes depends on breeding cycles, impacting annual crops differently than trees or vegetatively propagated plants. Strategies to minimize off-target effects and identification methods are crucial for safe genome-editing applications.

18.14 Unintended Pleiotropic Effects in Genetic Modification

Certain genetic modifications, including new breeding approaches, may lead to unintended secondary phenotypes known as pleiotropic effects. These effects, expressed alongside the desired traits, are inherent in the final breeding products and can occur with various breeding methods (Ladics *et al.*, 2015). An example is the modification of plants for increased disease resistance, resulting in pleiotropic effects such as yield decrease and susceptibility to other pathogens (van Esse *et al.*, 2020; Zhao *et al.*, 2022). Unintended effects can also arise from modifications near integrated DNA sites, particularly in cisgenesis, intragenesis, or SDN-3 applications. Understanding altered genomic sequences provides insights into potential unintended modifications and their phenotypic consequences. These modifications, once introduced, cannot be removed through segregation in subsequent breeding steps. Molecular characterization helps predict and assess the impact of unintended effects.

18.15 Unintended Changes in Plant Genome Editing: Transformation Effects

The introduction of exogenous effects or molecules during genome editing, such as via stable genetic transformation may lead to unintended genetic or epigenetic changes. These alterations can occur during the integration of genetic constructs into the plant genome, in processes like cisgenesis/intragenesis and certain genome-editing approaches. Random integration can disrupt functional genomic sequences or introduce unrelated DNA sequences, potentially impacting nearby endogenous genes. Additionally, constructs transiently introduced for method-related components might integrate into the recipient cells' genome, raising concerns about their complete removal from the final product. Whole-genome sequencing has revealed instances of multiple independent insertions of vector backbone sequences during the transformation of the CRISPR/Cas9 expression construct (Kubicek *et al.*, 2019). Continuous assessment and monitoring are crucial to ensuring the absence of unintended modifications in the final modified plant product.

18.16 Biosafety in Genome Editing: Risks and Future Prospective

Genome-editing technologies, notably CRISPR/Cas9, offer transformative possibilities in agriculture and medicine. However, biosafety remains a priority, demanding meticulous evaluation and adaptive regulations to address ethical, environmental, and health considerations associated with these powerful tools. Precision breeding through genome-editing results in enhanced food crop quality, increase in yield, and disease resistance, that is, presenting sustainable solutions while navigating ethical considerations for responsible advancement.

The regulatory landscape for genome-edited crops varies globally, with different countries adopting diverse approaches. Many countries have permitted certain types of genome editing, such as SDN 1 and 2, to be free of stringent regulation due to their similarity to conventional breeding methods (Table 18.2).

18.17 Ethical Issues Regarding Plant Genome Editing

In the last two decades, genome-editing technologies, including zinc-finger nucleases, meganucleases, transcription activator-like effector nucleases (TALENs), and the revolutionary CRISPR/Cas9 system, have evolved with increased efficiency. The emergence of synthetic gene drives has allowed rapid gene spread within populations. These technologies enable precise genome modifications in nonhuman animals, surpassing classical transgenesis with fewer off-target effects. Engineered nucleases facilitate genetic changes without foreign DNA, offering applications like enhancing livestock productivity, creating disease-resistant animals, and controlling vector-borne diseases. Despite the excitement surrounding these advancements, ethical and governance concerns have triggered debates in academic and public spheres.

18.18 Unintended Mutations Risk

The CRISPR/Cas9 system, while revolutionary, poses a risk of off-target mutations in the genome. Mosaicism, introduced during different embryonic development stages, can lead to unintended mutations, potentially causing cell death or transformation (Zhang *et al.*, 2014; Bortesi *et al.*, 2016). Ongoing improvements aim for safer and more efficient usage. Addressing challenges like enhancing the efficiency of knockout and homologous recombination is crucial (Eduardo, 2017). Scientists are developing methods to minimize off-target effects and mosaic mutations, thus working toward higher editing precision (Koo *et al.*, 2015; Kleinstiver *et al.*, 2016; Slaymaker *et al.*, 2016). Despite claims of highly efficient engineered Cas9 proteins, rigorous testing for each organism remains essential.

18.19 Ethical Concerns in Synthetic Biology

The field of synthetic biology, merging molecular biology with computer science, chemistry, and engineering, involves designing organisms or devices using biological materials (Cameron *et al.*, 2014). However, the capability to design and edit entire genomes raises ethical apprehensions, especially regarding potential military or terrorist applications. A significant concern is the misuse of CRISPR/Cas9 to synthesize and manipulate pathogens, enhancing their threat to human health. Experiments manipulating viruses, such as the H5N1 strain of influenza, pose risks to society (Herfst *et al.*, 2012). Some research publications gave the idea of new research lines aiming to generate harmful organisms. The accessibility of CRISPR/Cas9 kits to broader users beyond traditional biotechnological entities and research institutions highlights the need for effective governance and regulations.

18.20 Bioethics: Navigating Risks and Responsibilities

Bioethics demands transparent communication on the benefits and risks of genome editing, urging respect for experimental organisms, especially in the presence of sentient people. Debates range from anthropocentrism to bioconservatism, with a moderate stance acknowledging reciprocal impacts between humans and nature. The moral relevance of nonhuman organisms depends on their interconnectedness with other organisms at universal level to maintain equilibrium. Releasing the modified organisms raises concerns about upsetting the balance of nature. Discussions on genome modifications involve debates on interfering with intrinsic values and the ethical implications of technological interventions. The precautionary principle advocates anticipating potential and undesirable consequences and acknowledging uncertainties. Responsibility extends to preserving biodiversity and imposing ethical limits on human intervention. The power of genome editing demands ethical scrutiny and inclusive public discourse.

The rapid advancement of genome editing, particularly with CRISPR/Cas9, holds great promise but raises ethical concerns. Issues like unintended mutations and the potential misuse of synthetic biology tools require ongoing vigilance. Ethical considerations span personalized animal models, xenotransplantation, and the broader landscape of genome-editing applications. Bioethics, advocating for a balanced approach, emphasizes transparent communication and responsible practices to navigate the risks and responsibilities associated with the transformative era of genetic manipulation.

18.21 Future Directions

- Enhanced Precision: Advancements in CRISPR/Cas9 variants and novel nucleases such as Cpf1 aim for greater precision and minimize off-target effects.
- Integration of AI and Bioinformatics: Using AI and bioinformatics can improve the prediction of off-target effects and facilitate the designing of more accurate editing strategies, increasing efficiency and reliability.
- Resilient Crops: Genome editing holds promise in creating crop varieties that are resistant to diseases, pests, and environmental stresses, boosting yields and sustainability.
- Ethical and Regulatory Frameworks: Robust frameworks are crucial for addressing biosafety, environmental impact, and navigating societal implications effectively.
- Public Engagement and Communication: Transparent communication and education are essential to build trust and acceptance of genome-editing technologies.
- Global Collaboration: International cooperation is key to harmonizing regulatory standards and promoting equitable access to the benefits of genome editing.
- Broad Applications: Beyond agriculture, genome editing has diverse applications in biomedicine and industrial biotechnology, including biofuels and pharmaceuticals.

18.22 Conclusion

The advancements in genome-editing technologies, particularly CRISPR-Cas9, have revolutionized agriculture by enabling precise genetic modifications that enhance crop yields, disease resistance, and environmental adaptability. However, these advancements pose significant regulatory and ethical challenges. Current regulations, often outdated, vary globally and struggle to keep pace with technological progress. Regulatory frameworks worldwide must evolve to address the unique risks and benefits of these new technologies, balancing innovation with safety and public trust. This approach aims to streamline regulatory processes, promote innovation, and ensure safety. Ongoing public engagement and education are crucial to building trust and addressing ethical concerns. Ethical considerations, including potential unintended consequences and biosecurity risks, underscore the necessity for vigilant oversight and ongoing dialogue among scientists, policy makers, and the public.

References

Adenle, A.A., Morris, E.J. and Parayil, G. (2013) Status of development, regulation and adoption of GM Agriculture in Africa: Views and positions of stakeholder groups. *Food Policy* 43, 159–166.

AgBio Investor GM Monitor (2024) Global GM crop area 2023 review. Available at: https://gm.agbioinvestor.com/downloads (accessed 20 September 2024).

Bortesi, L., Zhu, C., Zischewski, J., Perez, L., Bassié, L. *et al.* (2016) Patterns of CRISPR/Cas9 activity in plants, animals and microbes. *Plant Biotechnology Journal* 14(12), 2203–2216.

Burke, D. (2004) GM food and crops: What went wrong in the UK? Many of the public's concerns have little to do with science. *EMBO Reports* 5(5), 432–436.

Cameron, D.E., Bashor, C.J. and Collins, J.J. (2014) A brief history of synthetic biology. *Nature Reviews Microbiology* 12(5), 381–390.

Caradus, J.R. (2023) Processes for regulating genetically modified and gene edited plants. *GM Crops and Food* 14(1), 1–41.

Eckerstorfer, M.F., Engelhard, M., Heissenberger, A., Simon, S. and Teichmann, H. (2019) Plants developed by new genetic modification techniques – comparison of existing regulatory frameworks in the EU and non-EU countries. *Frontiers in Bioengineering and Biotechnology* 7, 26.

Eduardo, R. (2017) Ethical issues in genome editing for non-human organisms using CRISPR/Cas9 system. *Journal of Clinical Research and Bioethics* 8(2), 1000300.

Ellens, K.W., Levac, D., Pearson, C., Savoie, A., Strand, N. *et al.* (2019) Canadian regulatory aspects of gene editing technologies. *Transgenic Research* 28, 165–168.

El-Mounadi, K., Morales-Floriano, M.L. and Garcia-Ruiz, H. (2020) Principles, applications, and biosafety of plant genome editing using CRISPR-Cas9. *Frontiers in Plant Science* 11, 498555.

Giraldo, P.A., Shinozuka, H., Spangenberg, G.C., Cogan, N.O.I. and Smith, K.F. (2019) *Frontiers in Plant Science* 10, 486827.

Gizaw, Z. (2019) Public health risks related to food safety issues in the food market: A systematic literature review. *Environmental Health and Preventive Medicine* 24, 1–21.

Gordon, D.R., Jaffe, G., Doane, M., Glaser, A., Gremillion, T.M. *et al.* (2021) Responsible governance of gene editing in agriculture and the environment. *Nature Biotechnology* 39(9), 1055–1057.

Herfst, S., Schrauwen, E.J.A., Linster, M., Chutinimitkul, S., De, E. *et al.* (2012) Airborne transmission of influenza A/H5N1 virus between ferrets. *Science* 336(6088), 1534–1541.

Hobbs, J.E., Kerr, W.A. and Smyth, S.J. (2014) The perils of zero tolerance: Technology management, supply chains and thwarted globalisation. *International Journal of Technology and Globalisation* 7(3), 203–216.

Hull, R., Bosse, M. and Tzotzos, G. (2010) Training for implementing risk assessment regulations for the release of GM crops. *Aspects Applied Biology* 96, 1–8.

Idris, S.H., Mat Jalaluddin, N.S. and Chang, L.W. (2023) Ethical and legal implications of gene editing in plant breeding: A systematic literature review. *Journal of Zhejiang University-SCIENCE B* 24(12), 1093–1105.

Jaffe, G. (2004) Regulating transgenic crops: A comparative analysis of different regulatory processes. *Transgenic Research* 13, 5–19.

Jenkins, D., Juba, N., Crawford, B., Worthington, M. and Hummel, A. (2023) Regulation of plants developed through new breeding techniques must ensure societal benefits. *Nature Plants* 9(5), 679–684.

Khan, S.H. (2019) Genome-editing technologies: Concept, pros, and cons of various genome-editing techniques and bioethical concerns for clinical application. *Molecular Therapy-Nucleic Acids* 16, 326–334.

Kleinstiver, B.P., Pattanayak, V., Prew, M.S., Tsai, S.Q., Nguyen, N.T. *et al.* (2016) High-fidelity CRISPR–Cas9 nucleases with no detectable genome-wide off-target effects. *Nature* 529(7587), 490–495.

Koo, T., Lee, J. and Kim, J.S. (2015) Measuring and reducing off-target activities of programmable nucleases including CRISPR-Cas9. *Molecules and Cells* 38(6), 475–481.

Kubicek, C.P., Steindorff, A.S., Chenthamara, K., Manganiello, G., Henrissat, B. *et al.* (2019) Evolution and comparative genomics of the most common *Trichoderma species*. *BMC Genomics* 20(1), 1–24.

Ladics, G.S., Bartholomaeus, A., Bregitzer, P., Doerrer, N.G., Gray, A. *et al.* (2015) Genetic basis and detection of unintended effects in genetically modified crop plants. *Transgenic Research* 24, 587–603.

Lassoued, R., Macall, D.M., Smyth, S.J., Phillips, P.W.B. and Hesseln, H. (2020) How should we regulate products of new breeding techniques? Opinion of surveyed experts in plant biotechnology. *Biotechnology Reports* 26, e00460.

Liang, Z., Chen, K., Zhang, Y., Liu, J., Yin, K. *et al.* (2018) Genome editing of bread wheat using biolistic delivery of CRISPR/Cas9 *in vitro* transcripts or ribonucleoproteins. *Nature Protocols* 13(3), 413–430.

Menz, J., Modrzejewski, D., Hartung, F., Wilhelm, R. and Sprink, T. (2020) Genome edited crops touch the market: A view on the global development and regulatory environment. *Frontiers in Plant Science* 11, 586027.

Naeem, M. and Alkhnbashi, O.S. (2023) Current bioinformatics tools to optimize CRISPR/Cas9 experiments to reduce off-target effects. *International Journal of Molecular Sciences* 24(7), 6261.

Nordlee, J.A., Taylor, S.L., Townsend, J.A., Thomas, L.A. and Bush, R.K. (1996) Identification of a Brazil-nut allergen in transgenic soybeans. *New England Journal of Medicine* 334(11), 688–692.

Oh, J. and Ezezika, O.C. (2014) To label or not to label: Balancing the risks, benefits and costs of mandatory labelling of GM food in Africa. *Agriculture and Food Security* 3, 1–8.

Paull, J. and Hennig, B. (2019) New world map of Genetically Modified Organism (GMO) agriculture: North and South America is 85. *Acres Australia* 101, 59–60.

Pixley, K.V., Falck-Zepeda, J.B., Paarlberg, R.L., Phillips, P.W.B., Slamet-Loedin, I.H. *et al*. (2022) Genome-edited crops for improved food security of smallholder farmers. *Nature Genetics* 54(4), 364–367.

Rozas, P., Kessi-Pérez, E.I. and Martínez, C. (2022) Genetically modified organisms: Adapting regulatory frameworks for evolving genome editing technologies. *Biological Research* 55(1), 31.

Shimatani, Z., Kashojiya, S., Takayama, M., Terada, R., Arazoe, T. *et al*. (2017) Targeted base editing in rice and tomato using a CRISPR-Cas9 cytidine deaminase fusion. *Nature Biotechnology* 35(5), 441–443.

Slaymaker, I.M., Gao, L., Zetsche, B., Scott, D.A., Yan, W.X. *et al*. (2016) Rationally engineered Cas9 nucleases with improved specificity. *Science* 351(6268), 84–88.

Turnbull, C., Lillemo, M. and Hvoslef-Eide, T.A.K. (2021) Global regulation of genetically modified crops amid the gene edited crop boom–a review. *Frontiers in Plant Science* 12, 630396.

van Esse, H.P., Reuber, T.L. and van der Does, D. (2020) Genetic modification to improve disease resistance in crops. *New Phytologist* 225(1), 70–86.

Zhang, F., Wen, Y. and Guo, X. (2014) CRISPR/Cas9 for genome editing: Progress, implications and challenges. *Human Molecular Genetics* 23(R1), R40–R46.

Zhao, Y., Zhu, X., Chen, X. and Zhou, J.M. (2022) From plant immunity to crop disease resistance. *Journal of Genetics and Genomics* 49(8), 693–703.

Index